Sustainable Materials and Technology

Sustainable Materials and Technology (SMT) book series publishes research monographs (both edited and authored volumes) showcasing the latest developments in the field and comprehensively covering topics such as:

- Recycling of waste into useful material and their energy applications
- Catalytic action of Nano oxides for efficient carbon reforming process
- Sustainable technologies for plastic transformation
- Bifunctional nanoparticles for sustainable water splitting applications
- Sustainable dying and printing
- New materials from waste
- Sustainable Manure Management and Technology: Potentials, Uses and limitations
- Sustainable Mechanical Engineering Approach
- Sustainable biochemistry for the improvement of health
- Sustainable development of Mechanical recycling of automotive components
- Sustainable-waste recycling and conversion in useful materials for different applications
- Sustainable development of inexpensive Nano-photo catalysts
- Sustainable development of recycling of discarded lithium ion batteries
- Modern sustainable cement and concrete
- Sustainable adsorbent for hazardous removal
- Sustainable superior electromagnetic shielding materials
- Excellent sustainable nanostructured materials for energy storage device
- Sustainable development of heavy metal detoxification from water
- Carbon dioxide utilization for sustainable energy
- Sustainable development in green syntheses of materials
- Environment friendly and sustainable cloth for garments application
- Sustainable design and application of eco-materials
- Nanoparticles for sustainable environment applications
- Sustainable remediation of industrial contaminated water towards potential industrial applications
- Biomaterials for sustainable bioremediations

Sajid Maqsood · Muhammad Nihal Naseer ·
Soottawat Benjakul · Asad A. Zaidi
Editors

Fish Waste to Valuable Products

Springer

Editors
Sajid Maqsood
Department of Food Science
College of Agriculture and Veterinary Medicine
United Arab Emirates University
Al Ain, United Arab Emirates

Soottawat Benjakul
Department of Food Technology
Faculty of Agro-Industry
Prince of Songkla University
Songkhla, Thailand

Muhammad Nihal Naseer
Department of Mechanical Engineering
Seoul National University
Seoul, Korea (Republic of)

Asad A. Zaidi
Department of Mechanical Engineering
Hamdard University
Karachi, Pakistan

ISSN 2731-0426 ISSN 2731-0434 (electronic)
Sustainable Materials and Technology
ISBN 978-981-99-8592-0 ISBN 978-981-99-8593-7 (eBook)
https://doi.org/10.1007/978-981-99-8593-7

This Springer imprint is published by the registered company Springer Nature Singapore Pte Ltd.
The registered company address is: 152 Beach Road, #21-01/04 Gateway East, Singapore 189721, Singapore

Paper in this product is recyclable.

Contents

Fish Waste Composition and Classification

Gokulprasanth Murugan, Kamalii Ahilan, Vaisshali Prakash Arul Prakasam, Joshna Malreddy, Soottawat Benjakul, and Muralidharan Nagarajan

Abstrasct Globally, food waste is a menacing issue, which is primely focused on Sustainable Development Goal (SDG) Target 12.3, aiming to cut down wastage by 2030. Thirty-five percent of the fish caught in fisheries and aquaculture are thrown away as fish waste and by-products. Fish processing waste accounts for about 70% of them, which can be turned into a valuable resource through a circular bioeconomy. The utilization of fish waste can also assuage the pressure on the fishery's resources and nurture the sustainability of the realm. The valorization of fish waste and by-products is gaining significant attention as a source of nutrition besides being used as fodder or thrown away as waste. The by-products of fish typically include heads (9–12% of the total weight of the fish), bones (9–15%), scales (around 5%), viscera (12–18%), and skin (1–3%). Indeed, fish waste is justified to be a good source of protein (58% dry matter), fat (19%), and minerals. The potential techniques are being utilized to treat fish waste and recover valuable bioactive compounds. Therefore, this chapter underlines the enormous power of fish waste and its by-products.

A Chapter Submitted to Fish Waste to Valuable Products: Recent Applications and Research Update

G. Murugan · M. Nagarajan (✉)
Department of Fish Processing Technology, Tamil Nadu Dr. J Jayalalithaa Fisheries University, Dr. M.G.R Fisheries College and Research Institute, Ponneri, Tamil Nadu 601 204, India
e-mail: murali@tnfu.ac.in

K. Ahilan · J. Malreddy
Department of Aquaculture, Tamil Nadu Dr. J Jayalalithaa Fisheries University, Dr. M.G.R Fisheries College and Research Institute, Ponneri, Tamil Nadu 601 204, India

V. P. A. Prakasam
Department of Fish Processing Technology, Kerala University of Fisheries and Ocean Studies, Panangad, Kochi, Kerala 682 506, India

S. Benjakul
International Center of Excellence in Seafood Science and Innovation (ICE-SSI), Faculty of Agro-Industry, Prince of Songkla University, Hat Yai, Songkhla 90110, Thailand

S. Maqsood et al. (eds.), *Fish Waste to Valuable Products*, Sustainable Materials and Technology, https://doi.org/10.1007/978-981-99-8593-7_1

Keywords Bioactive compounds · By-products · Circular bioeconomy · Enzymes · Fish waste treatment · Nutrition · Potential utilization · Protein · Sustainability · Underutilization

1 Introduction

Nowadays, the fish production has tremendously increased all over the world. However, this negatively entails an increased fish waste generation. The fish waste and by-products usually consist of heads, bones, viscera, scales, and skin. The utilization of fish waste is low profitable due to the lack of proper knowledge about the final product or partly thrown away (Guillen et al. 2018). The better fish waste management paves the way for a circular bioeconomy. The circular bioeconomy aims at attaining the environment and terms of resources. The fundamental role of bioeconomy is bio-waste valorization. Fish waste is one of the potential starting materials due to its eminent nutritional composition and gains a promising market value among bio-waste (Shahidi et al. 2019). Hence, commercializing fish waste boosts economic development and reduces environmental issues. The ideology of the composition of fish waste entails the prior valorization and potential usage of fish waste. Therefore, this chapter provides a clear view of fish waste and its composition.

2 Fish Waste and Its Composition

A tremendous quantity of fish waste is generated as the fish processing industry expands, accounting for up to processed fish (70%) and total fish mass (¾) (Rustad et al. 2011). Fish by-products decompose quickly unless properly processed or placed in suitable circumstances due to heavier bacterial load and endogenic enzyme content, which can create severe environmental and food-technical concerns (Chalamaiah et al. 2012).

Stevens et al. (2018) stated that by-products are the leftovers (edible or non-edible) after manufacturing primary products. Heads, skins, frames, trimmings, viscera, and blood are common leftovers of finfish (Table 1).

Underutilization of Fish Waste

Scales, skin, bones, and fins are common fish processing waste. They are normally dumped on land and into the ocean. This practice results in the underutilization of raw materials and has an impact on the utilization of available resources. The disposal of these wastes causes several challenges for the environment (Fig. 1 and Table 2).

Table 1 Body components of different fish species

Fish species	Fillets/ meat	Head with gills	Bones	Skin with scales	Fins	Viscera	References
Hake	47.5	15.2	–	4.3	2.3	–	Wheaton and Lawson (1985)
Common carp	42.2	19.3	15.2	10.3	3.6	4.7	Javeed and Mahendrakar (1995)
Catla	63.4	17.3	4.5	9.0	2.9	3.9	Sachindra and Mahendrakar (2014)
Cod	–	20.2	9.7	4.2	–	5.6	Falch et al. (2006)
Haddock	–	18.9	10.6	4.5	–	6.2	Sachindra and Mahendrakar (2014)
Cage cultured channel catfish	–	59.4	–	6.4	–	7.6	Sachindra and Mahendrakar (2014)

Source Sachindra and Mahendrakar (2014)

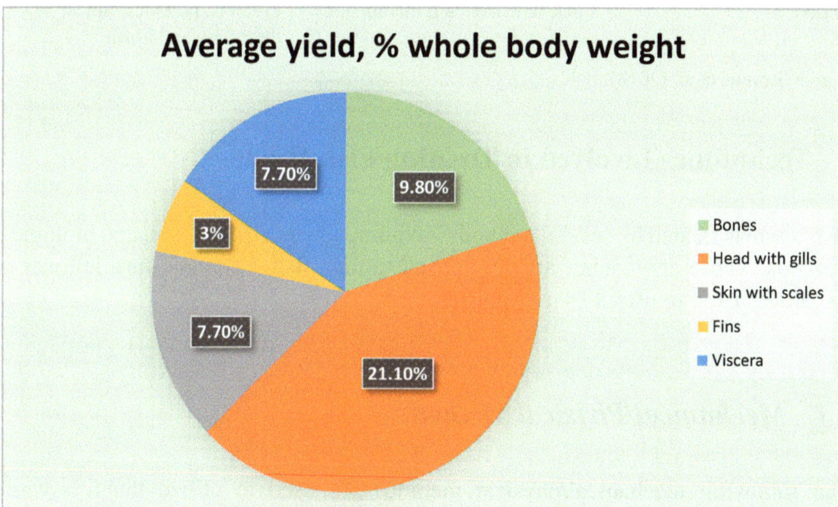

Fig. 1 Estimation production of by-products from finfish processing FAO (2013). *Source* Sachindra and Mahendrakar (2014)

Table 2 Valuable components and the potential use of fish by-products

Type of fish waste	Possible utilization (Currently being utilized as)	Valuable components
Fish head waste	Food, food-grade hydrolysates, fish oil, fish meal, animal-grade hydrolysates, nutraceuticals, pet food, and cosmetics	Proteins, peptides, lipids, collagen, gelatin, minerals including calcium, and flavoring agents/substances
Fish visceral waste	Food-grade hydrolysates, animal-grade hydrolysates, fish meal, fish oil, fuel, and fertilizers	Proteins, peptides, lipids, and enzymes such as lipases
Fish skin waste (with belly flap)	Fish meal, fish oil, cosmetics, food, fish meal, nutraceuticals, cosmetics, leather, fuel, and fertilizers	Collagen, gelatin, lipids, proteins, peptides, minerals and flavoring agents
Fishbone waste–Frames (bones, flesh, fins)	Food, fish meal, fish oil, food-grade hydrolysates, calcium powder, animal-grade, pet food nutraceuticals, and cosmetics	Proteins, peptides, lipids, hydroxy appetite, collagen, gelatin, minerals including calcium, and flavoring agents
Fish scale waste	Pearl essence, dietary supplements, and biomedical applications	Guanine, chitin, protein, minerals, collagen, and gelatin
Trimmings	Food, fish oil, fish meal, pet food, and animal-grade hydrolysates	Proteins, peptides, and lipids
Blood	Fuel, fertilizer, and therapeutics	Proteins, peptides, lipids, thrombin, and fibrin

Source Stevens et al. (2018)

3 Techniques Involved in Treating Fish Waste

Fish waste is treated by several methods including physical, chemical, and biological methods. These approaches are being used to treat fish waste and turn leftover or discarded fish into useful by-products.

3.1 Mechanical/Physical Method

The following mechanical/physical methods are used to utilize the fish waste (Table 3).

3.2 Chemical Method

The following chemical methods are used to utilize the fish waste (Table 4).

Table 3 Mechanical/physical method of treating fish waste

Type of waste	Processing method	Final product	References
Low-value fish species or unmarketed fishes	Breaking press cake using a hammer mill, pressing, grinding, and pulverizing	Fish meal	Gopakumar, (2002)
By-product from fish meal	Cooking fish offal during fish meal production	Fish solubles and stick water	Hertrampf and Piedad-Pascual (2000)
Fish liver	Cooking, pressing, and refining	Fish oil/shark liver oil	Balachandran (2001)
Swim bladder	Drying	Fish maws	Gopakumar (2002)
Swim bladder	Drying and compressed by ribbon rollers	Isinglass	Gopakumar, (2002)
Salted or fresh skins	Cooking, evaporation, and drying	Fish glue	Akter et al. (2016)
Shark skin	Drying	Shark leather	Balachandran (2001)
Trash fishes	Cooking, pressing, and drying	Fish flake	Govindan (1982)
Trash fishes	Cooking, pressing, blending, and drying	Fish soup powder (C.I.F.T)	Govindan (1982)
Fish bones	Separation of bones and drying	Shark cartilage	Balachandran (2001)
Shark fins	Drying	Shark fin rays	Balachandran (2001)
Head, tail, and body shells of prawn	Drying	Prawn manure	Balachandran (2001)

3.3 Biological/Enzymatic Method

The following biological/enzymatic methods are used to utilize the fish waste (Table 5).

4 Potential Utilization of Fish Waste

4.1 Fish Silage

Fish silage is defined as the process of preserving entire fish by the addition of acid (chemical silage) or by the fermentation with sugar and microbial culture (biological silage). Acid/chemical silage method has been employed on a commercial scale in Denmark for more than 30 years. The fish silage production is popularly being

Table 4 Chemical method of treating fish waste

Type of waste	Processing method	Final product	References
Skins, scales, and bones	Acid (acetic acid) and alkali treatment (sodium hydroxide)	Gelatin/ collagen	Karim and Bhat (2009)
Processing waste and neglected raw materials—fins, viscera, gills, head, etc	Acid fermentation—Formic acid	Fish silage	Gopakumar (2002)
Fish scales	Acid treatment—Acetic acid Preservative—Salicylic acid	Pearl essence	Balachandran (2001)
Head, gut, and trimmings of fish	Acid treatment—Hydrochloric acid	Fish protein hydrolysate	Petrova et al. (2018)
Head, shell, viscera, and gills of crustaceans	Demineralization—Hydrochloric acid Deproteinization—Sodium hydroxide	Chitin	Synowiecki and Al-Khateeb (2003)
Head, shell, viscera, and gills of crustaceans	Alkali treatment—Sodium hydroxide	Chitosan	Varun et al. (2017)
Fish liver	Alkali digestion—Hydrochloric acid	Fish oil/shark liver oil	Balachandran (2001)
Salted or fresh skins	Washing—Sodium hydroxide Hydrolysis—Acetic acid (catalyst)	Fish glue	Akter et al. (2016)
Gall bladder	Preservation—Hydrochloric acid	Insulin	Balachandran (2001)
Processing waste of fish	Partial hydrolysis—Acetic acid	Fish albumin	Balachandran (2001)
Processing waste of fish	Acid hydrolysis—Hydrochloric or sulfuric acid Alkali hydrolysis—Sodium hydroxide	Peptone	Gildberg (2004)
Shark skin	Softening—Calcium hydroxide containing some sodium sulfide	Shark leather	Balachandran (2001)
Shark fins	Soaking—Glacial acetic acid	Shark fin rays	Balachandran (2001)
Fish bones	Alkali treatment—Calcium carbonate	Fish calcium	Balachandran (2001)
Fish offal	Digestion—Sulfuric acid	Fertilizer	Balachandran (2001)
Head, shell, viscera, and gills of crustaceans	Acid treatment—Hydrochloric acid	Glucosamine hydrochloride	Balachandran (2001)
Shark teeth	Alkali treatment—Sodium hydroxide Bleaching—Hydrogen peroxide	Shark teeth (ornaments)	Gopakumar (2002)

Table 5 Biological/enzymatic method of treating fish waste

Type of waste	Processing method	Final product	References
Fish liver	Enzymatic digestion—Papain	Fish oil/shark liver oil	Balachandran (2001)
Fish scales	Enzyme—Papain	Pearl essence	Balachandran (2001)
Processing waste of fish	Hydrolysis—Peptic/tryptic enzyme	Peptone	Balachandran (2001)
Processing waste and neglected raw materials—fins, viscera, gills, and head	Fermentation by bacterial action-*Lactobacillus plantarum*	Fish silage	Gopakumar (2002)
Shark skin	Hydrolysis—Trypsin	Shark leather	Balachandran (2001)
Processing waste of fish	Fermentation by enzymes	Pit fish manure	Govindan (1982)
Head, gut, and trimmings of fish	Enzymatic treatment—Bromelain/papain	Fish protein hydrolysate	Zamora-Sillero et al. (2018)
Processing waste	Enzymatic treatment—Protease enzymes	Foliar spray	Naung and Hpa-an (2019)

utilized in Southeast Asia countries for the utilization of waste and contributes to the circular economy. Fish silage has an excellent protein value composed of hydrolyzed fish protein and micronutrients. They mostly consist of free amino acids and peptides such as methionine, histidine, glycine, alanine, and tyrosine (Xavier et al. 2017).

The raw material selection (Table 6) plays a pivotal role in arbitrating the quality of the silage. The raw material should be fresh and natural, ideally only after several hours of fish being processed into the by-products. However, the low-quality raw material can be utilized as fertilizer and not for feed preparation.

The main aim of fish silage production is to dually preserve the raw materials and augment the bioavailability of the nutrients. Thus, the fish silage paves an excellent pathway to reduce waste and convert them into a valuable product dually in terms of economy and nutrition. The microbial spoilage in the silage is prevented by lowering

Table 6 Raw materials used in fish silage

Raw material for fish silage	Sources	References
Low-value fishes of different species composition	Trawling operation by-catches	Raa et al. (1982)
Processing waste (Shrimp peeling, fish fileting, and surimi products)	Processing plants	Raghuanth and Gopakumar (2002)
Neglected raw materials (fins, viscera, gills, and head)	Retail markets	Gopakumar (2002)

the pH, which is done by the addition of acid or in situ acid production. The presence of fermentable sugar initially prevents the deamination of amino acids by bacteria and in the case of biological silage. Later, the lactic acid bacteria predominate suppressing or killing the spoilage bacteria by lowering the pH and producing numerous antibiotic substances called bacteriocins.

The common ways involved in fish silage production are chemical silage and biological silage. In chemical silage, the preservation is carried out either by the addition of organic acids (formic, citric, and propionic acids), inorganic acids (hydrochloric, sulfuric, and phosphoric acid), or a mixture of both. Fish silage prepared using inorganic acid requires neutralization utilizing chalk or high salt levels, which is not nutritionally satisfactory. Therefore, organic acids like formic acid at 2–3% are more suitable for storing in a tropical climate. The biological silage is an alternative process in which a slurry is prepared using minced fish, 5–10% by weight of molasses, and 30% by weight of water to which *Lactobacillus plantarum* of 18–22 h old culture is inoculated, mixed well and allowed for fermentation process for 72 h. LAB helps in transforming the sugar molecules to lactic acid and resulted with the lowered pH (4.5), which in turn reduces the growth of spoilage bacteria (Xavier et al. 2017).

Fish Silage Production

Fish silage production is a simple process that does not require sophisticated equipment. However, proper and regular cleaning, checking, and maintenance of the equipment are vital to obtaining good quality silage. The equipment is differentiated from low-cost manual units to higher automated plants. The various components involved are a grinder, pumps, mixing tank, and storage tank. The grinder helps to grind the product to less than 1 mm particle size which ensures the penetration of preservative into the core particle. The pump assists in the proper circulation of the product to expose the particles to acid and enzymes which enables the transformation of fish into silage and the pumps are also used in transferring the silage from one place to another. The mixing tank which is made of acid-resistant material is used for mixing fresh raw material, acid, and antioxidant. The producers use sulfite (e.g., Potassium metabisulphite) to control the oxidation and growth of fungi. The storage tank should be resistant to corrosion and a weekly check of the contents in the storage tank (check of pH) should be carried out to prevent rotting across the tank pockets (Fig. 2).

The protective face shields and safety glasses should be worn throughout the process. Acid-resistant gloves and rubber boots should be worn during handling and mixing of acids with silage. The production of silage requires the knowledge of raw material and silage. This knowledge helps in predicting the behavior of each substance. The substantial physical properties of silage are viscosity, termination of hydrolysis, and heat convection coefficients (Arason 1994).

The optimal temperature between 5 and 40 °C should be maintained which has a good impact on hydrolyzing the proteins. High temperature leads to enzyme inactivation, while low temperature prolongs the process time. The undissolved bones at the bottom should be removed to arrest the pH rise which could lead to a rotting process of silage. Decantation of silage should be carried out to separate the oil.

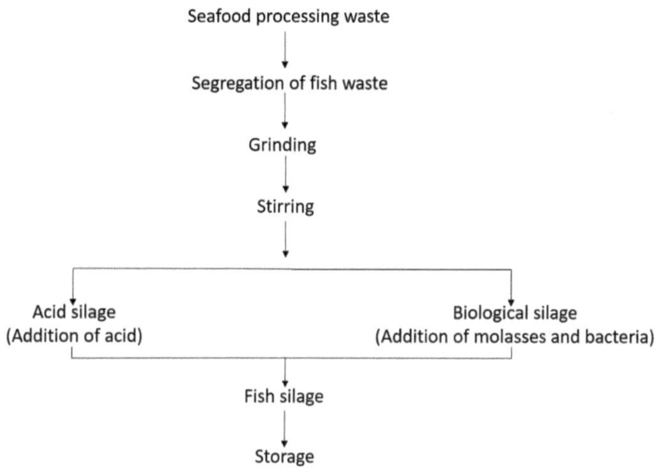

Fig. 2 Fish silage production

Antioxidants should be added along with acids to maintain the silage's quality. The fish silage could be stored for years without any prominent reduction in its nutritional quality. However, regular stirring and pH maintenance are essential.

The fish silage is rich in nutrients and has higher digestibility due to hydrolyzed proteins. The low acidity level of silage makes it suitable to be directly used as feed for pigs without any prior treatment resulting in higher growth rates. The inclusion of silage in feed ingredients produced stronger pellets thereby reducing the wastage during feeding and dust during transportation (Table 7).

Table 7 Pros and cons of fish silage

Pros of fish silage	Cons of fish silage
No expensive equipment is required	High water content makes it difficult for long-distance transportation
It reduces environmental impact by recycling the fish waste	Gases like CO_2 and H_2S produced in the tank residues can lower the oxygen levels causing fatal effects on the workers
It provides nutrients, bioactive components, and essential supplements to improve animal health	
It economically supports by converting waste into a valuable product that could promptly replace expensive feed ingredients	

4.2 *Animal Feed*

Animal feed from food/fish waste has gained increasing attention due to its multiple benefits to the human and environment also reducing the cost of feed production (Westendorf 2000). Fish/fishery waste has proven to possess essential protein, fat, and minerals. Fish waste has a plentiful number of fatty acids (monosaturated acids, oleic acids, and palmitic acid) and a low concentration of harmful substances (such as Hg, Pb, Cd, and As). The utilization of food waste as animal feed (protein source) has been studied and it has also been used as poultry and swine feed.

Fish wastes from white fish (haddock, cod, and plaice) are used to prepare intermediate moisture animal feed (Pig feed). The fish silage has a good amino acid profile which is favorable for monogastric animals, though it has less amount of lysine, methionine, and tryptophan when compared with the fish meal (Kjos et al. 1999). Therefore, they have been directly fed to pigs without any prior mixing and treatment. The pigs fed with fish silage have higher growth rates, good health, and reduced mortality.

Currently, fish waste is being used as a substrate for rearing insects (*Hermetia illucens*), which is further used as aquafeed ingredients for Asian sea bass (Chaklader et al. 2021).

4.3 *Fish Sauce*

In the past, the consumption of fermented fishery products has been on-trend. Fish sauce is a clear liquid (brown in color) seasoning that is utilized in Southeast Asia. It is basically produced by a combination of fish and salt (2:1 marine water fish, 3:1—freshwater fish) and allowed to fermentation for a duration of more than 6 months (30–35 °C). The resultant product is rich in flavor-active compounds which are developed during the progressive fermentation process (Zang et al. 2020).

The fish sauce is popular in Southeast Asia having different names depending on the country it has been manufactured (Table 8). For instance: Burma-Ngapi, China-Yeesui.

Thai fish (Nam pla) sauce is very popular in Western countries, specifically in the US when compared to other fish sauces. The fish sauce mostly produced in Thailand has expanded its domestic production to an international level over the last 5 decades.

The two methods involved in fish sauce production are the traditional and other is enzymatic methods. Traditionally, the fish is grounded and placed in jars made up of clay with layers of salt (fish to salt in the ratio of 3:1), this ratio depending on the country (Beddows et al., 1998). Then, they are buried in the ground and left to ferment for several months. The addition of proteolytic enzymes before fermentation is carried out in the enzymatic method. However, the price of proteolytic enzymes is very expensive which becomes a constraint for fish sauce production. In that case, alternative papaya sap which contains papain, a proteolytic enzyme, or young

Table 8 Fish sauce manufactured in various countries

Country	Fish sauce	Commonly used species
Malaysia	Budu	*Stolephorus spp.*
Philippines	Patis	*Clupea spp.* *Stolephorus spp.* *Leiognathus spp.* *Decapterus spp.*
Indonesia	Ketjap-ikan	*Clupea spp.* *Stolephorus spp.* *Osteochilus spp. (freshwater fish)* *Leiognathus spp.*
Vietnam and Cambodia	Nouc-nam/Nouc-mam	*Stolephorus spp.* *Rastrelliger spp.* *Engraulis spp.* *Decapterus spp.*
Thailand	Nam pla	*Rastrelliger spp.* *Cirrhinus spp.* *Stolephorus spp.*
Japan	Ishiru/Shottsuru	*Astroscopus japonicus* *Clupea pilchardus*
India and Pakistan	Colombo-cure	*Rastrelliger spp.* *Cybium spp.* *Clupea spp.*
Korea	Aekjeot	*Engraulis japonica* *Astroscopus japonicus*

pineapples containing bromelain can be utilized. The enzymatically produced fish sauce has a different color and odor however the same nutritional content as traditional fish sauce. The filtration is done and the filtrate is further ripened under the sun (2–4 weeks) after the fermentation process to get top-grade fish sauce. Ultrafiltration (UF) has currently been used for recovering antioxidant peptides (Najafian and Babji 2019) in fish sauce production, whereas electrodialysis has been carried out for desalination (Ratanasanya et al. 2018).

Fish sauce is tremendously affected by the five major factors which include salt, fish, fish and salt ratio, oxygen level, and minor ingredients. The unique flavor and aroma of fish sauce are mainly caused due to the presence of volatile compounds, aspartic acid and glutamic acid (amino acids), nucleotides, peptides, and succinic acid (organic acid). The three major contributory factors of fish sauce aroma are ammoniacal, cheesy, and meaty notes. Fish sauce consists of halophilic bacteria due to the high level of salt content. The major roles of these bacteria are protein degradation and flavor-aroma development.

The quality of fish sauce varies based on the type of processing method and fish species. The minerals found in the fish sauce are potassium, sulfur, phosphorus, magnesium, calcium, and iron, etc. In addition, water-soluble vitamins such as B_{6}, B_{12}, thiamine, niacin, and riboflavin are presented in the sauce.

Table 9 Pros and cons of fish sauce

Pros of fish sauce	Cons of fish sauce
It is used as an ingredient in cooking	The distinctive odor limits its usage in food preparation
It consists of all essential amino acids	
The production technique is simple and cheap	

The chemical composition (pH, nitrogen content, and volatile acids) varies depending on manufacturing conditions. In general, the amino acid content increases due to polypeptide nitrogen breakdown (Table 9).

4.4 *Biofuel*

The fish waste consists of a good amount of fish oil. The yield generally depends on the fat level of fish species used, almost 50% of the body weight is utilized as waste. It possesses better quality fish oil, which can be utilized for the production of biofuel.

Nowadays, one of the greatest challenges is to maintain the global energy supply security. In that case, biofuel plays a viable role as an alternative replacement for conventional fuels. The biofuels obtained from the wastes are classified into two useful products: biodiesel and biogas (bio-methane). In chemical terms, biodiesel contains alkyl esters (long-chain fatty acids), whereas biogas is composed of methane and carbon dioxide. Biodiesel can be extracted using other methods such as trans-esterification, single-step trans-esterification using an alkaline catalyst, microwave-assisted lipid extraction, separation process, two-stage reaction process, and conventional processes including wet rendering and dry rendering. Biogas is produced during the anaerobic digestion of fish waste. Anaerobic digestion is a process in which various microorganism are involved and undergoes biodegradation. The fats, proteins, and carbohydrates present in a fish waste substrate have biologically degraded to biogas during this condition. Therefore, the yield of biogas depends on the content of the organic compounds. The microbes involved in the process are mostly acetogens and methanogens. The four steps involved in the decomposition process are hydrolysis, acidogenesis, acetogenesis, and methanogenesis.

In the hydrolysis step, the enzymes like amylases, lipases, proteases, cellulases, and hemicelluloses help in converting polysaccharides into small monomers, which paves the way for easier digestion for bacteria. This step is followed by acidogenesis, acetogenesis, and methanogenesis, where the methanogen bacteria convert the acetogenesis products to methane and carbon dioxide. The methane (by-product) is used as biogas. Several varieties of reactors are used in biogas production from fish waste.

Table 10 Pros and cons of biogas

Pros of biogas	Cons of biogas
It is used as an alternative source of energy	Risk of gas poisoning
Biogas production is eco-friendly and cost-effective	It requires a large land area for the establishment
It reduces greenhouse gas (GHG) emissions and the recovery of methane from the biogas can provide a cost-effective source of renewable energy	Huge investment needed for setting up of biogas plant

Though fish waste has high valued organic carbon for methane production, the presence of a high amount of N_2 and NH_3 is limited. The anaerobic digestion of fish waste is feasible with co-digestion, in which the fish waste is mixed with slurry or other wastes hence biogas is produced. The limitation of co-digestion is maintaining the balance of macro and micronutrients (Yuvaraj et al. 2019) (Table 10).

4.5 Cosmetics

The regaining of chemical compounds from fish waste is a promising field of study to develop valuable by-products (cosmetics, nutraceuticals, and pharmaceuticals). Chitosan is generally extracted from shrimp and crab shells. However in recent times, the extraction of chitosan from scales is being extensively carried out (Kumari and Rath 2014). The chitosan has the ability to form films and poly oxy salts and possesses optical characteristics that make them attractive for utilization in cosmetic industries. The cationic and humectant properties of chitosan make a wiser application in skin and hair care products. In addition, chitosan is an excellent flocculent and adheres easily to hair and skin. Therefore, it has tremendous usage in the field of cosmetics. Although, the extraction of chitosan and chitin from fish scales has promising results, only limited study has been carried out for the characterization of these compounds.

Collagen obtained from fish waste has good antioxidant properties making it suitable for the cosmetic industry. Collagen peptides, an emerging substance obtained from fish waste prevent skin wrinkles and improve the elasticity of the skin. Moreover, a collagen drink prepared from fish is examined for improvement in the aging of the skin. The results showcased that collagen drink improved the cell viability of ultraviolet A irradiated human skin fibroblast and reduced reactive oxygen species production. In addition, it improved elastin and collagen production (Lin et al. 2020).

Gelatin is a derivative of its parent compound collagen. Gelatin is obtained by the partial hydrolysis of collagen. Gelatin is utilized as a gelling agent in face creams, bath salts, body lotions, and shampoos in the cosmetic industry (Elgadir et al. 2013). It protects against UV radiation by boosting immunity, thereby increasing hydroxyproline in the skin, which makes fish gelatin a novel ingredient for future skin anti-aging products.

4.6 *Value-Added Products*

4.6.1 Fish Protein Hydrolysate

FPH has prepared from fish/fish waste using the protein hydrolyzation process (the breaking down of proteins). Generally, FPH has a range of improved qualities including better functional and bioactive properties than anti-oxidative or anti-hypertensive activity (He et al. 2013). FPH has recently been employed as cryoprotectants for frozen fish items.

FPH is prepared in two types: liquid and dry. Liquid FPH is a watery combination of hydrolyzed proteins (90% moisture). FPH in the liquid state is exceedingly unstable for prolonged storage and is also difficult to transport. Therefore, dried FPH is preferred since it has a better shelf life than liquid FPH and is also easier to store and transport. However, in liquid FPH, the removal of large volumes of water is a difficult and expensive task, which is one of the problems faced by dried FPH manufacturers. Thus, the FPH has enough protein sources for human consumption. However, the process of dehydration requires a large amount of energy and is thus quite expensive (Petrova et al. 2018).

Extraction of FPH

FPH is produced by various techniques such as autolysis, chemical hydrolysis (acid and alkaline hydrolysis), enzymatic hydrolysis, and bacterial fermentation. Chemical hydrolysis and enzymatic hydrolysis are the most often utilized procedures due to their numerous advantages. The chemical hydrolysis technique is affordable, quick, and yields high protein recovery. Nevertheless, there is limited control over the uniformity of hydrolyzed products with considerable variability in the free amino acid profile due to non-specific peptide bond breaking (Siddik et al. 2021).

Enzymatic hydrolysis is extensively used to manufacture precise hydrolysates that maintain the nutritional content of the source protein in order to make high-grade FPH (Zamora-Sillero et al. 2018). This method requires minimal reaction time and is effective in targeting certain peptide bonds and amino acids with optimal activity under certain circumstances. Furthermore, the final product of enzymatic hydrolysis does not include any leftover organic solvents or hazardous compounds (Najafian and Babji 2012).

Debittering of FPHs: Fish protein hydrolysate is quite bitter in taste. It is plausibly caused by certain peptides present in it. An enzyme (acidic serine carboxypeptidase) isolated from the hepatopancreas of the squid can reduce the bitterness of FPH by eliminating certain amino acids in the carboxy terminal (Komai et al. 2007) (Fig. 3).

Commercial Significance of FPH

FPH is used as a flavor enhancer in dietic foods such as macaroni, meringues, souffles, or bread. However, in the preparation of fish paste, fish soup, and shellfish analogs, which is used as the source of small amino acids and peptides (Muzaddadi et al. 2016). FPH is reported to be an effective growth promoter for different kinds

Fig. 3 Extraction of FPH

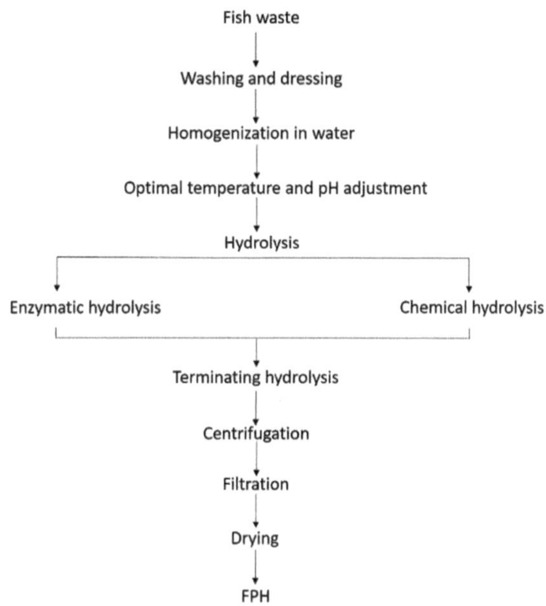

of bacteria, yeasts, and molds due to the high quantity of amino acids, peptones, and other important elements (Guerard et al. 2001). FPH has improved antioxidant, anticancer, immunomodulatory, anticoagulant, anti-obesity, and anti-hypertensive properties and is widely employed in nutraceuticals. Vasotensin, Amizate, Protizen, Seacure, and Molval are commercial nutraceutical products of FPH marketed in several countries (Elavarasan 2019) (Table 11).

Table 11 Pros and cons of FPH

Pros of FPH	Cons of FPH
By chemical hydrolysis, – High recovery yields – Fast process – Process cost is low	By chemical hydrolysis, – Removal of cysteine and tryptophan – Partial elimination of serine, threonine, and tyrosine
By enzymatic hydrolysis, – High recovery yields – Having higher nutritive value – Contamination of waste is low	By enzymatic hydrolysis, – High-cost enzymes (except autolysis) – Bitterness – Long-time process
Easily digestible and highly nutritious	
Used as a flavor enhancer, stabilizer in beverages, milk replacers, and protein supplements	

Source Pedziwiatr (2017)

4.6.2 Collagen/gelatin

Collagen is the fibrous protein and presents in the extracellular matrix of the body's connective tissues (i.e., bones, skin, tendons, ligaments, and cartilage) (Muller et al. 2003). It is the only protein that is abundant in the animal kingdom. Plants and unicellular creatures lack it, as polysaccharides and cellulose fill the void. Generally, collagen is found in the body walls and cuticles of invertebrates (Silvipriya et al. 2015). Collagen's fundamental structural unit is made up of three polypeptide chains. Each chain is 300 nm long and includes 1050 amino acids twisted around one another in a typical right-handed helical shape. It has a molecular weight of about 290 kDa and a diameter of around 1.5 nm (Silvipriya et al. 2015). In general, collagen is extracted from the skin, bones, scales, fins, and swim bladder of fishes.

Collagen from fish has two α-chain variants such as α_1 and α_2 (Gómez-Guillén et al. 2002). Changes in the amino acid composition of α chain types, end up with small differences in hydrophobicity. Despite having the same molecular weight (95,000 Da), these chain variants can be separated by SDS-PAGE because of their differing affinity.

Gelatin—A derivative of collagen

Gelatin is eventually obtained by heating the collagen over the transition temperature of the triple-helix structure commonly known as "superhelix." A three-peptide strand with a helix structure forms the superhelix (Ward and Courts 1977). The collagen molecule's helical shape collapses; subsequently, the molecular chains unfurl and decrease in molecular weight and such changes occur in a very narrow range of temperatures (Rahman et al. 2008). This treatment is required to destabilize the protein's non-covalent bonding and results in sufficient swelling and rupture of inter- and intra-molecular bonds, which leads to collagen solubilization (Gómez-Guillén et al. 2002) and gelatin conversion with improved hydration capacity (Mohtar et al. 2010).

Extraction of gelatin

Commercial significance of gelatin

Gelatin is a significant biodegradable polymer and is extensively used in the food, pharmaceutical, cosmetics, and photographical industries due to its effectiveness and technological capabilities. However, gelatin is not a naturally occurring protein. This heterogeneous protein has outstanding film-forming capability, high flexibility, good processability, effective gas barrier properties, high availability, and cheap cost (Mousazadeh et al. 2021) (Fig. 4).

In the food industry, fish gelatin is a highly adaptable food additive generally employed to increase the stability, consistency, and elasticity of food and is utilized in the confectionery, packaging, and the dairy processing industry. Gelatin is typically advised to be used in foods to lower the carbohydrate content and increase the protein content in diets designed for diabetes people because it has minimal calories.

Fig. 4 Extraction of gelatin

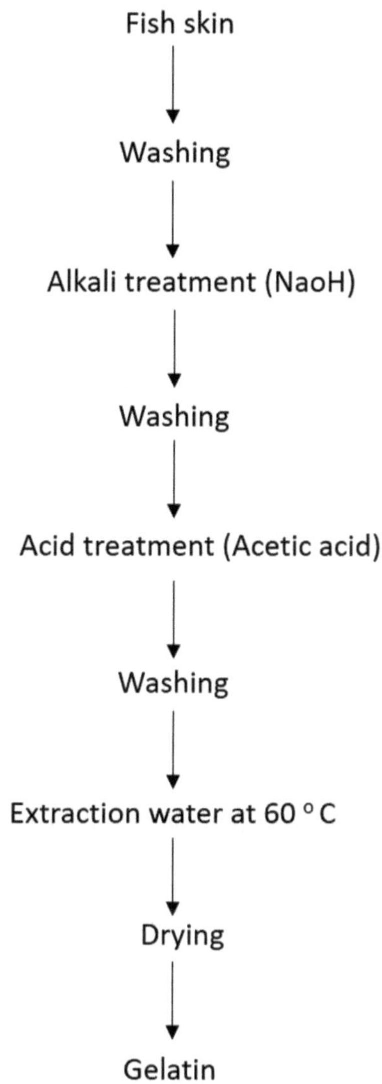

In the pharmaceutical industry, fish gelatin is widely used for the production of numerous products such as ointments, cosmetics, capsules, tablet coatings, and emulsions. In addition, fish gelatin can be used in the production of microencapsulated foods and other pharmaceutical additives.

4.6.3 Proteins, Minerals, and Enzymes

Proteins

Muscle proteins are highly nutritious and readily digestible and abundant in fish frames. As a result, rather than being dumped as trash, proteins from fish waste may be recovered using several techniques. Over the years, three techniques have been used to recover protein from fish processing waste. They have aimed to break the peptide bonds from the encrypted mother proteins or isolate the protein using pH changes. Li-Chan (2015) used the strategy to produce peptides and bioactive protein hydrolysates in one of the following three ways, namely chemical, enzymatic, or fermentation techniques. The recovery method used for fish protein is influenced by the product cost, quality, and time (Ananey-Obiri et al. 2019).

Enzymatic Hydrolysis: The most prevalent technique of recovering protein from seafood processing waste is usually by enzymatic or proteolytic hydrolysis. This process is carried out naturally with proteolytic enzymes or expedited and regulated by exogenous enzymes.

Chemical Hydrolysis: The chemical approach of recovering fish protein entails breaking down the proteins into peptides. In general, this process of recovering proteins is quite easy and affordable. However, it is a challenging process to manage due to the undefined peptide breakage during hydrolysis. In acid hydrolysis, HCl or H_2SO_4 is used to hydrolyze the proteins at high temperatures and also at high pressure (Kristinsson and Rasco 2000).

Fermentation Hydrolysis: Microorganisms are used in fermentation to break down fish proteins into amino acids and peptides. A variety of microbes have been found and employed to extract fish protein. The functionality of FPHs recovered by this technique might vary due to changes in microbes used in the fermentation culture (Daliri et al. 2017). In this method, protein hydrolysates have been produced by using lactic acid bacteria (LAB) *Pediococcus acidilactici* NCIM5368 (Chakka et al. 2015) and *Enterococcus faecium* NCIM5335 (Balakrishnan et al. 2011).

Minerals

Fish bones have generally been separated after the removal of muscle proteins from the frames. Minerals such as calcium, hydroxyapatite, and phosphorus are found in 60–70% of fish bones (Kim and Mendis 2006). Generally, a lower amount of calcium is present in most people's diets. Therefore, the consumption of smaller fishes can help to increase the level of calcium in the human body. The fish bones obtained from the processing industry can be used to improve calcium levels in our body. Bones should be converted into edible form before being used as a fortified meal (Ishikawa et al. 1990). Fish bones are an excellent source of $Ca_{10}(PO_4)6(OH)_2$ (hydroxyapatite) that can be used for the medical application. In the past, bone fractures are repaired using autografts, xenografts, and allografts. However, these materials are proven to be inefficient owing to mechanical instability and incompatibility.

Enzymes

Fishery waste such as head, liver, viscera, and shell are abundant in enzymes such as proteases, alkaline phosphatase, chitinase, and hyaluronidase. These enzymes are mostly abundant in the intestines, followed by pancreatic tissues, pyloric ceca, hepatopancreas, shell, and other waste components (Sriket 2014). It must be essential in relatively large quantities and acceptable quality for enzyme recovery. The commercially viable enzyme separation techniques from crude waste extracts are ultrafiltration, precipitation by salts and polyacrylic acids, isoelectric solubilization/precipitation, pH shift, overcooled acetone extraction, flocculation and membrane filtration (Benjakul et al. 2010).

Proteases are the digestive enzymes usually found in the digestive system of marine animals. It can be classified as endopeptidases or exopeptidases based on the hydrolysis of a protein molecule. Other digestive enzymes including trypsin, pepsin, gastricsin, chymotrypsin, elastase, carboxypeptidase, collagenase, and carboxyl esterase are also found in marine animals.

Lipases are a diverse group of enzymes that catalyze the hydrolysis of ester bonds in various substrates including phospholipids, triglycerides, vitamin esters, and cholesteryl esters (Venugopal 2016). Hydrolytic enzymes known as chitinases are capable of dissolving the glycosidic linkages found in chitin polymers. True chitinases are divided into chitinase and chitobiase and these enzymes are commonly found in species that must either remodel their own chitin or digest chitinous material consumed as food. Animals that feed on worms and arthropods require chitinases as their primary food source to digest their diet (Kim and Dewapriya 2014).

Other enzymes: Hyaluronidase, an animal tissue constituent that breaks hyaluronate to improve tissue permeability is recovered from shrimp shells which found in shallow tropical seas. Alkaline phosphatase, nucleoside phosphorylase, 5′-nucleotidase, urease, xanthine oxidase, trimethylamine oxide (TMAO) reductase, and TMAO-demethylase are enzymes that have been used in seafood quality evaluation (Benjakul et al. 2010) (Table 12).

4.6.4 Bioactive Compounds

Fish is a vital source of nutraceuticals due to its unique combination of bioactive compounds such as long-chain PUFAs (DHA and EPA), ω-3 PUFAs, peptides, protein hydrolysates, minerals, amino acids, vitamins, collagen, gelatin, fish bone, fish oil and fat-soluble vitamins (Kundam et al. 2018). Numerous nutraceuticals are used for preventative reasons in several sectors of medicine due to the high therapeutic value of seafood by-products. They can be used to reduce complications or treat a variety of disorders, including cardiovascular disease, cancer, rickets, viral infections, hypertension, dermatologic issues, particularly during pregnancy, and parasite infections (Siddiqui et al. 2015). In addition to these advantages, fish-derived nutraceuticals have anti-inflammatory, anticoagulant, and antioxidant properties (Tables 13 and 14).

Table 12 Seafood enzymes and their applications

Applications	Enzymes involved	Process	References
Recovery of proteins from fish processing waste	Proteases—Pepsin, trypsin, and pyloric caeca	Collagen is extracted from the fins, skins, scales, bones, head, and swim bladders of bighead carp using collagenases, pepsin from tuna, or trypsin from cod or tuna pyloric caeca	Ahmad and Benjakul (2010)
Fish protein hydrolysates	Trypsin, alcalase, chymotrypsin, and pepsin	FPH is produced by treating fish flesh with alcalase, trypsin, pepsin, chymotrypsin, or other enzymes under optimum pH and temperature	Venugopal (2016)
Debittering of FPHs	Acidic serine carboxypeptidase	An acidic serine carboxypeptidase isolated from squid hepatopancreas can reduce bitterness by eliminating certain amino acids in the carboxy terminal	Komai et al. (2007)
Fish sauce	Proteolytic enzymes	Fish sauce is a popular condiment due to its distinct flavor and taste. It is made by autolyzing fish with in situ proteolytic enzymes. Exogenous proteases can speed up the manufacturing process	Lopetcharat et al., 2001
Ripening of fermented fishery products	Exogenous proteases	Ripening leads to the formation of a distinctive taste and soft texture in fermented fishery products during storage, which may be hastened by exogenous proteases	Bekhit (2011)
Meat tenderization	Proteases	Shrimp protease can tenderize meat. The harsh texture of squid rings can be loosened by protease treatment after removing the rough outer proteoglycans layer of the rings	Aoki et al. (2004) and Melendo et al. (1997)
Caviar production	Collagenases	Caviar is cured fish eggs from species such as salmon, white sturgeon, and trout. Manually removing roe sacks of eggs to obtain caviar is time-consuming. Hence proteases, notable collagenases, can help with the easy removal of supporting tissue	Gildberg (1993)
Production of PUFAs	Lipases	Lipases can be used for the enrichment of DHA and EPA in fish oils	Venugopal (2016)

(continued)

Table 12 (continued)

Applications	Enzymes involved	Process	References
Flavor enhancement	Lipases	Lipases enhance the flavor together with calpains, muscle cathepsins, aminopeptidases, and peptidases	Bekhit (2011)
Shelf-life extension of fishery products	Lysozyme, glucose oxidase, and catalase	Dipping fresh shrimp in lysozyme at concentrations up to 150 g/mL inhibits the growth of spoilage microorganisms due to the enzyme's impact on the mucopeptide structure of microbial cell walls. The combination of glucose oxidase, lysozyme, and catalase can promote fish quality	Myrnes and Johansen (1994)
Color retention in cooked and frozen shrimp	Glucose oxidase	The characteristic yellow color of pre-cooked frozen shrimp or crab caused by oxidation can be minimized by immersing cooked shrimp in glucose oxidase	Venugopal (2016)

Table 13 Finfish processing waste and possible bioactive components

Percentage of waste from whole fish	Valuable bioactive components
Head (14–20%)	Gelatin, collagen, lipid, protein hydrolysate, protein, squalene, bioactive peptides, cartilage, flavor, and calcium
Gut (15–20%)	Protein hydrolysate, lipid, protein, squalene, bioactive peptides, enzymes, and flavor
Skin (1–3%)	Gelatin, collagen, elastin, and cartilage
Bones (10–16%)	Gelatin, collagen, cartilage, calcium, chondroitin sulfate, and other minerals
Trimmings (1–5%)	Protein hydrolysate, protein, and bioactive peptides

Source Chandrasekaran (2012)

- Intestine waste of fish species such as mackerel, black halibut, spiny dogfish, arctic char, salmon, fried calamari, sardine, and anchovies are rich in bioactive components nutraceuticals such as ω-3 and ω-6 fatty acids. It has a myriad of medical applications particularly therapeutic and clinical applications such as improving insulin sensitivity, anti-inflammatory agent, reducing the risk of cardiovascular problems, increasing vascular adhesion molecule-1 expression, and improving eyesight (Ashraf et al., 2020).
- Chitin and chitosan are recovered from the crustacean. It accelerates the healing of wounds, lowers blood cholesterol, acts as an anti-ulcer and anti-aging agent, and is used in cosmetics and ophthalmology.
- Chondroitin recovered from shark cartilage reduces the risk of osteoarthritis and supports the nutritional supplement.

Table 14 Shellfish processing waste and possible bioactive components

Percentage of waste from whole shellfish		Valuable bioactive components
Crustaceans		
Shrimp/prawn	Head and shell (65–85%)	Chitosan oligosaccharides, chitosan, chitin, N-acetyl chitooligosaccharides, protein hydrolysate, D-glucosamine, pigment protein, flavor, and enzymes
Crab	Back shell, viscera, gills, and claws shell (60–70%)	Chitin, pigment
Lobster	Head and shell (up to 60%)	Pigment, chitin, and flavor
Krill	Head and shell (7–74%)	Pigment, chitin, oil, and protein hydrolysate
Crayfish	Head and shell (up to 85%)	Pigment, chitin, oil, and flavor
Mollusks and cephalopods		
Scallop, clam, oyster, mussel, etc.	Shell and non-edible part (60–80%)	Protein hydrolysate, enzyme, and flavor
Squid	Gladius or pen, ink bag, liver, and other organs (8–20%)	Bioactive peptides, collagen, gelatin, chitin and enzymes
Octopus	Intestine, eyes, and mouth apparatus (25–32%)	Gelatin and collagen
Coelenterate and echinoderm		
Sea urchin	Shell and viscera (10–20%)	Gelatin and collagen

Source Chandrasekaran (2012)

- Glucosamine recovered from shellfish shells acts as an anti-inflammatory agent, reduces the risk of osteoarthritis, and supports the dietary supplement.
- Body walls and cuticles of invertebrates are rich in collagen. It reduces the risk of osteoarthritis, reduces hypertension, plays a vital role in tissue engineering, and acts as an antioxidant, anti-hypertensive, and anti-skin aging.
- Hyaluronidase is an endo-enzyme that is extracted from shellfish waste and has enormous therapeutic applications.
- Squalene extracted from shark liver reduces the risk of cardiovascular problems and acts as an antioxidant, anti-bacterial, antifungal, and anti-cancerous agent.
- Squalamine extracted from the liver, gallbladder, and intestine waste of sharks is effective against bacteria and fungi and plays a vital role in tumor angiogenesis activities.
- Astaxanthin, β-carotene, lutein, and Zeaxanthin are rich in freshwater fish, red fishes, and other fishes. It has antioxidant and anti-atherogenic properties, kills cancer cells, reduces the risk of neurological, cardiovascular problems, and psoriasis, and also plays a vital role in cosmetics.

5 Conclusion

The success of any food realm depends on sensible waste management. Fish being a highly perishable product needs more attention for waste management in the fishery industry. Fish wastes are rich in a number of novel components which include bioactive compounds, peptides, enzymes, etc. The effective utilization of the resourceful fish waste and its by-products enables the cut-off in dumping the wastes in aquatic environments which leads to eutrophication. Fish waste valorization plays a major role in achieving one of the Sustainable Development Goals' 2030 – 'zero discards.' It also paves the way to attain a sustainable circular economy by eminent utilization of other discarded fish wastes and by-products in the food chain. To date, there are several technologies to convert fish waste into valuable resources used in a myriad of industries paving the way to attaining a green approach in the fisheries industry. In a nutshell, the conversion of fish wastes into high-value products is promising and the intervention of food industries can bring a way forward in the future, in terms of product diversification, resource mobilization, and sustainable growth.

References

Ahmad M, Benjakul S (2010) Extraction and characterisation of pepsin-solubilised collagen from the skin of unicorn leatherjacket (*Aluterus monocerous*). Food Chem 120(3):817–824

Akter S, Rahman MA, Bhowmik S, Haidar MI, Alam AN (2016) Assessment of fishery wastes and suitability of its utilization in the manufacture of fish glue. Am J Food Nutrition 6:77–81

Ananey-Obiri D, Matthews LG, Tahergorabi R (2019) Proteins from fish processing by-products. In: *Proteins: Sustainable source, processing and applications*. Academic Press, pp 163–191

Aoki H, Ahsan MN, Matsuo K, Hagiwara T, Watabe S (2004) Partial purification of proteases that are generated by processing of the Northern shrimp *Pandalus borealis* and which can tenderize beef. Int J Food Sci Technol 39(5):471–480

Arason S (1994) Production of fish silage. In: *Fisheries processing*. Springer, Boston, MA, pp 244–272

Ashraf SA, Adnan M, Patel M, Siddiqui AJ, Sachidanandan M, Snoussi M, Hadi S (2020) Fish-based bioactives as potent nutraceuticals: exploring the therapeutic perspective of sustainable food from the sea. Mar Drugs 18(5):265

Balachandran KK (2001) Post-harvest technology of fish and fish products. Daya Books

Balakrishnan B, Prasad B, Rai AK, Velappan SP, Subbanna MN, Narayan B (2011) In vitro antioxidant and antibacterial properties of hydrolysed proteins of delimed tannery fleshings: comparison of acid hydrolysis and fermentation methods. Biodegradation 22(2):287–295

Beddows CG (1998) Fermented fish and fish products. In: Microbiology of fermented foods. Springer, Boston, MA, pp 416–440

Bekhit AEA (2011) Fish Roe: Fermentation. Encyclopedia of biotechnology in agriculture and food

Benjakul S, Klomklao S, Simpson BK (2010) 10 Enzymes in fish processing. Enzymes Food Technol 211

Chakka AK, Elias M, Jini R, Sakhare PZ, Bhaskar N (2015) In-vitro antioxidant and antibacterial properties of fermentatively and enzymatically prepared chicken liver protein hydrolysates. J Food Sci Technol 52(12):8059–8067

Chaklader MR, Howieson J, Foysal MJ, Fotedar R (2021) Transformation of fish waste protein to *Hermetia illucens* protein improves the efficacy of poultry by-products in the culture of juvenile barramundi, *Lates calcarifer*. Sci Total Environ 796:149045

Chalamaiah M, Hemalatha R, Jyothirmayi T (2012) Fish protein hydrolysates: proximate composition, amino acid composition, antioxidant activities and applications: a review. Food Chem 135(4):3020–3038

Chandrasekaran, M. ed., 2012. Valorization of food processing by-products. CRC press

Daliri EBM, Lee BH, Oh DH (2017) Current perspectives on antihypertensive probiotics. Probiotics Antimicrob Proteins 9(2):91–101

Elavarasan K (2019) Health benefits and potential applications of fish protein hydrolysate. ICAR-Central Institute of Fisheries Technology

Elgadir AM, Mirghani ME, Adam A (2013) Fish gelatin and its applications in selected pharmaceutical aspects as alternative source to pork gelatin. J Food Agric Environ 11:73–79

Falch E, Rustad T, Jonsdottir R, Shaw NB, Dumay J, Berge JP, Arason S, Kerry JP, Sandbakk M, Aursand M (2006) Geographical and seasonal differences in lipid composition and relative weight of by-products from gadiform species. J Food Compos Anal 19(6–7):727–736

FAO (2013) Fisheries and aquaculture information and statistics service. Food and Agricultural Organization, Rome, Italy http://www.fao.org/figis/servlet/TabSelector?tb_ds = Production and tb_mode = Table and tb_act = Select and tb_grp = Country

Gildberg A (1993) Enzymic processing of marine raw materials. Process Biochem 28(1):1–15

Gildberg AR (2004) Enzymes and bioactive peptides from fish waste related to fish silage, fish feed and fish sauce production. J Aquat Food Prod Technol 13(2):3–11

Gómez-Guillén MC, Turnay J, Fernández-Dıaz MD, Ulmo N, Lizarbe MA, Montero P (2002) Structural and physical properties of gelatin extracted from different marine species: a comparative study. Food Hydrocolloids 16(1):25–34

Gopakumar K (2002) Textbook of fish processing technology

Govidan TK (1982) Fish processing technology. CIFT, Kochi

Guerard F, Dufosse L, De La Broise D, Binet A (2001) Enzymatic hydrolysis of proteins from yellowfin tuna (*Thunnus albacares*) wastes using Alcalase. J Mol Catal B Enzym 11(4–6):1051–1059

Guillen J, Holmes SJ, Carvalho N, Casey J, Dörner H, Gibin M, Mannini A, Vasilakopoulos P, Zanzi A (2018) A review of the European Union landing obligation focusing on its implications for fisheries and the environment. Sustainability 10(4):900

He S, Franco C, Zhang W (2013) Functions, applications and production of protein hydrolysates from fish processing co-products (FPCP). Food Res Int 50(1):289–297

Hertrampf JW, Piedad-Pascual F (2000) Fish solubles (de-hydrated). In: Handbook on ingredients for aquaculture feeds. Springer, Dordrecht, pp 211–216

Ishikawa M, Kato M, Mihori T, Watanabe H, Sakai Y (1990) Effect of vapor pressure on the rate of softening of fish bone by super-heated steam cooking. Nion Suisan Gakkaishi 56(10):1687–1691

Javeed A, Mahendrakar NS (1995) Effect of different levels of molasses and salt on acid production and volume of fermenting mass during ensiling of tropical freshwater fish viscera. J Food Sci Technol 32(2):115–118

Karim AA, Bhat R (2009) Fish gelatin: properties, challenges, and prospects as an alternative to mammalian gelatins. Food Hydrocolloids 23(3):563–576

Kim SK, Dewapriya P (2014) Enzymes from fish processing waste materials and their commercial applications. In: Seafood processing by-products. Springer, New York, NY, pp. 183–196

Kim SK, Mendis E (2006) Bioactive compounds from marine processing byproducts–a review. Food Res Int 39(4):383–393

Kjos NP, Skrede A, Øverland M (1999) Effects of dietary fish silage and fish fat on growth performance and sensory quality of growing-finishing pigs. Can J Anim Sci 79(2):139–147

Komai T, Kawabata C, Tojo H, Gocho S, Ichishima E (2007) Purification of serine carboxypeptidase from the hepatopancreas of Japanese common squid *Todarodes pacificus* and its application for elimination of bitterness from bitter peptides. Fish Sci 73(2):404–411

Kristinsson HG, Rasco BA (2000) Fish protein hydrolysates: production, biochemical, and functional properties. Crit Rev Food Sci Nutr 40(1):43–81

Kumari S, Rath PK (2014) Extraction and characterization of chitin and chitosan from (*Labeo rohita*) fish scales. Proc Mater Sci 6:482–489

Kundam DN, Acham IO, Girgih AT (2018) Bioactive compounds in fish and their health benefits. Asian Food Sci. J 4(4):1–14

Li-Chan EC (2015) Bioactive peptides and protein hydrolysates: research trends and challenges for application as nutraceuticals and functional food ingredients. Curr Opin Food Sci 1:28–37

Lin P, Hua N, Hsu YC, Kan KW, Chen JH, Lin YH, Lin YH, Kuan CM (2020) Oral collagen drink for antiaging: antioxidation, facilitation of the increase of collagen synthesis, and improvement of protein folding and DNA repair in human skin whfibroblasts. Oxidat Med Cell Longevity

Lopetcharat K, Choi YJ, Park JW, Daeschel MA (2001) Fish sauce products and manufacturing: a review. Food Rev Intl 17(1):65–88

Melendo JA, Beltrán JA, Roncalés P (1997) Tenderization of squid (*Loligo vulgaris* and *Illex coindetii*) with bromelain and a bovine spleen lysosomal-enriched extract. Food Res Int 30(5):335–341

Mohtar NF, Perera C, Quek SY (2010) Optimisation of gelatine extraction from hoki (*Macruronus novaezelandiae*) skins and measurement of gel strength and SDS–PAGE. Food Chem 122(1):307–313

Mousazadeh S, Ehsani A, Kia EM, Ghasempour Z (2021) Zinc oxide nanoparticles and periodate oxidation in developing pH-sensitive packaging film based on modified gelatin. Food Packag Shelf Life 28:100654

Muller WE (2003) The origin of metazoan complexity: Porifera as integrated animals. Integr Comp Biol 43(1):3–10

Muzaddadi AU, Devatkal S, Oberoi HS (2016) Seafood enzymes and their application in food processing. In: Agro-industrial wastes as feedstock for enzyme production. Academic Press, pp. 201–232

Myrnes B, Johansen A (1994) Recovery of lysozyme from scallop waste. Prep Biochem Biotechnol 24(1):69–80

Najafian L, Babji AS (2012) A review of fish-derived antioxidant and antimicrobial peptides: their production, assessment, and applications. Peptides 33(1):178–185

Najafian L, Babji AS (2019) Purification and identification of antioxidant peptides from fermented fish sauce (Budu). J Aquat Food Prod Technol 28(1):14–24

Naung SHH, Hpa-an M (2019) Foliar fertilizer production from waste fish

Pedziwiatr P (2017) Aquaculture waste management. Acta Innov 22:20–29

Petrova I, Tolstorebrov I, Eikevik TM (2018) Production of fish protein hydrolysates step by step: technological aspects, equipment used, major energy costs and methods of their minimizing. Int Aquatic Res 10(3):223–241

Raa J, Gildberg A, Olley JN (1982) Fish silage: a review. Crit Rev Food Sci Nutr 16(4):383–419

Raghunath MR, Gopakumar K (2002) Trends in production and utilization of fish silage. J Food Sci Technol (Mysore) 39(2):103–110

Rahman MS, Al-Saidi GS, Guizani N (2008) Thermal characterisation of gelatin extracted from yellowfin tuna skin and commercial mammalian gelatin. Food Chem 108(2):472–481

Ratanasanya S, Chindapan N, Polvichai J, Sirinaovakul B, Devahastin S (2018) Particle swarm optimization as alternative tool to sensory evaluation to produce high-quality low-sodium fish sauce via electrodialysis. J Food Eng 228:84–90

Rustad T, Storrø I, Slizyte R (2011) Possibilities for the utilisation of marine by-products. Int J Food Sci Technol 46(10):2001–2014

Sachindra NM, Mahendrakar NS (2014) Fish processing byproducts (Quality Assessment And Applications). Studium Press

Shahidi F, Varatharajan V, Peng H, Senadheera R (2019) Utilization of marine by-products for the recovery of value-added products. J Food Bioactives 6

Siddik MA, Howieson J, Fotedar R, Partridge GJ (2021) Enzymatic fish protein hydrolysates in finfish aquaculture: a review. Rev Aquac 13(1):406–430
Siddiqui AJ, Bhardwaj J, Goyal M, Prakash K, Soni A, Tiwari V, Puri SK (2015) Assessment of real-time method to detect liver parasite burden under different experimental conditions in mice infected with Plasmodium yoelii sporozoites. Microb Pathog 89:35–42
Silvipriya KS, Kumar KK, Bhat AR, Kumar BD, John A (2015) Collagen: Animal sources and biomedical application. J Appl Pharmaceut Sci 5(3):123–127
Sriket C (2014) Proteases in fish and shellfish: Role on muscle softening and prevention. Int Food Res J 21(2)
Stevens JR, Newton RW, Tlusty M, Little DC (2018) The rise of aquaculture by-products: Increasing food production, value, and sustainability through strategic utilisation. Mar Policy 90:115–124
Synowiecki J, Al-Khateeb NA (2003) Production, properties, and some new applications of chitin and its derivatives
Varun TK, Senani S, Jayapal N, Chikkerur J, Roy S, Tekulapally VB, Gautam M, Kumar N (2017) Extraction of chitosan and its oligomers from shrimp shell waste, their characterization and antimicrobial effect. Veterinary World 10(2):170
Venugopal V (2016) Enzymes from seafood processing waste and their applications in seafood processing. Adv Food Nutr Res 78:47–69
Ward AG, Courts A (1977) Science and technology of gelatin. Academic press
Westendorf ML (2000) Food waste as animal feed: an introduction. Food waste to animal feed, pp 3–16
Wheaton FW, Lawson TB (1985) Processing aquatic food products. John Wiley and Sons, New York, USA
Xavier MKA, Geethalekshmi V, Senapati SR, Mathew PT, Joseph AC, Ramachandran Nair KG (2017) Valorization of squid processing waste as animal feed ingredient by acid ensilaging process. Waste Biomass Valoriz 8(6):2009–2015
Yuvaraj D, Bharathiraja B, Rithika J, Dhanasree S, Ezhilarasi V, Lavanya A, Praveenkumar R (2019) Production of biofuels from fish wastes: an overview. Biofuels 10(3):301–307
Zamora-Sillero J, Ramos P, Monserrat JM, Prentice C (2018) Evaluation of the antioxidant activity in vitro and in hippocampal HT-22 cells system of protein hydrolysates of common carp (*Cyprinus carpio*) by-product. J Aquat Food Prod Technol 27(1):21–34
Zang J, Xu Y, Xia W, Regenstein JM (2020) Quality, functionality, and microbiology of fermented fish: a review. Crit Rev Food Sci Nutr 60(7):1228–1242

Application of Emerging Technologies for Processing of Fish Waste and By-Products

Sandhiya Venkatesh, Soottawat Benjakul, and Muralidharan Nagarajan

Abstract The seafood industry is the most complex of all food industries across the world, mainly based on fish and shellfish, which are a unique source of nutrients. A wide range of processing technologies has been employed to obtain premium quality and meet the requirements of consumers. In addition, novel product development and innovative or emerging processing can be intensively implemented to increase the yield, and eating quality and to ensure the safety of seafood. However, 35% of the total fish catch in fisheries and aquaculture is referred to as fish waste and by-products, which can be turned into several value-added marketable products. Discards including non-target species, and fish processing waste from the world's fisheries surpass 20 million tons per year, accounting for 25% of total fish catch production. The by-products represent a significant potential source of bioactive compounds with nutraceutical properties that can be isolated and can be readily used as functional ingredients. The valorization of fish waste and by-products is gaining significant attention besides being used as fodder or thrown away as waste. The role of technologies in the seafood processing sector has evolved swiftly over the past decade for increased productivity, waste reduction, augmented recovery, and utilization, increased shelf-life, and improved food safety, thus facilitating exports. A number of innovative processing technologies have recently emerged. The fishery waste and bycatch have normally been utilized for fish meal production, nonetheless apart from fish meal there are copious numbers of products with high bioactivities that can be isolated by employing sustainable emerging technologies like Ultrasound-Assisted Extraction (UAE), Supercritical Fluid Extraction (SFE), Microwave-Assisted extraction (MAE), Membrane Technology, Nanotechnology, High Pressure Processing,

S. Venkatesh
Department of Post-Harvest Technology, ICAR-Central Institute of Fisheries Education, Mumbai, India

S. Benjakul
Department of Food Technology, Faculty of Agro-Industry, Prince of Songkla University, Hat YaiSongkhla 90112, Thailand

M. Nagarajan (✉)
Dr. MGR Fisheries College & Research Institute, Tamil Nadu Dr. J Jayalalithaa Fisheries University, Ponneri, Tamil Nadu 601 204, India
e-mail: murali@tnfu.ac.in

S. Maqsood et al. (eds.), *Fish Waste to Valuable Products*, Sustainable Materials and Technology, https://doi.org/10.1007/978-981-99-8593-7_2

Pulsed Electric Field and Cold plasma, etc. Therefore, this chapter will deal in detail with the emerging technologies that can be employed in fish waste and by-products valorization.

Keywords Seafood waste · Valorization · By-products · Emerging technologies · Supercritical fluid extraction · Assisted extraction · Assisted extraction · Pressurized solvent extraction · Pulsed electric field extraction

1 Introduction

Fish remains a healthy food consumed by three billion people all over the world which comprises 20% of their average per capita animal protein. Fish is considered the most nutritious food because of its balanced protein content with high digestibility, vitamins A and D, and some essential micro and macronutrients. The increase in the production of fisheries and aquaculture sector has resulted in enormous quantities of fish waste. The fish waste can be considered as the best raw materials (Rustad 2003). By-products can be produced from both the leftover materials after processing the fish and also from the excess catch. Conventionally, fish waste or excess catch was diverted to fishmeal and fish oil production. In fact, fish waste contains a good proportion of protein and lipid hence can be diverted to the development of various kinds of by-products with high commercial value. The only concern is the quality of fishery by-product however it is directly dependent on the quality of raw material, i.e., fish waste. The huge quantity of unused waste results in economic loss and environmental pollution. Thus, the production of any valuable fishery by-product is economical since the raw material cost is minimal. The main technical challenges in the processing of fish waste are greater microbial load and enzymatic action; hence proper storage and handling of the fish waste becomes an essential prerequisite. The focus can be shifted toward high-quality products that are prepared using environmentally friendly green extraction techniques, apart from fish meal and fish oil this would provide numerous benefits such as reduction in environmental pollution, effective utilization of the healthy components of fishery waste, value addition component in fishery waste utilization may generate income, supply of low-cost convenient products to the consumers. Fishery waste includes head, trimmings, viscera, scale, skin, fin, and bone. Each part of the waste can be used for a specific product. Fishery waste can be turned into animal feed supplements, human food uses, and biotechnological and pharmaceutical applications. Collagen constitutes a major portion of skin and scale (Blanco et al. 2017), and viscera serves as a source of lipids and enzymes (Villamil et al. 2017). There are enormous amounts of crustacean, bivalve, and squid processing waste that can also be converted into valuable by-products. Fishery by-products are divided into two groups such as highly degradable such as viscera, blood, and the stable one such as head, skin, and bone. Therefore, establishing effective and secure procedures for the extraction of desired nutrients and bioactive substances is essential for the successful and ethical exploitation of marine resources. In recent times, the idea of green/clean technology has evolved, implying the employment of more ecologically friendly ways for ingredient processing. Thus, an alternative clean extraction technology which is innovative such as supercritical fluid extraction (SFE), ultrasound-assisted extraction (UAE), pulsed electric fields (PEF), and microwave-assisted extraction (MAE), have been discovered.

2 The Need for New Extraction Techniques

The solid waste from fisheries includes head, frame, tail, skin, gut, fin, and skin. During fileting 30–50% meat, 4–5% skin, 21–25% head, and 24–34% bone are being wasted during fileting. constitute nearly 45% (Ghaly et al. 2013).

Fish scale constitutes 2% of fish body weight (Ferraro et al. 2010) from which hydroxyapatite, calcium carbonate, and type 1 collagen can be extracted. Fish bones are a rich source of hydroxyapatite. Fish viscera constitutes about 12–18% of the total fish body considered as a rich source of enzymes and probiotics (Caruso 2016). Collagen can be found in abundance in fish skin. The Soxhelt extraction method was developed by Soxhelt in 1879, since then it has been used for the extraction of lipids for several decades. However, there is an increased need for the development and optimization of new techniques. The new techniques should be in such a way that automation can be done, reduction in solvent consumption which will reduce the pollution resulting from solvent, reduction in sample preparation step, reduction in extraction time, and omission of the final concentration step (Wan et al. 1996). The experimented new techniques are as follows: supercritical fluid extraction, ultrasound-assisted extraction, microwave-assisted extraction, pressure-assisted extraction, cold plasma, pulsed electric field, membrane technology, etc., The improved new techniques are of great importance as the extraction time becomes very less as it operates under either high temperature or pressure (Eskilsson et al. 2000) (Fig. 1).

2.1 Solid–Liquid Extraction (SLE)

SLE is a relatively new term that describes the process of removing or extracting the soluble analyte from the solid matrix with the aid of suitable solvents. SLE basically consists of five steps: (Grosso et al. 2015).

1. Solvent infiltration into the solid matrix.
2. Solubilization of the analyte with the solvent.
3. Transportation of the analyte from the inner network to the outer surface of the matrix.
4. Escapement of the analyte from the outer surface of the matrix to the bulk solvent solution.

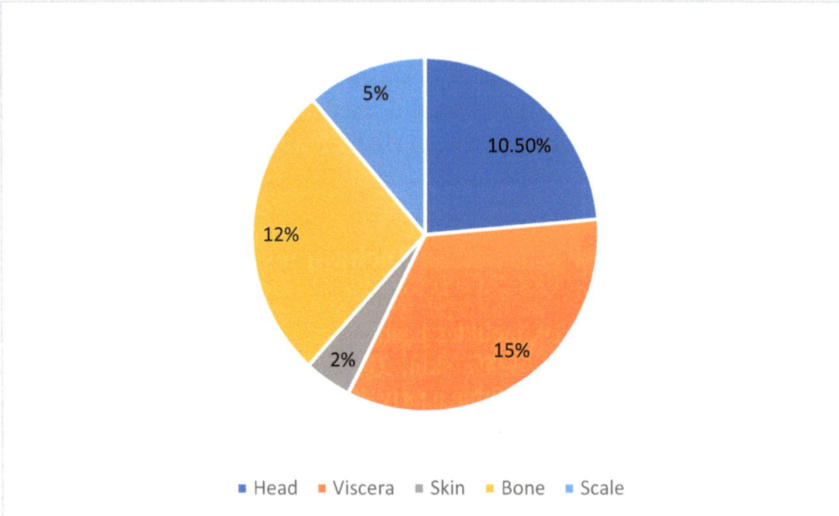

Fig. 1 Fish waste and their proportion, SOFIA, 2022

5. Concentration of the extract by removing the solvent.

The extraction of compounds from a complex solid matrix is an age-old technique that is used in the making of medicines, perfumes, foods, etc., SLE techniques can be divided into two categories—traditional and modern methods, traditional methods include Soxhlet, maceration, infusion, hydro distillation, pressing, percolation, etc., whereas modern techniques involve supercritical fluid extraction (SFE), ultrasound-assisted extraction (UAE), microwave-assisted extraction (MAE), pressurized solvent extraction, pulsed electric field assisted extraction (PFE), ionic solvents extraction, enzymes assisted extraction.

SLE depends on several factors that can be divided into solvent factors and the compound of interest factors. Solvent factors include purity, viscosity, toxicity, polarity, and volatility; compound of interest factors involves polarity and thermolability. These factors must be taken into account while extracting a particular compound with the SLE process (Fig. 2).

2.2 *Supercritical Fluid Extraction (SFE)*

2.2.1 Fundamentals

Extraction is the removal of soluble matter from insoluble compounds with the aid of solvent. The most commonly used organic solvents were petroleum solvents, namely, hexane, ethyl ether, toluene, benzene, xylene, etc., SCF is the method of separating

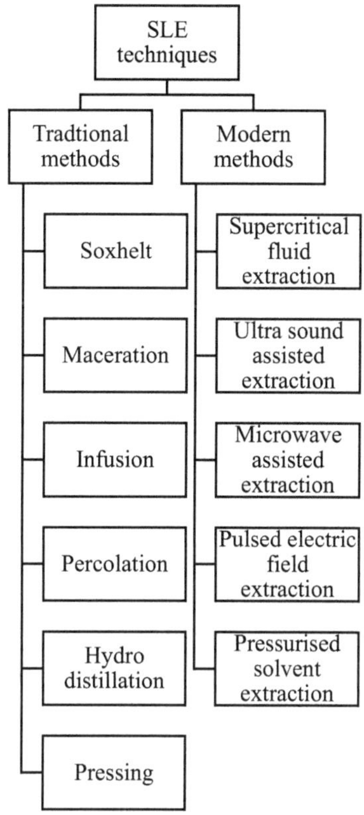

Fig. 2 Classification of SLE techniques

the analyte from its matrix with the use of supercritical fluid. The most common supercritical fluid used is supercritical CO_2 owing to its safety, cost-effectiveness, and availability.

Baron Cagniard de La tour (1822) first reported the supercritical phase by observing the disappearance of the gas–liquid boundary line for the first time when the temperature increased in certain materials. Hannah and Hogarth (1879), later demonstrated the solvating power of supercritical fluids.

Supercritical fluid (SCF) is any material that can be either gas or liquid and is used above its critical temperature and pressure where both the state (gas and liquid) co-exist. Therefore, supercritical fluids do not have any phase. The word fluid denotes the intermediate form of both the gas and liquid form. The fluid at the supercritical stage possesses the solvating power of liquid as well as the transporting capacity of the gas (Askin and Ötles 2005). Thus, supercritical fluid will have more efficiency than the normal solvents at atmospheric pressure. In other words, supercritical fluid can be defined as highly compressed gas having a similar density to its liquid phase (Saito 2013). Critical state or Critical point—The state at which any substance has its liquid and gaseous phase at the same density, i.e., the two phases have the same temperature, pressure, and volume.

Carbon dioxide (CO_2) is non-toxic, incombustible, easily available, and high purity, moreover, it is generally recognized as safe (GRAS) and hence it is widely used as supercritical fluid. It has a high diffusivity coefficient and low viscosity; hence, it can easily penetrate into the sample matrix, is chemically inert, and prevents the degradation of the analyte by providing non-oxidative environment, environment-friendly, odorless, and tasteless. The main advantage of using CO_2 is that it can be easily removed from the analyte by the reduction in pressure as they have low critical temperature and pressure of 31.1 °C and 7.4 MPa respectively. CO_2 can be used alone or can be combined with any other co-solvents. Cosolvents namely methanol, ethanol, ethyl ether, dichloromethane, acetone, acetonitrile, and water can be added to the SFE CO_2. Methanol is 20% miscible with CO_2 and so it is efficient but ethanol is the best choice of cosolvent as it is environmentally friendly.

SFE CO2 process is influenced by the following factors such as CO_2 density, CO_2 flow rate, cosolvent characteristics, pressure, extraction time, and temperature.

However, phenolics, alkaloids, and glycosidic compounds are poorly soluble in CO_2 and this limitation can be overcome by the addition of modifiers called co-solvents.

2.2.2 Instrumentation

SFE CO_2 contains the following components such as CO_2 cylinder, pump, heater, extraction cell, pressure controller, and extract collector. The sample is placed in the extraction cell in the heated chamber where the CO_2 is pumped. The analyte and the CO_2 mixture move to the separator where CO_2 converts to the gaseous phase and escapes from the analyte. The advantage of this method is the concentration step of the analyte by removing the solvent can be omitted due to the concentrated form of the analyte. The sample preparation for the SFE process involves a reduction of the moisture content to 20% for the effectual extraction (Rubio-Rodriguez et al. 2008). Plaza and Rodriguez-Meizoso (2013) reported that important variables including pressure, temperature, time, co-solvents, etc. have an impact on the extraction efficiency of SFE (Fig. 3).

2.2.3 Application in Fishery By-Products Processing

Application of the SFE includes extraction of high-value oils, isolation of lipid-soluble compounds, and extraction of natural aromas. The main by-product of fish is the fish oil which is rich in ω-3 fatty acids-DHA and EPA which can easily be extracted by the SFE CO_2 technique. Fish oil was usually obtained along with the fish meal by the application of the wet reduction method and the quality was substantially low. Due to the increase in the demand for fish oil owing to its health benefits, it is essential that it should be of good quality. In order to extract oil from fish by-products, Rubio-Rodriguez et al. (2012) examined four distinct extraction techniques: cold extraction, wet reduction, enzymatic extraction, and supercritical fluid extraction.

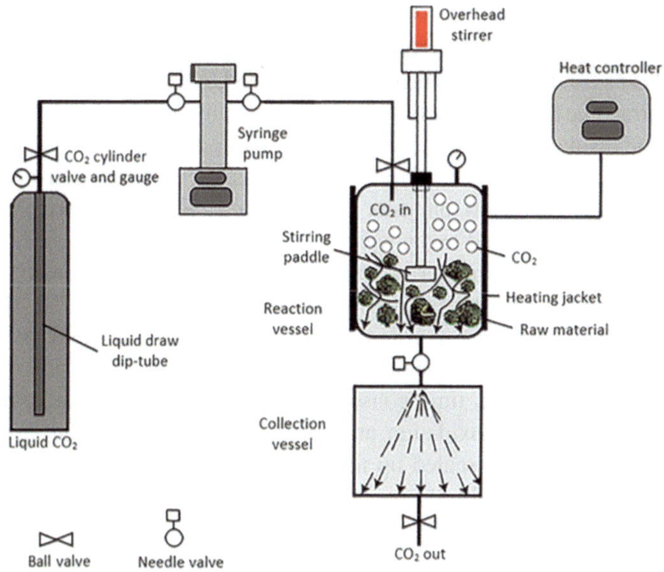

Fig. 3 Schematic representation of the SFE process (adopted from Khaw et al. 2017)

SFE CO_2 was found to yield good-quality oil as it has better oxidation stability and low arsenic content. Létisse et al. (2006) optimized the SFE CO_2 method for the extraction of sardine oil containing DHA and EPA at 75 °C, 300 bars pressure, time period of 45 min. It has been found that the SFE CO_2 method for the extraction of fish oil has more advantages over the quality point of view rather than the yield.

Astaxanthin, a lipophilic compound from shrimp shell waste that acts as an antioxidant has several pharmaceutical properties. Radzali et al. (2014) investigated the extraction of astaxanthin using solvent extraction method and SFE CO_2 with different co-solvents. SFE CO_2 with ethanol gave a higher yield than other co-solvents. López et al. (2004) optimized the SFE CO_2 process for the extraction of carotenoids from crustaceans and found that 200 bars pressure, 60 °C temperature, and 15% (v/v) ethanol were the suitable parameters for the extraction process. Hence, the by-products from fish waste can effectively be extracted by supercritical fluid extraction method.

2.3 *Ultrasound-Assisted Extraction (UAE)*

2.3.1 Fundamentals

UAE employs the principle of disruption of the cells in the sample matrix by the cavitation bubble formation, thus paving the way for easier and faster infiltration of the solvent into the sample matrix. This could be a result of the solvent's increased surface area of contact with the analyte-containing sample. The frequency of the ultrasound is beyond the human hearing range. The UAE process involves the application of ultrasonic waves into a liquid medium containing the sample matrix which facilitates the disintegration of the cells of the sample matrix (Chemat et al. 2017a).

Ultrasounds can be classified in terms of usage as low-intensity and high-intensity ultrasound frequencies. Low-intensity ultrasounds have a high frequency ranging from 100 kHz to 1 MHz and less power of 1 W/cm^2 which are generally utilized to study the physicochemical properties of the food material. High-intensity ultrasound has a low frequency ranging from 16 to 100 kHz and more power of 10–100 W/cm^2. This high-intensity ultrasound is commonly used for short-time rapid extraction and it usually changes the physicochemical properties of the food material.

UAE is simple and less time consuming, it can be operated at room temperature thus preventing the oxidation and further degradation of the analyte, cost-effective, however, can be scaled up to an industrial level, less solvent consumption, environmentally friendly (Louie et al. 2020). The three main considerations in ultrasound-assisted extraction are the solvent composition, extraction time, and input power.

The application of the ultrasound results in the cavitation bubble formation in the solvent in contact with the sample which enhances the mass transfer of the analyte to the solvent (Pacheco et al. 2020). UAE can easily be combined with other methods such as Soxhlet, microwave-assisted extraction, and supercritical fluid extraction (Chemat et al. 2017a).

2.3.2 Instrumentation

UAE assembly includes an ultrasound generator, transducer, ultrasound cylinder probe, thermocouple, and data recorder.

The ultrasonic effect can be produced by the sonicator. A water bath positioned beneath the extraction setup aids in temperature control. The sample-containing solvent will be submerged completely with the ultrasonic probe. The ultrasonic device could cause a cavitation with a bubble burst effect to aid in the extraction process (Fig. 4).

2.3.3 Application

Álvarez et al. (2017) utilized the sequential acid/alkaline extraction method combined with the UAE technique for the recovery of protein from the mackerel waste and found that it resulted in 100% recovery of protein. Siewe et al. (2020) used and optimized the ultrasound-assisted enzymatic extraction method for the extraction of umami compounds from the Rohu head waste. The gelatin was extracted by ultrasound-assisted extraction and microwave-assisted extraction on the molecular and physicochemical properties of the extracted gelatin were studied. However, the

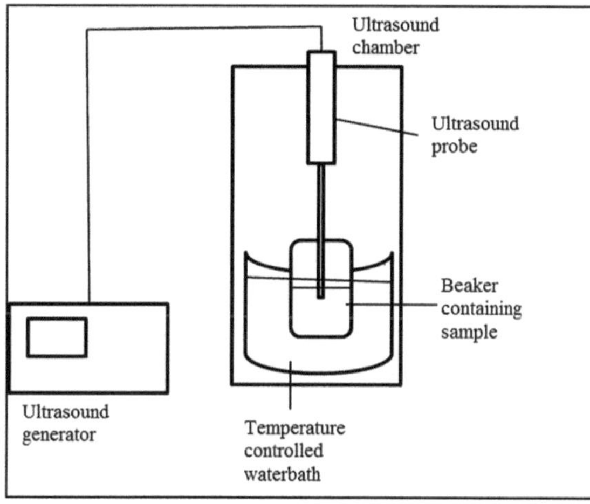

Fig. 4 Instrumentation of UAE (adopted from Zahari et al. 2020)

Table 1 Solvents commonly used in MAE with their dielectric constant at 20 °C

Solvent	Dielectric constant at 20 °C
Acetone	20.7
Acetonitrile	37.5
Ethanol	24.3
Hexane	1.89
Methanol	32.6
2-Propanol	19.9
Water	78.3

Source Jassie et al. (1997)

electrophoretic pattern of the gelatin showed higher gel strength of the UAE-extracted gelatin compared to MAE gelatin (Mirzapour-Kouhdasht et al. (2019). Petcharat et al. (2021) used an ultrasound-assisted extraction method for the collagen extraction from clown feather back which enhanced the extraction efficiency of the collagen. Thus, the UAE method can be used effectively in combination with other methods for the extraction of protein, collagen, aroma compounds, and bioactive compounds from the fishery by-products with a higher rate of recovery in less time (Table 1).

2.4 *Microwave-Assisted Extraction (MAE)*

2.4.1 Fundamentals

Microwaves are a short electromagnetic spectrum ranging from 300 MHz to 300 GHz and wavelength of 1 cm to 1 m (Mandal et al. 2007). Microwaves are composed of two perpendicular oscillating fields, i.e., magnetic and electric fields. A magnetron is a device that generates the microwave. The emission of microwaves results in a continuous movement of water molecules which is a dipole, this continuous oscillation of water molecules results in inter and intramolecular friction and collisions resulting in the production of heat. In simple words, microwaves cause dipole rotation of the organic molecules, and heating causes the breakage of hydrogen bonding (Zakkaria et al. 2021). The internal heating of the matrix occurs when the sample matrix is exposed to microwaves (Grosso et al. 2015). This in turn creates pressurized effects in the sample that induces the breakdown of the cells in the sample matrix. This facilitates the faster release of the analyte into the solvent (Kaufmann and Christen, 2002). UAE and MAE have majorly been used for polysaccharides extraction (Soria et al. 2012). The solvents used in MAE must have significant dielectric constant/relative permittivity (Table 2) (Vanderburg et al. 2000). MAE heats the extraction solvent using microwave radiation to improve solvent dispersion into the sample and speed up the partitioning of extractants from the sample to the solvent. The solvents must be polar in order to absorb the microwave energy, the commonly used solvents are methanol and ethyl acetate. MAE remains inappropriate for polyunsaturated fatty acids (PUFA) extraction due to its thermal nature as they are more prone to oxidation and degradation processes (Yousuf et al. 2018). Free radical generation is another major disadvantage of this process. However, most of the demerits can be overcome by using temperature-controlled microwave reactors, while it would be expensive on a large scale.

The extraction can be condensed into three steps:

1. The solutes dissociate from the active sites in the sample matrix as the temperature and pressure rise.
2. The solute is released to the solvent.
3. The solvent diffuses across the sample matrix.

The primary aim of microwave heating during MAE is the moisture content of the sample matrix. The pressure inside the cells rises as the water evaporation, ruptures the cell walls and accelerates the release of valuable chemicals.

2.4.2 Instrumentation

MAE can be done by two types of instruments such as open and closed. A closed microwave instrument is the most common one, in which the MAE is done at controlled pressure and temperature. However, in open microwave instruments, the extraction is based on atmospheric pressure (Kaufmann et al. 2002). Microwave energy, solvent, and extraction time are the deciding factors in the effective extraction in the MAE process. The volume and boiling point of the solvents basically determine the pressure inside the vessel. The solvent should be heated beyond its atmospheric pressure to increase the process efficiency. Paramount importance should be given to extraction vessels and MAE equipment as there is a possibility of occurrence of high pressure which may lead to explosion (Lehotay and Schenck 2000). An open

Table 2 Advantages and disadvantages of the emerging technologies

Methods	Advantages	Disadvantages
SFE	• No toxicity • Environment friendly • Low cost • Less time consuming • Eliminates the concentration step	• High installation cost • SFE CO_2 can be used only for non-polar analytes • High power consumption
UAE	• Reduction in time • Reduction in energy consumption • Reduction in solvent consumption • More infiltration of the solvent into the sample matrix results in greater extraction	• Sonolysis causes the formation of free radicals • Induce lipid oxidation • Scaling up process is difficult • Greater power consumption
MAE	• Faster extraction time • Less quantity of solvent is needed • Multiple extractions can be done • Loss of volatile compounds is minimum	• Final clean-up step is required • Time consumption in cooling of the extraction vessel • Solvent used must be able to absorb microwaves • Not suitable for highly volatile solvents
PSE	• Faster extraction (around 15 min) • Use of a limited amount of solvent (15–40 ml) • No need for further filtration of the extract • Automatic process and hence easy to use	• High cost of installation • Optimization of the variables
PEF	• Cost-effective • Low energy consumption • No need for further purification • Suitable for extraction of heat-sensitive substances • Mostly applicable to all kinds of solid and liquid samples • Retention of color, flavor, and nutrients • Shorter extraction period • Environmentally friendly • Effectively combined with other processing techniques	• High installation cost • Scaling up into a commercial unit is difficult

microwave system can be used for thermolabile substances as it uses low temperature. The particle size of the extracted material ranges from 100 μm to 2 nm (Eskilsson and Björklund 2006). Thus, powdered sample substrate increases the extraction efficiency as it increases the surface area.

The following are the components of the microwave instrument: (Mandal et al. 2007).

1. Magnetron—generates the microwaves.
2. Wave guide—to which the magnetron is attached, propagates the microwaves from the magnetron to microwave cavity.
3. Applicator—in which is sample is placed.
4. Circulator—to facilitate the microwave in a forward direction (Fig. 5).

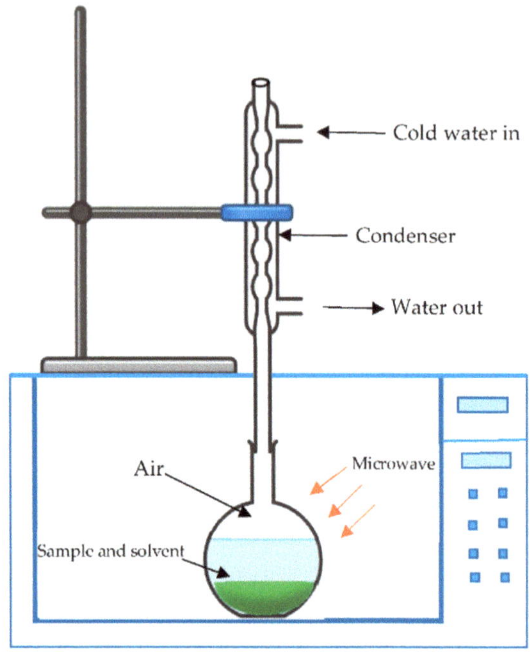

Fig. 5 Instrumentation of MAE (adopted from Saifullah et al. 2021)

2.4.3 Application of MAE in Fishery by-products Processing

MAE method of fish fat extraction showed a 90% reduction in the extraction time, less residues, negligible lipid oxidation compounds, and reduced solvent consumption with greater precision, accuracy, and robustness. Costa and Bragagnolo (2017) investigated the efficiency of MAE of fish lipids and found that MAE is an efficient method than other fat extraction methods such as Soxhelt, Bligh and Dyer method, and Folch method.

2.5 Pressurized Solvent Extraction (PSE)

2.5.1 Fundamentals

PSE is also known as accelerated solvent extraction, high pressure solvent extraction, pressurized liquid extraction, high temperature solvent extraction, pressurized hot solvent extraction, subcritical solvent extraction. PSE involves high temperature and pressure for the extraction process. Normally, the temperature of 50–200 °C and pressure of 50–3000 psi for a relatively short period of time of about 5–10 min are used in the process (De la and Armenta 2011). A rise in temperature aids in breaking down the sample matrix which increases the solubility of the target molecules and thus increases the process efficiency. The main advantage of PSE is that the relatively high pressure allows the solvent to be in a liquid state which enhances the extraction process. The presence of high moisture content hinders the extraction process as it lowers the contact between the solvent and the sample matrix. Hence, the main factor influencing the extraction process is the temperature, pressure, and the sample matrix. PSE was developed as an alternative to Soxhlet and other extraction methods, including SFE, UAE, and MAE in the middle of the 1990s. The rising temperature leads to increased extraction yields due to the improved distribution coefficient and kinetics. However, rising pressure increases the working temperature higher than the boiling point of the solvent (María 2018). The high temperature of the process aids in decreasing the viscosity of the solvent thereby enhancing the contact and flow of the solvent through the sample matrix, resulting in matrix swelling and faster analyte diffusion.

2.5.2 Instrumentation

PSE involves static, flow-through/dynamic, and mixed systems. A static system involves placing of sample inside a cell with a fixed volume of solvent. Flow through the system involves placing the sample in a cell in which continuous pumping of solvent occurs. Flow through the system increases the extraction efficiency, however, it has the greater disadvantage of using a large volume of solvent. A mixed system involves the combination of both static and dynamic systems.

The sample is mixed with desiccants such as diatomaceous earth and anhydrous sodium sulfate which is then placed in an extraction vessel containing the determined amount of solvent. The target molecules are collected after the complete extraction. The extraction cell is again washed with the solvent to remove trace target molecules that are attached to the cell. The extraction cell is purged with nitrogen gas (inert gas) to remove the memory effects at the end of the process. Dionex was the first firm to patent the PSE system.

PSE system consists of the following parts such as solvent reservoir, solvent controller, pump, gas tank, valves, extraction cell, collection vessel, and oven (Fig. 6).

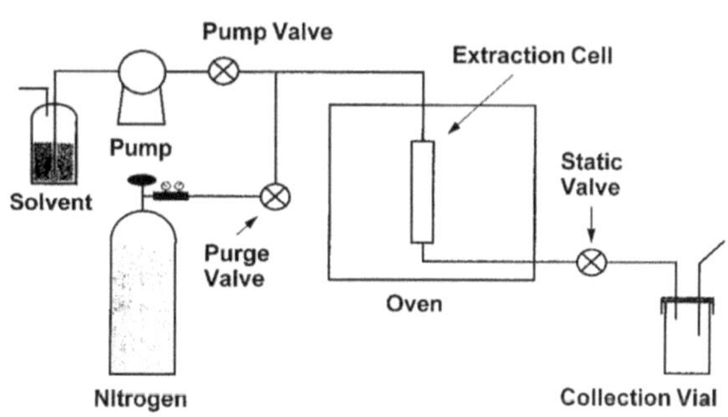

Fig. 6 Schematic representation of the PSE system (adopted from Richter et al. 1996)

2.5.3 Application of PSE in Fisheries by-product Processing

Wang et al. (2021) used PSE for the extraction of protein from rainbow trout fish processing waste at 45–55 °C for 15 min at pH 5.2–6.8 and a pressure of 103.4 bars. The result shows that it retains the antioxidant properties of the active substances.

2.6 Pulsed Electric Field Extraction (PEF)

2.6.1 Fundamentals

In recent years, the use of pulsed electric fields (PEF) for the extraction of functional components from food waste and by-products has received increased attention (Bansal et al. 2015). As the PEF method does not affect the nutritive quality of the end product, it is a good alternative for other extraction procedures such as boiling, MAE, UAE, etc., PEF processing involves exposing food held between two electrodes to high voltage pulses for a minimal period. PEF affects the cell structure by disrupting the cell membrane (Barba et al. 2015). The efficiency of the PEF system depends on parameters such as solvent nature and sample composition (pH, conductivity, size, and shape). There is a field threshold value of about 1–10 (kV/cm) depending on the kind of cell, i.e., microbial, plant, or animal. However, an excess level causes a local dielectric breakdown of the membrane which leads to the development of a pore that can subsequently function as a conductive channel (Kumar et al. 2015). PEF can be considered as non-thermal cell membrane permeabilization method.

2.6.2 Instrumentation

The PEF approach requires an electric field strength (EFS) between 100 and 300 V/cm in batch mode and between 20 and 80 kV/cm in continuous mode extraction (Ranjha et al. 2021). High-voltage pulse generator, a treatment chamber with a fluid management assembly, and a monitoring and control system make up a PEF unit (Alexandre et al. 2017). The sample is placed between the electrodes in the treatment chamber. The power source produces the energy which gets stored in the capacitor and gets discharged in the treatment chamber to produce an electric field. The factors important in the PEF process are the number of pulses, pulse duration, pulse width, pulse shape, pulse specific, and pulse frequency (Puértolas and Barba 2016) (Fig. 7).

2.6.3 Application of PEF in Fishery by-products Processing

Zhou et al. (2012) employed the PEF technique for the extraction of calcium from the fish bones, further, the authors compared PEF with UAE and found that PEF resulted in greater extraction efficiency in a relatively shorter period than UAE.

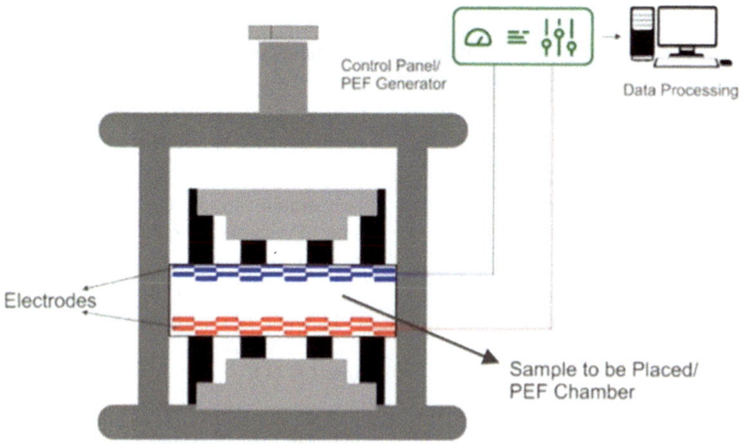

Fig. 7 Schematic representation of PEF system (adopted from Ranjha et al. 2021)

The chondroitin sulfate is extracted from fish bones using the PEF technique (He et al. 2014). The extraction procedure of collagen from fish bones was optimized using PEF and semi bionic method by He et al. (2017). Li et al. (2016) employed the PEF method for the isolation of abalone viscera protein and the highest yield was obtained with the solvent-to-material ratio of 1 to 4 when the PEF was applied as 600 μs at the intensity strength of 20 kV/cm. PEF method resulted in greater emulsifying properties of the protein in addition to the yield when compared to the conventional technique however viscosity and foaming properties were reduced. PEF can effectively be implemented as a pretreatment procedure for fishery waste for further processing (Gomez et al. 2019) (Table 3).

Table 3 Bioactive compounds extracted using green technologies

Bioactive compounds	Source	Method	Optimum conditions	Reference
Proteins	Mackerel (whole)	UAE	Frequency: 40 kHz, 60% amplitude, 0.1 M NaOH, time: 10 min	Álvarez et al. (2018)
	Rainbow trout and sole waste	PSE	Temperature: 45–55 °C, time:15 min, pH 5.2–6.8, pressure:103.4 bars	Wang et al. (2021)
	Rainbow trout and sole waste	PFE	Electric field strength: 1–3 kV/cm, 123–energy density 300 kJ/kg, time:15–24 h	Wang et al. (2021)
Collagen	Japanese seabass skin	UAE	Frequency: 20 kHz, 80% amplitude, 0.1 M acetic acid, time: 3 h	Kim et al. (2013)
Gelatin	Bighead carp scales	UAE	Temperature: 60, 70, and 80 °C for time: 1 h	Huang et al. (2017)
	Bighead carp scales	UAE	Temperature: 60 °C for time:1, 3, and 5 h	Tu et al. (2015)
Chitosan	Squid pens	SFE	Temperature: 35, 45, 50 °C and Pressure: 20, 30, 40 MPa for time:1 h, Co-solvent: 1 wt % acetic acid	Lee et al. (2012)
Carotenoids	Shrimp waste	SFE	Temperature: 50 °C, Pressure: 30 MPa, CO_2 flow with ethanol: 8.3 × 10 − 5 kg/s, Ethanol flow: 4.4 × 10 − 6 kg/s for 200 min	Sánchez et al. (2011)
Astaxanthin	Crawfish	SFE	Temperature: 50–70 °C, Pressure: 13.8–31.0 MPa CO_2 flow: 1.0–1.5 L/min Co-solvent: 10% ethanol	Charest et al. (2001)
PUFA	Rohu head waste	UAE	Temperature ≥ 40 °C, Pressure ≥ 25 Mpa, CO_2 flow ≥ 10 kg/h, Extraction time: 3 h 20 kHz, 40% amplitude, for 5, 10, and 15 min	Bruno et al. (2019) Taati et al. (2017) Kuvendziev et al. (2018) Sahena et al. (2014) Létisse et al. (2008) Haq et al. (2017) Rahimi et al., (2017) de la Fuente et al., (2022)

(continued)

Table 3 (continued)

Bioactive compounds	Source	Method	Optimum conditions	Reference
	Tuna waste	SFE	Temperature ≥ 40 °C, Pressure ≥ 25 MPa CO_2 flow ≥ 10 kg/h, Extraction time: 3 h	
	Common carp waste	SFE	40, 50, and 60 °C, Pressure: 200, 300, 350, and 400 bar, CO_2 flow: 0.194 kg/h Extraction time: 180 min	
	Tuna head	SFE	Temperature: 65 °C, Pressure: 40 Mpa, CO_2 flow with ethanol: 2.4 mL/min, Ethanol flow: 0.6 mL/min, Extraction time: 120 min	
	Sardine head waste	SFE	Temperature: 75 °C, Pressure: 300 bar, CO_2 flow: 2.5 mL/min, Extraction time: 45 min Temperature: 45 °C, Pressure: 250 bar, CO_2 Flow: 27 g/min, Extraction time: 3 h	
	Salmon waste	SFE	Temperature: 45 °C, Pressure: 15–25 Mpa, CO_2 flow: 27 g/min, Extraction time: 2 h	
	Sardine fish	MWE	Hexane + Isopropanol solvent mixture, Time:1–4 min, Power: 800 W	
	Salmon waste	MWE	Time: 14.6 min, power: 291.9 W, solid/liquid ratio: 80.1 g/L for backbones, 10.8 min, 50.0 W, 80.0 g/L for heads, and 14.3 min, 960.6 W, 99.5 g/L for viscera Maximum of 1000 W	
Vitamins	Mackerel flesh part	SFE	Temperature: 45 °C, Pressure: 15–25 Mpa, CO_2 flow: 27 g/min, Extraction time: 2 h	Haque et al. (2014)

3 Conclusion

The exploration of relatively new techniques that are environmentally friendly for the effective utilization of marine resources is the need of the hour. The field of functional component discovery and its application in nutraceuticals is expanding. In fact, there are many interesting substances found in fish bycatch and food waste, including bioactive peptides, polysaccharides, polyunsaturated fats, carotenoids, polyphenolic compounds, minerals, collagen, gelatin, saponins, phycobiliproteins, and phytosterols. In addition to their low prices of fish waste the main benefits of the waste-derived nutraceuticals include their easy availability as raw materials, high rates of recovery, and intriguing functional characteristics of the separated chemicals. The nutritional value and possible health advantages of bioactive substances found in by-products of fish and shellfish have greatly increased the interest in their extraction. The production of novel products as part of the strategy for valuing seafood by-products can result in a more environmentally responsible utilization of marine resources and good economic returns for the industry. Therefore, it is crucial to establish appropriate extraction technologies that permit minimal processing with good quality and yield, ensuring the safety of the final product, thus meeting the objective of sustainable waste management for expanding global human population.

References

Alexandre EM, Moreira SA, Pintado M, Saraiva JA (2017) Emergent extraction technologies to valorize fruit and vegetable residues. Agricultural Research Updates; Gorawala, P., Mandhatri, S., Eds, pp 37–79

Álvarez C, Lélu P, Lynch SA, Tiwari BK (2018) Optimised protein recovery from mackerel whole fish by using sequential acid/alkaline isoelectric solubilization precipitation (ISP) extraction assisted by ultrasound. LWT 88:210–216

Anas KK, Lekshmi RGK, Tejpal CS, Mathew S (2020) Supercritical fluid extraction of fish oil– recent perspectives

Askin R, Ötles S (2005) Supercritical fluids. Acta Sci Polon Technol Alimentaria 4(1):3–16

Bansal V, Sharma A, Ghanshyam C, Singla ML, Kim KH (2015) Influence of pulsed electric field and heat treatment on Emblica officinalis juice inoculated with Zygosaccharomyces bailii. Food Bioprod Process 95:146–154

Barba FJ, Parniakov O, Pereira SA, Wiktor A, Grimi N, Boussetta N, Saraiva JA, Raso J, Martin-Belloso O, Witrowa-Rajchert D, Lebovka N (2015) Current applications and new opportunities for the use of pulsed electric fields in food science and industry. Food Res Int 77:773–798

Blanco M, Vázquez JA, Pérez-Martín RI, Sotelo CG (2017) Hydrolysates of fish skin collagen: an opportunity for valorizing fish industry byproducts. Mar Drugs 15:131

Caruso G (2016) Fishery wastes and by-products: A resource to be valorised. J Fish Sci 10(1)

Charest DJ, Balaban MO, Marshall MR, Cornell JA (2001) Astaxanthin extraction from crawfish shells by supercritical CO_2 with ethanol as cosolvent. J Aquat Food Prod Technol 10(3):81–96

Chaves JO, De Souza MC, Da Silva LC, Lachos-Perez D, Torres-Mayanga PC, Machado APDF, Forster-Carneiro T, Vázquez-Espinosa M, González-de-Peredo AV, Barbero GF, Rostagno MA (2020) Extraction of flavonoids from natural sources using modern techniques. Front Chem 8:507887

Chemat F, Rombaut N, Sicaire AG, Meullemiestre A, Fabiano-Tixier AS, Abert-Vian M (2017a) Ultrasound assisted extraction of food and natural products. Mechanisms, techniques, combinations, protocols and applications. A review. Ultrason Sonochem 34:540–560
Chemat F, Rombaut N, Meullemiestre A, Turk M, Perino S, Fabiano-Tixier AS, Abert-Vian M (2017b) Review of Green Food Processing techniques. Preservation, transformation, and extraction. Innov Food Sci Emerg Technol 41:357–377
Costa DDSV, Bragagnolo N (2017) Development and validation of a novel microwave assisted extraction method for fish lipids. Eur J Lipid Sci Technol 119(3):1600108
De la Guardia M, Armenta S (2011) Greening sample treatments. In Comprehensive analytical chemistry, Vol. 57, pp 87–120. Elsevier
de la Fuente B, Pinela J, Mandim F, Heleno SA, Ferreira IC, Barba FJ, Berrada H, Caleja C, Barros L (2022) Nutritional and bioactive oils from salmon (Salmo salar) side streams obtained by Soxhlet and optimized microwave-assisted extraction. Food Chem 386:132778
Eskilsson CS, Björklund E (2000) Analytical-scale microwave-assisted extraction. J Chromatogr A 902(1):227–250
FAO (2022) The State of World Fisheries and Aquaculture 2022. Towards Blue Transformation. Rome, FAO
Ferraro V, Cruz IB, Jorge RF, Malcata FX, Pintado ME, Castro PM (2010) Valorisation of natural extracts from marine source focused on marine by-products: a review. Food Res Int 43(9):2221–2233
Ghaly AE, Ramakrishnan VV, Brooks MS, Budge SM, Dave D (2013) Fish processing wastes as a potential source of proteins. Amino acids and oils: A critical review. J Microb Biochem Technol 5(4):107–129
Gómez B, Munekata PE, Gavahian M, Barba FJ, Martí-Quijal FJ, Bolumar T, Campagnol PCB, Tomasevic I, Lorenzo JM (2019) Application of pulsed electric fields in meat and fish processing industries: an overview. Food Res Int 123:95–105
Grosso C, Valentão P, Ferreres F, Andrade PB (2015) Alternative and efficient extraction methods for marine-derived compounds. Mar Drugs 13(5):3182–3230
Haq M, Ahmed R, Cho YJ, Chun BS (2017) Quality properties and bio-potentiality of edible oils from Atlantic salmon by-products extracted by supercritial carbon dioxide and conventional methods. Waste Biomass Valoriz 8(6):1953–1967
Haque ASM, Asaduzzaman AKM, Chun BS (2014) Fatty acid composition and stability of extracted mackerel muscle oil and oil-polyethylene glycol particles formed by gas saturated solution process. Fisheries Aquatic Sci 17(1):67–73
He G, Yin Y, Yan X, Yu Q (2014) Optimisation extraction of chondroitin sulfate from fish bone by high intensity pulsed electric fields. Food Chem 164:205–210
He G, Yin Y, Yan X, Wang Y (2017) Semi-bionic extraction of effective ingredient from fishbone by high intensity pulsed electric fields. J Food Process Eng 40(2):e12392
Huang T, Tu ZC, Xinchen-Shangguan W, H., Zhang, L. and Sha, X.M. (2017) Rheological and structural properties of fish scales gelatin: Effects of conventional and ultrasound-assisted extraction. Int J Food Prop 20(sup2):1210–1220
Jassie L, Revesz R, Kierstead T, Hasty E, Metz S, Kingston HM, Haswell SJ (1997) Microwave-enhanced chemistry. Fundamentals, sample preparation and applications, p 569
Kaufmann B, Christen P (2002) Recent extraction techniques for natural products: microwave-assisted extraction and pressurised solvent extraction. Phytochem Anal 13:105–113
Khaw KY, Parat MO, Shaw PN, Falconer JR (2017) Solvent supercritical fluid technologies to extract bioactive compounds from natural sources: a review. Molecules 22(7):1186
Kim HK, Kim YH, Park HJ, Lee NH (2013) Application of ultrasonic treatment to extraction of collagen from the skins of sea bass Lateolabrax japonicus. Fish Sci 79(5):849–856
Kumar Y, Patel KK, Kumar V (2015) Pulsed electric field processing in food technology. Int J Eng Stud Tech Approach 1(2):6–17

Kuvendziev S, Lisichkov K, Zeković Z, Marinkovski M, Musliu ZH (2018) Supercritical fluid extraction of fish oil from common carp (Cyprinus carpio L.) tissues. J Supercrit Fluids 133:528–534
Lee WK, Tsai ML, Shieh YT (2012) Fractionation of chitosan by supercritical carbon dioxide/acetic acid aqueous solution. J Supercrit Fluids 71:86–91
Lehotay SJ, Schenck FJ (2000) Multi-residue methods: extraction. Encyclopedia of separation science, pp 3409–3415
Létisse M, Comeau L (2008) Enrichment of eicosapentaenoic acid and docosahexaenoic acid from sardine by-products by supercritical fluid fractionation. J Sep Sci 31(8):1374–1380
Létisse M, Rozières M, Hiol A, Sergent M, Comeau L (2006) Enrichment of EPA and DHA from sardine by supercritical fluid extraction without organic modifier: I. Optimization of extraction conditions. J Supercrit Fluids 38(1):27–36
Li M, Lin JIE, Chen J, Fang T (2016) Pulsed electric field-assisted enzymatic extraction of protein from abalone (Haliotis discus hannai Ino) viscera. J Food Process Eng 39(6):702–710
López M, Arce L, Garrido J, Rıos A, Valcárcel M (2004) Selective extraction of astaxanthin from crustaceans by use of supercritical carbon dioxide. Talanta 64(3):726–731
Louie KB, Kosina SM, Hu Y, Otani H, de Raad M, Kuftin AN, Mouncey NJ, Bowen BP, Northen TR (2020) Mass spectrometry for natural product discovery
Mandal V, Mohan Y, Hemalath S (2007) Microwave assisted extraction-an innovative andpromising extraction tool for medicinal plant research. Pharmcogn Rev 1(1):7–18
Mirzapour-Kouhdasht A, Sabzipour F, Taghizadeh MS, Moosavi-Nasab M (2019) Physicochemical, rheological, and molecular characterization of colloidal gelatin produced from Common carp by-products using microwave and ultrasound-assisted extraction. J Texture Stud 50(5):416–425
Pacheco-Fernández I, Pino V (2020) Extraction with ionic liquids-organic compounds. In Liquid-phase extraction, pp 499–537). Elsevier
Petcharat T, Benjakul S, Karnjanapratum S, Nalinanon S (2021) Ultrasound-assisted extraction of collagen from clown featherback (Chitala ornata) skin: yield and molecular characteristics. J Sci Food Agric 101(2):648–658
Plaza M, Rodríguez-Meizoso I (2013) Advanced extraction processes to obtain bioactives from marine foods. Bioactive compounds from marine foods: Plant and animal sources, pp 343–371
Puértolas E, Barba FJ (2016) Electrotechnologies applied to valorization of by-products from food industry: main findings, energy and economic cost of their industrialization. Food Bioprod Process 100:172–184
Radzali SA, Baharin BS, Othman R, Markom M, Rahman RA (2014) Co-solvent selection for supercritical fluid extraction of astaxanthin and other carotenoids from Penaeus monodon waste. J Oleo Sci p. ess13184
Rahimi MA, Omar R, Ethaib S, Mazlina MS, Biak DA, Aisyah RN (2017) Microwave-assisted extraction of lipid from fish waste. In: IOP Conference Series: Materials Science and Engineering (Vol. 206, No. 1, p. 012096). IOP Publishing
Ranjha MMA, Kanwal R, Shafique B, Arshad RN, Irfan S, Kieliszek M, Kowalczewski PŁ, Irfan M, Khalid MZ, Roobab U, Aadil RM (2021) A critical review on pulsed electric field: A novel technology for the extraction of phytoconstituents. Molecules 26(16):4893
Richter BE, Jones BA, Ezzell JL, Porter NL, Avdalovic N, Pohl C (1996) Accelerated solvent extraction: a technique for sample preparation. Anal Chem 68(6):1033–1039
Rubio-Rodríguez N, Sara M, Beltrán S, Jaime I, Sanz MT, Rovira J (2012) Supercritical fluid extraction of fish oil from fish by-products: a comparison with other extraction methods. J Food Eng 109(2):238–248
Rustad T (2003) Utilisation of marine by-products. Elec J Env Agricult Food Chem Title 2(4):458–463
Sahena F, Zaidul ISM, Norulaini NNA, Jinap S, Jahurul MHA, Omar MAK (2014) Storage stability and quality of polyunsaturated fatty acid rich oil fraction from Longtail tuna (Thunnus tonggol) head using supercritical extraction. CyTA-J Food 12(2):183–188

Saifullah M, McCullum R, Vuong QV (2021) Optimization of microwave-assisted extraction of polyphenols from Lemon Myrtle: Comparison of modern and conventional extraction techniques based on bioactivity and total polyphenols in dry extracts. Processes 9(12):2212

Saito M (2013) History of supercritical fluid chromatography: instrumental development. J Biosci Bioeng 115(6):590–599

Sánchez-Camargo AP, Meireles MÂA, Lopes BLF, Cabral FA (2011) Proximate composition and extraction of carotenoids and lipids from Brazilian redspotted shrimp waste (Farfantepenaeus paulensis). J Food Eng 102(1):87–93

Siewe FB, Kudre TG, Narayan B (2021) Optimisation of ultrasound-assisted enzymatic extraction conditions of umami compounds from fish by-products using the combination of fractional factorial design and central composite design. Food Chem 334:127498

Soria AC, Brokl M, Sanz ML, Martínez-Castro I (2012) Sample preparation for the determination of carbohydrates in food and beverages

Sosa-Hernández JE, Escobedo-Avellaneda Z, Iqbal HM, Welti-Chanes J (2018) State-of-the-art extraction methodologies for bioactive compounds from algal biome to meet bio-economy challenges and opportunities. Molecules 23(11):2953

Taati MM, Shabanpour B, Ojagh M (2017) Extraction of oil from tuna by-product by supercritical fluid extraction (SFE) and comparison with wet reduction method. Aquacult Aquarium Conserv Legislation 10(6):1546–1553

Tena MT (2019) Extraction | Pressurized Liquid Extraction,Editor(s): Paul Worsfold, Colin Poole, Alan Townshend, Manuel Miró, Encyclopedia of Analytical Science (Third Edition),Academic Press,2019,Pages 78–83,ISBN 9780081019849

Tu ZC, Huang T, Wang H, Sha XM, Shi Y, Huang XQ, Man ZZ, Li DJ (2015) Physico-chemical properties of gelatin from bighead carp (Hypophthalmichthys nobilis) scales by ultrasound-assisted extraction. J Food Sci Technol 52(4):2166–2174

Vandenburg, H.J., 2000. POLYMERS| Supercritical Fluid Extraction.

Villamil O, Váquiro H, Solanilla JF (2017) Fish viscera protein hydrolysates: Production, potential applications and functional and bioactive properties. Food Chem 224:160–171

Wang M, Zhou J, Collado MC, Barba FJ (2021) Accelerated solvent extraction and pulsed electric fields for valorization of Rainbow Trout (Oncorhynchus mykiss) and Sole (Dover sole) by-products: Protein content, molecular weight distribution and antioxidant potential of the extracts. Mar Drugs 19(4):207

Yousuf O, Gaibimei P, Singh A (2018) Ultrasound assisted extraction of oil from soybean. Int J Curr Microbiol App Sci 7:843–852

Zahari NAAR, Chong GH, Abdullah LC, Chua BL (2020) Ultrasonic-assisted extraction (UAE) process on thymol concentration from Plectranthus amboinicus leaves: Kinetic modeling and optimization. Processes 8(3):322

Zakaria N, Rahman R, Zaidel D, Dailin D, Jusoh M (2021) Microwave-assisted extraction of pectin from pineapple pee. Malaysian J Fund Appl Sci 17(1):33–38

Zhou Y, Sui S, Huang H, He G, Wang S, Yin Y, Ma Z (2012) Process optimization for extraction of fishbone calcium assisted by high intensity pulsed electric fields. Trans Chin Soc Agricult Eng 28(23):265–270

Emerging Nonthermal Technologies for the Processing of Fish Waste and By-Products

T. Jayasree Joshi, S. Sivaranjani, Nelluri Puja, Rahul Kumar Rout, and Tammineni Dushyanth Kumar

Abstract Fish production has steadily increased as a result of industrialization and advancements in fishing technology. Along with this, there has been a significant rise in the amount of fish waste produced, necessitating the disposal and recycling of this waste. Fish waste comprises the head, bones, skin, scales, blood, and viscera. Minerals, vitamins, proteins, and polyunsaturated fatty acids are highly concentrated in these. Conventional recovery methods for processing and extraction of bioactive components often use heat and are limited due to excessive energy consumption, loss of functional properties, poor stability of end-product, and manufacturing cost. Nonthermal processes like supercritical fluid technology, pulsed electric field, high-pressure processing, ozone technology, ultrasound treatment, membrane processing, cold plasma, etc., offer high processing efficiency with minimal loss of nutritional components. The chapter provides insight into the principles and current developments of novel nonthermal technologies that are employed in the processing of fish waste and by-products.

Keywords Fish waste · Nonthermal methods · HPP · PEF · Ozone · Cold plasma · Extraction techniques

1 Introduction

The fish production and processing sectors have expanded tremendously over the past three decades (Wangkheirakpam et al. 2019). According to FAO (2018), global consumption of fish surged from 67% in the 1960s to 87% in 2016. The amount of fish waste generated has also increased significantly. As a result, there is an urge to manage and recycle waste using eco-friendly and sustainable techniques. Fish waste includes the head, bones, skin, scales, blood, and viscera. These are rich in macro and micronutrients. With adequate treatment, bioactive components, lipids, protein

T. J. Joshi (✉) · S. Sivaranjani · N. Puja · R. K. Rout · T. D. Kumar
Agricultural and Food Engineering Department, Indian Institute of Technology, Kharagpur, West Bengal 721302, India
e-mail: jayasreejoshi48@gmail.com

S. Maqsood et al. (eds.), *Fish Waste to Valuable Products*, Sustainable Materials and Technology, https://doi.org/10.1007/978-981-99-8593-7_3

hydrolysates, and other nutrient components can be recovered from fish waste (Li et al., 2016; Bruno et al., 2019; Rodrigues, et al., 2021). Apart from food applications, the by-products generated could be exploited in the pharmaceutical, cosmetic, and nutraceutical sectors if adequately processed. Conventional methods used to process fish waste and by-products include solvent extraction, fermentation, wet pressing, enzymatic hydrolysis, distillation, and smoking (Tyson, et al., 2004; Aspevik et al., 2017). However, these approaches are limited because of significant energy consumption, loss of functional characteristics, poor stability of the product, and operating costs (Al Khawli et al. 2009; Ferraro et al., 2013). Novel nonthermal techniques can tackle most of these issues with high processing efficiency. Supercritical fluid (SCF) technology, Pulsed electric field (PEF), high-pressure processing (HPP), ozone technology, ultrasound (US) treatment, membrane processing, and cold plasma (CP) are the nonthermal technologies that have been used in the management of fish waste and by-products (Li et al. 2016; Jaswir, et al., 2017; Roslan et al. 2017). They have also shown huge potential in terms of preserving microbiological safety while preserving organoleptic properties. Ultrasound and enzyme-assisted extraction techniques have also demonstrated a high extraction rate, substantially reduced processing time, solvent selectivity, low energy demand, and low cost without damaging heat-sensitive components (Thirukumaran et al. 2022). The chapter delves into the fundamentals and recent advances of novel nonthermal methods used in the processing of fish waste and by-products. The challenges of implementing these technologies into practice at the industrial level are also discussed.

2 Fish Waste and By-Products

Processing of fish and fish products generates a large amount of waste, which includes the head, skin, bones, blood, scales, and viscera. From these by-products, protein hydrolysates, amino acids, collagen, lipids, gelatin, enzymes, and other bioactive components can be isolated (Esteban et al. 2007; Wangkheirakpam et al. 2019). Kasankala et al. (2007) extracted gelatin from the skin of grass carp and obtained a yield of 19.83%. The gelatin extracts recovered were also found to be high in proline and hydroxyproline. Gelatin was also recovered from the skins of Nile perch, Grass carp, Tilapia, and Dog sharks (Muyonga et al. 2004; Kasankala et al. 2007; Zeng et al. 2010; Shyni et al. 2014). Gelatin is the most widely extracted component from fish skins and bones. This is obtained by partially hydrolyzing animal collagen. Gelatin, unlike collagen, is soluble in water. Conversion of collagen to gelatin involves structural changes, which can be explained in three ways: (1) long peptide chain breakdown and rearrangement, (2) chain detachment by side bond breakage, and (3) chain reconfiguration (Boran et al. 2010). In the process of extracting gelatin, undesirable materials like fats and minerals are removed. Type A and Type B gelatins are obtained after the process. Type A gelatin is obtained by treating collagen with an acid, which has an isoelectric point at a pH range of 6–9. Type B gelatin is obtained by treating collagen with an alkali, having an isoelectric point at pH 5 (Karim and Bhat 2009).

After the extraction of type A or type B gelatin, the following stages are performed: hot water extraction, filtration, evaporation, and drying.

Fish lipids, like Omega-3 fatty acids (also called n-3 fatty acids), are high-value products obtained from fish waste and by-products. They require either chemical or physical interactions for separation from the matrix. Fishbone has been used as an excellent source of calcium. Fishbone powder is used in the formulation of food, feed, and fertilizer formulations (Asikin et al. 2019). Nawaz et al. 2020 improved the in vitro digestibility of fishbone powder using autoclave processing. Fish scales, gills, head, and gut have been used for the production of various enzymes (Ramakrishnan et al. 2013; Harikrishna et al. 2017; Murthy et al. 2018). Silva et al. (2014) isolated protein hydrolysates rich in essential amino acids and fatty acids from tilapia fish waste using enzyme-assisted extraction. Protein hydrolysates have been utilized in the treatment of cancer, heart disease, and immunological disorders, as well as in nutritional supplements (Ahn et al. 2012). Through enzymatic treatment, Hathwar et al. (2011) retrieved proteins and lipid waste from catla and rohu visceral waste. Using the fungal protease enzyme, they got a high lipid recovery of 74.9%, and the protein hydrolysate generated by the enzyme exhibited high antioxidant activity.

3 Technologies for the Processing of Fish Waste and By-Products

The following session discusses in detail the different conventional and novel nonthermal methods used in the processing of fish waste. Fig. 1 also gives a brief overview of the technologies currently used in fish waste valorization.

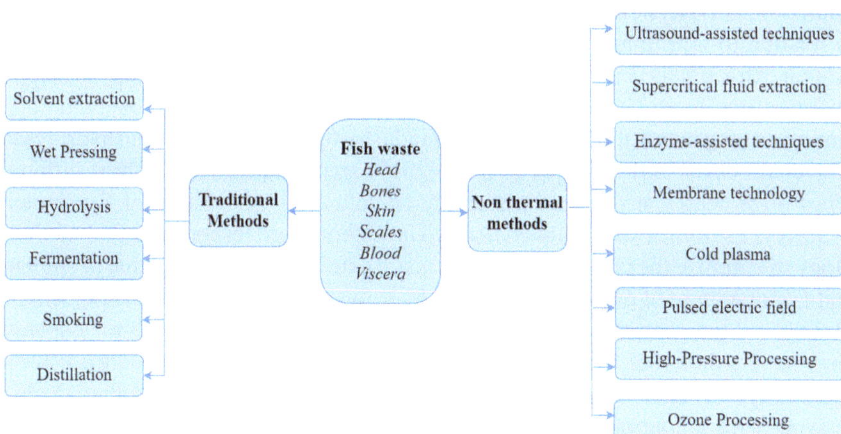

Fig. 1 Overview of the traditional and nonthermal technologies currently used in the fish waste valorization

3.1 Conventional Approaches

Fish waste contains a variety of beneficial compounds. Conventional approaches such as solvent extraction, smoking, wet pressing, distillation, fermentation, and esterification are used to valorize fish waste. The most popular approach for separating desired components from a material using a suitable solvent is solvent extraction. Organic solvents are used for the extraction of lipids and lipid-based compounds as lipids are readily soluble in them, giving a higher yield (Furse et al. 2015). Different organic solvents reported in studies include methylene chloride, mixture of chloroform and methanol, mixture of acetone and ethyl acetate, and hexane (Oterhals and Vogt 2013). Toyoshima et al. (2004) extracted oil from sardine waste using chloroform and methanol as solvents in a ratio of 2:1. They were able to recover 7 g/100 g of oil from the fish waste. Sathivel et al. (2002) obtained a yield of 34% of fat from catfish viscera using solvent extraction technique. Other conventional methods for extracting fish oil from fish tissue include the Goldfisch method, chloroform–methanol method, Bligh and Dyer method, and the acid digestion method (Shahidi 2003; Dave and Ghaly 2013). Fermentation is another approach, which is used in the production of oil, protein, silage, and enzymes from fish waste. Shabani et al. (2021) used Sardine wastes as a substrate for the production of silage using *Lactobacillus plantarum*. The silage obtained is supplemented in diet broiler chicken and they noticed a considerable increase in the weight of the chickens.

Dry and wet rendering is another traditional approach that is used in the processing of fish waste. Rendering refers to a process where protein coagulation takes place under high-pressure steam, giving a miscella of oil with bound water. From it, oil is recovered using filtration or centrifugation (Cretton et al. 2022). Wet rendering uses live steam, while dry rendering uses the heat of steam inside a jacketed vessel. Enzymatic processing of various biopolymers in foods is another important step for improving the physical, chemical, and organoleptic qualities of the original food in relation to nutritional value and intestine absorption characteristics. Enzymatic protein extraction is performed at under-regulated pH levels to ensure the nutritional quality of the food. Carbohydrases, proteases, and lipases are the most commonly employed enzymes in the food processing industry to prepare fish protein hydrolysate (Dave and Ghaly 2013). Vázquez et al. (2017) studied the effect of three different types of proteases at different pH and temperatures for extraction of fish protein hydrolysates from *Scyliorhinus canicula*. The authors used response surface methodology to optimize the conditions for enzymatic extractions and found that pH of 9.40 and a temperature of 64.6 °C are the best hydrolysis conditions for alcalase, and pH of 8.90 and temperature of 60.8 °C are the best hydrolysis conditions for Esperase enzyme. Furthermore, Araujo et al. (2021) utilized fish waste to produce fish oil, protein hydrolysates, and collagen simultaneously. The authors used the Alcalase enzyme with a controlled temperature (50 °C) and pH (8) for 180 min. Using a response surface technique, Nurdiyana et al. (2008) optimized the protein extraction conditions from freeze-dried fish waste.

3.2 Nonthermal Technologies for Processing of Fish Wastes and By-Products

The following sessions discuss in detail about the different nonthermal techniques used in the valorization of fish waste.

3.2.1 Extraction Techniques

Fish processing waste is often used to extract collagen, gelatin, hydroxyapatite, lipids, enzymes, protein hydrolysates, and bioactive components. The extraction characteristics and yield of these components are highly influenced by raw material type, composition, properties of the material, type of solvent, concentration of solvent, temperature, pH, pressure, and extraction time (Caruso et al. 2020). The constraints with conventional methods were low yield and longer extraction time. As a result, the emphasis was shifted to innovative extraction technologies with the intention of increasing yield while decreasing extraction time. Table 1 represents different extraction techniques followed for extracting valuable components from fish wastes and by-products. Different extraction techniques are also discussed in brief in the following sections.

Ultrasound-Assisted Extraction (UAE)

The US is one of the fastest and efficient methods of extraction, which is capturing industrial applications to a large extent. Applying the principle of "acoustic cavitation" results in the disruption of cell wall and an increase in the mass transfer of cell contents, thereby increasing the yield as a whole (Khanashyam et al. 2023). It also increases the functional characteristics of gelatin (Tu et al. 2015). Astaxanthin was extracted from shrimp shells under the optimized extraction conditions of 23.6% amplitude for 13.9 min at 26.3 °C (Sharayei et al. 2021). In UAE, the polarity of the solvents and the time of extraction play a vital role in extraction yield (Šimat et al. 2022). Kim et al. (2013) recovered collagen from the skin of sea bass using UAE. They used a sample: acetic acid ratio of 1:200 for the extraction. A rapid increase in collagen yield with amplitude and duration of the ultrasound was observed. The extraction solvent is also important in the yield. The extraction yield increased in accordance with the amount of acid added for acid concentrations between 1/5 and 1/50. The authors also observed that adding acid to the skin of sea bass resulted in modifications of collagen fiber structure. The cavitation effect of ultrasonic treatment also improved collagen extraction. According to Kim et al. (2012), ultrasonic treatment enhances extraction yield by 4.65 times due to the cavitation effect.

Table 1 Literatures on the extraction of valuable products from fish and crustacean wastes using different techniques

Fish waste/ by-product	Extraction technique	Component extracted	Major findings	References
Shrimp byproducts	High-pressure extraction	Astaxanthin	Increase in astaxanthin content up to 300 MPa; yield reduced beyond 300 MPa	Šimat et al. (2022)
Shrimp shells	UAE	Astaxanthin	Improvement in the extraction of astaxanthin as compared to conventional extraction techniques	Sharayei et al. (2021)
Crustacean by-products	SCFE	Astaxanthin	Increased yield of free and conjugated astaxanthins; increase in ethanol increased extraction yield; pressure above 400 bar hampered extraction	Sánchez-Camargo et al. (2012)
Carp scales	UAE	Gelatin	Enhanced yield observed and also the free amino group content in the extracted gelatin was high	Tu, et al. (2015)
Sea bass skin	UAE	Collagen	Improved yield with increasing treatment time observed	Kim et al. (2013)
Tuna head	SCFE	Oil	Obtained high yield and they separated the extracted oil into different fractions based on the molecular weight	Ferdosh, et al. (2016)
Fish skin	HPP	Gelatin	Resulted in higher yield and concentration, as well as a reduction in pretreatment time	Jaswir, et al. (2017)

(continued)

Table 1 (continued)

Fish waste/ by-product	Extraction technique	Component extracted	Major findings	References
Shrimp by-products	Combined US and PEF-assisted treatment	Lipids	Higher electric field gave better disintegration; ultrasound improved mass transfer due to electroporation; PEF was better than US treatment in terms of extraction yield	Gulzar and Benjakul (2020)
Bighead carp scales	EAE	Gelatin	Higher gel strength in enzymatic extraction as compared to other techniques	Muhammad et al. (2016)
Black and red tilapia skins	Hot water extraction (at 45 °C)	Gelatin	Black and red tilapia gave 5.39% and 7.81% gelatin, respectively	Jamilah and Harvinder (2002)
Nile perch skin and bone	Hot water extraction at 50 °C, 60 °C, and 70 °C, for 5 h each	Gelatin	Skin yielded 4.9% gelatin; Bone yield was 2.9% (db)	Muyonga et al. (2004)
Cuttlefish skin	Incubation for 4 h at 40 °C (continuous stirring), followed by centrifugation (10,000 g, 30 min, 4 °C)	Gelatin	Alkali-treated skin treated with pepsin; Gelatin yield of 6.61%, 7.72%, and 9.22% for 5, 10, and 15 U pepsin/g fish skin	Jridi et al. (2013)

Supercritical Fluid Extraction (SCFE)

SCFE is the technique of separating one component from another using supercritical fluids as the solvent. Rodrigues et al. (2021) extracted bioactive lipids from canned sardine waste using SCFE. They found that raising pressure at a constant temperature resulted in a considerable increase in yields, whereas increasing temperature at a given pressure had no influence on extraction yields. Létisse et al. (2006) also obtained comparable results. The optimized conditions for SCFE of free and conjugated astaxanthins are a pressure of 215.68 bar, and 1.89 mL/min flow rate at 58.66 °C (Radzali et al. 2016). The yield of free and conjugated astaxanthins obtained was 12.20 µg/g and 58.50 µg/g, respectively. In the same study, another observation was made, i.e., an increase in extraction yield (26–34.8 µg/g) with an increase in ethanol content (5–15%). This trend was observed due to higher mass transfer because of the higher amount of solvent, resulting in a higher yield (Sánchez-Camargo et al. 2012).

In another study, Ferdosh et al. (2016) extracted oil from tuna heads using the SCFE technique, and they obtained a yield of 35.6%. They found that the approach was extremely effective in extracting omega-3 fatty acid, omega-6 fatty acid, and DHA from tuna heads.

Enzyme-Assisted Extraction (EAE)

EAE is a novel, safe, and environmentally friendly method for extracting bioactives that employ specialized enzymes. This method is often combined with other methods to get improved yields. Pezeshk et al. (2019) extracted protein hydrolysate from tuna viscera using enzyme hydrolysis followed by ultrafiltration. They separated the protein hydrolysate into different molecular fractions and observed that fractions of <3 kDa have high antibacterial activity. Pepsin enzyme (547 U/g) was used to extract gelatin from carp scales (Usman et al. 2022). They found that, in comparison with other extraction methods, enzymatic extraction produced gelatin with greater gel strength. Pepsin was also used for the extraction of gelatin from the skins of Chinese sturgeon (Muhammad et al. 2016). Feng et al. (2013) extracted collagen from sturgeon skin by using the EAE technique. Pepsin was the enzyme utilized for the study. The optimum conditions for the extraction were found to be a solid–liquid ratio of 1:11.88, enzyme concentration of 2.42%, and an extraction period of 6.45 h. Under optimal conditions, a maximum yield of 86.69% pepsin-soluble collagen was obtained. They also revealed that increasing the enzyme concentration to a certain level enhanced collagen production, which thereafter declined.

3.3 *Membrane Technology*

Membrane technology is a separation process wherein selective particles are segregated from the rest of the mixture by passing through a semi-permeable membrane (Youravong & Marthosa 2017). Based on the pore size and operating principle, they are classified as microfiltration, ultrafiltration, nanofiltration, reverse osmosis, and electromembrane filtration (Cardoso and Nunes 2013). It is mainly employed for concentration, extraction, and fractionation of high-quality biomolecules, recovery of valuable components, and purification of food and wastewater streams. In recent days, membrane technology has been successfully used for the treatment and valorization of fish waste which has been reported by several studies. Pezeshk et al. (2019) fractionated fish protein hydrolysates obtained from yellowfin tuna (Thunnus albacores) viscera by using ultrafiltration membranes of four different molecular weights (<3, 3–10, 10–30, and >30 kDa). He found that the better quality peptide fractions were obtained from <3 kDa membrane, which can be substituted as antimicrobials and antioxidants. Roslan et al. (2017) reported that multilayer ultrafiltration membrane improved the yield of peptides from tilapia waste protein hydrolysates when compared to a single membrane. He also concluded that the specificity of biomolecules can be enhanced by altering the pore size and orientation of the membrane. It can also be used in conjunction with chromatography to isolate selective peptides (Cardoso and Nunes 2013).

This technology also helps in purifying the biofuels obtained from the fish waste by using a membrane-based gas permeation unit with an efficiency up to 99% for separating methane and carbon dioxide (Kratky and Zamazal 2020). Similar results (efficiency of 99.9%) were reported by Maghami et al. (2018) when a ceramic ultra-membrane was used for the separation of glycerin and biodiesel produced from waste fish oil. Moreover, it can also be used in combination with enzymatic hydrolysis in a bioreactor for the synthesis and isolation of high-quality peptides from fish waste which has been highlighted in a few studies (Abejón et al. 2018; Benhabiles et al. 2012; Cardoso and Nunes 2013). Membrane fouling and membrane flux play an important role in the membrane separation process as such they must be maintained within limits to ensure better performance and separation efficiency.

3.4 Cold Plasma (CP)

CP is another innovative nonthermal technology that employs plasma, the fourth state of matter, for processing of food products. CP has a wide range of applications like decontamination of food, toxin degradation, inactivation of enzymes, and modification of packaging materials (Pankaj and Keener 2017). Plasma-activated water produced by CP was also found to play a key role in surface decontamination and curing of meat (Zhao et al. 2020). Several studies (Chiper et al. 2011; Choi et al. 2017; Hajhoseini et al. 2020) reported the processing of fish by using CP but very limited studies are available on the valorization of fish waste by CP. The effect of atmospheric CP on crude protease extract and muscle protein of hairtail fish was investigated by Koddy et al. (2021). He found that CP is capable of minimizing the crude protease activity (0.035 units/mg protein) when treated at 50 kV for 240 s. The water retention capacity and color of the hair tail fish proteins were greatly increased by CP. Choi et al. (2016) reported the usage of CP for processing dried Alaska pollock shreds and found that it significantly inactivated most of the microbes. Treatment of dried Alaska pollock shreds for 2 min with CP was found to be optimal based on the sensory and physico-chemical attributes. Apart from this CP can be used for the modification of fish proteins to enhance their functional properties. It's also possible to employ CP in enhancing the mechanical properties of packaging films developed from fish waste. It can also enhance the quality of proteins developed from fish waste. However, there is a strong need for further research and a more detailed analysis in the valorization of fish waste by CP.

3.5 Pulsed Electric Field

PEF is one of the nonthermal techniques which use high-frequency electric pulses at high voltages. Low thermal heat generation and short processing times make PEF a unique technology for application in the fish waste industry. He et al. (2017) extracted various valuable compounds such as calcium, chondroitin sulfate, and collagen from fishbone using semi-bionic extraction paired with a pulsed electric field. The semi-bionic procedure involves maintaining the pH and adding enzymes like pepsin and trypsin to simulate the human digestive tract. Semi-bionic PEF extraction of substances such as calcium, chondroitin sulfate, and collagen has proven to be an efficient approach for increasing extraction yield. The authors reported calcium extraction yields of 19.8 mg/ml, chondroitin sulfate extraction yields of 39.268 mg/ml, and a collagen extraction of 3.5227%. Franco et al. (2020) used PEF to extract antioxidants from fish residues such as gills, bones, and heads from sea bass and sea bream. The authors observed the highest antioxidant activity from residues of sea bream species. They also observed that extracts from gills showed the highest antioxidant activity compared to other parts. Wang et al. (2021) observed the antioxidant activity of extracts from rainbow trout and sole residues by accelerated solvent extraction combined with PEF. Both the treatments considerably improved the antioxidant capacity of extracts.

High-intensity pulsed electric field (HI-PEF) employs higher intensity electric field pulses than normal PEF and has a shorter extraction time, higher yield, and lower operational temperature. Zhou et al. (2017) utilized HI-PEF to extract protein fractions from mussels. When compared to traditional methods, the authors observed that HI-PEF extracts protein at a much faster rate. The optimized conditions for maximum protein extraction of 77.08% from mussels were 20 kV/cm electric field with a pulse number of 8 and enzymolysis time of two hours. Li et al. (2016) studied the effect of PEF on protein extraction from Abalone viscera. The authors compared the functional properties of protein extracted with PEF and protein extracted by enzymatic extraction. The protein extracted from PEF showed better emulsifying properties compared to only enzymatic extraction. However, a reduction in foaming and viscosity was observed in PEF-treated samples.

3.6 High-Pressure Processing (HPP)

HPP has been successfully employed for the extension of the shelf life of marine products by improving physicochemical, microbial, and sensory attributes. Despite the fact that HPP has been used for the treatment of fish products, the significance of HPP on waste valorization has been highlighted by a few studies. High-pressure extraction (HPE) demonstrates high extraction efficiency, less processing time, improved quality, and a higher yield of bioactive components from natural materials (Huang et al. 2013). Astaxanthin belongs to the carotenoid group, which is a principal compound prevalent in marine animals. It has wide applications ranging from food, cosmetics, and pharmaceutical applications, as it has high antioxidant activity. High-pressure-assisted extraction of astaxanthin from shrimp waste viz., shell and head at varying pressure, holding time, and liquid-to-solid ratio on extraction yield was evaluated (Du et al. 2013; Li et al. 2017). Yield of astaxanthin from shrimp discards was found to be 89.15 μg/g at optimum conditions of 210 MPa pressure and 9.2 min pressure holding time whereas ultrasonic extraction recorded 6.5% lower yield than HPE (Du et al. 2013). Similar results were found for dried shrimp waste where ethanol was used as a solvent that shows highest yield of astaxanthin at all pressure levels. Also, HPE process recorded the lowest EC50 values than conventional extraction, standard Vitamin C and standard Vitamin E solutions by DPPH and superoxide anion radical scavenging activity that shows higher antioxidant capacity of astaxanthin extracted by the HPE process (Li et al. 2017). Irna et al. (2018) examined the six species of shrimp for its total carotenoid and astaxanthin yield by HPE and chemical extraction. Irrespective of shrimp species, HPE resulted in higher carotenoid content and astaxanthin yield with shorter extraction time. Compared to other species, *Penaeus monodon* contains the highest carotenoid content and yield of astaxanthin.

HPE application is also extended to gelatin extraction, which is obtained by the hydrolysis of collagen from fish waste. High-pressure results in protein denaturation due to breaking of non-covalent bonds, thereby enabling gelatin extraction (Chang et al. 2013). It is reported that HPP allows better acid penetration into the proteins, resulting in higher yield of gelatin (Sarupria et al. 2010). HPP treatment is generally given after acid or alkali pre-treatment, where the clean fish scales, skins, or bones are sealed in an LDPE pouch along with distilled or de-ionized water (Jaswir et al. 2017). It generally takes half the time for extraction as compared to conventional extraction. Jaswir et al. (2017) studied gelatin from four types of fish skin that was extracted by acid treatment followed by water extraction. High pressure influences protein denaturation, which distresses the non-covalent bond in the collagen. It was found that the yield of gelatin was higher when pressure was applied at acid treatment than at the extraction stage because high pressure induces more acid to penetrate into the cell membrane. Also, pH and gel strength do not show a significant difference between HPE and the conventional method. Similar results for yield were found by Gómez-Guillén et al. (2005) where skin from dover sole was used for gelatin extraction. Highpressure treatment not only influences the extraction of components but also assists the enzymatic hydrolysis of protein isolate and gelatin to form a hydrolysate. Fish protein hydrolysates were obtained from tilapia by-products, which effectively decreased the processing time and increased free amino acids in high-pressure processing (Hemker et al. 2020). High pressure promoted the enzyme hydrolysis with better antioxidant capacity, although this cannot be extended to all enzymes that are involved in the hydrolysis (Alemán et al. 2011).

3.7 *Ozone Processing*

Ozone is a nonthermal food processing technology alternative to thermal treatment (Sivaranjani et al. 2021). Ozone application in fish products has been explored in the production process of biofuel (Onishi et al. 2004). Due to the increasing demand for fossil fuels, there is a need for substitute resources, which can provide the same properties as fossil fuels. Biofuels are mainly produced from biomass obtained from vegetable and animal wastes. These biofuels serve as a renewable transportation fuel, also reducing carbon dioxide emissions (Demirbas 2007). Biofuel from fish processing waste can be recovered and purified by physical, chemical, biological, and thermal processes (Jayasinghe and Hawboldt 2012). However, the biodiesel from these processes has limited application due to the strong odor in the oil. Ozone treatment of fish oil in the biodiesel production process and property analysis has been observed by Kato et al. (2004).

Biodiesel production involves the conversion of fatty acids into hydrocarbons due to the action of ozone on the bonds of fatty acids in fish waste oil. The biodiesel production method begins with purifying fish oil with kaolin before depositing it in a reactor with iron oxide and calcium phosphate as a catalyst. The authors also carried out ozone treatment at a concentration of 8000 ppm at 5 g/h for 1 h at room temperature. Following the primary treatment, the mixture is filtered again using zeolite and then passes to a secondary treatment. The secondary treatment consists of treating fish oil in the same manner as the primary treatment but without the catalyst for 30 min (Kato et al. 2004; Onishi et al. 2004). The yield of the biodiesel obtained from ozone treatment (95–96%) was found to be higher when compared with transesterification process (90%) (Yahyaee et al. 2013). The ozone-treated biodiesel has better properties such as density (0.87 g/cm^3 at 15 °C), higher heating value (10,700 kcal/kg), flash (38 °C) and pour (−16 °C) points, almost similar to commercial diesel fuel (Kato et al. 2004). While the flash point of methyl-esterified fish oil and vegetable oil is 164 °C and 130 °C, respectively, which results in less combustible fuel (Kato et al. 2004; Yahyaee et al. 2013). In terms of water content and acid index, pyrolyzed fish oil has a higher value than ozone-treated biofuel, which indicates the presence of more free fatty acids (Meier et al. 2009). Biodiesel from ozone treatment contains low to high molecular weight hydrocarbons, which facilitates lower soot, polyaromatic compounds, and harmful material emissions such as carbon dioxide. Besides, this process doesn't require any waste treatment steps for the production of biodiesel (Kato et al. 2004). Thus, ozone-treated fish oil was found to be a suitable alternative to commercial diesel fuel in terms of fuel quality.

4 Future Perspectives and Challenges

Even though emerging nonthermal technologies offer better valorization of fish waste than conventional approaches, they do have significant downsides. Prolonged exposure time and intense cavitation have reduced gel strength and physicochemical and functional properties during UAE treatment of fish waste (Al Khawli et al. 2019). To preserve the functional characteristics of the end product, it is critical to optimize the processing conditions of novel extraction techniques. It is also vital to establish suitable extraction methods that allow for minimal processing with maximum quality and yield while assuring product safety. The high initial cost of HPP and PEF technologies, and the cost for the generation of CP also poses significant challenges. Future research is also needed to investigate the toxicological consequences of cold plasma on fish by-products (Thirukumaran et al. 2022). Despite the fact that most nonthermal techniques maintain food quality, they are still on the laboratory scale. Some of the key challenges preventing their industrial-scale deployment are machinery development, optimization and control of process parameters during processing, and regulatory approval by authorities (Ali et al. 2021). Hence, further investigations are required to optimize the manufacturing and processing processes.

5 Conclusion

Novel nonthermal technologies are prospective methods for valorizing fish processing owing to their high efficiency, solvent selectivity, shortened processing time, nontoxicity, low energy consumption, and ecofriendliness. Conventional techniques for processing fish waste and by-products lead to limited loss of functional properties and poor stability issues of the final product. This chapter provided an overview of the novel technologies utilized in the treatment of fish waste and by-products, including SCF technology, PEF, HPP, US, CP, ozone technology, and membrane processing. Compared to traditional processing processes, these technologies cause only minimal changes in sensory and nutritional quality. But, limited research is available on the optimization of the processing parameters and scale-up studies. More research is needed to enable industrial adoption of these technologies on a large scale.

References

Abejón R, Belleville MP, Sanchez-Marcano J, Garea A, Irabien A (2018) Optimal design of industrial scale continuous process for fractionation by membrane technologies of protein hydrolysate derived from fish wastes. Sep Purif Technol 197:137–146

Ahn CB, Je JY, Cho YS (2012) Antioxidant and anti-inflammatory peptide fraction from salmon byproduct protein hydrolysates by peptic hydrolysis. Food Res Int 49(1):92–98

Al Khawli F, Pateiro M, Domínguez R, Lorenzo JM, Gullón P, Kousoulaki K, Ferrer E, Berrada H, Barba FJ (2019) Innovative green technologies of intensification for valorization of seafood and their by-products. Mar Drugs 17(12):689
Alemán A, Giménez B, Gómez-Guillén MC, Montero P (2011) Enzymatic hydrolysis of fish gelatin under high pressure treatment. Int J Food Sci Technol 46(6):1129–1136
Ali A, Wei S, Liu Z, Fan X, Sun Q, Xia Q, Liu S, Hao J, Deng C (2021) Non-thermal processing technologies for the recovery of bioactive compounds from marine by-products. Lwt 147:111549
Araujo J, Sica P, Costa C, Márquez MC (2021) Enzymatic hydrolysis of fish waste as an alternative to produce high value-added products. Waste Biomass Valoriz 12(2):847–855
Asikin AN, Kusumaningrum I, Hidayat T (2019) Effect of knife-fish bone powder addition on characteristics of starch and seaweed kerupuk as calcium and crude fiber sources. Curr Res Nutr Food Sci 7(2):584
Aspevik T, Oterhals Å, Rønning SB, Altintzoglou T, Wubshet SG, Gildberg A, Afseth NK, Whitaker RD, Lindberg D (2017) Valorization of proteins from coand by-products from the fish and meat industry. In: Chemistry and chemical technologies in waste valorization. Springer, pp 123–150
Benhabiles MS, Abdi N, Drouiche N, Lounici H, Pauss A, Goosen MFA, Mameri N (2012) Fish protein hydrolysate production from sardine solid waste by crude pepsin enzymatic hydrolysis in a bioreactor coupled to an ultrafiltration unit. Mater Sci Eng C 32(4):922–928
Boran G, Mulvaney SJ, Regenstein JM (2010) Rheological properties of gelatin from silver carp skin compared to commercially available gelatins from different sources. J. Food Sci. 75(8). https://doi.org/10.1111/j.1750-3841.2010.01543.x
Bruno SF, Kudre TG, Bhaskar N (2019) Impact of pretreatment-assisted enzymatic extraction on recovery, physicochemical and rheological properties of oil from Labeo rohita head. J Food Process Eng 42(3):e12990
Cardoso C, Nunes ML (2013) Improved utilization of fish waste, discards, and by-products and low-value fish towards food and health products. Utiliz Fish Waste 26–58
Caruso G, Floris R, Serangeli C, Di Paola L (2020) Fishery wastes as a yet undiscovered treasure from the sea: Biomolecules sources, extraction methods and valorization. Mar Drugs 18(12):622
Chang HJ, Niu XY, Tang C, Wang Q (2013) Effect of high pressure processing on thermal characteristics of perimysium and endomysium collagen from beef semitendinosus muscle. 食品科学 34(13):15
Chiper AS, Chen W, Mejlholm O, Dalgaard P, Stamate E (2011) Atmospheric pressure plasma produced inside a closed package by a dielectric barrier discharge in Ar/CO_2 for bacterial inactivation of biological samples. Plasma Sources Sci Technol 20(2):025008
Choi S, Puligundla P, Mok C (2017) Effect of corona discharge plasma on microbial decontamination of dried squid shreds including physico-chemical and sensory evaluation. Lwt 75:323–328
Cretton M, Malanga G, Mazzuca Sobczuk T, Mazzuca M (2022) Marine lipids as a source of high-quality fatty acids and antioxidants. Food Rev Int 1–24
Dave D, Ghaly A (2013) Fish processing wastes as a potential source of proteins. Amino acids and oils: a critical review. J Microb Biochem Technol 5:107–129
Demirbas A (2007) Importance of biodiesel as transportation fuel. Energy Policy 35(9):4661–4670
Du J, He J, Yu Y, Zhu S, Li J (2013) Astaxanthin extracts from shrimp (Litopenaeusvannamei) discards assisted by high pressure processing. In: 2013 Kansas City, Missouri, July 21–24, 2013 (p. 1). American Society of Agricultural and Biological Engineers
Esteban MB, Garcia AJ, Ramos P, Marquez MC (2007) Evaluation of fruit–vegetable and fish wastes as alternative feedstuffs in pig diets. Waste Manage 27(2):193–200
FAO., (2018) The state of world fisheries and aquaculture 2018. Meeting the sustainable development goals, Rome
Feng W, Zhao T, Zhou Y, Li F, Zou Y, Bai S, Wang W, Yang L, Wu X (2013) Optimization of enzyme-assisted extraction and characterization of collagen from Chinese sturgeon (Acipenser sturio Linnaeus) skin. Pharmacogn Mag 9(Suppl 1):S32
Ferdosh S, Sarker MZI, Rahman NNNA, Akanda MJH, Ghafoor K, Kadir MOA (2016) Simultaneous extraction and fractionation of fish oil from tuna by-product using supercritical carbon

dioxide (SC-CO_2). J Aquat Food Prod Technol 25:230–239. https://doi.org/10.1080/10498850.2013.843629
Ferraro V, Carvalho AP, Piccirillo C, Santos MM, Castro L, Pintado E (2013) Extraction of high added value biological compounds from sardine, sardine-type fish and mackerel canning residues — a review. P.M. Mater Sci Eng C 33:3111–3120. https://doi.org/10.1016/j.msec.2013.04.003
Franco D, Munekata PE, Agregán R, Bermúdez R, López-Pedrouso M, Pateiro M, Lorenzo JM (2020) Application of pulsed electric fields for obtaining antioxidant extracts from fish residues. Antioxidants 9(2):90
Furse S, Egmond MR, Killian JA (2015) Isolation of lipids from biological samples. Mol Membr Biol. https://doi.org/10.3109/09687688.2015.1050468
Gómez-Guillén MC, Giménez B, Montero P (2005) Extraction of gelatin from fish skins by high pressure treatment. Food Hydrocolloids 19(5):923–928
Gulzar S, Benjakul S (2020) Impact of pulsed electric field pretreatment on yield and quality of lipid extracted from cephalothorax of Pacific white shrimp (Litopenaeusvannamei) by ultrasound-assisted process. Int J Food Sci Technol 55(2). https://doi.org/10.1111/ijfs.14316
Hajhoseini A, Sharifan A, Yousefi HR (2020) Effects of atmospheric cold plasma on microbial growth of Listeria innocua and Staphylococcus aureus in ready-to-eat fish products. Iran J Fish Sci 19(1):262–271
Harikrishna N, Mahalakshmi S, Kiran Kumar K, Reddy G (2017) Fish scales as potential substrate for production of alkaline protease and amino acid rich aqua hydrolyzate by *Bacillus altitudinis* GVC11. Indian J Microbiol 57(3):339–343
Hathwar SC, Bijinu B, Rai AK, Narayan B (2011) Simultaneous recovery of lipids and proteins by enzymatic hydrolysis of fish industry waste using different commercial proteases. Appl Biochem Biotechnol 164(1):115–124
He G, Yin Y, Yan X, Wang Y (2017) Semi-bionic extraction of effective ingredient from fishbone by high intensity pulsed electric fields. J Food Process Eng 40(2):e12392
Hemker AK, Nguyen LT, Karwe M, Salvi D (2020) Effects of pressure-assisted enzymatic hydrolysis on functional and bioactive properties of tilapia (*Oreochromis niloticus*) by-product protein hydrolysates. LWT 122:109003
Huang HW, Hsu CP, Yang BB, Wang CY (2013) Advances in the extraction of natural ingredients by high pressure extraction technology. Trends Food Sci Technol 33(1):54–62
Irna C, Jaswir I, Othman R, Jimat DN (2018) Comparison between high-pressure processing and chemical extraction: astaxanthin yield from six species of shrimp carapace. J Dietary Supplem 15(6):805–813
Jamilah B, Harvinder KG (2002) Properties of gelatins from skins of fish - Black tilapia (Oreochromis mossambicus) and red tilapia (Oreochromis nilotica). Food Chem 77(1). https://doi.org/10.1016/S0308-8146(01)00328-4
Jaswir I, Yusof N, Jamal P, Jami MS (2017) Novel method for gelatin extraction of various local fish using High Pressure Processing (HPP). Int Food Res J 24
Jayasinghe P, Hawboldt K (2012) A review of bio-oils from waste biomass: focus on fish processing waste. Renew Sustain Energy Rev 16(1):798–821
Jridi M, Nasri R, Lassoued I, Souissi N, Mbarek A, Barkia A, Nasri M (2013) Chemical and biophysical properties of gelatins extracted from alkali-pretreated skin of cuttlefish (*Sepia officinalis*) using pepsin. Food Res Int 54(2):1680–1687
Karim AA, Bhat R (2009) Fish gelatin: properties, challenges, and prospects as an alternative to mammalian gelatins. Food Hydrocolloids. https://doi.org/10.1016/j.foodhyd.2008.07.002
Kasankala LM, Xue Y, Weilong Y, Hong SD, He Q (2007) Optimization of gelatine extraction from grass carp (*Catenopharyngodon idella*) fish skin by response surface methodology. Biores Technol 98(17):3338–3343
Kato S, Kunisawa K, Kojima T, Murakami S (2004) Evaluation of ozone treated fish waste oil as a fuel for transportation. J Chem Eng Jpn 37(7):863–870

Khanashyam AC, Shanker MA, Thorakkattu P, Joshi TJ, Pandiselvam R, Kothakota A (2023) Applications of ultrasound technology for inactivation of food enzymes. In: Enzyme inactivation in food processing. Apple Academic Press, pp 201–237

Kim HK, Kim YH, Kim YJ, Park HJ, Lee NH (2012) Effects of ultrasonic treatment on collagen extraction from skins of the sea bass Lateolabrax japonicas. Fish Sci 78:485–490

Kim HK, Kim YH, Park HJ, Lee NH (2013) Application of ultrasonic treatment to extraction of collagen from the skins of sea bass Lateolabrax japonicus. Fish Sci 79:849–856. https://doi.org/10.1007/s12562-013-0648-z

Koddy JK, Miao W, Hatab S, Tang L, Xu H, Nyaisaba BM, Chen M, Deng S (2021) Understanding the role of atmospheric cold plasma (ACP) in maintaining the quality of hairtail (*Trichiurus Lepturus*). Food Chem 343:128418

Kratky L, Zamazal P (2020) Economic feasibility and sensitivity analysis of fish waste processing biorefinery. J Clean Prod 243:118677

Létisse, M., Rozières, M., Hiol, A., Sergent, M. and Comeau, L., 2006. Enrichment of EPA and DHA from sardine by supercritical fluid extraction without organic modifier: I. Optimization of extraction conditions. J Supercrit Fluids 38(1):27–36

Li M, Lin J, Chen J, Fang T (2016) Pulsed electric field-assisted enzymatic extraction of protein from abalone (Haliotis Discus Hannai Ino) viscera. J Food Process Eng 39:702–710. https://doi.org/10.1111/jfpe.12262

Li J, Sun W, Ramaswamy HS, Yu Y, Zhu S, Wang J, Li H (2017) High pressure extraction of astaxanthin from shrimp waste (Penaeus Vannamei Boone): effect on yield and antioxidant activity. J Food Process Eng 40(2):e12353

Maghami M, Sadrameli SM, Shamloo M (2018) Glycerin purification using asymmetric nano-structured ceramic membranes from production of waste fish oil biodiesel. Heat Mass Transf 54(9):2683–2690

Muhammad KML, Ariffin F, Abd Razak HKB, Sulaiman PDS (2016) Review of fish gelatin extraction, properties and packaging applications. Food Sci Nutr 56:47–59

Murthy LN, Phadke GG, Unnikrishnan P, Annamalai J, Joshy CG, Zynudheen AA, Ravishankar CN (2018) Valorization of fish viscera for crude proteases production and its use in bioactive protein hydrolysate preparation. Waste Biomass Valoriz 9(10):1735–1746

Muyonga JH, Cole CGB, Duodu KG (2004) Extraction and physico-chemical characterisation of Nile perch (*Lates niloticus*) skin and bone gelatin. Food Hydrocolloids 18(4):581–592

Nawaz A, Li E, Irshad S, Hammad HHM, Liu J, Shahbaz HM, Ahmed W, Regenstein JM (2020) Improved effect of autoclave processing on size reduction, chemical structure, nutritional, mechanical and in vitro digestibility properties of fish bone powder. Adv Powder Technol 31(6):2513–2520

Nurdiyana H, Siti Mazalina MK, Fadhilah SN, M. (2008) Optimization of protein extraction from freeze dried fish waste using response surface methodology (RSM). Int J Eng Technol 5(1):48–56

Onishi K, Kato S, Kunisawa K, Kojima T, Murakami S (2004) Production of biodiesel by ozone treatment. In: Asian Pacific Confederation of Chemical Engineering congress program and abstracts Asian Pacific Confederation of Chemical Engineers congress program and abstracts. The Society of Chemical Engineers, Japan, pp 565–565

Oterhals A, Vogt G (2013) Impact of extraction, refining and concentration stages on the stability of fish oil. In: Food Enrichment with Omega-3 Fatty Acids. https://doi.org/10.1533/9780857098863.2.111

Pankaj SK, Keener KM (2017) Cold plasma: background, applications and current trends. Curr Opin Food Sci 16:49–52

Pezeshk S, Ojagh SM, Rezaei M, Shabanpour B (2019) Fractionation of protein hydrolysates of fish waste using membrane ultrafiltration: investigation of antibacterial and antioxidant activities. Probiotics Antimicrob Proteins 11(3):1015–1022

Ramakrishnan VV, Ghaly AE, Brooks MS, Budge SM (2013) Extraction of proteins from mackerel fish processing waste using alcalase enzyme. Bioprocess Biotech 3(2):2–10

Rodrigues LA, Pereira CV, Partidario AMC, Gouveia LF, Simoes P, Paiva A, Matias AA (2021) Supercritical CO_2 extraction of bioactive lipids from canned sardine waste streams. J CO_2 Util 43:101359. https://doi.org/10.1016/j.jcou.2020.101359

Roslan J, Kamal SMM, Yunos KFM, Abdullah N (2017) Assessment on multilayer ultrafiltration membrane for fractionation of tilapia by-product protein hydrolysate with angiotensin I-converting enzyme (ACE) inhibitory activity. Sep Purif Technol 173:250–257

Sánchez-Camargo AP, Meireles MÂA, Ferreira AL, Saito E, Cabral FA (2012) Extraction of ω-3 fatty acids and astaxanthin from Brazilian redspotted shrimp waste using supercritical CO2+ ethanol mixtures. J Supercrit Fluids 61:71–77

Sarupria S, Ghosh T, García AE, Garde S (2010) Studying pressure denaturation of a protein by molecular dynamics simulations. Proteins Struct Funct Bioinform 78(7):1641–1651

Shabani A, Boldaji F, Dastar B, Ghoorchi T, Zerehdaran S, Ashayerizadeh A (2021) Evaluation of increasing concentrations of fish waste silage in diets on growth performance, gastrointestinal microbial population, and intestinal morphology of broiler chickens. Anim Feed Sci Technol 275:114874

Shahidi F (2003) Extraction and measurement of total lipids. In: Current protocols in food analytical chemistry

Sharayei P, Azarpazhooh E, Zomorodi S, Einafshar S, Ramaswamy HS (2021) Optimization of ultrasonic-assisted extraction of astaxanthin from green tiger (*Penaeus semisulcatus*) shrimp shell. Ultrason Sonochem 76:105666

Shyni K, Hema GS, Ninan G, Mathew S, Joshy CG, Lakshmanan PT (2014) Isolation and characterization of gelatin from the skins of skipjack tuna (*Katsuwonus pelamis*), dog shark (*Scoliodon sorrakowah*), and rohu (*Labeo rohita*). Food Hydrocolloids 39:68–76

Silva JFX, Ribeiro K, Silva JF, Cahú TB, Bezerra RS (2014) Utilization of tilapia processing waste for the production of fish protein hydrolysate. Anim Feed Sci Technol 196:96–106

Šimat V, Rathod NB, Čagalj M, Hamed I, Generalić Mekinić I (2022) Astaxanthin from crustaceans and their byproducts: a bioactive metabolite candidate for therapeutic application. Mar Drugs 20(3):206

Sivaranjani S, Prasath VA, Pandiselvam R, Kothakota A, Khaneghah AM (2021) Recent advances in applications of ozone in the cereal industry. LWT 146:111412

Thirukumaran R, Priya VKA, Krishnamoorthy S, Ramakrishnan P, Moses JA, Anandharamakrishnan C (2022) Resource recovery from fish waste: prospects and the usage of intensified extraction technologies. Chemosphere 299:134361

Toyoshima K, Noguchi R, Hosokawa M, Fukunaga K, Nishiyama T, Takahashi R, Miyashita K (2004) Separation of sardine oil without heating from surimi waste and its effect on lipid metabolism in rats. J Agric Food Chem 52(8):2372–2375

Tu ZC, Huang T, Wang H, Sha XM, Shi Y, Huang XQ, Man ZZ, Li DJ (2015) Physico-chemical properties of gelatin from bighead carp (*Hypophthalmichthys nobilis*) scales by ultrasound-assisted extraction. J Food Sci Technol 52(4):2166–2174

Tyson KS, Bozell J, Wallace R, Petersen E, Moens L (2004) *Biomass oil analysis: research needs and recommendations*. National Renewable Energy Lab Golden Co

Usman M, Sahar A, Inam-Ur-Raheem M, Rahman UU, Sameen A, Aadil RM (2022) Gelatin extraction from fish waste and potential applications in food sector. Int J Food Sci Technol 57(1):154–163

Vázquez JA, Blanco M, Massa AE, Amado IR, Pérez-Martín RI (2017) Production of fish protein hydrolysates from *Scyliorhinus canicula* discards with antihypertensive and antioxidant activities by enzymatic hydrolysis and mathematical optimization using response surface methodology. Mar Drugs 15(10):306

Wang M, Zhou J, Collado MC, Barba FJ (2021) Accelerated solvent extraction and pulsed electric fields for valorization of Rainbow Trout (*Oncorhynchus mykiss*) and Sole (*Dover sole*) by-products: Protein content, molecular weight distribution and antioxidant potential of the extracts. Mar Drugs 19(4):207

Wangkheirakpam MR, Mahanand SS, Majumdar RK, Sharma S, Hidangmayum DD, Netam S (2019) Fish waste utilization with reference to fish protein hydrolisate-a review. Fish Technol 56(3):169–178

Yahyaee R, Ghobadian B, Najafi G (2013) Waste fish oil biodiesel as a source of renewable fuel in Iran. Renew Sustain Energy Rev 17:312–319

Youravong W, Marthosa S (2017) Membrane technology in fish-processing waste utilization: some insights on sustainability. Sustain Challenge Agrofood Sector 22:575

Zeng S, Yan X, Cao W, Hong P, Zhang C, Li L (2010) Optimisation of extraction conditions and characteristics of skin gelatin from Nile tilapia (*Oreochromis niloticus*). Int J Food Sci Technol 45(9):1807–1813

Zhao YM, Patange A, Sun DW, Tiwari B (2020) Plasma-activated water: Physicochemical properties, microbial inactivation mechanisms, factors influencing antimicrobial effectiveness, and applications in the food industry. Compreh Rev Food Sci Food Safety 19(6):3951–3979

Zhou Y, He Q, Zhou D (2017) Optimization extraction of protein from mussel by high-intensity pulsed electric fields. J Food Process Preserv 41(3). https://doi.org/10.1111/jfpp.12962

Strategies to Recover Protein and Lipids from Fish Processing By-Products

S. Sivaranjani, Nelluri Puja, Rahul Kumar Rout, T. Jayasree Joshi, Shagolshem Mukta Singh, M. Indumathi, and Tammineni Dushyanth Kumar

Abstract The fish processing industry has an immense opportunity for converting fish by-products into value-added products. By-products have been shown to be an excellent source of bioactive components and have diverse functional characteristics. Proteins, oil, gelatin, collagen, chitin, chitosan, silage, as well as several other economically valuable items can be produced from fish waste. Proteins (15–30%) and lipids (0.5–25%) make up a large portion of fish waste, depending on age, species, sex, season, and the environment of growth. Numerous investigations have shown that these compounds have a multitude of uses in the pharmaceutical, food, and feed industry. Conventional approaches like acid/alkali hydrolysis and solvent extraction techniques are limited owing to their lower extraction efficiency and high processing time. Microwave, ultrasound, supercritical fluid, and pulsed electric field technology are some of the novel intensified techniques that have been applied to treat fish waste, and they have proved to enhance extraction yield and functional attributes. Isoelectric solubilization and precipitation techniques have also shown high protein and lipid recovery yields. Another recent technique that has gained popularity in the recovery of isolated fish protein is pH shift acid and alkaline solubilization. This chapter provides an overview of techniques for isolation of protein and lipids, recent advancements, challenges, and their applications.

Keywords Proteins · Lipids · Intensified extraction techniques · Enzymatic hydrolysis

S. Sivaranjani (✉) · N. Puja · R. K. Rout · T. J. Joshi · S. M. Singh · T. D. Kumar
Agricultural and Food Engineering Department, Indian Institute of Technology, Kharagpur, West Bengal 721302, India
e-mail: ssivaranjani13@gmail.com

M. Indumathi
Department of Food Technology, Excel Engineering College, Namakkal, Tamil Nadu 637303, India

S. Maqsood et al. (eds.), *Fish Waste to Valuable Products*, Sustainable Materials and Technology, https://doi.org/10.1007/978-981-99-8593-7_5

1 Introduction

The fish processing industry proves to be one of the major sectors in the world. Of the total fish captured, around 70% of fish is processed, of which 20–80% accounts for fish waste according to processing method and fish species (Ghaly et al. 2013). All these wastes are of major concern to the environment, which leads to the recovery of valuable products from fish waste. The valuable products are rich in nutrients, viz., proteins, lipids, amino acids, carotenoids, polysaccharides, and minerals. In seafood discards, proteins constitute around 58% (db) and are present in fish heads, bones, and trimmings (Sasidharan and Venugopal 2020). The protein recovery methods can be conventional or novel methods. The novel methods of protein extraction from fish waste include the use of membranes, ultrasound, and pulsed electric fields. This enables the high extraction efficiency and protein quality. Fish oil contains a range of fatty acids that include omega-3 fatty acids, viz., docosahexaenoic acid and eicosapentaenoic acid. The lipid extraction from fish waste employs novel techniques such as accelerated solvent, cold plasma, supercritical, microwave, pulsed electric field, and hybrid technology. The applications of fish proteins and lipids include agriculture, energy, pharmaceutical and biomedical, nutraceutical and functional foods, cosmetics, bioremediation, food industry, etc. This chapter deals with protein and lipid recovery from fish processing by-products and waste. It also describes the recent advances in extraction techniques and applications in various industries.

2 Proteins and Lipids in Fish By-Products

Fish oil contributes to around 2% of global fat and oil consumption (Adeoti and Hawboldt 2014). Fish by-products are rich in proteins and essential fatty acids (Bhaskar et al. 2010; Sachindra et al. 2010). The by-products are widely employed to produce fish protein hydrolysates through autolytic and enzymatic hydrolysis (Liaset et al. 2002). These protein hydrolysates can be utilized as an emulsifier in a variety of applications, including salad dressings and spreads (Saha and Hayashi 2001). In addition, they offer high nutritional, water-holding, gelling, textural, and foaming characteristics (Kristinsson 2007; Šližyte et al. 2005). Other proteinaceous components that can be recovered from fish by-products include collagen, gelatin, bioactive peptides, and enzymes. Collagen is produced from fish skin and bones and has a variety of therapeutic and cosmetic applications. Gelatin is a hydrolyzed form of collagen that is widely used as a food additive in the food processing industry to enhance the textural attributes and stability of food products. Several investigations have also revealed that peptides generated from fish proteins have antioxidative effects (Kristinsson 2007; Šližyte et al. 2009; Yang et al. 2008). Researchers have also recovered proteinaceous enzymes from fish by-products such as trypsin, chymotrypsin, pepsin, calpain, cathepsins, elastase, etc. (Bougatef 2013; Gildberg, 2004; Venugopal 2016).

Fish lipids are composed of fatty acids with varying lengths and degrees of unsaturation (Jobling et al. 2002). The majority of fish lipids are triacylglycerols, which exist as storage fat in hydrophobic aggregates. Fish lipids differ from other lipid sources due to high levels of omega-3 fatty acids such as eicosapentaenoic acid and docosahexaenoic acid (Rustad et al. 2011). These polyunsaturated fats have been proven to lower the risk of heart disease, hypertension, cancer, and immune and allergic diseases (Parolini 2019; König et al. 2005; Song et al. 2014). The major limiting factor that restricts the use of fish oil is its sensitivity to lipid oxidation, which can alter the flavour and reduce nutritional value. The oxidative stability of lipids is affected by external and internal elements such as gaseous composition, light, temperature, and chemical composition (Rustad et al. 2011).

3 Extraction Techniques for Protein

Fish waste is extremely high in protein content. These proteins are superior to plant proteins because they have a higher nutritional value. Protein extraction techniques are evolving, and new methods are being developed to improve yield and quality. This extracted protein is used in a variety of industries.

3.1 Conventional Approaches

Conventional approaches include the use of solvents and enzymes that are employed especially for the extraction of protein and lipids from fish.

3.1.1 Chemical Method

Proteins are chemically hydrolyzed by cleaving peptide bonds with either an acid or a base. Methods for extracting proteins with varying qualities and efficiencies include solvent, physical, enzymatic, and acid hydrolysis (Al Khawli et al. 2019). With solvent and physical methods, the structure of recovered protein molecules can be preserved (Hadnađev et al. 2017). Long-chain proteins in seafood, on the other hand, can be broken down, resulting in highly valuable short-chain peptides, i.e., essential amino acids, through enzymatic and acid hydrolysis (Cao et al. 2010). There have been numerous processes proposed for the acid or alkaline hydrolysis of fish protein. Because of its low cost and simple methodology, it has proven to be a popular method in the industry (Tu et al. 2015).

3.1.2 Acid Hydrolysis

Acid hydrolysis is considered a pre-treatment step prior to enzymatic hydrolysis of protein from the fish tissue (Manninen 2004). The protein structure can be denatured under the action of strong acids such as HCl and H_2SO_4 to produce lactic acid as the final product. Once the acid hydrolysis is neutralized, the hydrolysates contain a huge quantity of salt, resulting in the enhancement of taste and flavour. Under controlled conditions, the acid hydrolyzed seafood protein hydrolysate indicates a viable product with high nutritive value (Fernandes 2016). After distillation, the acid hydrolysate obtained from seafood can be used as a specific medium for microbial growth. Acid hydrolysis activity is affected by acid activity, temperature, and pressure. Protein concentration and acid ratio are inversely related. Acid hydrolysis can be used as a pre-treatment step for enzymatic hydrolysis in the next step to improve fish protein yield for food consideration (Ramakrishnan et al. 2013).

3.1.3 Enzymatic Hydrolysis

Protease enzymes cleave the special chemical bonds of the long-chain protein to create mixtures of free amino acids and oligopeptides to impart a better functional property for health (Prihanto et al. 2019). The processing conditions and enzyme–substrate ratio can influence the yield and quality of protein. Some of the exogenous enzymes can increase the rate of fermentation in fish and produce a protein hydrolysate like fish sauce (Laohakunjit et al. 2014). The use of low-temperature processing (less than 60 °C) and neutral media which is an optimum condition for the production of desired enzymes is an easy process to extract the enzymatic proteins. In the enzymatic hydrolysis of protein, the hydrolysis rapidly increases in the initial phase and remains constant with the increase in time. With increased hydrolysis time, the abundance of free amino acids such as Pro, Leu, Ala, and Trp, and antioxidant activity from enzymatic seafood hydrolysis increases (Gao et al. 2006). Enzymatic extraction yielded a higher yield and extracted at a faster rate than ultrasonic extraction. Hydrolysis represents a sustainable method for extracting proteins from various sources (plant, animal, and algae) in food processing, resulting in protein hydrolysates.

Traditional protein extraction methods are difficult to control and almost always result in products with variable chemical composition and functional properties. Protein hydrolysis with strong chemicals and solvents is carried out at high temperatures and pH levels, yielding products with low nutritional value, poor functionality, and limited use as flavour enhancers. Because of the high functionality and nutritive value of protein hydrolysates, chemical hydrolysis is an alternative method for the food industry. However, low reaction rate, high cost, low media yield, and difficult enzyme recovery are drawbacks that force the development of new novel methods for extracting fish protein.

3.2 Novel Techniques

3.2.1 Pulsed Electric Field

Enzymatic extraction has incomplete hydrolysis and this can be overcome by assisting pulsed electric field (PEF) in the extraction method. Complete hydrolysis gives a higher abalone viscera protein recovery yield with better emulsifying properties (Li et al. 2016). Franco et al. (2020) mixed 50 mg of chopped fish residues (bones, heads, and gills) with 50 ml of distilled water and then crushed and homogenized completely. For PEF treatment, the sample or the homogenates were placed in between the two electrodes with a 5 cm gap and 1.8 cm height. The PEF of different potential differences, pulse width, frequency, and pulse number by using semiconductor-based positive Marx modulator Epulsus-PM1-10 were applied based on the electrical conductivity of the homogenates (bones, heads, and gills). The entire procedure was carried out by protecting from the sunlight. The ISO (International Organization for Standardization) standards were used for measuring the protein content of the homogenates and resulted in an average of 15.43% protein obtained from the fish residues. The study also mentioned that the gills produced more amount of protein.

3.2.2 Ultrasound-Assisted Extraction

Ultrasound induces instant and strong cavitation to the biomass or fish discards or fish processing by-products. The instant formation and collapsed of the cavitation bubbles interrupted the cell walls and membranes of the discards and creates new micro-cavities which assist solvent penetration. This changes the structure of the materials and increases the mass transfer which leads to increasing yield and reduces processing time and solvent used. The adoption of ultrasound-intensification into the supercritical carbon-dioxide ($SCCO_2$) extraction process will increase the yields, functionality and will also provide a sustainable method (greener technologies) for recovery of protein from the discards or fish processing by-products or underutilized marine bioresources (Nguyen et al. 2020). Song et al. (2018) reported that the collagen extraction yield can be increased by 2 times than the commercial production from the skins of flatfish by using 0.05 M acetic acid solution and treating for 3 h in an ultrasound system and collected by centrifugation. It also reduces extraction time. A similar method was used by Kim et al. (2012) which showed that ultrasonic treatment increases the extraction of protein from the sea bass skins and the rate of extraction increases with an increase in ultrasonic amplitudes. Álvarez and Tiwari (2015) recovered 97% of protein by using 60% amplitude of ultrasound for 10 min in 0.1 M NaOH buffer from fish processing by-products. The recovered proteins contain high amounts of essential and non-essential amino acids, and thus can also be used for food purposes.

3.2.3 PH Shifting

pH shift process also known as isoelectric solubilization precipitation (ISP) includes the early homogenization of the fish processing by-products with either dilute alkali (10.8–11.5 pH) or dilute acid (2.5–3.5 pH). During the process, the insoluble impurities like oil, skin, membranes, bones, etc., are removed, while the sarcoplasmic and myofibrillar proteins are dissolved. The treatment has been installed to recover 90% of proteins in the form of fish protein isolated (FPI) from the discarded by-products. Up to a maximum of 90% of proteins can be recovered from the discards by altering the pH of the discard solution to the isoelectric pH of the protein (5.2–6 pH). The enzymatic degradation and denaturation of the proteins recovered from the discards can be evaded by performing the treatment below 10 °C (Sasisharan and Venugopal 2020a).

The fish protein recovered from the discards is divided into two types, i.e., Type A Fish Protein Concentration and Type B Fish Protein Concentration (FPC). Type A protein produced by solvent extraction or by alkaline method is odorless and contains less than 1% fat. The pH shifting process produced type B FPC by drying and grinding fish mince. It contains a higher quantity of lipids and thus has a fishy odour. The process recovered a higher amount of protein without any prior treatment to the fish but metal corrosion arises due to acid (Sanmartín et al. 2009).

3.2.4 Ultrafiltration

Protein can be recovered from the discards or fish processing by-products by membrane separation by means of ultrafiltration (UF) and is driven by pressure gradient. The pore sizes of UF membranes of size around 0.1–0.01 μm. UF membranes are taken as the most suitable for protein extraction from meat waste. UF membranes of pore size 10–30 kDa are efficient for extracting micro and macro solutes which also include protein hydrolysates (Castro-Muñoz et al. 2021).

The UF membrane arrangement is generally cross flow and the larger proteins are retained and concentrated. The method is affected by membrane type, operating temperature and pressure, and liquid pre-treatment. The process produced high quality of permeate but is time consuming and the membranes are expensive (Sanmartín et al. 2009). Franco et al. (2020) also used ultrasound to improve the extraction yield from fish by-products (bones, heads, and gills). Chabeaud et al. (2009) extracted 90% of protein hydrolysates from the fish fillet co-products with 4 kDa UF membranes.

4 Extraction Techniques for Lipids

Fish oil extraction methods encompass traditional methods such as wet pressing and solvent extraction to novel green approaches such as supercritical carbon dioxide extraction, high-pressure extraction, microwave-assisted extraction, and enzyme-assisted extraction. The different techniques that are employed in the extraction of fish lipids are detailed in the following sessions.

4.1 *Physical Method*

The physical process of lipid extraction involves wet pressing rendered fish by-products. The processes involved in the physical extraction of fish oil from by-products are shown in Fig. 1. The hydraulic press is the most commonly employed device for the extraction of oils. Manual loading of raw materials into perforated and horizontal hydraulic press boxes was done. The boxes were pressed together due to the ram's upward hydraulic pressure. The liquid portion and filter cake are separated using a centrifuge from the feed during the pressing process. The effectiveness of separation is determined by both the design of the centrifuge and its mode of operation. Effective centrifugation also necessitates precise temperature control (nearly 60 °C) (Adeoti, and Hawboldt 2014). To make the extracted oil safe for human consumption, refining procedures such as degumming, alkali refining, bleaching, deodorization, and antioxidant addition are necessary.

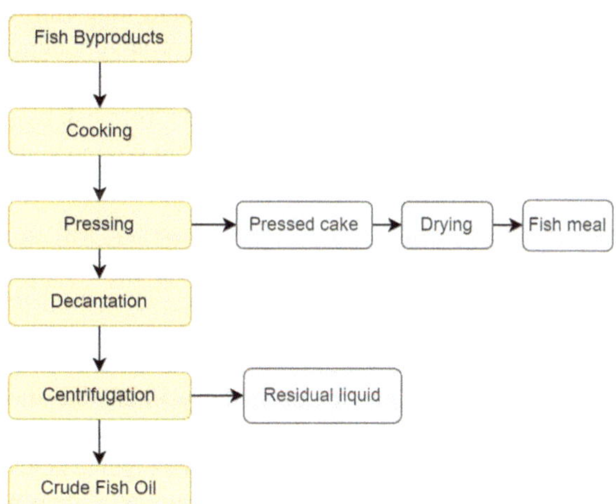

Fig. 1 Processes involved in the physical extraction of fish oil from by-products

4.2 Chemical Method

Chemical extraction procedures entail the use of several organic solvents. Traditional methods like Solvent extraction, Soxhlet method, and Bligh and Dyer (BD) method are commonly used for the extraction of lipids from fish. Hexane, cyclohexane, acetone, benzene, and chloroform are the most commonly used solvents for lipid extraction. The solvents work by rupturing cellular structures or altering the contact forces between lipids and the cell matrix (Mercer and Armenta 2011; Xiao 2010). The selection of solvent is critical as the extraction efficiency is highly contingent on the polarity of the lipids and solvents (Ramluckan et al. 2014). BD technique for lipid extraction utilizes a specific ratio of polar and non-polar solvents. The solvents include chloroform, methanol, and water. The approach yields excellent recovery with minimal solvent usage (Kumari et al. 2011). The main downside of this process is that it generates significant amounts of waste solvents, which makes solvent recycling expensive and also raises concerns about the safety of handling organic solvents (Norziah et al. 2009; Adeoti, and Hawboldt 2014).

The Soxhlet method is a long-established method for extracting lipids from food products. The extraction setup consists of a flask, an extraction chamber, and a condenser (Hewavitharana et al. 2020). The method is based on the principle of solid–liquid extraction. The solvent extracts the lipids and transports them into the flask as it passes through the sample. After the extraction is complete, the solvent is evaporated and the amount of residual lipid is calculated. On average, this extraction process takes 6–24 h (Nielsen 2017, Señoráns and Luna 2012). The major solvents used for the process include acetone, hexane, toluene, petroleum ether, etc. The approach produces lower yields than the Bligh and Dyer method. The requirement of large quantities of solvents and long run time are two drawbacks of the method. Furthermore, the outcomes are largely dependent on operational variables (Adeoti, and Hawboldt 2014). Acid–alkali aided extraction is another technique that is employed for the extraction of lipids from fish by-products. The process involves hydrolysis of the sample to release lipid components, followed by extraction using suitable solvents (Xiao 2010).

4.3 Biological Method

Fermentation and enzymatic hydrolysis are the most commonly used biological approaches for the extraction of lipids. For fermentation, minced fish waste is combined with carbohydrate sources or organic acids and then inoculated with suitable microbes. The centrifugation process is then used to separate the oil generated after fermentation (Jayasinghe and Hawboldt 2012). However, there is limited research available on the fermentation and enzymatic hydrolysis of fish waste. Rai et al. 2010, investigated the impact of lactic acid fermentation on the recovery of lipids from fish viscera and observed that fermentation could recover more than

85% of the oil contained in the viscera. Šližyte et al. (2005), compared the enzymatic hydrolysis of cod fish by-products using Flavourzyme and Neutrase enzyme. They observed that the use of Neutrase yielded more oil (6–16%) than Flavourzyme. By using a fermentation strategy, it would also be able to recover other functional elements such as protein hydrolysate, collagen, etc., from the raw material (Rai et al. 2010). Hence there is a lot of opportunity for further research in this field.

4.4 Emerging Extraction Techniques for Lipids

Compared to conventional methods, these extraction intensification techniques follow sustainable green chemistry and proved to be efficient and feasible in extracting lipids from fish wastes and by-products. The potential of each extraction technique is measured by the time and energy consumption, along with the quality and quantity of extracted products (Thirukumaran et al. 2022). The non-conventional techniques reported in the literature for extraction of lipids from fish wastes include supercritical fluid extraction, pulsed electric field, ultrasound-assisted extraction, and non-thermal cold plasma, to name a few. Each technique is discussed in detail under the following subsections.

4.4.1 Supercritical Fluid Extraction (SCFE)

This method is suitable for better recovery of a desirable extract with minimum environmental implications. The most common solvent used here is carbon dioxide (CO_2), which is also called the supercritical fluid for its dual behaviour; at 31.1 °C and 7.4 MPa, it shows diffusivity in the range of gas, while viscosity and flow behaviour of that of liquids (Fang et al. 2018). SCFE is suitable for extracting lipid-soluble compounds having antioxidative properties from fish wastes, and the process does not disturb their functional properties. It is especially suitable for the extraction of omega-3 fatty acids, as the extraction provides an oxygen-free environment with moderate temperature application, thus preventing omega-3 fatty acid oxidation and thermal degradation (Melgosa et al. 2021). SCFE can also remove polar impurities, heavy metals, inorganic impurities, and pollutants like lead, arsenobetaine, cadmium, etc., owing to the low polarity of CO_2. This helps in the elimination of subsequent fish oil refining (Melgosa et al. 2020). It was also observed that SCFE-based fish oil shows a longer shelf life with higher oxidative stability as compared to other extraction techniques (Thirukumaran et al. 2022).

4.4.2 Subcritical Water Hydrolysis (SWH)

This is yet another novel and environmentally friendly method for extracting valuable ingredients from fish wastes and by-products. Factors responsible for better extraction in this technique include ideal hydrolysis of sub-critical water, strong catalytic behaviour, and high dissolution effect (Thirukumaran et al. 2022). The ideal thermodynamic condition for SWH is a temperature of 100–380 °C and a pressure up to 22 MPa (di Domenico Ziero et al. 2020). Water can dissolve non-polar compounds under these conditions. Studies were carried on with SWH for extraction of amino acids, fish collagen peptides, hydrolysates, etc. However, studies on lipid extraction are scarce and need to be focused in this domain as it shows potential for efficient extraction for the same.

4.4.3 Ultrasound-Assisted Extraction (UAE)

UAE also follows green chemistry for the extraction of bioactive compounds, having key advantages like minimum energy usage, scalability convenience, and less time of processing over other established methods (Thirukumaran et al. 2022). Acoustic cavitation facilitates the retrieval of bioactive components from fish processing wastes and by-products. It was reported for extracting heat-sensitive lipids like polyunsaturated fatty acids (PUFA) from fish wastes (Alfio et al. 2021). Bruno et al. (2019b) studied the effect of enzymatic extraction coupled with UAE for extracting fish oil from rohu (*Labeo rohita*) heads and reported that UAE helped achieve more oxidative stability, quality of mono-unsaturated and polyunsaturated fatty acids (MUFA and PUFA), and viscosity of the oil as compared to those extracted without UAE. Another study focused on the extraction of squid oil from fishery by-catch (non-target fishery species, which are often discarded) using ultrasound-assisted aqueous enzymatic method (Liu et al. 2020). The results demonstrated a synergistic effect of increased extraction rate and economic value with little to no secondary pollutant generation.

4.4.4 Microwave-Assisted Extraction (MAE)

Microwave is best for its quick heating phenomena, owing to ionic polarization and dipole rotation. It has the potential to affect the cellular structure of the material significantly (Nouri et al. 2016). Rahimi et al. (2017) carried out a comparative study of oil extraction from sardine fish wastes by means of novel MAE and conventional soxhlet extraction. MAE yielded 80.5 mg lipid per gram of fish waste, which was about twice the yield achieved in soxhlet extraction (46.6 mg oil/g fish waste). Moreover, the extraction time for MAE was 10 min, which was about 24 times less than that of soxhlet extraction (4 h). Another study focused on the extraction of ω-3 fatty acid-rich fish oil from *Pangasianodon gigas* wastes using MAE and conventional methods. The yield was 9% of raw material for MAE, which is higher than the

4.5% yield of oil in the conventional method. The extract of MAE was also economical and superior as compared to that of the conventional method (Chimsook and Wannalangka 2015).

4.4.5 Pulsed Electric Field (PEF) Extraction

PEF is a non-thermal extraction technique with the potential of complete product valorization. It works on the principle of very high energy electric pulses ranging from 10 to 80 kV/cm applied over the samples for a very short time (of the order of milliseconds), which generates phenomena like electroporation and electro-permeabilization, thus resulting in efficient extraction (Arshad et al. 2021). Additionally, PEF can be used as a pre-treatment to accelerate the extraction process (Gulzar & Benjakul 2020). In this study, PEF was used as a pre-treatment for solvent extraction, it showed reduced peroxide value, which might be due to inactivation of oxidative enzymes during PEF treatment. Moreover, the lipid extracts were rich in PUFA and carotenoids (like β-carotene, astaxanthin, and canthaxanthin among others).

4.4.6 Accelerated Solvent Extraction

Accelerated solvent extraction (ASE) is an extraction technique that uses solvents commonly used in extraction techniques such as Soxhlet extraction at elevated temperatures and pressures. ASE was first described by Richter et al. (1996) as a sample preparation technique to reduce the time and solvent required to prepare the sample. To reduce the cost of extracted compounds from the fish waste, several researchers employed ASE to recover several useful compounds from fish tissue. Dodds et al. (2004) extracted lipids from fish tissue using ASE. The authors used three different types of solvent systems such as Methanol–Chloroform, Hexane–Isopropanol, and Methylene chloride to extract lipids from salmon halibut tissues. The authors found the extraction efficiencies of lipids through ASE have increased and sample preparation has also been decreased. This method can also be used to extract lipids from by-products of fish processing.

Despite the fact that the above experiments confirmed ASE as a viable lipid recovery technology, more research and development in large-scale recovery and feasibility are required. Because high-volume lipid recovery with this approach has been unusual, oil quality for biofuel application must still be pursued. One of the major constraints will almost certainly be environmental concerns about the massive amount of remaining solvent.

4.4.7 Cold Plasma

Cold plasma (CP) is a novel, non-thermal technique that has gained popularity in solvent biorefinery. It is preferred for enzyme inactivation, microbial decontamination, detoxification, structural modification, disinfection of instruments, etc. (Pankaj et al. 2018). With a continuous supply of energy over a very small surface area, a material will change its state from solid to liquid, and then to gas. When the gaseous state is further provided with intensified energy, ionization occurs which results in the generation of free electrons and ions. This state is called 'plasma', and since the process takes place at ambient temperature, it is called 'cold plasma'. A study reported the application of CP for preventing lipid oxidation and microbial inactivation of mackerel fish discards. The process did not show a significant effect in preventing lipid oxidation, as the end products were still susceptible after CP treatment, but showed promising results in reducing microbial load. The physicochemical properties were also minimally hampered (Albertos et al. 2017). Another study was conducted to check the lipid peroxidation level of Alaska pollock shreds. The lipid quality index (LQI) was found to be below 8 mg malondialdehyde/g dried shreds, which was safe for human consumption. This was obtained after 2 min of CP treatment, which was also sufficient for microbial reduction (Choi et al. 2016). Though CP has promising results in the valorization of fish wastes and by-products, the studies are scarce. Hence, in-depth studies are needed to understand the efficacy and feasibility of CP for a broad variety of fish and aquatic waste utilization.

4.4.8 Combined Approaches (Hurdle Technology)

Despite having advantages in terms of extraction yield, extraction time, and quality of extracts among others, a single extraction technique cannot fulfil all the possible benefits. Every novel extraction technique has certain shortcomings, e.g., poor conducting capacity of fish wastes and low volume application in PEF, intense cavitation for prolonged extraction period in UAE, localized heating in MAE, etc., have been reported as the drawbacks in previous studies (Gómez et al. 2019). These shortcomings pave the path for combining two or more novel extraction techniques, so that one technique can mask the demerit of the other, and can give a synergistic effect towards better extraction in terms of quality, yield, and time. Gulzar and Benjakul (2020) used a combination of PEF and UAE to extract lipids from the cephalothorax of Pacific white shrimp. The extraction yield was 30.34 g oil/100 g solid waste, which was about 50% higher yield than that of the conventional solvent extraction technique. The reason was the synergistic effect of electroporation due to PEF and acoustic cavitation because of UAE, which facilitated better solvent penetration in the matrix of fish wastes, thereby accelerating the process with higher yield. The combination studies are recently practiced, and the trends show that it can actually prove effective towards synergistic extraction of valuable compounds from fish wastes and by-products (Table 1).

Table 1 Novel extraction techniques for lipids from fish wastes and by-products

Fish wastes and by-products	Extraction method	Salient findings	References
1. Supercritical fluid extraction (SCFE)			
Off-cuts of Orange roughy (*Hoplostethus atlanticus*) and Atlantic salmon (*Salmo salar*)	SCFE	PUFA in extracted oil was highly pure with minimum oxidation	Fang et al. (2019)
Sardine heads and tails	SCFE	Extraction conditions: 30 MPa; 75 °C; flow rate of CO_2 = 2.5 $mL.min^{-1}$; 45 min extraction; improved concentration of eicosapentaenoic acid (EPA) and docosahexaenoic acid (DHA) with selective isolation	Zhang et al. (2019)
Longtail fish (*Thunnus tonggol*)	SCFE	Extraction of superior nutritional polar and neutral lipids	Ferdosh et al. (2016)
Indian mackerel skin tissues	SCFE and Soxhlet extraction	Extraction conditions: SCFE—45–75 °C at a pressure of 20–35 MPa; Soxhlet extraction at 78 °C; Oil yield was more than twice in SCFE than soxhlet extraction	Bruno et al. (2019a)
Hake species (*Merluccius capensis; Merluccius paradoxus*) by-products	SCFE	Extraction conditions: 25 MPa at 40 °C for 3 h; Yield of fish oil was around 96% of total oil	Pateiro et al. (2021)
Hake by-products	SCFE	Oil recovery of about 20 g/100 g by-products dry weight (expressed in dry basis)	Rubio-Rodríguez et al. (2012)
Tuna by-products	SCFE	20 g oil for every 100 g fish waste (db)	Ferdosh et al. (2015)
Indian mackerel by-products	SCFE	Fish oil recovery of 50 g per100 g dry waste	Hajeb et al. (2015)
African catfish wastes	SCFE	Very high oil recovery; about 67 g fish oil/100 g waste (db)	Lin et al. (2012)
2. Ultrasound-assisted extraction (UAE)			
Rohu (*Labeo rohita*) head	UAE coupled with enzymatic extraction	Extraction conditions: 20 kHz frequency; 40% amplitude; 15 min extraction time; 68.08% extraction yield of fish oil in UAE as compared to 39.03% in conventional method; enhancement of MUFA and PUFA in the extracted oil	Bruno et al. (2019b)

(continued)

Table 1 (continued)

Fish wastes and by-products	Extraction method	Salient findings	References
Fishery by-catch wastes	UAE assisted with enzymatic extraction	Extraction conditions: Ultrasonication time—1.5 h; stomach protease enzyme—0.1%; increased extraction rate with minimum to no secondary pollutant generation	Liu et al. (2020)
3. Microwave-assisted extraction (MAE)			
Sardine fish waste	MAE and soxhlet extraction	Extraction conditions: MAE time—10 min, solvents—distilled water and a mixture of hexane and isopropanol; Soxhlet time—4 h, solvent—hexane; lipid yield was twofold higher in case of MAE (80.5 mg/g) as compared to soxhlet extraction (46.6 mg/g)	Rahimi et al. (2017)
Pangasianodon gigas waste	MAE	Extraction conditions: microwave power—110 W; time—60 s; yield was 9% of raw material for MAE as compared to 4.5% for conventional method; quality of extract was superior in MAE	Chimsook and Wannalangka (2015)
4. Pulsed electric field (PEF)			
Pacific white shrimp cephalothorax	PEF pre-treatment followed by solvent extraction	Pre-treatment improved the quality of oil by reducing its peroxide value; the extracted oil was rich in PUFA and carotenoids	Gulzar & Benjakul (2020)
Multiple fish wastes	PEF	Extraction conditions: 55–60 kV/cm PEF for 1–2 h (pulsed effect); improvement in extraction rate; Increase in EPA and DHA content in fish oil	Lu et al. (2018)
5. Cold plasma (CP)			
Mackerel fish discards	Cold plasma (CP)	No significant effect on reducing lipid oxidation; significant reduction in microbial load	Albertos et al. (2017)
Alaska pollock shreds	CP	Treatment time—2 min; Lipid quality index (LQI) <8 mg malondialdehyde/kg dried shreds, which was safe for human consumption	Choi et al. (2016)

(continued)

Table 1 (continued)

Fish wastes and by-products	Extraction method	Salient findings	References
6. Combination approach			
Pacific white shrimp cephalothorax	PEF and UAE	Extraction conditions: PEF at 16 kV/cm at a rate of 240 number; UAE at 80% amplitude for 25 min; Oil yield of 30.34 g per 100 g solid waste (50% higher than solvent extraction method); minimum lipid oxidation as suggested by thiobarbituric acid reactive substances (TBARS) value	Gulzar & Benjakul (2020)

5 Practical Applications of Fish Proteins and Lipids

Fish proteins and lipids derived from fish waste have potential applications in food, pharmaceutical, and cosmetics industries which are summarized briefly below. Fish proteins include bioactive peptides, protein hydrolysates, collagen, gelatin, and other essential amino acids. Fish lipids like omega-3 fatty acids and others can be used for producing fish oils and biofuels from fish wastes.

5.1 Food Industry Applications

Fish proteins like collagen and gelatin obtained from fish waste like skin, scales, and bones can be used as a suitable gelling, emulsifier, and stabilizing agent in different product developments. Many studies have reported the usage of gelatin obtained from fish waste in food applications. Yin et al. (2021) reported the usage of xanthan gum-modified fish gelatin from tilapia fish skin to improve the rheological properties of low-fat yogurt. He found that the addition of xanthan gum and fish gelatin significantly improved the water-holding capacity and consistency of low-fat yogurt when compared to the mammalian porcine gelatin. Bhagwat and Dandge (2016) disclosed that fish collagen obtained from Cyprinus carpio fish scales can be used as an additive in paneer production. It was found that collagen improved the protein content of paneer with no profound alteration in its sensory and texture characteristics. Collagen and gelatin can also be used in developing functional foods and as an additive in many foods (Duan et al. 2018; Chotphruethipong et al. 2021). Fish gelatin can also be used for developing packaging films with good barrier properties (Etxabide et al. 2017; Hanani et al. 2019).

Bioactive peptides obtained from fish proteins can be employed as an ingredient in formulating functional and convenience foods (Caruso et al. 2020). Hydrolysates from fish protein can act as an emulsifier, foaming agent, and dispersing agent. Owing to their essential amino acid content, they can also be used as a dietary supplement

(Gao et al. 2021). They can also be used as antioxidants to extend the shelf life and improve the quality of food products. According to Dekkers et al. (2011), tilapia fish protein hydrolysates improved the oxidative stability of mahi mahi red fish fillets by reducing the formation of lipid hydroperoxides and thiobarbutric acid reactive substances. Similarly, fish protein powder can be used for formulating several food products and also for developing edible films (Shaviklo 2015). Omega 3 fatty acids produced from fish waste have a high potential for nutraceutical applications (Alfio et al. 2021). They can also be employed in food fortification.

5.2 *Biofuel Applications*

Fish waste is a good source of fish oil which can be further processed to fish biodiesel. Fish oil is generally produced from the lipids obtained from fish waste. The oil obtained is purified further to obtain pure biodiesel. Biodiesel produced from fish waste is generally used in combination with regular diesel (Jayasinghe and Hawboldt 2012). Fish biodiesel generally has higher or similar calorific value when compared to other available biodiesels (Martins et al. 2015). Fish biodiesel can also be used in diesel engines and internal combustion engines (Steigers 2003). Many studies have highlighted the production of biodiesel from fish waste and their efficiency as biofuel. Samat et al. (2018) emphasized that biodiesel can be manufactured from fish oil obtained from fish waste by an economically friendly two-step esterification process. He found that reaction time and temperature, solvent to fish oil molar ratio positively affected the yield of fish biodiesel. Martins et al. (2015) reported the production of biodiesel by transesterification from tilapia fish oil and found that the calorific value of fish oil is on par with diesel along with other properties like flash point, oxidation stability, combustibility, and kinematic viscosity. Another study disclosed that ozone-treated fish oil yielded biodiesel with better properties like lowered pour and flash points, and improved combustibility and ignition which can be used as a regular diesel (Kato et al. 2004). Preto et al. (2008) evaluated the performance efficiency of crude fish oil by comparing it with fuel oil and blends of fish oil. He found that the blend of fish oil and fuel oil exhibited lower viscosity, CO emissions, and higher flash points which can be used conventionally in combustors and furnaces.

5.3 *Pharmaceutical and Cosmeceutical Applications*

Many bioactive peptides obtained from fish waste were reported to possess antioxidant, antihypertensive, anticancer, and antimicrobial properties which can be exploited by employing them in pharmaceuticals. Peptides obtained from fish waste can be incorporated as an active component in cosmetic formulations (Lintner and Peschard 2000). Fish waste-derived collagen can be included as an ingredient in cosmetics owing to its improved bioactivity, mechanical strength, moisture-retaining capacity, and mild odour (Venkatesan et al. 2017). Xhauflaire-Uhoda et al. (2008) found that serum characterized by fish collagen has a better hydrating and firming effect on the skin following usage for a brief time. Currently, fish collagen has a high demand in tissue engineering for manufacturing scaffolds pertaining to its superior biocompatibility, biosafety, and minimal antigenicity (Lin et al. 2020). Gelatin obtained from fish waste can be used for encapsulation and coatings of certain compounds in pharmaceutical industries. It can also be used for the formation of bones and reduces the brittleness of bones (Nomura et al. 2005; Noma et al. 2017). Gelatin can also be employed in cosmetics as a gelling agent, antioxidant, and anti-ageing in skin care products (Al-Nimry et al. 2021).

Fish oil and its fatty acids like omega-3 and omega-6 have numerous benefits on skin like anti-photoaging, skin whitening, reduced risk of skin cancer, and others (Huang et al. 2018). Fortes et al. (2008) found that consumption of omega-3 fatty acids enriched fish oil by melanoma patients diminished the risk of melanoma. Similar results were reported by Hakim et al. (2000) in squamous cell carcinoma patients. Some others reported that fatty acids like DHA, EPA, oleic acid, LA, and ALA possess fast and quick wound healing properties (McDaniel et al. 2008; Rodrigues et al. 2016; Peng et al. 2018). Usage of fish oil was also found to decrease photoaging and hyperpigmentation in the experimental group (Shigeta et al. 2004; Puglia et al. 2005). Although fish oil has many benefits on skin, clinical studies are still required for enhanced cure.

5.4 Future Perspectives and Challenges

Despite the availability of raw materials and processing technology for fish waste, there is still room for advancement and improvement. This chapter describes protein and lipid extraction techniques and their applications in various fields. These recovery techniques could be commercialized and formulated into various ingredients to improve functional properties. Protease inhibitors are required to inhibit proteolytic activity during the heat-induced gelation method of protein extraction. This necessitates the use of additives (Chen and Jaczynski 2007). There are methods for simultaneously extracting protein and lipids from fish waste that could have a variety of functionalities (Hathwar et al. 2011). Therefore, hybrid techniques overcome the shortcomings of individual techniques. Forthcoming research trends should also be on a way to efficiently obtain pure, stable, and bioavailable products that are resistant to functional changes, while clear communication of progress in sustainability as well as the safety of waste-derived products is critical to consumer acceptability.

6 Conclusion

The efficient lowering and utilization of fish waste, as well as its transformation into high-value products such as proteins and lipids, is an appealing solution from both an environmental and economic standpoint. Dietary supplements obtained from fish oil are widely marketed around the world, and this encourages the development of responsible strategies for using renewable resources. Seafood protein products demand their usage in beauty and sports nutrition that shows remarkable properties. In conclusion, utilizing fish wastes and by-products reduces environmental pollution also empowers the need for protein and oil-derived products for a variety of purposes.

References

Adeoti IA, Hawboldt K (2014) A review of lipid extraction from fish processing by-product for use as a biofuel. Biomass Bioenerg 63:330–340

Al Khawli F, Pateiro M, Domínguez R, Lorenzo JM, Gullón P, Kousoulaki K, Ferrer E, Berrada H, Barba FJ (2019) Innovative green technologies of intensification for valorization of seafood and their by-products. Mar Drugs 17(12):689

Albertos I, Martín-Diana A, Cullen PJ, Tiwari BK, Ojha SK, Bourke P, Álvarez C, Rico D (2017) Effects of dielectric barrier discharge (DBD) generated plasma on microbial reduction and quality parameters of fresh mackerel (Scomber scombrus) fillets. Innov Food Sci Emerg Technol 44:117–122

Alfio VG, Manzo C, Micillo R (2021) From fish waste to value: An overview of the sustainable recovery of omega-3 for food supplements. Molecules 26(4):1002

Al-Nimry S, Dayah AA, Hasan I, Daghmash R (2021) Cosmetic, biomedical and pharmaceutical applications of fish gelatin/hydrolysates. Mar Drugs 19(3):145

Álvarez C, Tiwari BK (2015) Ultrasound assisted extraction of proteins from fish processing by-products. institute of food technologist. Chicago, US

Arshad RN, Abdul-Malek Z, Roobab U, Qureshi MI, Khan N, Ahmad MH, Liu ZW, Aadil RM (2021) Effective valorization of food wastes and by-products through pulsed electric field: A systematic review. J Food Process Eng 44(3):e13629

Bhagwat PK, Dandge PB (2016) Isolation, characterization and valorizable applications of fish scale collagen in food and agriculture industries. Biocatal Agric Biotechnol 7:234–240

Bhaskar N, Sachindra NM, Suresh PV, Mahendrakar NS (2010) Microbial reclamation of fish industry bi-products. In: Montet D, Ray RC (eds) Aquaculture microbiology, pp. 248–275. Science, Enfield

Bougatef A (2013) Trypsins from fish processing waste: Characteristics and biotechnological applications–comprehensive review. J Clean Prod 57:257–265

Bruno SF, Ekorong FJAA, Karkal SS, Cathrine MSB, Kudre TG (2019a) Green and innovative techniques for recovery of valuable compounds from seafood by-products and discards: A review. Trends Food Sci Technol 85

Bruno SF, Kudre TG, Bhaskar N (2019b). Impact of pretreatment-assisted enzymatic extraction on recovery, physicochemical and rheological properties of oil from Labeo rohita head. J Food Process Eng 42(3)

Cao X, Li C, Wen H, Gu Z (2010) Extraction technique and characteristics of soluble protein in germinated brown rice. Int J Food Prop 13(4):810–820

Caruso G, Floris R, Serangeli C, Di Paola L (2020) Fishery wastes as a yet undiscovered treasure from the sea: Biomolecules sources, extraction methods and valorization. Mar Drugs 18(12):622

Castro-Muñoz R, García-Depraect O, León-Becerril E, Cassano A, Conidi C, Fíla V (2021) Recovery of protein-based compounds from meat by-products by membrane-assisted separations: A review. J Chem Technol Biotechnol 96(11):3025–3042

Chabeaud A, Vandanjon L, Bourseau P, Jaouen P, Chaplain-Derouiniot M, Guérard F (2009) Performances of ultrafiltration membranes for fractionating a fish protein hydrolysate: Application to the refining of bioactive peptidic fractions. Sep Purif Technol 66(3):463–471

Chen YC, Jaczynski J (2007) Protein recovery from rainbow trout (Oncorhynchus mykiss) processing byproducts via isoelectric solubilization/precipitation and its gelation properties as affected by functional additives. J Agric Food Chem 55(22):9079–9088

Chimsook T, Wannalangka W (2015) Effect of microwave pretreatment on extraction yield and quality of catfish oil in Northern Thailand. MATEC Web of Conferences, 35.

Choi S, Puligundla P, Mok C (2016). Microbial decontamination of dried alaska pollock shreds using corona discharge plasma jet: Effects on physicochemical and sensory characteristics. J Food Sci 81(4)

Chotphruethipong L, Binlateh T, Hutamekalin P, Sukketsiri W, Aluko RE, Benjakul S (2021) Hydrolyzed collagen from defatted sea bass skin and its conjugate with epigallocatechin gallate: in vitro antioxidant, anti-inflammatory, wound-healing and anti-obesity activities. Food Biosci 43:101303

Dekkers E, Raghavan S, Kristinsson HG, Marshall MR (2011) Oxidative stability of mahi mahi red muscle dipped in tilapia protein hydrolysates. Food Chem 124(2):640-645

Dodds ED, McCoy MR, Geldenhuys A, Rea LD, Kennish JM (2004) Microscale recovery of total lipids from fish tissue by accelerated solvent extraction. J Am Oil Chem Soc 81(9):835–840

di Domenico Ziero H, Buller LS, Mudhoo A, Ampese LC, Mussatto SI, Carneiro TF (2020) An overview of subcritical and supercritical water treatment of different biomasses for protein and amino acids production and recovery. J Environ Chem Eng 8(5)

Duan R, Zhang J, Liu L, Cui W, Regenstein JM (2018) The functional properties and application of gelatin derived from the skin of channel catfish (*Ictalurus punctatus*). Food Chem 239:464–469

Etxabide A, Uranga J, Guerrero P, De la Caba K (2017) Development of active gelatin films by means of valorisation of food processing waste: A review. Food Hydrocolloids 68:192–198

Fang Y, Gu S, Liu S, Zhang J, Ding Y, Liu J (2018) Extraction of oil from high-moisture tuna liver by subcritical dimethyl ether: Feasibility and optimization by the response surface method. RSC Adv 8(5)

Fang Y, Liu S, Hu W, Zhang J, Ding Y, Liu J (2019) Extraction of oil from high-moisture tuna livers by subcritical dimethyl ether: A comparison with different extraction methods. Eur J Lipid Sci Technol 121(2)

Ferdosh S, Sarker ZI, Norulaini N, Oliveira A, Yunus K, Chowdury AJ, Akanda J, Omar M (2015) Quality of tuna fish oils extracted from processing the by-products of three species of neritic tuna using supercritical carbon dioxide. J Food Process Preserv 39(4)

Ferdosh S, Sarker MZI, Rahman NNNA, Akanda MJH, Ghafoor K, Kadir MOA (2016) Simultaneous extraction and fractionation of fish oil from tuna by-product using supercritical carbon dioxide (SC-CO_2). J Aquat Food Prod Technol 25(2)

Fernandes P (2016) Enzymes in fish and seafood processing. Front Bioeng Biotechnol 4:59

Fortes C, Mastroeni S, Melchi F, Pilla MA, Antonelli G, Camaioni D, Alotto M, Pasquini P (2008) A protective effect of the Mediterranean diet for cutaneous melanoma. Int J Epidemiol 37(5):1018–1029

Franco D, Munekata PE, Agregán R, Bermúdez R, López-Pedrouso M, Pateiro M, Lorenzo JM (2020) Application of pulsed electric fields for obtaining antioxidant extracts from fish residues. Antioxidants 9(2):90

Gao MT, Hirata M, Toorisaka E, Hano T (2006) Acid-hydrolysis of fish wastes for lactic acid fermentation. Biores Technol 97(18):2414–2420

Gao R, Yu Q, Shen Y, Chu Q, Chen G, Fen S, Yang M, Yuan L, McClements DJ, Sun Q (2021) Production, bioactive properties, and potential applications of fish protein hydrolysates: Developments and challenges. Trends Food Sci Technol 110:687–699

Ghaly AE, Ramakrishnan VV, Brooks MS, Budge SM, Dave D (2013) Fish processing wastes as a potential source of proteins. Amino acids and oils: A critical review. J Microb Biochem Technol 5(4): 107–129

Gildberg AR (2004) Enzymes and bioactive peptides from fish waste related to fish silage, fish feed and fish sauce production. J Aquat Food Prod Technol 13(2):3–11

Gómez B, Munekata PES, Gavahian M, Barba FJ, Martí-Quijal FJ, Bolumar T, Campagnol PCB, Tomasevic I, Lorenzo JM (2019) Application of pulsed electric fields in meat and fish processing industries: An overview. Food Res Int 123

Gulzar S, Benjakul S (2020) Impact of pulsed electric field pretreatment on yield and quality of lipid extracted from cephalothorax of Pacific white shrimp (Litopenaeus vannamei) by ultrasound-assisted process. Int J Food Sci Technol 55(2)

Hadnađev MS, Hadnađev-Dapčević T, Pojić MM, Šarić BM, Mišan AČ, Jovanov PT, Sakač MB (2017) Progress in vegetable proteins isolation techniques: A review. Food Feed Res 44(1):11–21

Hajeb P, Selamat J, Afsah-Hejri L, Mahyudin NA, Shakibazadeh S, Sarker MZI (2015) Effect of supercritical fluid extraction on the reduction of toxic elements in fish oil compared with other extraction methods. J Food Prot 78(1)

Hakim IA, Harris RB, Ritenbaugh C (2000) Fat intake and risk of squamous cell carcinoma of the skin. Nutr Cancer 36(2):155–162

Hanani ZN, Yee FC, Nor-Khaizura MAR (2019) Effect of pomegranate (Punica granatum L.) peel powder on the antioxidant and antimicrobial properties of fish gelatin films as active packaging. Food Hydrocolloids 89:253–259

Hathwar SC, Bijinu B, Rai AK, Narayan B (2011) Simultaneous recovery of lipids and proteins by enzymatic hydrolysis of fish industry waste using different commercial proteases. Appl Biochem Biotechnol 164(1):115–124

Hewavitharana GG, Perera DN, Navaratne SB, Wickramasinghe I (2020) Extraction methods of fat from food samples and preparation of fatty acid methyl esters for gas chromatography: A review. Arab J Chem 13(8):6865–6875

Huang TH, Wang PW, Yang SC, Chou WL, Fang JY (2018) Cosmetic and therapeutic applications of fish oil's fatty acids on the skin. Mar Drugs 16(8):256

Jayasinghe P, Hawboldt K (2012) A review of bio-oils from waste biomass: Focus on fish processing waste. Renew Sustain Energy Rev 16(1):798–821

Jobling M, Larsen AV, Andreassen B, Sigholt T, Olsen RL (2002) Influence of a dietary shift on temporal changes in fat deposition and fatty acid composition of Atlantic salmon post-smolt during the early phase of seawater rearing. Aquac Res 33(11):875–889

Kato S, Kunisawa K, Kojima T, Murakami S (2004) Evaluation of ozone treated fish waste oil as a fuel for transportation. J Chem Eng Japan 37(7):863–870

Kim HK, Kim YH, Kim YJ, Park HJ, Lee NH (2012) Effects of ultrasonic treatment on collagen extraction from skins of the sea bass Lateolabrax japonicus. Fish Sci 78(2):485–490

König A, Bouzan C, Cohen JT, Connor WE, Kris-Etherton PM, Gray GM, Lawrence RS, Savitz DA, Teutsch SM (2005) A quantitative analysis of fish consumption and coronary heart disease mortality. Am J Prev Med 29(4):335–346

Kristinsson HG (2007) Aquatic food protein hydrolysates. Maximising the value of marine by-products, pp. 229–248. Woodhead Publishing

Kumari P, Reddy CRK, Jha B (2011) Comparative evaluation and selection of a method for lipid and fatty acid extraction from macroalgae. Anal Biochem 415(2):134–144

Laohakunjit N, Selamassakul O, Kerdchoechuen O (2014) Seafood-like flavour obtained from the enzymatic hydrolysis of the protein by-products of seaweed (Gracilaria sp.). Food Chem 158:162–170

Li M, Lin JIE, Chen J, Fang T (2016) Pulsed electric field-assisted enzymatic extraction of protein from abalone (Haliotis discus hannai Ino) viscera. J Food Process Eng 39(6):702–710

Liaset B, Nortvedt R, Lied E, Espe M (2002) Studies on the nitrogen recovery in enzymic hydrolysis of Atlantic salmon (*Salmo salar, L.*) frames by Protamex™ protease. Process Biochem 37(11): 1263–1269

Lin L, Xu Y, Li Y, Gong X, Wei M, Zhang W, Zhang X, Xu Y (2020) Nanofibrous Wharton's jelly scaffold in combination with adipose-derived stem cells for cartilage engineering. Mater Des 186:108216

Lin LL, Huang HC, Ogihara S, Wang JT, Wu MC, McNeil PL, Chen CN, Juan HF (2012) Helicobacter pylori disrupts host cell membranes, initiating a repair response and cell proliferation. Int J Mol Sci 13(8)

Lintner K, Peschard O (2000) Biologically active peptides: from a laboratory bench curiosity to a functional skin care product. Int J Cosmet Sci 22(3):207–218

Liu S, Zhang Y, Sun Q, Wei S, Zhao Y, Ji H (2020) Fish oil extraction method and application. China Patent: CN111139126A

Lu Z, Wei C, Huang Y (2018) Fish oil extraction technology based on high voltage pulsed electric field. China Patent: CN108117922A

Manninen AH (2004) Protein hydrolysates in sports and exercise: A brief review. J Sports Sci Med 3(2):60

Martins GI, Secco D, Rosa HA, Bariccatti RA, Dolci BD, de Souza SNM, Santos RF, da Silva TRB, Gurgacz F (2015) Physical and chemical properties of fish oil biodiesel produced in Brazil. Renew Sustain Energy Rev 42:154–157

McDaniel JC, Belury M, Ahijevych K, Blakely W (2008) Omega-3 fatty acids effect on wound healing. Wound Repair Regen 16(3):337–345

Melgosa R, Trigueros E, Sanz MT, Cardeira M, Rodrigues L, Fernández N, Matias AA, Bronze MR, Marques M, Paiva A, Simões P (2020) Supercritical CO_2 and subcritical water technologies for the production of bioactive extracts from sardine (Sardina pilchardus) waste. J Supercrit Fluids 164

Melgosa R, Sanz MT, Beltrán S (2021) Supercritical CO_2 processing of omega-3 polyunsaturated fatty acids—Towards a biorefinery for fish waste valorization. J Supercrit Fluids 169

Mercer P, Armenta RE (2011) Developments in oil extraction from microalgae. Eur J Lipid Sci Technol 113(5):539–547

Nguyen TT, Heimann K, Zhang W (2020) Protein recovery from underutilised marine bioresources for product development with nutraceutical and pharmaceutical bioactivities. Mar Drugs 18(8):391

Nielsen SS (2017) Food analysis laboratory manual. Springer

Noma T, Takasugi S, Shioyama M, Yamaji T, Itou H, Suzuki Y, Sakuraba K, Sawaki K (2017) Effects of dietary gelatin hydrolysates on bone mineral density in magnesium-deficient rats. BMC Musculoskelet Disord 18(1):1–6

Nomura Y, Oohashi K, Watanabe M, Kasugai S (2005) Increase in bone mineral density through oral administration of shark gelatin to ovariectomized rats. Nutrition 21(11–12):1120–1126

Norziah MH, Nuraini J, Lee KY (2009) Studies on the extraction and characterization of fish oil from wastes of seafood processing industry. Asian J Food Agro-Ind 2(4):959–973

Nouri M, Khodaiyan F, Razavi SH, Mousavi M (2016) Improvement of chitosan production from Persian Gulf shrimp waste by response surface methodology. Food Hydrocolloids 59

Pankaj SK, Wan Z, Keener KM (2018) Effects of cold plasma on food quality: A review. Foods 7(1)

Parolini C (2019) Effects of fish n-3 PUFAs on intestinal microbiota and immune system. Mar Drugs 17(6):374

Pateiro M, Domínguez R, Varzakas T, Munekata PES, Movilla Fierro E, Lorenzo JM (2021) Omega-3-rich oils from marine side streams and their potential application in food. Marine Drugs 19(5)

Peng YC, Yang FL, Subeq YM, Tien CC, Chao YFC, Lee RP (2018) Lipid emulsion enriched in omega-3 PUFA accelerates wound healing: A placebo-controlled animal study. World J Surg 42(6):1714–1720

Preto F, Zhang F, Wang J (2008) A study on using fish oil as an alternative fuel for conventional combustors. Fuel 87(10–11):2258–2268
Prihanto AA, Nurdiani R, Bagus AD (2019) Production and characteristics of fish protein hydrolysate from parrotfish (Chlorurus sordidus) head. PeerJ 7:e8297
Puglia C, Tropea S, Rizza L, Santagati NA, Bonina F (2005) In vitro percutaneous absorption studies and in vivo evaluation of anti-inflammatory activity of essential fatty acids (EFA) from fish oil extracts. Int J Pharm 299(1–2):41–48
Rahimi MA, Omar R, Ethaib S, Siti Mazlina MK, Awang Biak DR, Nor Aisyah R (2017) Microwave-assisted extraction of lipid from fish waste. IOP Conf Ser: Mater Sci Eng 206(1)
Rai AK, Swapna HC, Bhaskar N, Halami PM, Sachindra NM (2010) Effect of fermentation ensilaging on recovery of oil from fresh water fish viscera. Enzyme Microb Technol 46(1):9–13
Ramakrishnan VV, Ghaly AE, Brooks MS, Budge SM (2013) Extraction of proteins from mackerel fish processing waste using alcalase enzyme. Bioprocess Biotech 3(2):2–10
Ramluckan K, Moodley KG, Bux F (2014) An evaluation of the efficacy of using selected solvents for the extraction of lipids from algal biomass by the soxhlet extraction method. Fuel 116:103–108
Richter BE, Jones BA, Ezzell JL, Porter NL, Avdalovic N, Pohl C (1996) Accelerated solvent extraction: A technique for sample preparation. Anal Chem 68(6):1033–1039
Rodrigues HG, Vinolo MA, Sato FT, Magdalon J, Kuhl CM, Yamagata AS, Pessoa AFM, Malheiros G, Dos Santos MF, Lima C, Farsky SH (2016) Oral administration of linoleic acid induces new vessel formation and improves skin wound healing in diabetic rats. PLoS ONE 11(10):e0165115
Rubio-Rodríguez N, de Diego SM, Beltrán S, Jaime I, Sanz MT, Rovira J (2012) Supercritical fluid extraction of fish oil from fish by-products: A comparison with other extraction methods. J Food Eng 109(2)
Rustad T, Storrø I, Slizyte R (2011) Possibilities for the utilisation of marine by-products. Int J Food Sci Technol 46(10):2001–2014
Sachindra NM, Bhaskar N, Hosokawa M, Miyashita K (2010) Value addition to fish processing by-products. In: Alasalvar C, Miyashita K, Wanasundara U, Shahidi F (eds) Seafood quality, safety and health applications. Blackwell, UK, pp 390–401
Saha BC, Hayashi K (2001) Debittering of protein hydrolyzates. Biotechnol Adv 19(5):355–370
Samat AF, Muhamad NAS, Abd Rasib NA, Hassan SAM, Sohaimi KSA, Iberahim NI (2018) The potential of biodiesel production derived from fish waste. IOP Conf Ser: Mater Sci Eng 318(1): 012017. IOP Publishing
Sanmartín E, Arboleya JC, Villamiel M, Moreno FJ (2009) Recent advances in the recovery and improvement of functional proteins from fish processing by-products: Use of protein Glycation as an alternative method. Compr Rev Food Sci Food Saf 8(4):332–344
Sasidharan A, Venugopal V (2020a) Recovery of functional proteins from seafood processing discards and their applications. AFSTI Newsletter. Mumbai chapter. pp:4–6
Sasidharan A, Venugopal V (2020b) Proteins and co-products from seafood processing discards: Their recovery, functional properties and applications. Waste Biomass Valorization 11(11):5647–5663
Señoráns FJ, Luna P (2012) Sample preparation techniques for the determination of fats in food. Compr. Sampl. Sample Preparat 4:203–211
Shaviklo AR (2015) Development of fish protein powder as an ingredient for food applications: A review. J Food Sci Technol 52(2):648–661
Shigeta Y, Imanaka H, Ando H, Ryu A, Oku N, Baba N, Makino T (2004) Skin whitening effect of linoleic acid is enhanced by liposomal formulations. Biol Pharm Bull 27(4):591–594
Šližyte R, Daukšas E, Falch E, Storrø I, Rustad T (2005) Yield and composition of different fractions obtained after enzymatic hydrolysis of cod (Gadus morhua) by-products. Process Biochem 40(3–4):1415–1424
Šližytė R, Mozuraitytė R, Martínez-Alvarez O, Falch E, Fouchereau-Peron M, Rustad T (2009) Functional, bioactive and antioxidative properties of hydrolysates obtained from cod (Gadus morhua) backbones. Process Biochem 44(6):668–677

Song M, Chan AT, Fuchs CS, Ogino S, Hu FB, Mozaffarian D, Ma J, Willett WC, Giovannucci EL, Wu K (2014) Dietary intake of fish, ω-3 and ω-6 fatty acids and risk of colorectal cancer: A prospective study in US men and women. Int J Cancer 135(10):2413–2423

Song KM, Jung SK, Kim YH, Kim YE, Lee NH (2018) Development of industrial ultrasound system for mass production of collagen and biochemical characteristics of extracted collagen. Food Bioprod Process 110:96–103

Steigers JA (2003) Demonstrating the use of fish oil in a large stationary diesel engine. Steigers Corporation, Littleton, CO, USA

Thirukumaran R, Anu Priya VK, Krishnamoorthy S, Ramakrishnan P, Moses JA, Anandharamakrishnan C (2022) Resource recovery from fish waste: Prospects and the usage of intensified extraction technologies. Chemosphere 299

Tu GL, Bui THN, Tran TTT, Ton NMN, Le VVM (2015) Comparison of enzymatic and ultrasonic extraction of albumin from defatted pumpkin (Cucurbita pepo) seed powder. Food Technol Biotechnol 53(4):479–487

Venkatesan J, Anil S, Kim SK, Shim MS (2017) Marine fish proteins and peptides for cosmeceuticals: A review. Mar Drugs 15(5):143

Venugopal V (2016) Enzymes from seafood processing waste and their applications in seafood processing. Adv Food Nutr Res 78:47–69

Venugopal V (2021) Valorization of seafood processing discards: Bioconversion and bio-refinery approaches. Front Sustain Food Syst 5:611835

Xhauflaire-Uhoda E, Fontaine K, Piérard GE (2008) Kinetics of moisturizing and firming effects of cosmetic formulations. Int J Cosmet Sci 30(2):131–138

Xiao L (2010) Evaluation of extraction methods for recovery of fatty acids from marine products (Doctoral dissertation)

Yang JI, Ho HY, Chu YJ, Chow CJ (2008) Characteristic and antioxidant activity of retorted gelatin hydrolysates from cobia (Rachycentron canadum) skin. Food Chem 110(1):128–136

Yin M, Yang D, Lai S, Yang H (2021) Rheological properties of xanthan-modified fish gelatin and its potential to replace mammalian gelatin in low-fat stirred yogurt. LWT 147:111643

Zhang J, Tao N, Zhao Y, Wang X, Wang M (2019) Comparison of the fatty acid and triglyceride profiles of big eye tuna (Thunnus obesus), atlantic salmon (salmo salar) and bighead carp (Aristichthysnobilis) Heads. Molecules 24(21)

Collagen and Gelatin from Fish Processing By-Products for Biomedical Applications

Sonali Jana, Piyali Das, Prabal Ranjan Ghosh, and Samit Kumar Nandi

Abstract Fish discards that otherwise constitute a threat to environmental health are also a reservoir of bioactive molecules, peptides, and polymers that hold immense potential for biomedical applications. Fish discards, including fish scales, skin, fins, tails, etc., are largely collagen proteins that can be easily isolated from these discards by simple isolation protocols. Endowed with several advantageous characteristics such as limited immunogenic properties, easy extractability, lower risk of zoonosis transmission, and biocompatibility, fish products-derived collagen and gelatin have emerged as an appropriate alternative for their mammalian counterparts. Using simple extraction techniques, fish-derived collagen and gelatin can be turned into scaffolds and constructs using cutting-edge technologies like 3D printing and electrospinning, among others based on the therapeutic demands of the concerned tissue for various tissue engineering applications. Although these two natural polymers made from fish also have weak mechanical qualities, these flaws have been painstakingly fixed as a result of the latest technical breakthroughs, maximizing their utility. The role of fish collagen and gelatin in drug delivery wound healing and therapeutics is indispensable which signifies their importance in the commercial aspect. Entwined with technology, these discards or by-products could be viably transformed into value-added products that can immensely contribute to the biomedical sector simultaneously abating the burden on the marine and soil environment.

Keywords Fish wastes · Collagen · Gelatin · Wound healing · Tissue engineering · Biomedical applications

---

S. Jana · P. R. Ghosh
Department of Veterinary Physiology, West Bengal University of Animal and Fishery Sciences, Kolkata 700037, India

P. Das
Department of Biological Sciences, School of Life Sciences and Biotechnology, Adamas University, Barasat, West Bengal 700126, India

S. K. Nandi (✉)
Department of Veterinary Surgery and Radiology, West Bengal University of Animal and Fishery Sciences, Kolkata 700037, India
e-mail: samitnandi1967@gmail.com

S. Maqsood et al. (eds.), *Fish Waste to Valuable Products*, Sustainable Materials and Technology, https://doi.org/10.1007/978-981-99-8593-7_6

1 Introduction

With the global rise in fish production, there has been a proportionate increase in the number of biowastes from the fish processing industries consisting of fish's head, skin, bones, scales, tail, and viscera. These wastes form about 50–70% of the total catch (Chalamaiah et al. 2012). The discards often being dumped in landfills and into water bodies contributes to environmental pollution thus raising concerns about environmental health. The notable hazards from the dumping of fish discards in the ocean have an adverse effect on the oxygen balance of the ocean with an increase in BOD and COD leading to disturbance in the pH of water and destruction of oceanic biota. When dumped in landfills, these wastes emit harmful gases such as methane, hydrogen sulfide, and ammonia upon decomposition. Researchers have also raised concerns over the loss of nutrients through these wastes as they contain abundant proteins and essential fatty acids (Venugopal 2021). The conversion of the discards into fish meal, plant fertilizers, etc., is not an economically attractive avenue, thus valorization of these wastes into value-added products, simultaneously reducing the burden on the environment can lead to a sustainable flourishing of the fish processing industries (Ideia et al. 2020). These fish discards are also a valuable source of bioactive molecules, protein hydrolysates, peptides, enzymes, proteins, and oils that are potential raw materials in the pharmaceutical and biomedical sectors. Fish waste-derived collagen and gelatin are two macromolecules that have been intensively investigated in the biomedical field (Fig. 1). These have emerged as potential substitutes for mammalian-origin collagen and gelatin due to several factors. Bovine and porcine-derived collagen although of commercial importance has found restricted usage in recent times as they are known to incite allergic reactions as well as tend to spread zoonotic diseases including foot and mouth and bovine spongiform encephalopathy, scrapie, etc. Moreover, there are religious concerns attached to the usage of bovine and porcine-derived products. Thus, fish-derived collagen has emerged as a potent substitute for mammalian collagen, and is also safe (GRAS; Generally recognized as safe) as per FDA recommendations (Kulkarni and Maniyar 2020).

In the extracellular matrix (ECM) of living animals' tissues, collagen is a protein that is present in large amounts. This fibrous protein makes up around 25–35% of the body's total protein. It is present in bones, cartilage, aponeuroses, skin, tendons, ligaments, and other connective tissues. According to their structural characteristics, collagen has been divided into 29 varieties, with type I being the most prevalent in the body followed by types II, III, and IV. Collagen mostly contains the amino acids glycine, proline, and hydroxyproline (Lim et al. 2019). The skin of fish and mammals primarily contains type I collagen, which has a triple helix structure and various chains. The majority of fish wastes such as skin, scales, and bones have been used to extract collagen for commercial and therapeutic purposes (Ahn et al. 2021).

Gelatin, a biopolymer is widely used in the food sector along with in pharmaceutical and biomedical fields due to its unique properties. The majority of it is taken out of bones and skin from pigs and cows. But in recent years, there has been a lot of interest in the usage of gelatin made from fish waste such as fish skin. This is mostly

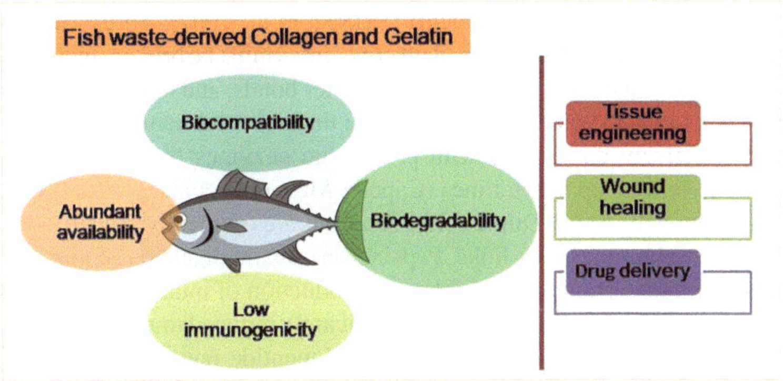

Fig. 1 Graphical illustration of biomedical use of fish waste-derived collagen and gelatin

due to religious constraints and the possibility of transmission of zoonotic diseases. Collagen is partially hydrolyzed to produce gelatin, which causes the collagen to become denaturized and lose its hydrogen bonds, causing the chains to separate and change from a triple helical conformation to a coiled shape. However, the amino acid makeup is still much the same as that of the original collagen and is made up of triplets of Gly-Proline-Hydroxyproline (Karim and Bhat 2009). Many freshwater and marine fish species have been studied to obtain collagen and gelatin. This includes catfish, Nile perch, Rohu, Catla, Eel, Red Snapper, Tilapia, and Leather Jacket Cod fish (Jana et al. 2022a; Muralidharan et al. 2013; Muyonga et al. 2004; Niu et al. 2013; Pati et al. 2010; Sanaei et al. 2013; Shakila et al. 2012).

2 Extraction of Collagen and Gelatin from Fish Wastes

Collagen must be extracted from fish wastes using several consecutive procedures (Figs. 2A–D and 3). It can be taken out of the fish's skin, scales, bones, muscles, and even swim bladder. While different extraction methodologies can be applied depending on the raw materials, the initial steps are common to all the starting materials. This includes a thorough cleaning of dirt and debris when collagen is to be extracted from the skin and scales. The next step is chopping the raw materials into very small pieces that will ease the overall handling of the materials during the extraction process. It is also essential to eliminate non-collagenous parts of the raw material to improve extraction efficacy. This pretreatment is done by immersing the raw material in either dilute acids or an alkali solution that facilitates the breaking down of solid cross-linkages in the collagen triple helix. A mild solution of NaOH is frequently used as an alkali for this purpose. Further alcoholic treatment with butanol or ethanol is often done to wash away the remnant fats within the raw material (Lim et al. 2019). Hard tissue, typically composed of various minerals such as the bones and scales is demineralized by treatment with ethylenediaminetetraacetic acid or strong acid (hydrochloric acid) (Skierka et al. 2007). To extract collagen, several intra- and intermolecular covalent cross-links must be broken, including those involving lysine and hydroxylysine residues, ester bonds, and other interactions with saccharides (Ran and Wang 2014). The more unstable and mature ketamine cross-links are effectively dissociated with proteolytic enzymes like pepsin, increasing the collagen yield by up to ten times (Capella-Monsonís et al. 2018; Delgado et al. 2017; Zeugolis et al. 2008). Dilute acidic solutions are frequently used to break down intermolecular aldimine cross-links. Pepsin's enzymatic digestion of non-helical N- and C-telopeptide portions also enables the production of intact mono triple helices known as atelocollagen, which causes a significantly reduced immune response since the antigenic P-determinant sequence at the telopeptide regions is eliminated. The various methods used to extract it have given the final collagen the labels acid-soluble collagen (ASC), pepsin-soluble collagen (PSC), and salt-soluble collagen (SSC). Recently, there have also been reports of other techniques such as the deep eutectic solvent (DES) method, the extrusion and ultrasound-assisted approach, and supercritical fluid (SF) extractions (Jafari et al. 2020).

Collagen is subjected to chemical or enzymatic treatment to produce gelatin, which results in the degradation of the tertiary, secondary, and partial collagen primary structures (Karayannakidis and Zotos 2016) (Fig. 2E). Out of the 18 distinct amino acids that make up gelatin, glycine, proline, and hydroxyl-proline are the most common. Type A and Type B of gelatin can be distinguished from one another; with Type A being acquired using an acid extraction method and Type B being obtained through an alkaline procedure, respectively (Kamble et al. 2014). Gelatin is the term for partially hydrolyzed collagen, which is also water-soluble. The procedure for extracting gelatin is similar to that used to extract collagen, and then the collagen is heated to create gelatin. Dinçer et al. (2016) extracted gelatin from fish scales in three main stages, i.e., removal of non-collagen proteins with the use of 5% NaCl and 0.4% NaOH solution then removal of lipids by 10% Isobutyl alcohol and then de-mineralization in 0.5 N EDTA solutions (pH 7. 66) with different periods. The residual scales after filtration were treated with a dilute acid solution of 0.05 M acetic acid for 3 h and then filtered and heated at 60 °C with water for 12 h to obtain gelatin. Fish bones must be pretreated with a concentrated acid solution (3% HCl) for around 9–12 days to completely remove all hard minerals before gelatin can be extracted from them. The remaining soft matrix was boiled in water to extract gelatin (Muyonga et al. 2004). The gel strength of gelatin rests on several reasons like gelatin concentration, the composition of amino acids, temperature, time, pH during extraction, and importantly its content of imino acid and glycine (Kamble et al. 2014). Gelatin can also be extracted by using an enzyme such as pepsin (pH 2.0) after pretreatment with alkali. Increasing the concentration of pepsin decreased the gel strength even though the extraction yield increased (Jridi et al. 2013). Table 1 gives summarized information on the different extraction protocols employed for collagen and gelatin extraction from fish wastes.

The yield of collagen is known to be significantly dependent on several aspects, together with temperature, pH, and the concentration of the extraction solution. The characteristics of extracted gelatin are similarly influenced by these variables

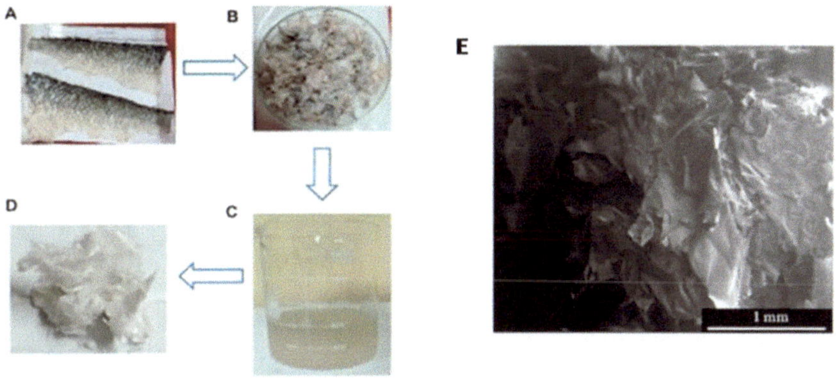

Fig. 2 Various stages of extraction of Rohu fish skin collagen. Skin is cleaned (**A**), chopped into pieces (**B**), treated with weak acid and pepsin to extract collagen (**C**), and lyophilized to form collagen sponges (**D**) (Replicated with permission from the American Chemical Society (Ref. Jana et al. 2022a, b)). SEM image of Tilapia fish scale gelatin (**E**). (Replicated with permission from the European Polymer Journal (Sghayyar et al. 2020))

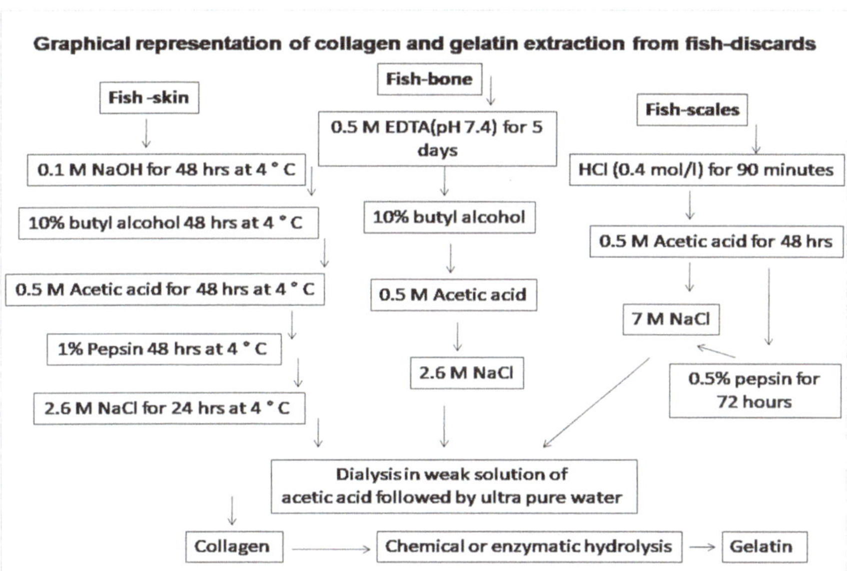

Fig. 3 Graphical representation of collagen and gelatin extraction from fish discards

(Gómez-guilløen and Montero 2001). Therefore, it is crucial to maximize these variables to improve the yield and quality of the collagen and gelatin extracted.

Table 1 Summary of various fish wastes' collagen/gelatin extraction protocols

Fish	Type of waste	Collagen/gelatin extraction protocol	Characteristics	References
1. Carp fish (*Cyprinus carpio*)	Scales	Scales were demineralized by treating them with a solution of HCl (0.4 mol/l) for 90 min. For two to three days, acid-soluble collagen (ASC) was extracted from the sample using 0.5 M acetic acid. Centrifugation was used to separate the soluble collagen after precipitation with a 0.7 M salt solution. To extract the pepsin-soluble collagen, Next, 0.5 M acetic acid and 0.5% (w/v) pepsin were applied to the acid-insoluble part for 72 h (PSC). It was separated by salt treatment similarly	Type 1 collagen	Fengxiang et al. (2011)

(continued)

Table 1 (continued)

Fish	Type of waste	Collagen/gelatin extraction protocol	Characteristics	References
2. Tilapia (*Oreochromis sp.*)	Scales (extrusion–hydro-extraction process)	The fish scales were completely cleaned with water to a neutral pH after being pretreated with 0.1 N NaOH. The scales were finally dried and processed into powder using a mill. In a fixed ratio, powered scales were combined with double-distilled water, 1.26% citric acid (pH 2), and extrusion cooking was carried out in a screw extruder at a high temperature and constant speed. The extrudate of fish scales was dried, powdered, and combined with double-distilled water before being shaken for an hour at a temperature of 25 or 50 °C. To get soluble collagen, the mixture was centrifuged and then lyophilized	Type 1 collagen	Huang et al. (2016)

(continued)

Table 1 (continued)

Fish	Type of waste	Collagen/gelatin extraction protocol	Characteristics	References
3. Eel fish	Skin	The cleaned fish skin was continuously agitated in a 0.5 M acetic acid solution for 42 h at 4 °C . Salting out with 0.9 M NaCl precipitated the filtered fraction. By using centrifugation, the acid-soluble collagen (ASC) was isolated. To create pepsin-soluble collagen, this ASC underwent further treatment with 1% pepsin in 0.5 M of acetic acid for 24 h (PSC). The PSC was separated by centrifugation and salting out with NaCl. Finally, the precipitated collagen was dialyzed to obtain lyophilized salt-free collagen	Type 1 collagen	Govindharaj et al. (2019)

(continued)

Table 1 (continued)

Fish	Type of waste	Collagen/gelatin extraction protocol	Characteristics	References
4. Japanese sea bass (*Lateolabrax japonicas*), Skipjack tuna (*Katsuwonus pelamis*), ayu (*Plecoglossusaltivelis*), chub mackerel (*Scomber japonicas*), yellow sea bream (*Dentex tumi frons*), horse mackerel (*Trachurus japonicas*) and bullhead shark (*Heterodontus japonicas*)	Bones	To decalcify and de-fatten the bones, they were initially immersed in 0.5 M ethylenediaminetetraacetic acid (pH 7.4) for five days, with daily replacements of the fresh solution, before being treated with 10% butyl alcohol. After washing, the residue was frozen and dried. Multiple treatments with 0.5 M acetic acid were used for collagen extraction. By salting out with NaCl (2.6 M) containing TRIS HCl, collagen was precipitated (0.05 M, pH 7.5). Using centrifugation, this was further separated, dialyzed, and lyophilized	Type 1 collagen	Nagai and Suzuki (2000)
5. Grass carp (*Ctenopharyngodonidella*)	Swim bladder	The diced swim bladders were repeatedly rinsed with distilled water before being continuously stirred in 0.5 M acetic acid for 12 h. The goopy collagen was salted out with NaCl and centrifuged. Centrifugation was used to recover the precipitated collagen, which was then dialyzed and lyophilized to produce the dry collagen	Type 1 collagen	Li et al. (2022)

(continued)

Table 1 (continued)

Fish	Type of waste	Collagen/gelatin extraction protocol	Characteristics	References
6. Rainbow trout (*Oncorhynchuson-corhynchusmykiss*)	Skin	The skin was diced and submerged in a cold alkali solution after being rinsed in chilled water (0.01–0.21 N, NaOH). After washing, the skin was submerged in a cold acetic acid solution (0.01–0.21 N, at 7 °C), drained of the acid, and then washed in cold water. Three times the soaking took place. By immersing the samples in distilled water and heating them at 50 °C for 16 ± 2 h, gelatin was extracted. Gelatin extract was obtained by filtration with a cloth cheesecloth		Tabarestani et al. (2010)
7. Catfish (*Clarias gariepinus*)	Bones	The bones were cleaned, chopped into pieces, and demineralized by treating them with various concentrations of hydrochloric acid at 4 °C for several hours. A thorough wash with water was done until all the acids were neutralized. Gelatin extraction was carried out under high temperatures (30–90 °C) by mixing the bones with water and heating them for 2–8 h. The extracted gelatin was obtained by filtering and centrifugation		Sanaei et al. (2013)

3 Characteristics of Fish-Derived Collagen and Gelatin

The physicochemical behavior of fish-derived collagen differs in various aspects when compared to mammalian-origin collagen. Fish collagen has poor thermal stability and thus undergoes rapid denaturation under elevated temperatures. While higher thermal stability is positively correlated with a higher content of hydroxyproline; this temperature-dependent denaturation is also influenced by the water content and cross-linking intensity (Gauza-Włodarczyk et al. 2017). A larger concentration of hydroxyproline promotes stable interconnections in the collagen triple helix by encouraging hydrogen bonding between the molecules and improving physical stability (Ahn et al. 2021). Fish-derived collagen's thermal stability is directly correlated with both the temperature of the fish's body and the water in which it lives (Rigby 1968). The thermal stability of collagen is an important parameter that is considered for its varied biomedical applications; unfortunately, fish-derived collagen finds limited usage because it is less thermally stable than mammalian collagen (Nagai and Suzuki 2000). Collagen with improved mechanical characteristics has been isolated from fish wastes in recent studies. In this context, Pati et al. (2010) successfully isolated type 1 collagen from *Labeo rohita* (Rohu) and *Catla catla* (Catla) scales with higher denaturation temperature (36.5 °C) and considering its proximity to mammalian-origin collagen may find potential application in the biomedical sector. Collagen from cold-water fish has higher quantities of amino acids like serine and threonine than that of mammalian species (Muralidharan et al. 2013).

The content of the amino acids and the molecular weight distribution have a significant impact on the physical and chemical features of the extracted gelatin, including the gel's strength, viscosity, melting point, and setting behavior. With an increase in the amount of amino acids, gelatin's gel strength and melting point rise correspondingly (proline and hydroxyproline). Because it contains more proline and hydroxyproline than fish gelatin, mammalian gelatin has a stronger gel than fish gelatin (See et al. 2010). Additionally, it has been noted that fish living in warm water have higher imino acid concentrations than fish living in cold water. Additionally, when the number of chains in the extracted gelatin grows, the gel strength does as well (Karim and Bhat 2009). A distinct quality (Yamaguchi et al. 1976).

The main hurdle in making tissue engineering constructs from fish collagen and gelatin for biomedical purposes is their poor mechanical qualities, such as lower denaturation temperatures. Considering the huge underutilized resource from fish discards, the fish-derived collagen can be favorably transformed for suitable applications by various means such as crosslinking by physical or chemical means. In this regard, the application of UV and gamma radiation, or chemicals such as glutaraldehyde and carbodiimide has resulted in remarkable improvement in the mechanical stability of the fabricated constructs.

4 Recent Technological Innovations for the Fabrication of Scaffolds/Constructs from Fish-Derived Collagen and Gelatin

Typically, mammalian tissues (such as pig or bovine dermis and Achilles tendons of horse) are used to extract collagen for therapeutic purposes. Scales, skin, bones, and fins of freshwater and marine fish have also been used to extract type I collagen (Kozlowska et al. 2015). It is crucial to remember that fish scales and the inorganic component that makes them up make an ideal combination that has promise for use in the long term. Collagen-based products are frequently used in pharmacy and reconstructive medicine, but their qualities depend on the origin of the collagen as well as how it is purified and treated (Sionkowska et al. 2017). Both three-dimensional (3D) forms and thin films of collagen can be employed. By air-drying a molded collagen solution, collagen films with a thickness of 0.01–0.5 mm are created. Collagen is mostly researched in film shape as a substance for the healing process and as a barrier membrane (Sionkowska et al. 2017). Freeze-drying, 3D printing, electrospinning, thermal-induced phase separation, laser treatment, and crystal precipitation can all be used to create 3D scaffolds for tissue engineering (Sionkowska and Kozłowska 2013). The gelation procedure may be used to create collagen 3D sponges (freeze-drying technique). Depending on the inorganic particles' thickness, one may produce collagen composite or nanocomposite using these sponges. Numerous crosslinking techniques are employed to enhance physical characteristics. Physical crosslinking can be accomplished using dehydrothermal treatment (DHT) and UV radiation, while chemical crosslinking can be accomplished using N-(3-dimethylamino propyl)-N′-ethyl carbodiimide hydrochloride and N-hydroxyl-l succinimide (Sionkowska and Kozłowska 2013). The creation of mixes depending on collagen and chitosan is of particular interest to the research community. Using electrospinning methods, alginate, chitosan, collagen, and hydroxyapatite scaffold with promise for bone regeneration purposes was created (Yu et al. 2013). Adding collagen to the liquid phase of tricalcium phosphate cement, either in soluble or fibrillated form and possibly using a self-assembling technique that begins with collagen gel and hydroxyapatite precursors by an in vitro customized solubilization procedure, allows one to create collagen/hydroxyapatite composite (Kozłowska and Sionkowska 2015). Fish scales may be properly prepared to create novel composites with intriguing characteristics that can be used in biological applications. In one such study, hydroxyapatite from fish scales was investigated for use in bone regeneration. Collagen and similar polymers might be combined in advanced materials to create a new range of renewable polymeric substances that could be used as packaging material.

Over the past 10 years, several possible gelatin-based technologies have been created using this method. For instance, a biocomponent-based hydrogel made of gelatin and hyaluronic acid was first presented to give different targeted regions flexibility and encourage vascularization (Eke et al. 2017). Platelet-loaded chitosan–gelatin composite hydrogel was an innovative method for enhancing bone healing in a rat model with bilateral critical-sized radial bone lesions (Oryan et al. 2017). Furthermore, it was shown that angiogenic maturation markers, osteogenic factors, and alkaline phosphatase had increased mRNA levels. Although gelatin-based scaffolds for bone repair purposes have not yet been made commercially available, the coupling of gelatin/calcium phosphate ceramics and alternative polymer blends represents a viable alternative (Echave et al. 2019). Creating biomaterials with similar complicated topologies to those found in bone lesions is one of the toughest problems here. Injectable methods (Echave et al. 2019) and 3D-printed implantable constructions (Heo et al. 2017) that make it easier to create specially built composites that can properly fit into bone fractures have been researched as ways to get around this constraint.

5 Biomedical Use of Fish-Derived Collagen and Gelatin

A recent and quickly developing field of the biological sciences is tissue engineering and regenerative medicine. Employing technology and biological concepts to produce bioinspired organs and tissues from natural samples has recently become a popular notion and contentious issue among scientists. Due to MDC's high biocompatibility, there is speculation that it may be used to create biomaterial substrates for tissue engineering and regenerative medicine. This article goes into great detail about the multiple applications of collagen, a strong biomaterial, in the field of health (Fig. 4). It is important in the biomedical area since it appears in numerous forms. There have lately been several endeavors to utilize type I collagen as a biomaterial. Biochemical therapies offer remarkable sturdiness to the collagen matrix, but they might lead to cytotoxicity or poor biocompatibility (Yamada et al. 2014), while physical therapies contribute enough durability without any cytotoxicity. To organize cells radially, provide environmental signals, and control site-specific cellular regulation during tissue engineering, it is necessary to create a highly permeable, biodegradable scaffolding that resembles the natural extracellular matrix (Wang et al. 2006). Regarding cell seeding, development, migration, and the generation of new tissues, it is commonly acknowledged that the surface area, pore number, pore size, and pore wall architecture of scaffolds employed in bioengineering are crucial characteristics.

The use of jellyfish collagen comprising telopeptides increases IgM synthesis in the human hybridoma cell line HB4C5, along with IgM and IgG expression in human peripheral blood cells (Sugahara et al. 2006). Furthermore, enhanced transcription and translation brought on by jellyfish collagen leads to increased production of antibodies and cytokines (Nishimoto et al. 2008). Atelocollagen has been investigated in comparison to bovine collagen in vivo reactions of jellyfish in terms of safety for biological purposes (Song et al. 2006). The created substrate has an incredibly porous and interconnected pore structure that works well for increased cell seeding and provides the cells grown in the 3D matrix with outstanding feeding and oxygen delivery.

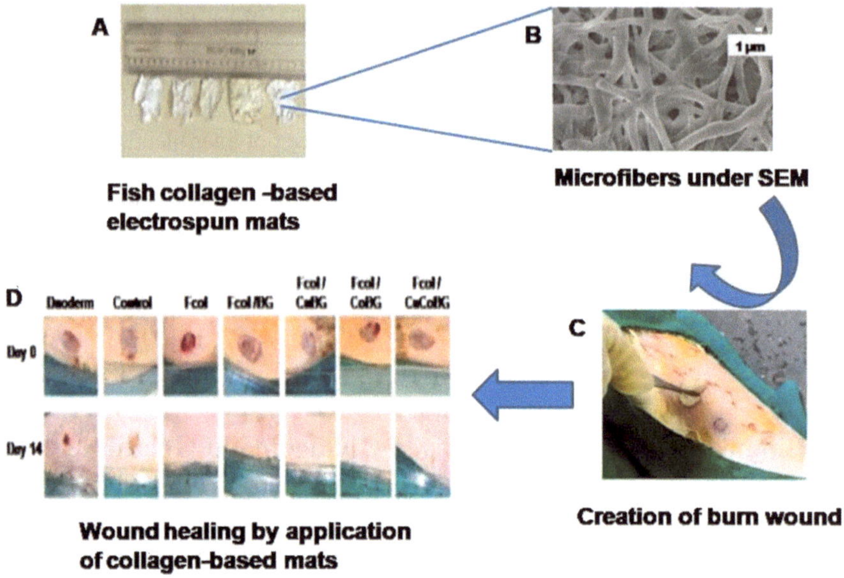

Fig. 4 Application of fish skin-derived collagen-based biomaterial for biomedical application (**A**). (Ref. Jana et al. 2022a). Application of electrospinning technology to create microfibrous mats with collagen as a base material (**B**). (Ref. Jana et al. 2022b). Burn-wound healing with collagen-based mats (**C**, **D**). (Ref. Kayal et al. 2023)

5.1 Bone Tissue Engineering

Fracture healing and restoration are postoperative regenerative mechanisms that retrace several aspects of bone growth and are types of tissue renewal. Therefore, bone restoration is a sophisticated physiological bone remodeling process that includes both bone creation and bone resorption. Avascular necrosis, osteoporosis, and atrophic non-unions are a few complicated diagnostic situations in which significant amounts of bone regeneration are necessary, such as for those of the skeletal restructuring of long bones caused by a defect, and injury induced by removal of bone tumor, infection, or skeletal irregularities (Dimitriou et al. 2011). The methods used in a wide range of pertinent clinical situations to repair or rejuvenate bones, manufacture bone grafts, or create bone replacements are now being revolutionized by advancements in tissue engineering technology. Functional bone transplants are currently in extremely high demand all around the world. Over half a million people in the US have bone repairs every year, at a cost of more than \$2.5 billion. Over the past ten years, analyses have shown that fish collagen-based scaffolds have high osteogenic activity. Fish collagen-based peptides produced from tilapia scales were found to promote diverse differentiation in primary bone marrow mesenchymal stem cells generated from rats. Fish collagen-based peptides not only enhance cellular survival, but also markedly increased the articulation of osteogenic and endothelial indicators, indicating that fish collagen-based peptides have the prospects to vigorously encourage osteogenic and endothelial differentiation. Adipogenic and chondrogenic marker expression was notably reported to be reduced by fish collagen-based peptides (Liu and Sun 2014). In their 2013b study, Yamada et al. investigated the biological effects of fish collagen peptides derived from cod bones and skin on human osteoblastic cells (NOS-1). For bone tissue engineering, it was discovered that treating osteoblast cell lines with peptides derived from fish collagen improved cellular proliferation, mineralization, expression of osteogenic markers, and alkaline phosphatase activity. In a mouse osteoblastic cell growth system (MC3T3-E1), the impact of fish collagen-based peptides isolated from Gadiformes and Pleuronectidae on collagen synthesis, purity, and calcification was revealed (Yamada et al. 2013). The study's findings emphasized the potential use of fish collagen-based peptides in wound healing by demonstrating that collagen therapy accelerated matrix remodeling and calcification in vitro and greatly increased the gene expression of various collagen-modifying enzymes.

Multiple investigations have confirmed the utility of fish collagen-based biomaterial scaffolds as enhancements and alternatives for bone tissue regeneration and renewal. The main goal of bone tissue engineering is to generate scaffolds that encourage cellular proliferation within them (Elango et al. 2018; Liu and Sun 2014). For this purpose, a nano/microfibrous scaffold was created using a freeze-drying technique and utilizing collagen taken from the scales of freshwater fish, *Catla catla* and *Labeo rohita* (Pati et al. 2012). In this study, mice were given the collagen solution, which just slightly induced an inflammatory reaction. The nano/microfibrous scaffolds demonstrated excellent biocompatibility and suggested potential uses in tissue engineering by considerably increasing the growth rates of human osteoblast-like cells (MG-63) and mouse fibroblasts (NIH3T3).

Biphasic constructs of biomimetic mineralized salmon collagen and fibrotic jellyfish collagen were produced using lyophilization and cross-linking methods, and they demonstrated their capacity to promote chondroblast and osteogenic growth of hMSCs in vitro (Bernhardt et al. 2018). Additionally, shark dermal collagen promotes osteoblast growth and collagen synthesis in bone cells (Calejo et al. 2012). This collagen may promote the adhesion and proliferation of osteoblast-like cells when combined with shark tooth calcium phosphate to create a 3D composite scaffold (Diogo et al. 2018). Furthermore, studies have demonstrated that the collagen polypeptide generated from the scales of two species of Sephareid fish may encourage osteoblast production and inhibit mature osteoclast growth, which can be used to prevent osteoporosis and encourage bone healing (Hu et al. 2016).

Future research by MDC could offer further options for bone restoration and transplantation. Fish swim bladder collagen (SBC) from sturgeons was used by scientists to create a unique collagen fiber wikestone hydrogel based on the dual network (DN) concept. The gel has exceptional biomechanical properties in vivo after being inserted into the osteochondral defect of the rabbit knee joint. It is advantageous to combine bone with wrapped DN gel and hydroxyapatite. Future implantable devices like artificial cartilage could be built using this special collagen matrix composite DN gel, which combines with a bone to create a soft, elastic ceramic material and has exceptional physical qualities (Mredha et al. 2017). Bioactive nano-hydroxyapatite (N-HA)-reinforced polylactide glycogen (PLGA) nanofiber membranes with low immunogenicity fish collagen protein were created for electrospinning-guided bone regeneration (GBR). It was established that the coating was compatible with human gingival fibroblasts and bone marrow mesenchymal stem cells (BMSCs) (HGF). The results of the experiment showed that the synthetic fibrous membrane is extremely effective in controlling bone or tissue regeneration (Jin et al. 2019). These results support the idea that scaffolds made of fish collagen play an increasingly important role in bone tissue creation and repair.

5.2 *Cartilage Tissue Engineering*

Degradation of the articular cartilage is one of the conditions that is most frequently diagnosed, and if not treated effectively, it can lead to complete joint replacement. Cartilage has a relatively restricted potential to renew itself due to its avascular nature. As a result, articular cartilage treatment has received attention and has great potential because of the study of cartilage tissue engineering, which seeks to restore, revitalize, and/or enhance the qualities of diseased or damaged articular cartilage. Fish collagen-based composites promote chondroblast development in cartilage tissue. For instance, the study contrasted the chondrogenic growth of human bone marrow mesenchymal stem cells (hMSCs) on tilapia-size collagen fibers with that of pig collagen and uncoated culture dishes. The findings showed that tilapia collagen fibrils in the chondrogenic media specifically stimulated HMSC cartilage formation. Haddock, cod, and pollock dermis-derived fish collagen polypeptides have been shown to enhance the chondrogenic growth of primary adipose-derived stromal cells (Raabe et al. 2010). Oral administration of glucosamine and fish collagen polypeptides derived from Gadiformes skins constructively managed cartilage degeneration in osteoarthritis-induced rabbit models, suggesting that fish collagen polypeptides may be useful as a disease-modifying drug in the treatment of deteriorated joint disorders (Ohnishi et al. 2013). In keeping with these findings, fish collagen polypeptides have been demonstrated to stimulate chondrogenesis in hMSCs, which is similar to its influence on osteogenic and osteoblastic differentiation in hMSCs (Hsu et al. 2016). Hsu et al. (2016) discovered that hMSCs cultivated with fish collagen polypeptides expressed more chondrogenic markers and fewer osteogenic markers, suggesting that fish collagen polypeptides might offer the optimal environment for in vitro chondrogenic differentiation of hMSCs. These findings lend credence to the use of fish collagen in cartilage repair. The proper expansion of primary human and rat nasal septum chondrocytes in a three-dimensional environment where chondrocytes attached to scaffolds made of fish collagen polypeptides and produced cartilaginous matrix proteins was demonstrated to benefit from the use of fish collagen polypeptide-based composites made using the freeze-drying method and collagen derived from the jellyfish (*Rhopilema esculentum*) (Bermueller et al. 2013). In the current work, fish collagen polypeptide scaffolds were found to be effective at preventing septal perforations in an orthotopic rat model, proving their appropriateness for the regeneration of nasal cartilage. The implantation of rabbit auricle chondrocytes into porous fish collagen was shown to enhance the formation of chondrospecific extracellular matrix (ECM) both in vivo and in vitro, which in turn accelerated the creation of cartilage under the rabbit's skin. When coupled with fish collagen (ADSCs), TGF-1 may successfully induce the chondrogenesis of adipocyte stromal cells (Elias 2007). In studies of rat nasal cartilage regeneration employing in situ models, the fish collagen polypeptide matrix demonstrates good functionality for cartilage tissue engineering (Bermueller et al. 2013). It also possesses cartilage-protective properties (Ohnishi et al. 2013). Indicating that the hybrid structure encourages chondrogenic development in hMSCs and provides a fresh approach for researching cartilage regeneration, the mixed scaffolds boosted the expression of chondrogenic hMSCs. Clinical trials showed that fish collagen polypeptides increased cartilage matrix production and decreased inflammation in osteoarthritic patients, rendering them a suitable choice for treatments against human osteoporosis and osteoarthritis (Cheung et al. 2015).

5.3 Dental Tissue Engineering

The cohabitation of soft and hard tissues, with the hard tissue encompassing the soft tissue and the dental pulp, is an unusual aspect of a tooth's structure (Aurrekoetxea et al. 2015). Regenerative dentistry is a growing profession as a result of the considerable advancements made in the recent generation in the field of dental tissue engineering toward fresh emerging concepts of the compensating approach (Chang et al. 2017). Fish collagen is relevant in dentistry for use in dentin-pulp restoration because it enhances cell survival and attachment, increases alkaline phosphatase activity, overexpresses the bone sialoprotein gene in rat odontoblast-like cells (MDPC-23), and also increases composite calcification. These effects are comparable to those of porcine skin (Tang and Saito 2015).

A hybrid nanofiber membrane made of chitosan, bioactive glass, and isolated tilapia collagen was created using an electrospinning technique. The nanocomposite promoted cell survival and osteogenic gene activity in human periodontal ligament cells in the study employing the canine class II infection deficiency model (HPDLCs). The blood levels of the peptides osteopontin (OPN) and Runt-related transcription factor 2 were also increased (RUNX-2) (Xu et al. 2021). Finally, the use of fish collagen polypeptide in tooth restoration has enormous promise. A biomimetic electrospun tilapia fish collagen/bioactive glass/chitosan nanofiber membrane with antibacterial action against *Staphylococcus mutants* improved hPDL cells' cell adhesion, stability, and osteogenic gene expression (Zhou et al. 2017a). In periodontal defective dog models, this scaffold also encouraged bone regeneration, suggesting its feasibility for use in the clinic as a controlled material or bone restoration substrate for promoting periodontal tissue formation.

5.4 *Corneal Tissue Engineering*

Corneal injury is the main cause of blindness around the globe (Lim et al. 2019). The only option left after conventional procedures fail to improve vision is corneal transplantation (keratoplasty). The most frequent solid tissue transplant surgery worldwide is corneal transplantation (Lim et al. 2019). Scar prevention and corneal healing were accomplished using collagen-based biomaterials (Chae et al. 2015). Human donor corneal tissue was substituted with a collagen matrix obtained from decellularized and decalcified fish scales (Zhao et al. 2018). In a rat anterior lamellar keratoplasty model, the ocular implantation of the fish collagen-based matrix demonstrated great cytocompatibility, suitable light transmission, believable light-scattering values, and the potential to be employed in keratoplasty. These results suggest that fish collagen may be used as a corneal substitute to help with the scarcity of corneal donor tissue. When tested against a denuded human amniotic membrane, fish collagen-based scaffolds produced from fish scales by freeze-drying demonstrated good physicomechanical properties. Additionally, they had increased microbial resistance, a better swelling ratio, and superior growth, survival, multiplication, and migration of limbal epithelial cells from limbal explants. These results promote fish collagen as a probable biodegradable polymer recipient for corneal restoration or transplantation (Lim et al. 2019). To comprehend its immunogenicity better, more research is required. The first version of the fish scale-derived collagen matrix (FSCM) shares this characteristic with human corneal tissue, according to statistics on light scattering and transmission (van Essen et al. 2013).

5.5 *Vascular Tissue Engineering*

The ability to comprehend and manage vasculature growth and differentiation as well as to construct vascular grafts or prosthetic blood and lymphatic arteries is imperative given the population of patients needing vascular entry due to a variety of illnesses, including coronary heart disease, peripheral arterial disease, and ischemia. The ability to comprehend and control vasculature growth and differentiation as well as to create vascular grafts or prosthetic blood and lymphatic arteries is imperative given the population of patients needing vascular entry due to a variety of illnesses, including coronary heart disease, peripheral arterial disease, and ischemia. Vascular tissue engineering has advanced significantly during the previous ten years (Serbo and Gerecht 2013). Fish collagen-based materials which are already extensively acknowledged and employed in a variety of therapeutic purposes have proven to be a viable substitute scaffold material for the making of vascular tissue. A collagen patch made from snakehead fish scales was produced by Wang et al. (2017) using lyophilization, cold pressing, and 1,4-butanediol diglycidyl ether (BDE), a well-researched bifunctional cross-linking epoxy chemical. This study provided evidence of the promising need for fish scale-derived collagen as a scaffolding material for vascular genesis and various other biological uses by demonstrating the satisfactory assimilation of the collagen patches into the surrounding tissues, with good invasion of cells, capillaries, and lymphatic vessels. For use as vascular grafts, tubular constructions made of PLGA and marine source collagen from jellyfish were electrospuned and freeze-dried (Jeong et al. 2007). This MC/PLGA composite construction significantly improved spatial cell orientation, simulating native vasculature in vivo. It also stimulated endothelial cell advancement and supported the proliferation of primary rabbit smooth muscle cells and endothelial cells. This suggests MC/PLGA scaffolds have a promising role in tissue-engineered vascular grafts.

5.6 Wound Dressing and Skin Tissue Engineering

Damage to a wound occurs from a disruption in the skin's defensive membrane activity. The goal of treating wound damage is to quickly return its structure and functionality to those of healthy tissue. Conventional treatment has been used to treat skin injuries since ancient times, although the wound cannot be healed quickly using classical medical dressings (Lim et al. 2019). Therefore, discovering cutting-edge bioengineering materials that can act as a transient barrier in injured skin and prevent the wound from becoming contaminated is a top priority. In this background, the fish collagen as wound care dressings has received a lot of interest (Fig. 5). A multipurpose, biocompatible skin wound dressing made of tilapia skin collagen and bioactive glass nanofiber was electrospun to prevent infection, hasten wound healing, and increase skin regeneration (Zhou et al. 2017b). The prospect of incorporating such a composite nanofiber as a useful skin wound dressing arises from its outstanding antibacterial behavior against *Staphylococcus aureus*, promotion of human keratinocyte adhesion, multiplication, and mobility (HaCaT), initiation of collagen type I and VEGF extravasation by skin fibroblasts, stimulation of the propagation of human vascular endothelial cells, and acceleration of rat dermal wound renewal and regeneration. Utilizing the inherent horny exoskeleton of marine sponges, innovative bio-based therapeutic compositions like powder or polymeric film have been created (Porifera, Dictyoceratida).

In the meantime, there is also another study group that, for instance, used a porous collagen sponge to study burn injuries in mice (Muthukumar et al. 2014), jellyfish collagen polypeptide to assess oral scars, and salmon skin damage in rats (Xu et al. 2021), and ethylene amine and fish scale collagen to assess wound healing (Xu et al. 2021), almost everything shows that fish collagen derivatives can encourage wound healing. Additionally, a wound healing investigation using a rat model revealed its effectiveness as a dermal alternative (Pal et al. 2016). The scientists discovered that fish collagen-based composites might swiftly and efficiently encourage rat wound healing (Xu et al. 2021). It can also accelerate the healing progression of wounds if it is formed into nanofibers or scaffolds. In vivo tests using MDC scaffolds performed by Catfish revealed that the scaffolds encouraged NIH/3T3 fibroblast migration and proliferation as well as tissue repair and regeneration (Evans et al. 2021). On a

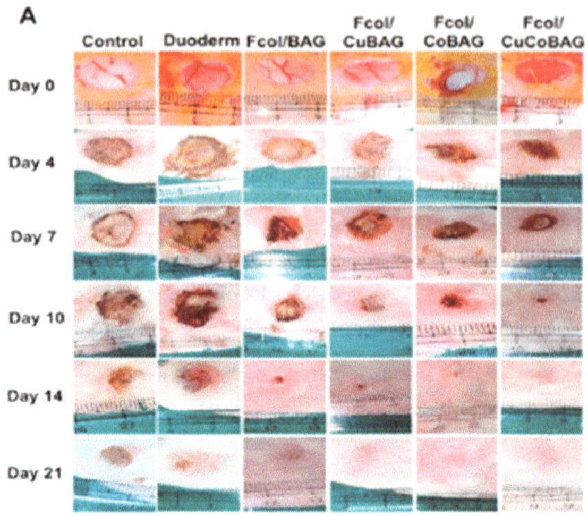

Fig. 5 Wound healing with electrospun wound dressing composed of Rohu fish dermal collagen and ion-doped bioactive glass composites. Replicated with authorization from the American Chemical Society (Ref. Jana et al. (2022a, b))

complex 3D collagen gel, fibroblasts from tiny hamster kidneys (BHK21) were transplanted. As a result of the findings, fish collagen-based polypeptides might be employed as a possible biodegradable polymer isolate for the clinical field since it might stimulate the propagation of BHK-21 cells (El-Rashidy et al. 2015). When human umbilical vein endothelial cells (HUVECs) are arranged in a monolayer, jellyfish collagen encourages the creation of simulated wounds (Xu et al. 2021). Additionally, the researchers found that uterine scar tension, uterine collapse risk, and uterine wound recovery in rats following cesarean section could all be significantly improved by oligopeptide compounds made from marine fish peptides (MFPs) (CS). It is hypothesized that its boosting impact is connected to the development and restoration of collagen fiber and smooth muscles, as well as the production of functional capillaries in scar tissue (Peng et al. 2020). Additionally, this discovery raises the possibility of using fish collagen-based composites in medicine and cosmetic combinations for skin injury or photoaging restoration. Therefore, marine collagen-based composites might be employed as the best coverings for wound treatment and therapy because of their two-fold capabilities of the healing process and antimicrobial activity.

5.7 Drug Delivery

Substantial obstacles to target-specific distribution and therapeutic efficacy, in vivo destabilization, poor accessibility, dispersion, and poor penetration into living tissue are presented by the use of huge components in drug delivery. Thus, using cutting-edge drug delivery devices to target medications at specific body parts may be an alternative approach to solving these crucial issues. Since the controlled release and dispersion of these pharmaceuticals have been so successful, nanotechnology is crucial in the creation of contemporary drugs and medicines (Patra et al. 2018). According to numerous sorts of research, collagen may be used as a carrier in different medication delivery methods. The lyophilized silver carp dermal collagen/chitosan/chondroitin sulfate composites including bFGF-laden PLGA nanoparticles, which were evenly dispersed inside the marine collagen-based loaded scaffolds; it was possible to create a scaffold-controlled discharge mechanism for skin tissue engineering. This marine collagen-based filled prosthesis can be used in sustained discharge targeted delivery as well as for the healing process and skin tissue construction, as demonstrated by the composites, which also demonstrated good bioactivity, a controllable and optimized protein discharge percentage, and the capacity to stimulate fibroblast cell growth and skin tissue formation (Cao et al. 2015). The scientists describe a modest method for producing collagenous peptide-chelated calcium (CPCC) from marine fish scales as well as a brand-new calcium supplement nanoparticle that is enhanced with CPCC. The CPCC and core–shell CPCC nanoparticles are the ideal options for calcium fortification since studies have shown that they greatly enhance calcium levels and bone mineralization strength in the rat femur (Xu et al. 2021). Fish-derived collagen polypeptides may also be used to distribute microproteins. The collagen taken from the jellyfish *Catostylus tagi* was used as a polymeric framework to create the micro granular protein delivery system. Cross-linking also boosted the stability of CMP in water, according to in vitro experiments, which allowed the release of microgranular proteins gradually. These demonstrate how fish-derived collagen polypeptides may be used to create microparticles that release bioactive molecules progressively (Nicklas et al. 2009).

Gelatin-based biopolymers are of special significance in the engineering of therapeutics systems that allow regulated, prolonged, and/or localized release of functional compounds while boosting accessibility and medicinal benefits. Medical image processing, anti-cancer therapy, tissue repair, and wound healing are just a few of the medical uses for this cutting technology. An intriguing example is the combination of gelatin and poly (-caprolactone) (PCL) to produce an electrospun fibrous scaffold that inhibits scarring and contains a TGF-1 inhibitor (Zhao et al. 2017). This approach effectively prevented scarring in vivo and effectively reduced fibroblast over-proliferation in a rabbit ear wound model. Lastly, gelatin-based microspheres and nanostructures are shown as possibilities as strong, bendable polymers for the delivery of growth factors. These traits are paving the way for novel therapeutic approaches in several clinical specialties, including cancer therapy, neuroprotection following post-ischemic brain damage (Kim et al. 2016), and vaccination. For instance, gelatin nanoparticles were employed by Sabet S. and colleagues as a non-viral hepatitis C vaccine and gene delivery technology to deliver the nonstructural protein 2 genes into bacterial cells.

6 Conclusion and Future Outlook

Bestowed with innumerable beneficial characteristics, such as biocompatibility, low immunogenicity, biodegradability, and abundant availability, fish waste-derived collagen and gelatin are indeed indispensable raw materials for the biomedical sector. With the advancement of technologies, these two biomolecules have found versatile usage in varied forms, especially in the tissue engineering sector for example skin, bone, nerves, cartilage, dental, and corneal tissue regeneration. Considering the drawbacks associated with mammalian-origin collagen and gelatin, and the flexibility and abundance of fish waste-derived collagen and gelatin the latter must soon reduce the dependability on these mammalian-derived biomolecules. In the future, we need to devise protocols and technologies that can achieve faster extraction, reducing the water requirement, i.e., tuning into completely green technology.

Acknowledgements The authors gratefully acknowledged the support of the Vice-Chancellor of Adamas University in Barasat, North 24 Parganas, and West Bengal University of Animal and Fishery Sciences in Kolkata, India. The authors also acknowledge the kind support of the ICAR National Professor project of the Indian Council of Agricultural Research, New Delhi.

References

Ahn H, Gong DJ, Lee HH, Seo JY, Song K-M, Eom SJ, Yeo SY (2021) Mechanical properties of porcine and fish skin-based collagen and conjugated collagen fibers. Polymers 13:2151

Aurrekoetxea M, Garcia-Gallastegui P, Irastorza I, Luzuriaga J, Uribe-Etxebarria V, Unda F, Ibarretxe G (2015) Dental pulp stem cells as a multifaceted tool for bioengineering and the regeneration of craniomaxillofacial tissues. Front Physiol 6:289

Bernhardt A, Paul B, Gelinsky M (2018) Biphasic scaffolds from marine collagens for regeneration of osteochondral defects. Mar Drugs 16:91

Bermueller C, Schwarz S, Elsaesser AF, Sewing J, Baur N, von Bomhard A, Scheithauer M, Notbohm H, Rotter N (2013) Marine collagen scaffolds for nasal cartilage repair: prevention of nasal septal perforations in a new orthotopic rat model using tissue engineering techniques. Tissue Eng Part A 19:2201–2214. https://doi.org/10.1089/ten.TEA.2012.0650

Calejo MT, Almeida AJ, Fernandes AI (2012) Exploring a new jellyfish collagen in the production of microparticles for protein delivery. J Microencapsul 29:520–531. https://doi.org/10.3109/02652048.2012.665089

Cao H, Chen M-M, Liu Y, Liu Y-Y, Huang Y-Q, Wang J-H, Chen J-D, Zhang Q-Q (2015) Fish collagen-based scaffold containing PLGA microspheres for controlled growth factor delivery in skin tissue engineering. Colloids Surf B 136:1098–1106

Capella-Monsonís H, Coentro JQ, Graceffa V, Wu Z, Zeugolis DI (2018) An experimental toolbox for characterization of mammalian collagen type I in biological specimens. Nat Protoc 13:507–529

Chae JJ, Ambrose WM, Espinoza FA, Mulreany DG, Ng S, Takezawa T, Trexler MM, Schein OD, Chuck RS, Elisseeff JH (2015) Regeneration of corneal epithelium utilizing a collagen vitriol membrane in rabbit models for corneal stromal wound and limbal stem cell deficiency. Acta Ophthalmol 93:e57-66. https://doi.org/10.1111/aos.12503

Chalamaiah M, Dinesh Kumar B, Hemalatha R, Jyothirmayi T (2012) Fish protein hydrolysates: proximate composition, amino acid composition, antioxidant activities and applications: a review. Food Chem 135:3020–3038. https://doi.org/10.1016/j.foodchem.2012.06.100

Chang B, Ahuja N, Ma C, Liu X (2017) Injectable scaffolds: preparation and application in dental and craniofacial regeneration. Mater Sci Eng R Rep 111:1–26

Cheung RCF, Ng TB, Wong JH (2015) Marine peptides: bioactivities and applications. Mar Drugs 13:4006–4043. https://doi.org/10.3390/md13074006

Delgado LM, Shologu N, Fuller K, Zeugolis DI (2017) Acetic acid and pepsin result in high yield, high purity and low macrophage response collagen for biomedical applications. Biomed Mater 12:065009

Dimitriou R, Jones E, McGonagle D, Giannoudis PV (2011) Bone regeneration: current concepts and future directions. BMC Med 9:66. https://doi.org/10.1186/1741-7015-9-66

Dinçer MT, Erdem ÖA, Kalkan H, Üçok MÇ (2016) Comparison of recovered carp scales (Cyprinus carpio) gelatin and commercial calf and pork skin gelatins Sazan pulu (Cyprinus carpio) kullanılarak elde edilen jelatin ve ticari dana ve domuz derisi jelatinlerinin kıyaslanması. Ege J Fish Aquat Sci 33:335–341

Diogo GS, López-Senra EL, Pirraco RP, Canadas RF, Fernandes EM, Serra J, Pérez-Martín RI, Sotelo CG, Marques AP, González P (2018) Marine collagen/apatite composite scaffolds envisaging hard tissue applications. Mar Drugs 16:269

Echave MC, Hernáez-Moya R, Iturriaga L, Pedraz JL, Lakshminarayanan R, Dolatshahi-Pirouz A, Taebnia N, Orive G (2019) Recent advances in gelatin-based therapeutics. Expert Opin Biol Ther 19:773–779. https://doi.org/10.1080/14712598.2019.1610383

Eke G, Mangir N, Hasirci N, MacNeil S, Hasirci V (2017) Development of a UV crosslinked biodegradable hydrogel containing adipose derived stem cells to promote vascularization for skin wounds and tissue engineering. Biomaterials 129:188–198. https://doi.org/10.1016/j.biomaterials.2017.03.021

Elango J, Lee JW, Wang S, Henrotin Y, de Val JEMS, Regenstein JM, Lim SY, Bao B, Wu W (2018) Evaluation of differentiated bone cells proliferation by blue shark skin collagen via biochemical for bone tissue engineering. Mar Drugs 16:E350. https://doi.org/10.3390/md16100350

Elias PM (2007) The skin barrier as an innate immune element. Semin Immunopathol 29:3–14

El-Rashidy AA, Gad A, Abu-Hussein AE-HG, Habib SI, Badr NA, Hashem AA (2015) Chemical and biological evaluation of Egyptian Nile Tilapia (Oreochromis niloticas) fish scale collagen. Int J Biol Macromol 79:618–626. https://doi.org/10.1016/j.ijbiomac.2015.05.019

Evans M, Lewis ED, Zakaria N, Pelipyagina T, Guthrie N (2021) A randomized, triple-blind, placebo-controlled, parallel study to evaluate the efficacy of a freshwater marine collagen on skin wrinkles and elasticity. J Cosmet Dermatol 20:825–834. https://doi.org/10.1111/jocd.13676

Fengxiang Z, Anning W, Zhihua L, Shengwen H, Lijun S (2011) Preparation and characterisation of collagen from freshwater fish scales. Food Nutr Sci 2(8):818–823

Gauza-Włodarczyk M, Kubisz L, Mielcarek S, Włodarczyk D (2017) Comparison of thermal properties of fish collagen and bovine collagen in the temperature range 298–670 K. Mater Sci Eng, C 80:468–471

Govindharaj M, Roopavath UK, Rath SN (2019) Valorization of discarded Marine Eel fish skin for collagen extraction as a 3D printable blue biomaterial for tissue engineering. J Clean Prod 230:412–419

Gómez-guilløen MC, Montero P (2001) Extraction of gelatin from megrim (Lepidorhombus boscii) skins with several organic acids. J Food Sci 66:213–216

Heo DN, Castro NJ, Lee S-J, Noh H, Zhu W, Zhang LG (2017) Enhanced bone tissue regeneration using a 3D printed microstructure incorporated with a hybrid nano hydrogel. Nanoscale 9:5055–5062. https://doi.org/10.1039/C6NR09652B

Hsu H-H, Uemura T, Yamaguchi I, Ikoma T, Tanaka J (2016) Chondrogenic differentiation of human mesenchymal stem cells on fish scale collagen. J Biosci Bioeng 122:219–225. https://doi.org/10.1016/j.jbiosc.2016.01.001

Hu C-H, Yao C-H, Chan T-M, Huang T-L, Sen Y, Huang C-Y, Ho C-Y (2016) Effects of different concentrations of collagenous peptide from fish scales on osteoblast proliferation and osteoclast resorption. Chin J Physiol 59:191–201. https://doi.org/10.4077/CJP.2016.BAE398
Huang C-Y, Kuo J-M, Wu S-J, Tsai H-T (2016) Isolation and characterization of fish scale collagen from tilapia (Oreochromis sp.) by a novel extrusion–hydro-extraction process. Food Chem 190:997–1006
Ideia P, Pinto J, Ferreira R, Figueiredo L, Spínola V, Castilho PC (2020) Fish processing industry residues: a review of valuable products extraction and characterization methods. Waste Biomass Valorization 11:3223–3246
Jafari H, Lista A, Siekapen MM, Ghaffari-Bohlouli P, Nie L, Alimoradi H, Shavandi A (2020) Fish collagen: extraction, characterization, and applications for biomaterials engineering. Polymers 12:2230
Jana S, Datta P, Das H, Jaiswal S, Ghosh PR, Lahiri D, Kundu B, Nandi SK (2022b) Copper and cobalt doped bioactive glass-fish dermal collagen electrospun mat triggers key events of diabetic wound healing in full-thickness skin defect model. J Mech Behav Biomed Mater 134:105414
Jana S, Datta P, Das H, Ghosh PR, Kundu B, Nandi SK (2022a) Engineering vascularizing electrospun dermal grafts by integrating fish collagen and ion-doped bioactive glass. ACS Biomaterials Science & Engineering
Jeong SI, Kim SY, Cho SK, Chong MS, Kim KS, Kim H, Lee SB, Lee YM (2007) Tissue-engineered vascular grafts composed of marine collagen and PLGA fibers using pulsatile perfusion bioreactors. Biomaterials 28:1115–1122. https://doi.org/10.1016/j.biomaterials.2006.10.025
Jin S, Sun F, Zou Q, Huang J, Zuo Y, Li Y, Wang S, Cheng L, Man Y, Yang F (2019) Fish collagen and hydroxyapatite reinforced poly (lactide-co-glycolide) fibrous membrane for guided bone regeneration. Biomacromolecules 20:2058–2067
Jridi M, Nasri R, Lassoued I, Souissi N, Mbarek A, Barkia A, Nasri M (2013) Chemical and biophysical properties of gelatins extracted from alkali-pretreated skin of cuttlefish (Sepia officinalis) using pepsin. Food Res Int 54:1680–1687
Kamble R, Shrangdher ST, Koli JM (2014) Physico-chemical properties of gelatin extracted from catla skin (Catla Catla) (Hamilton, 1822). Indian J Fundam Appl Life Sci 4:328–337
Karayannakidis PD, Zotos A (2016) Fish processing by-products as a potential source of gelatin: a review. J Aquat Food Prod Technol 25:65–92
Karim AA, Bhat R (2009) Fish gelatin: properties, challenges, and prospects as an alternative to mammalian gelatins. Food Hydrocoll 23:563–576
Kayal P, Jana S, Datta P, Das H, Kundu B, Nandi SK (2023) Microfibers of fish waste-derived collagen and ion-doped bioactive glass in stimulating the healing sequences in full-thickness cutaneous burn injury. J Drug Deliv Sci Technol 83:104429. https://doi.org/10.1016/j.jddst.2023.104429
Kim I-D, Sawicki E, Lee H-K, Lee E-H, Park HJ, Han P-L, Kim KK, Choi H, Lee J-K (2016) Robust neuroprotective effects of intranasally delivered iNOS siRNA encapsulated in gelatin nanoparticles in the postischemic brain. Nanomedicine 12:1219–1229. https://doi.org/10.1016/j.nano.2016.01.002
Kozlowska J, Sionkowska A, Skopinska-Wisniewska J, Piechowicz K (2015) Northern pike (Esox lucius) collagen: extraction, characterization and potential application. Int J Biol Macromol 81:220–227. https://doi.org/10.1016/j.ijbiomac.2015.08.002
Kozłowska J, Sionkowska A (2015) Effects of different crosslinking methods on the properties of collagen-calcium phosphate composite materials. Int J Biol Macromol 74:397–403. https://doi.org/10.1016/j.ijbiomac.2014.12.023
Kulkarni P, Maniyar M (2020) Utilization of fish collagen in pharmaceutical and biomedical industries: waste to wealth creation. Life Sci Inf Publ 6:11
Li Y, Yang L, Wu S, Chen J, Lin H (2022) Structural, functional, rheological, and biological properties of the swim bladder collagen extracted from grass carp (Ctenopharyngodon idella). LWT—Food Sci Technol 153:112518

Lim Y-S, Ok Y-J, Hwang S-Y, Kwak J-Y, Yoon S (2019) Marine collagen as a promising biomaterial for biomedical applications. Mar Drugs 17:467

Liu C, Sun J (2014) Potential application of hydrolyzed fish collagen for inducing the multi-directional differentiation of rat bone marrow mesenchymal stem cells. Biomacromolecules 15:436–443. https://doi.org/10.1021/bm401780v

Mredha MTI, Kitamura N, Nonoyama T, Wada S, Goto K, Zhang X, Nakajima T, Kurokawa T, Takagi Y, Yasuda K (2017) Anisotropic tough double network hydrogel from fish collagen and its spontaneous in vivo bonding to bone. Biomaterials 132:85–95

Muralidharan N, Jeya Shakila R, Sukumar D, Jeyasekaran G (2013) Skin, bone and muscle collagen extraction from the trash fish, leather jacket (Odonus niger) and their characterization. J Food Sci Technol 50:1106–1113. https://doi.org/10.1007/s13197-011-0440-y

Muthukumar T, Prabu P, Ghosh K, Sastry TP (2014) Fish scale collagen sponge incorporated with Macrotyloma uniflorum plant extract as a possible wound/burn dressing material. Colloids Surf B Biointerfaces 113:207–212. https://doi.org/10.1016/j.colsurfb.2013.09.019

Muyonga JH, Cole CGB, Duodu KG (2004) Extraction and physico-chemical characterisation of Nile perch (Lates niloticus) skin and bone gelatin. Food Hydrocoll 18:581–592

Nagai T, Suzuki N (2000) Isolation of collagen from fish waste material—skin, bone and fins. Food Chem 68:277–281

Nicklas M, Schatton W, Heinemann S, Hanke T, Kreuter J (2009) Preparation and characterization of marine sponge collagen nanoparticles and employment for the transdermal delivery of 17beta-estradiol-hemihydrate. Drug Dev Ind Pharm 35:1035–1042. https://doi.org/10.1080/036390409 02755213

Nishimoto S, Goto Y, Morishige H, Shiraishi R, Doi M, Akiyama K, Yamauchi S, Sugahara T (2008) Mode of action of the immunostimulatory effect of collagen from jellyfish. Biosci Biotechnol Biochem 72:2806–2814

Niu L, Zhou X, Yuan C, Bai Y, Lai K, Yang F, Huang Y (2013) Characterization of tilapia (Oreochromis niloticus) skin gelatin extracted with alkaline and different acid pretreatments. Food Hydrocolloids 33:336–341

Ohnishi A, Osaki T, Matahira Y, Tsuka T, Imagawa T, Okamoto Y, Minami S (2013) Evaluation of the chondroprotective effects of glucosamine and fish collagen peptide on a rabbit ACLT model using serum biomarkers. J Vet Med Sci 75:421–429. https://doi.org/10.1292/jvms.12-0240

Oryan A, Alemzadeh E, Moshiri A (2017) Burn wound healing: present concepts, treatment strategies and future directions. J Wound Care 26:5–19. https://doi.org/10.12968/jowc.2017.26.1.5

Pal P, Srivas PK, Dadhich P, Das B, Maity PP, Moulik D, Dhara S (2016) Accelerating full thickness wound healing using collagen sponge of mrigal fish (Cirrhinus cirrhosus) scale origin. Int J Biol Macromol 93:1507–1518. https://doi.org/10.1016/j.ijbiomac.2016.04.032

Pati F, Adhikari B, Dhara S (2010) Isolation and characterization of fish scale collagen of higher thermal stability. Biores Technol 101:3737–3742. https://doi.org/10.1016/j.biortech.2009.12.133

Pati F, Datta P, Adhikari B, Dhara S, Ghosh K, Das Mohapatra PK (2012) Collagen scaffolds derived from fresh water fish origin and their biocompatibility. J Biomed Mater Res A 100:1068–1079. https://doi.org/10.1002/jbm.a.33280

Patra JK, Das G, Fraceto LF, Campos EVR, Rodriguez-Torres MP, Acosta-Torres LS, Diaz-Torres LA, Grillo R, Swamy MK, Sharma S, Habtemariam S, Shin H-S (2018) Nano based drug delivery systems: recent developments and future prospects. J Nanobiotech 16:71. https://doi.org/10.1186/s12951-018-0392-8

Peng X, Xu J, Tian Y, Liu W, Peng B (2020) Marine fish peptides (collagen peptides) compound intake promotes wound healing in rats after cesarean section. Food Nutr Res. https://doi.org/10.29219/fnr.v64.4247

Raabe O, Reich C, Wenisch S, Hild A, Burg-Roderfeld M, Siebert HC, Arnhold S (2010) Hydrolyzed fish collagen induced chondrogenic diVerentiation of equine adipose tissue-derived stromal cells. Histochem Cell Biol 134:545–554

Ran X-G, Wang L-Y (2014) Use of ultrasonic and pepsin treatment in tandem for collagen extraction from meat industry by-products. J Sci Food Agric 94:585–590

Rigby BJ (1968) Amino-acid composition and thermal stability of the skin collagen of the Antarctic ice-fish. Nature 219:166–167

Sanaei AV, Mahmoodani F, See SF, Yusop SM, Babji AS (2013) Optimization of gelatin extraction and physico-chemical properties of catfish (Clarias gariepinus) bone gelatin. Int Food Res J 20:423

See SF, Hong PK, Ng KL, Wan Aida WM, Babji AS (2010) Physicochemical properties of gelatins extracted from skins of different freshwater fish species. Int Food Res J 17:809–816

Serbo JV, Gerecht S (2013) Vascular tissue engineering: biodegradable scaffold platforms to promote angiogenesis. Stem Cell Res Ther 4:8. https://doi.org/10.1186/scrt156

Sghayyar HN, Lim SS, Ahmed I, Lai JY, Cheong XY, Chong ZW, Lim AFX, Loh H-S (2020) Fish biowaste gelatin coated phosphate-glass fibres for wound-healing application. Eur Polymer J 122:109386

Shakila RJ, Jeevithan E, Varatharajakumar A, Jeyasekaran G, Sukumar D (2012) Functional characterization of gelatin extracted from bones of red snapper and grouper in comparison with mammalian gelatin. LWT-Food Sci Technol 48:30–36

Sionkowska A, Kozłowska J (2013) Properties and modification of porous 3-D collagen/hydroxyapatite composites. Int J Biol Macromol 52:250–259. https://doi.org/10.1016/j.ijbiomac.2012.10.002

Sionkowska A, Skrzyński S, Śmiechowski K, Kołodziejczak A (2017) The review of versatile application of collagen. Polym Adv Technol 28:4–9. https://doi.org/10.1002/pat.3842

Skierka E, Sadowska M, Karwowska A (2007) Optimization of condition for demineralization Baltic cod (Gadus morhua) backbone. Food Chem 105:215–218

Song E, Kim SY, Chun T, Byun H-J, Lee YM (2006) Collagen scaffolds derived from a marine source and their biocompatibility. Biomaterials 27:2951–2961

Sugahara T, Ueno M, Goto Y, Shiraishi R, Doi M, Akiyama K, Yamauchi S (2006) Immunostimulation effect of jellyfish collagen. Biosci Biotechnol Biochem 70:2131–2137. https://doi.org/10.1271/bbb.60076

Tang J, Saito T (2015) Biocompatibility of novel type I collagen purified from tilapia fish scale: an in vitro comparative study. Biomed Res Int 2015:e139476. https://doi.org/10.1155/2015/139476

van Essen TH, Lin CC, Hussain AK, Maas S, Lai HJ, Linnartz H, van den Berg TJTP, Salvatori DCF, Luyten GPM, Jager MJ (2013) A fish scale-derived collagen matrix as artificial cornea in rats: properties and potential. Invest Ophthalmol Vis Sci 54:3224–3233. https://doi.org/10.1167/iovs.13-11799

Venugopal V (2021) Valorization of seafood processing discards: bioconversion and bio-refinery approaches. Front Sustain Food Syst 5

Wang J-H, Hung C-H, Young T-H (2006) Proliferation and differentiation of neural stem cells on lysine–alanine sequential polymer substrates. Biomaterials 27:3441–3450. https://doi.org/10.1016/j.biomaterials.2006.02.002

Xu N, Peng X-L, Li H-R, Liu J-X, Cheng J-S-Y, Qi X-Y, Ye S-J, Gong H-L, Zhao X-H, Yu J, Xu G, Wei D-X (2021) Marine-derived collagen as biomaterials for human health. Front Nutr 8:702108. https://doi.org/10.3389/fnut.2021.702108

Yamada S, Nagaoka H, Terajima M, Tsuda N, Hayashi Y, Yamauchi M (2013) Effects of fish collagen peptides on collagen post-translational modifications and mineralization in an osteoblastic cell culture system. Dent Mater J 32:88–95. https://doi.org/10.4012/dmj.2012-220

Yamada S, Yamamoto K, Ikeda T, Yanagiguchi K, Hayashi Y (2014) Potency of fish collagen as a scaffold for regenerative medicine. BioMed Res Int 2014

Yamaguchi K, Lavéty J, Love RM (1976) The connective tissues of fish VIII. Comparative studies on hake, cod and catfish collagens. Int J Food Sci Technol 11:389–399. https://doi.org/10.1111/j.1365-2621.1976.tb00737.x

Yu C-C, Chang J-J, Lee Y-H, Lin Y-C, Wu M-H, Yang M-C, Chien C-T (2013) Electrospun scaffolds composing of alginate, chitosan, collagen and hydroxyapatite for applying in bone tissue engineering. Mater Lett 93:133–136. https://doi.org/10.1016/j.matlet.2012.11.040

Zeugolis DI, Paul RG, Attenburrow G (2008) Factors influencing the properties of reconstituted collagen fibers prior to self-assembly: animal species and collagen extraction method. J Biomed Mater Res Part A: Off J Soc Biomater Jpn Soc Biomater Aust Soc Biomater Korean Soc Biomater 86:892–904

Zhao X, Song W, Chen Y, Liu S, Ren L (2018) Collagen-based materials combined with microRNA for repairing cornea wounds and inhibiting scar formation. Biomater Sci 7:51–62. https://doi.org/10.1039/C8BM01054D

Zhou T, Liu X, Sui B, Liu C, Mo X, Sun J (2017a) Development of fish collagen/bioactive glass/chitosan composite nanofibers as a GTR/GBR membrane for inducing periodontal tissue regeneration. Biomed Mater 12:055004. https://doi.org/10.1088/1748-605X/aa7b55

Zhou T, Sui B, Mo X, Sun J (2017b) Multifunctional and biomimetic fish collagen/bioactive glass nanofibers: fabrication, antibacterial activity and inducing skin regeneration in vitro and in vivo. Int J Nanomedicine 12:3495–3507. https://doi.org/10.2147/IJN.S132459

Production of Antidiabetic Peptides from Fish Waste

Fernando Rivero-Pino, F. Javier Espejo-Carpio, Pedro J. García-Moreno, Raúl Pérez-Gálvez, Antonio Guadix, and Emilia M. Guadix

Abstract The underutilized fishery resources, such as non-commercial species and by-products generated on board and in land (e.g. heads, trimmings, skin and others), are an opportunity to obtain food products for human consumption, in line with the Sustainable Development Goals. Fish contains high-quality protein, with a content on a wet basis normally ranging from 15 to 25%, an adequate amino acid profile and good digestibility. Enzymatic hydrolysis of proteins releases peptides that may exert several physiological effects by interacting with human metabolism. Fish waste materials have been reported as a compelling source of antidiabetic peptides, which could be employed as ingredients to fortify foodstuffs intended for nutraceutical applications. This chapter provides a review of the production and characterization of antidiabetic peptides from fish waste protein, the mechanisms explaining their antidiabetic effect and their potential incorporation in food matrices as functional ingredients.

Keywords Fish by-products · Fish discards · Enzymatic hydrolysis · Amylase · Diabetes · Dipeptidyl peptidase IV · DPP-IV · Food matrix · Glucose · Glucosidase · Inhibitory peptides · Insulin

---

F. Rivero-Pino · F. J. Espejo-Carpio (✉) · P. J. García-Moreno · R. Pérez-Gálvez · A. Guadix · E. M. Guadix
Department of Chemical Engineering, University of Granada, 18071 Granada, Spain
e-mail: fjespejo@ugr.es

F. Rivero-Pino
e-mail: frivero@ugr.es

P. J. García-Moreno
e-mail: pjgarcia@ugr.es

R. Pérez-Gálvez
e-mail: rperezga@ugr.es

A. Guadix
e-mail: aguadix@ugr.es

E. M. Guadix
e-mail: eguadix@ugr.es

S. Maqsood et al. (eds.), *Fish Waste to Valuable Products*, Sustainable Materials and Technology, https://doi.org/10.1007/978-981-99-8593-7_7

1 Introduction

1.1 Fish Waste

The estimated population increase for the next decades would require a complete restructuring of the food system, including the search for new sources of macronutrients that supply the food demand in a sustainable way, without compromising the health of human beings, and ideally, preventing the development of diseases. A sustainable management of all the available natural resources should become a priority in modern societies (Thirukumaran et al. 2022), including the valorization of fish waste in agreement with the circular economy model. Aquaculture and capture fisheries account for production of almost 180 million tonnes in 2018, showing an increasing compounded annual growth rate of 2–3.5% (FAO 2020). Most of these aquatic resources are dedicated to food (87.2%) but only 50–60% of these materials are finally consumed by humans (Wu et al. 2021). Therefore, a considerable amount of by-products is generated, which currently end up as low-value products or waste. The latter implies the need to develop upgrading processes for the production of high-value compounds (e.g. bioactive compounds) (Abdollahi et al. 2021). Nevertheless, the potential use of fish waste as a source for the production of bioactive compounds requires a safety assessment due to different risks associated with fish consumption. This evaluation should be based on a case-by-case analysis, searching for specific analytes prone to accumulate in fish flesh such as heavy metals or polychlorinated biphenyl (Castro-González and Méndez-Armenta 2008; Judd et al. 2004).

In line with the principles of circular economy, fish waste is regarded as a sustainable source to obtain high-value products. The use of fish waste, including bones, head, skin, scales, fins, tails, gut, and viscera, as raw material to produce bioactive compounds has been proposed as a strategy to provide added value to fish waste materials. Similarly, non-targeted fish species, which are commonly discarded or underutilized due to their low commercial value, can be an adequate source of bioactive compounds (Despoti et al. 2021; Jawad 2021; Więcaszek et al. 2015). Nevertheless, the use of non-commercial species is constrained by the lack of knowledge on their compositional, safety and distribution, as well as the environmental impact that its exploitation would entail. In general, fishes are a source of macronutrients, where proteins are of special interest. Several factors influence the protein content of a species such as sex, age, sexual maturity, nutritive conditions and others. In addition, the variation of protein content can be rhythmic, periodic or non-periodic (Lozano and Hardisson 2003; Maschmeyer et al. 2020). Overall, the protein content does not present wide seasonal variations, as is the case of lipids. In previous studies covering both non-targeted and commercial undersized fish species from the Mediterranean Sea, the proximate composition of the species was monitored within the period 2011–2013, reporting a protein content from 15.5 to 23.1 wt% (Table 1) (García-Moreno et al. 2013; Morales-Medina et al. 2015). Protein content presented minor variations throughout the seasons even for non-fatty and semi-fatty fish species (e.g. axillary seabream, small-spotted catshark and bogue), where lipids may not be sufficient to provide the required energy. It is worth noting that small-spotted catshark showed the highest protein values (>20%), which is attributed to the higher presence of non-protein nitrogen compounds in elasmobranchs such as trimethylamine oxide (TMAO), ammonia and urea (García-Moreno et al. 2013; Morales-Medina et al. 2015; Morales-Medina et al. 2016a).

The use of fish waste for human consumption is aligned with the Sustainable Development Goals (SDGs). Among these SGDs listed by the United Nations, the number 14 is "Life Below Water". The aim of this goal is the conservation and responsible use of marine resources, which can be split into different objectives such as control of marine pollution, sustainable management of marine ecosystems or regulations on fishing exploitation. The oceans are crucial for maintaining life on Earth. Therefore, marine protected areas must be managed effectively, establishing conservation measures on at least 10% of coastal and marine areas.

In this framework, the valorization of fish by-products and waste materials is needed to achieve this goal (Nirmal et al. 2022a). In fisheries and aquaculture, it is estimated that 35% of the catch is lost or wasted each year worldwide and 12% of total fishery production is used for non-food purposes; in spite of the continuous efforts and the legislative measures taken in different areas. The amount of by-products generated by fishing and fish processing industry presents large variations among species. According to the estimation of by-products generated by commercial trade, fish trade and fish transformation activities in France (Fig. 1) represent between 33 and 64% of the total mass of fish catches (Perez-Galvez and Berge 2008).

This situation poses an economic and environmental problem that has to be tackled from different approaches. Reducing fish loss and waste can lead to reduced pressure on fish stocks and help improve resource sustainability as well as food security. As an example, in Europe, "Regulation (EU) 2020/852 on the establishment of a framework to facilitate sustainable investment" establishes the criteria for determining whether an economic activity qualifies as environmentally sustainable. Marine resources can address other social, economic and environmental challenges currently affecting our food system, by (i) providing high-quality proteins and fatty acids sources, in order to combat poverty (SDG1); (ii) provide safe and healthy foodstuff, ensuring food security and improved health (SDG2 and SGD3); (iii) to address climate change by introducing into the food chain products initially considered worthless (SDG13), among others.

Table 1 Average protein content of Mediterranean discarded species (years 2011–2013)

Species	Protein, wt% (wet basis)			
	Autumn	Winter	Spring	Summer
Axillary seabream (*Pagellus acarne*)	17.2	18.3	18.3	18.7
Small-spotted catshark (*Scyliorhinus canicula*)	19.2	20.9	21.1	22.8
Sardine (*Sardina pilchardus*)	16.5	17.0	16.9	19.2
Horse mackerel (*Trachurus mediterraneus*)	17.1	17.3	17.5	18.8
Bogue (*Boops boops*)	16.4	17.8	17.2	18.0

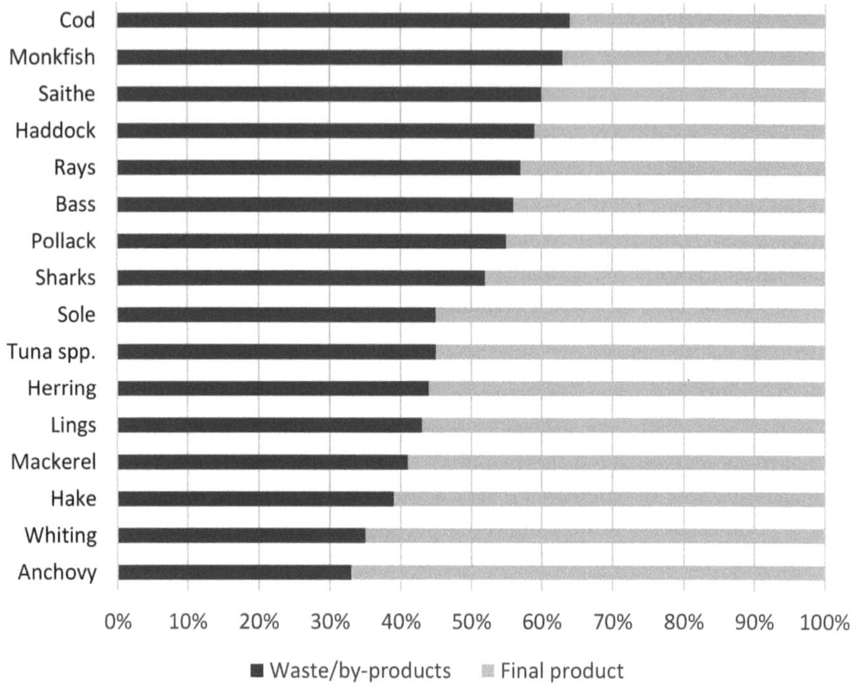

Fig. 1 Estimation of by-products generated by trade, canning and smoking industries in France, based on a survey from the French Institute for Marine Research and Exploitation (IFREMER) on 2004–2005

On the other hand, the twenty-first century has brought about a change in the lifestyle of human beings, with new consumption habits due to the development of products thanks to the technological revolution that has occurred in the food industry. This has led to a large part of the population consuming high levels of substances such as sugar or salts which may cause health disorders. This situation, added to the sedentary habits established in modern western society, has increased the prevalence of diseases related to the cardiovascular system or metabolic syndrome, with forecasts of even greater growth in the near future. To avoid this, current research in nutrition and food technology tries to develop products that meet the needs of consumers, at the same time, they produce a health beneficial effect. It has been shown that the consumption of certain foods can help prevent or pre-treat certain diseases, and this strategy could imply a reduction of economic cost, compared to those associated with the treatment of the disease, as well as the welfare of humans (Hewage et al. 2020; Li et al. 2010; Mata-Cases et al. 2020).

Consequently, the use of fish waste as a source of high-value products should be regarded as a strategy with a double purpose: (i) to make use of these currently wasted resources and (ii) to contribute to the development of a healthier society by turning this waste into products suitable for human consumption that can exert beneficial physiological effects on the human.

1.2 Prevention and Pre-treatment of Diabetes

Diabetes mellitus type II is a metabolic disease causing elevated blood glucose levels in humans (impaired glucose homeostasis). It is characterized by insulin resistance, which is an insufficient or inefficient production of this hormone—responsible for the glucose uptake in the muscle, or the lack of activity of it. There are several risk factors for prediabetes and type II diabetes, including sedentary behaviours or food consumption patterns. The consequences of this hyperglycaemia can lead to several health problems of different degrees of severity (e.g. from renal or kidney failure to nerve damage and cardiovascular disorders) ("IDF Diabetes Atlas" 2017; Rivero-Pino et al. 2020a).

There are numerous strategies currently available to overcome this metabolic disorder. Insulin injection is the most commonly known, but it cannot be orally administered. Medicaments with different targets are also available to treat the disease, such as dipeptidyl peptidase IV (DPP-IV) inhibitors, sodium-glucose cotransporter 2 (SGLT2) inhibitors, glucosidase inhibitors, amylases inhibitors, insulin sensitizers, insulin secretagogues, glucagon-like peptide (GLP-1) mimetics or glizofins. Nonetheless, these synthetic drugs may pose a risk for patients due to the side effects that can occur upon its consumption (Hira et al. 2021; Patil et al. 2015; Tahrani et al. 2011).

These adverse side effects have promoted the search for antidiabetic compounds of natural origin, such as bioactive peptides. Peptides and amino acids from food proteins play an essential role in the regulation of glucose homeostasis. These molecules can exert several activities, such as GLP-1 regulation, the inhibition of digestion-related enzymes (as described for the medicaments above), enhancement of cholecystokinin levels, a gut hormone regulating food intake, or as signalling molecules in enteroendocrine cells, among others. Thus, protein intake has been correlated with body fat loss, insulin secretion and glycaemia reduction. However, the mechanisms behind these physiological changes are still under research (Patil et al. 2015; Rivero-Pino et al. 2020a; Tahrani et al. 2011).

The digestion of carbohydrates (i.e. mainly consisting of polysaccharide molecules) releases glucose molecules, which increase their concentration in the bloodstream. A potential approach could be limiting the hydrolysis of such polymers by enzyme inhibitors i.e. amylase and glucosidase, that can occur at different levels of digestion, effectively avoiding postprandial hyperglycaemia (Ibrahim et al. 2017; Patil et al. 2015). In addition, delayed carbohydrate absorption can help stimulate GLP-1 secretion, leading to the incretin effect. Examples of compounds inhibiting these enzymes are acarbose, miglitol and voglibose, correlated with numerous side effects, such as diarrhoea or abdominal pain between others (Inzucchi et al. 2012). In this regard, some food-derived peptides have shown inhibitory activity against digestive enzymes, being an adequate strategy for diabetes prevention as well as for complementing the traditional treatments without negative effects.

Other treatments to tackle diabetes are based on the modulation of incretins. Food intake causes the release of hormones called incretins (GLP-1 and GIP), in concentrations ranging based on physiological needs and regulated by the enzyme DPP-IV, being this latter enzyme responsible for their degradation (e.g. DPP-IV inactivates more than 95% of GLP-1). Thus, it was discovered that the inhibition of DPP-IV would have an impact on the concentrations of incretins, increasing the half-life of these, and consequently, their incretin effect (Curran et al. 2019). Examples of these inhibitors, commonly known as gliptins, are sitagliptin or metformin, which have been showed to exert some adverse effects upon its consumption such as headache, dermatitis and the risk of acute pancreatitis, and long-term effects are not well characterized. Similarly, for the hydrolase inhibitors, the ability of food-derived peptides to interact with this enzyme, and exert the same inhibitory effect avoiding the negative effects is currently seen as an adequate strategy to prevent the development of diabetes.

Currently, in literature, the information regarding antidiabetic peptides from food-derived sources is related to the ability of these sequences to inhibit amylase, glucosidase and/or DPP-IV. At a physiological level, the most relevant would be the inhibition of DPP-IV, considering the mechanism of action of this enzyme.

1.3 Protein, Protein Hydrolysates and Peptides

Proteins are one of the most important components of the human diet, consisting of chains of amino acids linked (Nelson and Cox 2008). The intake of these macromolecules leads to the release of peptides during digestion that can have beneficial effects due to their similarity to the structure of human regulatory peptides, since they can interact with some enzymes and receptors involved in human metabolism. Furthermore, protein hydrolysis provides the amino acids necessary for the synthesis of endogenous proteins in ribosomes. Protein content can vary depending on fish species but, as mentioned before, within the same species the level of protein is considerably stable throughout the year. Fish waste can encompass diverse parts, including muscles, showing high digestibility and well-balanced amino acid composition (Table 2) and according to their solubility index, they can be classified as myofibrillar, sarcoplasmic, and alkali-soluble proteins. Fish skins, which account for 30% w/w of by-products, are a source of collagen, from which gelatin can be obtained by thermal denaturation or partial hydrolysis (Nirmal et al. 2022b).

Fish protein can be processed to manufacture protein hydrolysates, which are the resulting pool of peptides obtained by cleaving the native protein, by means of different techniques. The most common techniques to obtain peptides from protein are enzymatic hydrolysis, fermentation or acid/alkali hydrolysis (Kristinsson and Rasco 2000; Marti-Quijal et al. 2020; Nongonierma and FitzGerald 2017).

Table 2 Amino acid content of sardine (*Sardina pilchardus*), horse mackerel (*Trachurus mediterraneus*), small-spotted catshark (*Scyliorhinus canicula*) and blue-whiting (*Micromesistius poutassou*)

Amino acids composition, wt%	Sardine[a]	Horse mackerel[b]	Small-spotted catshark[a]	Blue-whiting[c]
Lys	8.5	8.8	6.3	7.4
Ala	10.6	4.4	11.6	5.4
Arg	2.6	5.5	3.0	5.9
Cys	0.5	n.m.	0.4	0.6
Met	2.4	2.2	2.3	1.9
Leu	7.3	6.2	7.4	5.8
Tyr	2.9	2.8	2.3	3.2
Thr	5.4	3.3	5.4	3.7
Pro	3.3	2.3	4.1	3.4
Phe	4.6	3.1	4.1	2.9
Ser	3.9	2.5	4.5	4.1
Asp	11.2	29.4	10.7	8.6
Val	3.6	4.3	2.7	3.9
Hyp	1.3	n.m.	3.8	0.8
Glu	20.9	15.7	20.1	12.5
His	3.4	2.3	1.6	1.5
Ile	3.4	3.9	3.1	3.4
Gly	4.3	3.5	6.7	5.5

n.m. Not measured
[a] García-Moreno et al. (2016)
[b] Morales-Medina et al. (2016b)
[b] Harnedy-Rothwell et al. (2021a, b)

Enzymatic hydrolysis is the most studied technique in literature to obtain protein hydrolysates and peptides, showing technological advantages compared to the other techniques. Indeed, according to a simple search in the scientific database Scopus, in the last 10 years, the number of studies dealing with protein hydrolysates was around 3500 studies of which 68% would be related to enzymatic hydrolysis.

The enzymatic hydrolysis reaction works under moderate conditions of temperature and pH, leading to a controllable reaction, and thus, a specific product, due to the action of proteases. It is widely reported that peptides exerting bioactivity usually contain from two to twenty amino acid residues. Different hydrolysates have been produced at a pilot or semi-pilot plant scale from fish discards (Harnedy-Rothwell et al. 2020; Remme et al. 2022; Vázquez et al. 2020) for their valorization, where the authors also carried out a sustainability assessment (Vázquez et al. 2020), and concluded that fish protein hydrolysates technology minimizes impact when compared with operating fish meal plants. The enzymes able to cleave proteins to produce the hydrolysates are called proteases (EC 3.4.X.X), classified either as endo-

or exopeptidases. Their active site determines its substrate specificity and therefore the length and composition of the peptides released (Tavano et al. 2018). Further information on the proteases employed to produce specific types of antidiabetic peptides can be found elsewhere (Rivero-Pino et al. 2020a).

The enzymatic hydrolysis of protein can be conducted employing the fish waste directly or stabilized by different pre-treatments. Commonly, a high-temperature thermal pretreatment is applied in order to deactivate the endogenous enzymes contained in the material, to avoid the uncontrolled hydrolysis of the material, obtaining a potentially uncharacterized substrate (Derouiche Ben Maiz et al. 2019). In addition, defatting the raw material, in order to obtain a high-concentrated protein product, allows for optimizing the release of peptides as well as avoiding the potential lipid oxidation of the fat content contained in the fish.

It has been also described the application of non-thermal pre-treatments to modify the structure of the protein, aiming to improve the release of bioactive peptides (Rivero-Pino et al. 2020e; Thirukumaran et al. 2022). One such example is the application of microwave before the hydrolysis carried out by Ketnawa et al. (2018a, b) on rainbow trout (*Oncorhynchus mykiss*) by-products, aiming to obtain DPP-IV inhibitory peptides. Similarly, ultrasound (Petcharat et al. 2021) or high pressure (Hemker et al. 2020; Rivero-Pino et al. 2020b) treatments have been employed on fish waste, assessing the effect over biological properties (e.g. antioxidant) or technologies properties (water holding capacity, oil binding capacity, emulsification, among others). This behaviour is caused by the modification of the tertiary and quaternary protein structures, denaturation, aggregation or precipitation.

The use of fish products as raw material to manufacture protein hydrolysates encompass several challenges, such as: (i) the reproducibility when manufacturing the final product on an industrial scale; (ii) the manufacturing of a product odourless, flavourless and visually appealing to the consumer (Harnedy-Rothwell et al. 2021a, b); (iii) microbiological contamination and lipid peroxidation; (iv) the feed of the fishes can affect the content of amino acids to some extent and (v) the effect of endogenous enzymes in releasing peptides before the enzymatic treatment has to be taken into consideration.

Furthermore, the safety of the manufactured product has to be evaluated, always considering hygiene measures, quality control, labelling requirements and relevant legislation according to the region where the product wants to be marketed.

2 Bioactivity of Peptides

Several peptide sequences have been identified in fish protein hydrolysates exerting inhibitory activity linked to an antidiabetic effect. The hierarchy of assays currently used to assess peptide activity begins with in vitro assays. These analyses should be completed with further studies on in vivo models, able to evaluate the interaction of bioactive compounds with other food components and their bioavailability after ingestion.

The inclusion of bioactive peptides within nutraceutical preparations should be accompanied by scientific statements on their potential health benefits. Health claims refer to how the intake of a food component implies a reduction of disease risk. A comprehensive and specific regulation applies to health-related claims in the scope of food, including their labelling, advertising, or any kind of communication activity. Each country/region has its own regulatory framework (Domínguez Díaz et al. 2020). Nowadays, there is no standard method to state a health claim for foodstuff but overall, the requirements in scientific terms are similar (Chalamaiah et al. 2019).

In the United States, food companies must make a request to the Food and Drug Administration, and the health claim depends on a *Significant Scientific Agreement* or on an authoritative statement from an appropriate scientific body of the US Government or the National Academy of Sciences or any of its subdivisions. In Europe, food business operators must submit their request to the European Commission, since health claims are regulated by Regulation (EC) No 1924/2006 on nutrition and health claims made on food, and the scientific evaluation is carried out by the European Food Safety. In Japan, the Minister of Consumer Affairs Agency of the Government of Japan is in charge of categorizing the products as *Food for Specified Health Use* (FOSHU).

2.1 *In Vitro Studies*

As mentioned above, antidiabetic peptides from food-derived proteins act as inhibitors of some of the proteases involved in carbohydrate metabolism (e.g. α-amylase, α-glucosidase and dipeptidyl peptidase IV). The effect of α-amylase and α-glucosidase inhibitors obtained from proteins has been reported to a lesser extent, compared to DPP-IV inhibition. A report evaluating the amylase inhibitory activity of fish water-soluble protein hydrolysates was conducted by Liu et al. (2013), using Response Surface Methodology to optimize the hydrolysis conditions for maximal release of amylase inhibitory peptides. Medenieks and Vasiljevic (2008) evaluated three species of underutilized fish (i.e. silver warehou, barracouta and Australian salmon) subjected to different enzymatic treatments as sources for bioactive peptides. Overall, no α-glucosidase inhibition and low amylase inhibition were observed. Many studies employed collagen as a substrate for obtaining inhibitory peptides. Kumar et al. (2019) carried out the hydrolysis of unicorn Leatherjacket (*Aluterus Monoceros*) collagen, aiming to unravel the potential of the peptides released in inhibiting α-amylase, reporting average half-maximal inhibitory concentration (IC_{50}) values ranging from 1.17 to 2.65 mg mL^{-1}, depending on the reaction temperature. Natsir et al. (2019) reported α-glucosidase inhibition of 24.47% as the highest activity exerted by the different samples analysed, from a collagen hydrolysate from yellowfin tuna (*Thunnus albacares*) bone hydrolysed using a bacterial collagenase. Similarly, Nur et al. (2021) evaluated the α-glucosidase inhibitory properties of a collagen hydrolysate from Lamuru (*Caranx ignobilis*) fishbone hydrolysed by collagenase enzyme from *Clostridium histolyticum*, reporting an IC_{50} value of 0.574 mg mL^{-1}. To the author's knowledge, there are no studies to date identifying α-amylase inhibiting peptides obtained from marine waste. However, Nguyen et al. (2022) were able to produce and identify hemi-pyocyanin, a potent α-amylase inhibitor, by microbial fermentation of fish-derived chitinous discards. Regarding α-glucosidase inhibitors, Matsui et al. (1999) identified two peptides with glucosidase inhibitory activity, VW (IC_{50} = 22.6 mM) and YYPL (IC_{50} = 3.7 mM) from sardine muscle hydrolysed with alkaline protease from *Bacillus licheniformis*.

DPP-IV inhibitory activity of peptides has been widely studied and identified in fish waste (Table 3) as well as other by-products from animal origin (Henriques et al. 2021; Rivero-Pino et al. 2020d). Most of the DPP-IV inhibitory peptides reported in the literature were produced by commercial proteases such as subtilisin, trypsin, papain or Flavourzyme. A common feature of DPP-IV inhibiting peptides is their short chain length. In this regard, Ketnawa et al. (2018a, b) hydrolysed rainbow trout (*Oncorhynchus mykiss*) frames and found that the most potent inhibitors were peptides with molecular sizes between 300 and 500 Da. Gelatin from skin and bones are one of the main fish waste substrate employed for the production of DPP-IV inhibitors. Jin et al. (2020) used several proteases to hydrolyse salmon skin, identifying the potent DPP-IV inhibitor, LDKVFR, in the fraction of molecular weight below 3 kDa of a trypsin hydrolysate. Similarly, Neves et al. (2017) hydrolysed salmon gelatin using different enzymes (Alcalase 2.4L, Flavourzyme 500L, Corolase PP, Promod 144MG and Brewer's Clarex). The authors found that Corolase PP and the combination of Alcalase and Flavourzyme yielded hydrolysates with high DPP-IV inhibitory activity (IC_{50} = 0.8 mg mL^{-1}). Both hydrolysates presented a significant proportion of peptides with molecular weight below 1000 Da. Moreover, this fraction allowed the identification of a number of active sequences in the Corolase PP hydrolysate (Table 3). The use of Flavourzyme as the sole enzyme is uncommon, since it is an enzyme mixture with predominant exopeptidase activity, which leads to a low extent of hydrolysis. Nevertheless, several studies confirm its capacity for releasing DPP-IV inhibitors from protein substrates. Atma et al. (2019) found that 4-h hydrolysis of bone gelatin using Flavourzyme can produce DPP-IV inhibitory hydrolysates, reporting maximal activity for the peptide fraction above 3 kDa. In this line, Sila et al. (2016) and Gui et al. (2022) obtained different sequences from fish skin, where the shorter peptides (GPAE and GPGA) showed very potent DPP-IV inhibitory capacity. Other studies identified sequences of inhibitory peptides in the hydrolysates of fish discards and fish cooking juice (Huang et al. 2012; Rivero-Pino et al. 2020d).

2.2 *Cell-Based Studies*

In situ cell-based and/or ex vivo studies are useful to reduce the gap between the in vitro assays and the in vivo studies, since they provide a higher quality analysis concerning the potential effect these compounds may have. Thus, cell culture assays

Table 3 Studies including identified DPP-IV inhibitory peptides obtained from fish waste

Source	Protease	Peptide	IC_{50}	Refs.
Carp (*Cyprinus carpio*) roe	Papain	IPNVAVD	0.777 mM	Zhang et al. (2020)
Atlantic salmon (*Salmo salar*) skin	Trypsin	LDKVFR	0.129 mM	Jin et al. (2020)
Salmon (*Salmo salar*) gelatin	Corolase PP	GGPAGPAV GPVA PP GF R (Free amino acid) Y (Free amino acid)	–	Neves et al. (2017)
Barbel (*Barbus callensis*) skin gelatine	Alcalase	WSG FSD	–	Sila et al. (2016)
Sardine (*Sardine pilchardus*) discards	Alcalase, trypsin, flavourzyme	NAPNPR YACSVR	–	Rivero-Pino et al. (2020d)
Sturgeon (*Acipenser schrencki*) skin	Flavourzyme	GPAGERGEGGPR SPGPDGKTGPR	2.14 mM 2.61 mM	Gui et al. (2022)
Tuna (*Thunnus tonggol*) cooking juice	Protease XXIII and orientase	PGVGGPLGPIGPCYE CAYQWQRPVDRIR PACGGFYISGRPG	0.116 mM 0.078 mM 0.096 mM	Huang et al. (2012)
Atlantic salmon (*Salmo salar*) skin gelatin	Flavourzyme	GPAE 49.6 μM GPGA 41.9 μM	49.6 μM 41.9 μM	Li-Chan et al. (2012)

are a strong tool to evaluate the antidiabetic effects of protein hydrolysates by evaluating different parameters such as: (i) insulin-secreting cells (pancreatic β cells) following glucose exposition; (ii) release of GLP-1 (intestinal hormone dependant on the DPP-IV enzyme); (iii) glucose absorption by fluorimetry; (iv) expression and secretion of gut hormones following a physiological stimuli. Beyond that, studies in animals provide scientific rationale on the mechanisms that these peptides can have, because it allows to evaluate the stability during gastrointestinal digestion, absorption in the lumen and their reaching to the target organ and the actual effect exerted.

In the last years, some authors have reported the efficacy of salmon and boarfish by-products as a source of antidiabetic peptides, by means of cell culture analysis and animal studies (Harnedy et al. 2018; Parthasarathy et al. 2018). In these reports, authors employed protein hydrolysates obtained with food-grade proteases (e.g. Alcalase and Flavourzyme) to obtain a product that can be resistant to gastrointestinal digestion and that exerts an effect on the models employed to evaluate the antidiabetic activity.

Zhang et al. (2020) analysed the potential of peptides derived from common carp (*Cyprinus carpio*) roe protein, employing four different proteases. The papain-hydrolysed with sample effectively inhibited DPP-IV, maintaining the activity after digestion. Analyses were carried out in cell models, and the identified peptide (IPNVAVD) had no cytotoxicity and reduced the amount of DPP-IV that Caco-2 cells released, as well as increased the absorption of glucose by insulin-resistant HepG2 cells.

Harnedy-Rothwell et al. (2020) identified that the boarfish-derived peptide reported in vitro as DPP-IV inhibitor with an IC_{50} value of 0.022 mM, showed an IC_{50} value of 0.004 mM in an in situ DPP-IV inhibition assay employing Caco-2 cells, highlighting the challenges in comparing and extrapolating the results among different types of analyses. In addition, this peptide was also responsible for a potent insulin secretory activity reported in pancreatic BRIN-BD11 cells. Moreover, recently, Heffernan et al. (2022) also demonstrated that blue whiting (*Micromesistius poutassou*) hydrolysates obtained by enzymatic hydrolysis were able to promote GLP-1 secretion and proglucagon production in murine enteroendocrine STC-1 cells. Theysgeur et al. (2020) recently identified peptides obtained from Tilapia (*Oreochromis niloticus*) able to exert inhibition over the enzyme DPP-IV, as well as able to regulate intestinal hormones (stimulation of CCK and GLP-1 secretion) on the enteroendocrine cells in an in vitro canine gastrointestinal simulated digestion model.

In terms of ex-vivo analysis, Hjorth et al. (2022) compared how the consumption of whey or salmon protein would affect cultured HepG2 cells by evaluating the transcriptomic profiling by using RNA sequencing, by exposing them to the serum of healthy male subjects, having intake 5.2 g of these samples. Authors concluded no difference in the nutrigenomic effects of the fishmeal and whey, though considering the whole scenario of food sustainability, fish by-products account for a more environmentally friendly source that could be used as a replacer.

On top of that, different reports employing animals can be also found concerning fish-derived peptides exerting antidiabetic effects. For instance, Siala et al. (2016) reported an inhibition of α-amylase exerted by grey triggerfish (*Balistes capriscus*) protein hydrolysates, obtained with different proteases, in alloxan-induced diabetic rats (AIDR), along with a decrease of blood glucose and glycated haemoglobin levels in the rats. In addition, other lipid-related parameters were affected by the intake of these peptides. Ben Slama-Ben Salem et al. (2018) evaluated the effect of peptides obtained from octopus (*Octopus vulgaris*) muscle proteins on AIDRs, showing increased α-amylase activity (in plasma, pancreas and intestine) and blood glucose, as well as a notable reduction in the levels of plasma insulin, among other modified parameters. The daily administration of the testing material for one month improved the glucose tolerance test, the glycaemic status of the animals and corrected the lipid profiles, including an attenuation of the increased activities of the plasma enzymes produced by diabetes.

Recently, Takahashi et al. (2021a) proved that a salmon milt peptide (*Oncorhynchus keta*), considered as an unused fish-processing by-product, is able to inhibit DPP-IV and has an antidiabetic effect on Sprague Dawley rats. Authors were able to identify 12 peptides and to determine that the one highly contributing to the inhibitory action was the sequence Ile-Pro, accounting for 1.3% of the total activity.

2.3 Human Studies

Regarding human studies, there has been recently an increase in the number of reports in order to clearly demonstrate the potential employment of these protein hydrolysates as a nutritional, health-promoting ingredient for human consumption. As evidence to really define whether a peptide (or protein hydrolysate) can exert a real effect in a population, well-defined human studies with a significant population size and with physiological changes clearly proved are needed. Nonetheless, these clinical trials are usually very expensive, and sometimes, conclusions are hard to draw. In this sense, randomized and controlled clinical intervention trials have been determined as the most reliable, and the number needed is variable. The reports must be of high quality in terms of methodology and reporting.

Dale et al. (2018) employed a sustainable marine resource (Atlantic cod (*Gadus morhua*)) as testing material in a double-blind cross-over trial to evaluate the glucose metabolism after meals in healthy subjects ($n = 41$, mean age 51). The study consisted of two study days including a wash-out in between, where the dose ingested was 20 mg kg^{-1} bw. The main parameter analysed was post-prandial response in glucose metabolism and GLP-1 concentrations, although no differences were observed between the testing item and the control. On the other hand, the insulin concentration after meals was significantly lower when the fish hydrolysate was ingested compared with the control. However, the limitations of the study make it impossible to draw any conclusions on the real effect of these protein hydrolysates as antidiabetic agents. Further studies in order to unravel the mechanistic behind are required to understand the physiological changes observed.

Crowe et al. (2018) reported the assessment of how peptides obtained from wasted boarfish (*Capros aper*) can affect postprandial glycaemic management in human participants. In a randomized controlled intervention crossover study, healthy subjects ($n = 20$) ingested 3.5 g of the sample, leading to an area under the curve (AUC) for insulin or glucose not different from the control, for 3 h. The design of this study is limited by the criteria assessed as well as the short time during which the metabolic-related parameters were evaluated.

Jensen et al. (2019) evaluated what effects the supplementation with cod protein peptides in older adults has on postprandial glucose metabolism. The experimental design consisted of a double-blind cross-over trial, where subjects ($n = 31$) received, during 1 week, a dose of the sample. Doses were different among groups, from 10 to 40 mg kg^{-1} bw. The measured postprandial response in glucose metabolism and in plasma, GLP-1 led to no significant differences among different dose groups; whereas, serum glucose and insulin levels decreased as the dose was increased. Later on, these authors also reported the effects of cod protein hydrolysate supplementation (dose of 4 g) during 8 weeks on glucose metabolism, lipid profile and body composition of Adults with Metabolic Syndrome ($n = 30$). The experimental design was a double-blind, randomized intervention study with a parallel-group design. Among the different parameters measured, no significant differences were found in postprandial insulin, post-prandial glucose of GLP-1, thus, not implying any physiological difference for the parameters assessed (Jensen et al. 2020).

Hovland et al. (2020) evaluated the efficacy of fish peptides (from fresh rest residual materials of herring (*Clupea harengus*), salmon (*Salmo salar*) or cod (*Gadus morhua*), a dose of 2.5 g per day, during 8 weeks) in a randomized, double-blind study ($n = 77$ overweight adults). Parameters analysed including assessing how it would affect glucose regulation and insulin sensitivity. The intake of the different samples affected diverse parameters in different extents, and authors concluded that effects were most pronounced following the supplementation with cod and herring samples, compared to salmon.

Vildmyren et al. (2020) reported how the ingestion of cod residual protein would have an impact on markers of glucose regulation in lean adults ($n = 50$, mean age 28) by a randomized double-blind study. The dose (8.1 g per day of protein) was ingested during 8 weeks. The intake of the protein did not have an effect on fasting glucose or insulin concentrations, but affected the levels of ketone bodies negatively in the subjects as well as an increase of trimethylamine N-oxide concentration was observed in plasma and urine. These are parameters related to impaired glucose metabolism, thus, further studies are needed to unravel the interactions occurring in the metabolism of the ingested protein.

Hustad et al. (2021) in a randomized controlled trial during 8 weeks with adults prone to suffer from T2DM, evaluating the effect of salmon protein supplement obtained from by-products (5.2 g of protein per day) on 2-h glucose, leading to no significant modifications compared to the control. However, it should be noted that in this study, the native protein was employed as a test item, and not a protein hydrolysate, even though peptides are expected to be in the material ingested.

Takahashi et al. (2021a, b) evaluated the effect of the administration of DPP-IV inhibitory peptides during 7 days (dose of 500 mg per day) from chum salmon milt (unused processing byproduct) on postprandial blood glucose level. The experimental design was a randomized, placebo-controlled, double-blind, crossover, pilot clinical trial with healthy Japanese subjects ($n = 15$). The main parameters evaluated were reduced blood glucose and insulin levels. According to the authors, no clear suppressive effect of the peptides on elevated postprandial blood glucose levels was observed. However, authors indicate the limitations from the study and the need of continuing investigating the effects, since positive results were observed in the animal studies.

Overall, the studies aiming to demonstrate the effect of fish-derived peptides on potential antidiabetic effects are lacking quality experimental design. A more in-depth analysis of the toxicokinetics parameters, as well as glucose-metabolism-related parameters are required in order to have an overview of the physiological changes occurring. In addition, further studies with human subjects in a state of pre-diabetes (exhibiting a result of 100–125 mg dL^{-1} on the Fasting Plasma Glucose Test;

or a level of 5.7–6.4% on the A1c test) are required in order to establish (always on a case-by-case basis) the functionality of these food-derived peptides in real patients. The impaired metabolism of patients is not necessarily representative of the status of healthy patients. Furthermore, differences among populations due to metabotyping have to be considered, because intervariability can occur.

3 Stability and Functionality in Food Matrices

Concerning the structure of the peptides, its maintenance after being included in food matrices and during processing and storage is essential to state them as bioactive ingredients (Rivero-Pino 2023). Firstly, the food matrix composition where fish-derived products are intended to be used has to be deeply characterized, considering that chemical reactions could lead to bioactivity changes (Capuano et al. 2017). Food matrix might also have an impact on the stability through and after the processing, while at the same time, it can be responsible for masking the bitterness that characterize the peptides (Aguilera 2018). During processing, heat treatments are commonly employed with several objectives, including the elimination of microbial activity or the modification of food structure, for instance, by producing Maillard compounds. These treatments may denature or provoke aggregation of proteins and peptides, depending on different parameters (López-Sánchez et al. 2016). Therefore, it would be necessary to investigate how thermal treatment (taking into account the time of the process) can imply modifications in terms of safety and health-promoting properties of fish-derived products. On the other hand, non-thermal treatments currently employed in the food industry are ultrasound, high hydrostatic pressure, pulsed electric field, ultraviolet application, etc. These treatments are also usually employed to change the structure of the components, eliminate microorganisms or deactivate enzymes, or promote the liberation of bioactive peptides. Overall, these processes strengthen food safety while not implying loss of the nutritional and sensorial characteristics of food, which is undoubtedly an adequate alternative with several applications and opportunities in the valorization of fish waste in the upcoming years.

Apart from processing, the chemical interactions with matrix compounds should be also evaluated during storage time, considering the temperature and the time (Rivero-Pino 2023). There are different studies investigating, for example, how heat treatment and long-term storage of a fish soup fortified with peptides can modify the structure, correlated with antioxidant and antihypertensive peptides (Rivero-Pino et al. 2020c). However, further research with diverse food matrices has to be done, including the investigation of how thermal and non-thermal treatments would have an impact on the bioactivity of peptides.

There are not many studies about food fortification with peptides in the literature. For example, Ayati et al. (2022) sought to create a yoghurt enhanced with three fish collagen-derived bioactive peptides, including GPLGAAGP, GRDGEP and MTGTQGEAGR at various doses (up to 1.0 mg mL^{-1}). The samples' DPP-IV inhibitory activity was assessed together with other factors, and the results showed high bioactive values at the concentrations examined. When compared to the control sample, the yoghurts enriched with bioactive peptides did not exhibit any physico-chemical or sensory modifications. On the other hand, Harnedy-Rothwell et al. (2021a, b) investigated two different treatments (conditions being (i) 90 °C during 1 min; (ii) 121 °C during 42 s) and subsequently, storage of tomato-based products fortified with protein hydrolysates derived from boarfish (*Capros aper*) on the antidiabetic activity. The DPP-IV inhibitory activity of the peptides was maintained after heat treatments, and in addition, no changes were observed during storage.

The effects of fish protein hydrolysate made from *Sind sardine* interacting with pistachio green hull extract were recently assessed by Amini Sarteshnizi et al. (2021). Due to the high concentration of polyphenols in the pistachio green hull extract, several interactions could take place and change the effect. These interactions would depend on the chemical groups exposed and the structure of the polyphenols. The inhibition of the enzymes α-glucosidase, α-amylase and DPP-IV was examined among other bioactivities. While the addition of the pistachio extract had no effect on the bioactivity for the inhibition of DPP-IV, it did result in a reduction in the inhibition of the other two enzymes related to antidiabetic effects.

There is very little information in literature on whether and to which extent heat treatments might modify the properties of α-glucosidase inhibitory peptides derived from fish waste or from other sources. Only Rivero-Pino (2021) reported that a commercial soup supplemented with the insect *Tenebrio molitor* peptides led to a significant loss of bioactivity after a heat treatment of 121 °C for 21 min, but maintained during storage.

As it has been indicated before, the interactions of peptides with different chemical structures can modify their potential bioactivity as well as their bioavailability. Wu et al. (2015) investigated the formation of stable colloidal complex nanoparticles by self-assembling tannic acid with fish scale gelatin hydrolysates. The tannic acid shows amylase inhibitory activity, but after the nanoparticle formation, this bioactivity decreased compared to that of free tannic acid. The author also evaluated how the interaction created would affect the bioavailability of the complex, indicating a capacity to inhibit Cu^{2+} ion-induced barrier impairment and hyperpermeability of Caco-2 intestinal epithelial tight junction by the nanoparticles.

Recently, Wu et al. (2022) aimed to convert inorganic chromium into organic chromium—claimed to have higher bioavailability and to act as antidiabetic agent—by bonding to the amine group of a fish scale collagen (from *Oreochromis niloticus*) extracted by enzymatic hydrolysis. Authors reported that the final structure was obtained and properly characterized, though no formal analysis on the antidiabetic activity was performed.

Overall, there is an evident lack of research concerning the interaction of fish waste-peptides with different food matrices, thermal and non-thermal processing consequences, and employment as ingredients that will need to be addressed in the following years.

4 Bioavailability of Peptides

The bioavailability and safety of fish peptides have to be analysed by means of animal and human studies specifically designed for that purpose, ideally according to harmonized guidelines of the Organisation for Economic Co-operation and Development (OECD). The analysis of toxicokinetics parameters could help define the actual bioavailability of these peptides and how these can modulate specific metabolic pathways, and thus, resulting in a modification of the physiological status of the individual.

The studies available reporting these parameters are scarce, and thus, there is not an exact reporting about how peptides can achieve target organs. Considering the complexity of the hydrolysates, if used as such as an ingredient, it is a challenge to establish the fate of each peptide, and whether the gastrointestinal digestion cleaves the peptides and leads to a modification of the bioactivity. Furthermore, as indicated previously, the food matrix can impact the bioavailability of the peptides. For instance, there are studies reporting how fish waste-derived peptides with proven antioxidant and antihypertensive activity are incorporated into fish-based liquid matrices, and the bioactivity was maintained after simulated gastrointestinal digestion (Rivero-Pino et al. 2020c).

Regarding antidiabetic peptides, Karimi et al. (2021) investigated whether maize herm peptides would increase the bioactive potential of bread before and after digestion, as well as the inhibition of the enzyme α-amylase during digestion. However, there is a lack of studies reporting fish-derived peptides with antidiabetic peptides used as functional ingredients. Harnedy-Rothwell et al. (2021a, b) fortified tomato-based products with peptides derived from boarfish (*Capros aper*), reporting maintained or enhanced bioactivity after simulated gastrointestinal digestion.

Given the low likelihood of chemical interactions occurring and the fact that it would prevent the bitter taste of peptides, boosting the sensory acceptance, high-fibre food matrices appear to be an appropriate vehicle to transport bioactive peptides (Sun et al. 2020). To encourage their employment in the food business for human consumption and stop the emergence of certain diseases, further study is required.

Zhang et al. (2018) demonstrated in vitro that IADHFL, a DPP-IV inhibitory peptide identified from bighead carp (*Hypophthalmichthys nobilis*) remained stable after simulated gastrointestinal digestion and inhibited DPP-IV enzyme in vitro, in cell cultures analysis. However, it should be noted that these analyses done with a purified peptide under in vitro conditions are not representative of the physiological conditions of the humans and of the products intended to be marketed. Purification of peptides in a sufficient amount extracted from fish-by products is still a technological challenge that needs to be addressed.

Ayati et al. (2022) as reported previously, manufactured a functional yoghurt with fish collagen-derived bioactive peptides. These samples were also subjected to in vitro gastrointestinal digestion, showing no significant difference in the bioactivity of two of the three samples analysed, indicating that these two peptides are not hydrolysed by digestive proteases.

In fact, the food matrix in which the antidiabetic food-derived peptides are included would have an impact on bioavailability, based on the pH value modifications. These parameters will influence the net charge, structure, and ability of them to interact with different targets, such as enzymes active sites (Rivero-Pino 2023), thus modulating their capacity of inhibiting them (Marcolini et al. 2015). During digestion, the amino acid content and sequence would affect the resistance to gastrointestinal digestion, with special attention to the terminal sites (Karaś 2019). In addition, it has been suggested that low steric hindrance values and high amphipathicity are correlated with an improved bioactivity exerted by the peptide, since it would stabilize the enzyme conformation (Mora et al. 2020).

One strategy that is currently being studied to overcome the main limitation of peptides, which is a low bioavailability as they are prone to be cleaved during digestion, is their encapsulation (Aguilar-Toala et al. 2022). The encapsulation of bioactive compounds is a way to ensure they reach the target organ and can exert their function. Although encapsulation of antidiabetic peptides has not been extensively studied, Cian et al. (2019) reported that encapsulation by spray drying prevented the reduction in DPP-IV inhibitory activity during the digestion of Lima bean (*Phaseolus lunatus*) protein hydrolysate.

Furthermore, Li et al. (2015) encapsulated antidiabetic Atlantic salmon peptides (molecular weight <1000 Da) in chitosan-coated liposomes obtained from milk fat globule membrane derived phospholipids. The physical stability of the nanocapsules during freeze-drying, freeze-thawing, and long-term storage was analysed, resulting in a prolonged release of the peptides in simulated biological fluids because of the chitosan coating.

Other authors have also reported the encapsulation of fish-derived peptides, without evaluating the antidiabetic potential of these, but other biological properties and bioavailability (Hanachi et al. 2022; Lima et al. 2021; Subara et al. 2018).

5 Fish Protein Hydrolysates Market

The valorization potential of fish waste is high. However, the number of patents focusing on the production of antidiabetic peptides from fish is limited. There are no patents describing the production of α-amylase or α-glucosidase inhibitory peptides from fish proteins. Only a few patents deal with the production of DPP-IV inhibitory peptides from fish species such as hairtail, silver carp, salmon, herring, cod, bonito (Jin and Y 2013; Shangwu and Ying 2014; Takahashi et al. 2014). The process usually consists of obtaining a uniform fish flesh pulp prior to carrying out enzymatic hydrolysis by proteases (Alcalase, trypsin, bromelain, Flavourzyme, papain among others). Afterwards, the hydrolysate produced is centrifuged and filtered through ultrafiltration membranes for purifying the most active fraction.

The patents regarding the use of fish waste as a source for peptides with DPP-IV inhibitory activity are even more limited. All the current patents employ fish skin as substrate. Particularly, the use of Atlantic salmon (*Salmo salar*) skin (Kuo-Chiang et al. 2013) implies a preliminary process to extract gelatin. Then, gelatin is hydrolysed by Alcalase, bromelain and Flavourzyme. Hydrolysates should be ultrafiltrated to recover the <1 kDa fraction, where the most potent DPP-IV inhibitors are found. Some of the active peptides identified in that fraction include four amino acid residues having an amino acid sequence of Gly-Pro-X_3-X_4, in which X_3 is alanine or glycine, and X_4 is glutamate or alanine. A similar process is proposed for the production of DPP-IV inhibitors from the skin of silver carp (*Hypophthalmichthys molitrix*) and tilapia (*Oreochromis niloticus*) (Yongkang 2017). In this process, after cleaning the skin, a protein slurry is generated using hot water and ultrasound treatment. After centrifugation, skin proteins (obtained in supernatant) are subjected to two sequential hydrolysis steps. Initially, skin proteins are hydrolysed by alkaline protease and trypsin during 1–2 h and then bromelain and Flavourzyme are added and kept hydrolysing for 1–2 additional hours. After hydrolysis, enzymes are deactivated thermally and the product is ultrafiltered by ceramic membranes. The filtration is carried out in two steps, first filtration through 5 kDa and then by 3 kDa membranes. The resulting permeate is purified by gel filtration (Sephadex G-25 gel). The resulting fractions are concentrated by freeze-drying. The invention also provides the application of the fish skin protein peptide in medicines, health-care foods and food additives. A slightly different process is described in CN109182435A (Yunping 2018), where croaker (*Micropogonias undulatus*) skin is cleaned, freeze-dried and crushed to a partial size is 25–50 μm. Then, the fish skin powder is hydrolysed with pepsin at pH 3–5 to extract collagen. Collagen is hydrolysed with photoresponse gelator and papain under ultraviolet lamp to obtain the active peptides. In all cases, fish waste hydrolysates, purified fractions or isolated peptides with DPP-IV inhibitory activity could be potentially added to medicines, health-care foods and food additives.

According to Global Market Insights, the Fish Protein Hydrolysates Market size is estimated to grow at over 4.5% compound annual growth rate between 2020 and 2026. The report indicates that the industry of enzymatically produced hydrolysate will achieve a value of USD 475 million by 2026, highlighting sources such as, tilapia, tuna, sardine, Atlantic salmon and cod fish, the same sources as those mostly under research as stated in the previous sections (Global Market Insight 2020).

Considering antidiabetic products, one example of commercially available preparation based on fish-derived protein hydrolysates is Nutripeptin™, a fine powder obtained from cod, claimed to reduce blood sugar levels after meals. Similarly, a white fish autolysate (*Molva Molva*) branded as Fortidium Liquamen®, is claimed to reduce the glycemic index, among other potential bioactivities. Similarly, Hydro MN Peptide is a marine cartilage extract with a mix of hydrolysed proteins (mainly Collagen) and polysaccharides, claimed by the company to reduce the glycaemic index and subsequently, alleviating symptoms of type II diabetes. In addition, Wellnex®, food-grade type-D collagen is a hydrolysate obtained from Tilapia (*Oreochromis* sp.) scale, claimed to exert antidiabetic effect. (Abachi et al. 2022; Nirmal et al. 2022a).

However, as indicated before, the health claims stated for these products vary depending on the country or region intended to be marketed on, and depend on the managers evaluating the data supporting these evidences. Therefore, and based on the scientific requirements mandatory in each region, a product can be considered to exert an effect or not, depending on the market.

6 Conclusions and Perspectives

A redefinition of our current food system is urgently needed, following the global situation affected by climate change and the increase of disease rate in humans. The close relationship that exists between diet and health makes human consumption habits participate in the prevention of the development of diseases. On the other hand, the full use of the resources available in nature is a key element to achieve a circular economy model. In this framework, fish-derived peptides obtained from waste (including non-commercial species and by-products from the industry) have proved not only to be safe, but also nutritionally advantageous upon its consumption, potentially exerting a beneficial effect in different biomarkers. The amount of high-quality protein, as well as the peptide sequences embedded in the native protein, that can be released by enzymatic hydrolysis processing, makes this raw material an adequate source of antidiabetic peptides. There are still some technological challenges that need to be addressed in the manufacturing of these products, from the obtaining of the peptides (unravelling the effect of novel technologies prior or during the release of peptides) to the formulation of the final product (stability, maintained bioactivity, consumer's acceptance). Furthermore, well-designed human studies, where the physiological effect is demonstrated, are essential to be carried out for each product intended to be marketed with potential health claims, in order not to mislead the consumer.

In addition, standardization of regulatory requirements would help companies aiming to sell fish protein hydrolysates products with health claims, to expand more rapidly, and to achieve a larger audience.

Acknowledgements This research was funded by the project B-AGR-400-UGR20 (Producción, concentración y estabilización de biopéptidos moduladores del índice glucémico (BioPepGI)) from University of Granada

References

Abachi S, Pilon G, Marette A, Bazinet L, Beaulieu L (2022) Beneficial effects of fish and fish peptides on main metabolic syndrome associated risk factors: diabetes, obesity and lipemia. Crit Rev Food Sci Nutr 1–49. https://doi.org/10.1080/10408398.2022.2052261

Abdollahi M, Wu H, Undeland I (2021) Impact of processing technology on macro- and micronutrient profile of protein-enriched products from fish backbones. Foods 10(5). https://doi.org/10.3390/foods10050950

Aguilar-Toala JE, Quintanar-Guerrero D, Liceaga AM, Zambrano-Zaragoza ML (2022) Encapsulation of bioactive peptides: a strategy to improve the stability, protect the nutraceutical bioactivity and support their food applications. RSC Adv 12(11):6449–6458. https://doi.org/10.1039/d1ra08590e

Aguilera JM (2018) The food matrix: implications in processing, nutrition and health. Crit Rev Food Sci Nutr. https://doi.org/10.1080/10408398.2018.1502743

AminiSarteshnizi R, Sahari MA, Ahmadi Gavlighi H, Regenstein JM, Nikoo M, Udenigwe CC (2021) Influence of fish protein hydrolysate-pistachio green hull extract interactions on antioxidant activity and inhibition of α-glucosidase, α-amylase, and DPP-IV enzymes. LWT 142:111019. https://doi.org/10.1016/j.lwt.2021.111019

Atma Y, Lioe HN, Prangdimurti E, Seftiono H, Taufik M, Mustopa AZ (2019) Dipeptidyl peptidase IV (DPP-IV) inhibitory activity of ultrafiltration and gel filtration fraction of gelatin hydrolyaste derived from bone of fish for antidiabetes. Int J Adv Sci Eng Inf Technol 9(6):2096–2103. https://doi.org/10.18517/ijaseit.9.6.10658

Ayati S, Eun JB, Atoub N, Mirzapour-Kouhdasht A (2022) Functional yogurt fortified with fish collagen-derived bioactive peptides: antioxidant capacity, ACE and DPP-IV inhibitory. J Food Process Preserv 46(1):1–10. https://doi.org/10.1111/jfpp.16208

Capuano E, Oliviero T, van Boekel MAJ (2017) Modelling food matrix effects on chemical reactivity: challenges and perspectives. Crit Rev Food Sci Nutr 58:2814–2828. https://doi.org/10.1080/10408398.2017.1342595

Castro-González MI, Méndez-Armenta M (2008) Heavy metals: implications associated to fish consumption. Environ Toxicol Pharmacol 26(3):263–271. https://doi.org/10.1016/j.etap.2008.06.001

Chalamaiah M, Keskin Ulug S, Hong H, Wu J (2019) Regulatory requirements of bioactive peptides (protein hydrolysates) from food proteins. J Funct Foods 58:123–129. https://doi.org/10.1016/J.JFF.2019.04.050

Cian R, Campos-Soldini A, Guerrero L, Drago S, Betancur D (2019) Bioactive Phaseolus lunatus peptides release from maltodextrin/gum Arabic microcapsules obtained by spray drying after simulated gastrointestinal digestion. Int J Food Sci Technol 54:2002–2009. https://doi.org/10.1111/ijfs.14031

Crowe W, McLaughlin CM, Allsopp PJ, Slevin MM, Harnedy PA, Cassidy Y, Baird J, Devaney M, Fitzgerald RJ, O'Harte FPM, McSorley EM (2018) The effect of boarfish protein hydrolysate on postprandial glycaemic response and satiety in healthy adults. Proc Nutr Soc 77(OCE3). https://doi.org/10.1017/s002966511800109x

Curran AM, Horner K, O'Sullivan V, Nongonierma AB, Le Maux S, Murphy E, Kelly P, Fitzgerald RJ, Brennan L (2019) Variable glycemic responses to intact and hydrolyzed milk proteins in overweight and obese adults reveal the need for precision nutrition. J Nutr 149(1):88–97. https://doi.org/10.1093/jn/nxy226

Dale HF, Jensen C, Hausken T, Lied E, Hatlebakk JG, Brønstad I, Hoff DAL, Lied GA (2018) Effect of a cod protein hydrolysate on postprandial glucose metabolism in healthy subjects: a double-blind cross-over trial. J Nutr Sci 7:1–9. https://doi.org/10.1017/jns.2018.23

Derouiche Ben Maiz H, Guadix EM, Guadix A, Gargouri M (2019) Valorisation of tuna viscera by endogenous enzymatic treatment. Int J Food Sci Technol 54(4):1100–1108. https://doi.org/10.1111/ijfs.14009

Despoti S, Stergiou KI, Machias A, Vassilopoulou V, Tsagarakis K, Valavanis V, Adamidou A, Giannoulaki M (2021) Assessing the spatial distribution of five non-commercial fish species in the Aegean Sea (Greece, eastern Mediterranean Sea) based on discards data. Reg Stud Mar Sci 44:101736. https://doi.org/10.1016/J.RSMA.2021.101736

Domínguez Díaz L, Fernández-Ruiz V, Cámara M (2020) An international regulatory review of food health-related claims in functional food products labeling. J Funct Foods 68:103896. https://doi.org/10.1016/j.jff.2020.103896

FAO (2020) The state of world fisheries and aquaculture 2020. In brief. Sustainability in action. https://doi.org/10.4060/ca9231en

García-Moreno PJ, Pérez-Gálvez R, Morales-Medina R, Guadix A, Guadix EM (2013) Discarded species in the west Mediterranean sea as sources of omega-3 PUFA. Eur J Lipid Sci Technol 115(9):982–989. https://doi.org/10.1002/ejlt.201300021

García-Moreno PJ, Guadix A, Guadix EM, Jacobsen C (2016) Physical and oxidative stability of fish oil-in-water emulsions stabilized with fish protein hydrolysates. Food Chem 203:124–135. https://doi.org/10.1016/j.foodchem.2016.02.073

Global Market Insight (2020) Fish protein hydrolysate market. https://www.gminsights.com/industry-analysis/fish-protein-hydrolysate-market

Gui M, Gao L, Rao L, Li P, Zhang Y, Han J-W, Li J (2022) Bioactive peptides identified from enzymatic hydrolysates of sturgeon skin. J Sci Food Agric 102:1948–1957. https://doi.org/10.1002/jsfa.11532

Hanachi A, Bianchi A, Kahn CJF, Velot E, Arab-Tehrany E, Cakir-Kiefer C, Linder M (2022) Encapsulation of salmon peptides in marine liposomes: physico-chemical properties, antiradical activities and biocompatibility assays. Mar Drugs 20(4). https://doi.org/10.3390/md20040249

Harnedy PA, Parthsarathy V, McLaughlin CM, O'Keeffe MB, Allsopp PJ, McSorley EM, O'Harte FPM, FitzGerald RJ (2018) Blue whiting (Micromesistius poutassou) muscle protein hydrolysate with in vitro and in vivo antidiabetic properties. J Funct Foods 40:137–145. https://doi.org/10.1016/j.jff.2017.10.045

Harnedy-Rothwell PA, McLaughlin CM, O'Keeffe MB, Le Gouic AV, Allsopp PJ, Mcsorley EM, Sharkey S, Whooley J, McGovern B, O'Harte FPM, Fitzgerald RJ (2020) Identification and characterisation of peptides from a boarfish (Capros aper) protein hydrolysate displaying in vitro dipeptidyl peptidase-IV (DPP-IV) inhibitory and insulinotropic activity. Food Res Int 131:108989. https://doi.org/10.1016/j.foodres.2020.108989

Harnedy-Rothwell PA, Khatib N, Sharkey S, Lafferty RA, Gite S, Whooley J, O'harte FPM, Fitzgerald RJ (2021a) Physicochemical, nutritional and in vitro antidiabetic characterisation of blue whiting (Micromesistius poutassou) protein hydrolysates. Mar Drugs 19(7). https://doi.org/10.3390/md19070383

Harnedy-Rothwell PA, McLaughlin CM, Crowe W, Allsopp PJ, McSorley EM, Devaney M, Whooley J, McGovern B, Parthsarathy V, O'Harte FPM, FitzGerald RJ (2021b) Stability to thermal treatment of dipeptidyl peptidase IV (DPP-IV) inhibitory activity of a boarfish (Capros aper) protein hydrolysate when incorporated into tomato-based products. Int J Food Sci Technol 56:158–165. https://doi.org/10.1111/ijfs.14615

Heffernan S, Nunn L, Gite S, Whooley J, Giblin L, Fitzgerald RJ, Brien NMO (2022) Blue whiting (Micromesistius poutassou) protein hydrolysates co-culture integrity. Mar Drugs 20:112

Hemker AK, Nguyen LT, Karwe M, Salvi D (2020) Effects of pressure-assisted enzymatic hydrolysis on functional and bioactive properties of tilapia (Oreochromis niloticus) by-product protein hydrolysates. LWT 122:109003. https://doi.org/10.1016/j.lwt.2019.109003

Henriques A, Vázquez JA, Valcarcel J, Mendes R, Bandarra NM, Pires C (2021) Characterization of protein hydrolysates from fish discards and by-products from the north-west spain fishing fleet as potential sources of bioactive peptides. Mar Drugs 19(6). https://doi.org/10.3390/md19060338

Hewage SS, Wu S, Neelakantan N, Yoong J (2020) Systematic review of effectiveness and cost-effectiveness of lifestyle interventions to improve clinical diabetes outcome measures in women with a history of GDM. Clin Nutr ESPEN 35:20–29. https://doi.org/10.1016/j.clnesp.2019.10.011

Hira T, Trakooncharoenvit A, Taguchi H, Hara H (2021) Improvement of glucose tolerance by food factors having glucagon-like peptide-1 releasing activity. Int J Mol Sci 22(12). https://doi.org/10.3390/ijms22126623

Hjorth M, Galigniana NM, Ween O, Ulven SM, Holven KB, Dalen KT, Sæther T (2022) Postprandial effects of Salmon fishmeal and whey on metabolic markers in serum and gene expression in liver cells. Nutrients 14(8). https://doi.org/10.3390/nu14081593

Hovland IH, Leikanger IS, Stokkeland O, Waage KH, Mjøs SA, Brokstad KA, McCann A, Ueland PM, Slizyte R, Carvajal A, Mellgren G, Remman T, Høgøy I, Gudbrandsen OA (2020) Effects

of low doses of fish and milk proteins on glucose regulation and markers of insulin sensitivity in overweight adults: a randomised, double blind study. Eur J Nutr 59(3):1013–1029. https://doi.org/10.1007/s00394-019-01963-0
Huang SL, Jao CL, Ho KP, Hsu KC (2012) Dipeptidyl-peptidase IV inhibitory activity of peptides derived from tuna cooking juice hydrolysates. Peptides 35:114–121. https://doi.org/10.1016/j.peptides.2012.03.006
Hustad KS, Ottestad I, Hjorth M, Dalen KT, Sæther T, Sheikh NA, Strømnes M, Ulven SM, Holven KB (2021) No effect of salmon fish protein on 2-h glucose in adults with increased risk of type 2 diabetes: a randomised controlled trial. Br J Nutr 126(9):1304–1313. https://doi.org/10.1017/S0007114521000040
Ibrahim MA, Bester MJ, Neitz AWH, Gaspar ARM (2017) Structural properties of bioactive peptides with α-glucosidase inhibitory activity. Chem Biol Drug Des 91:370–379. https://doi.org/10.1111/cbdd.13105
IDF Diabetes Atlas (2017) Federación Internacional de Diabetes, vol 8. International Diabetes Federation
Inzucchi SE, Bergenstal RM, Buse JB, Diamant M, Ferrannini E, Nauck M, Peters AL, Tsapas A, Wender R, Matthews DR (2012) Management of hyperglycemia in type 2 diabetes: a patient-centered approach. Diabetes Care 35(6):1364–1379. https://doi.org/10.2337/dc12-0413
Jawad LA (2021) The importance of non-commercial and small-sized fish species: a proposal for an additional revenue to Iraq BT—Tigris and Euphrates Rivers: their environment from headwaters to mouth. In: Jawad LA (Ed) Tigris and Euphrates Rivers: their environment from headwaters to mouth. Springer International Publishing, pp 1611–1623. https://doi.org/10.1007/978-3-030-57570-0_84
Jensen C, Dale HF, Hausken T, Hatlebakk JG, Brønstad I, Lied GA, Hoff DAL (2020) Supplementation with low doses of a cod protein hydrolysate on glucose regulation and lipid metabolism in adults with metabolic syndrome: a randomized, double-blind study. Nutrients 12(7):1–15. https://doi.org/10.3390/nu12071991
Jensen C, Dale HF, Hausken T, Lied E, Hatlebakk JG, Brønstad I, Lied GA, Hoff DAL (2019) Supplementation with cod protein hydrolysate in older adults: a dose range cross-over study. J Nutr Sci 4–11. https://doi.org/10.1017/jns.2019.37
Jin R, Teng X, Shang J, Wang D, Liu N (2020) Identification of novel DPP–IV inhibitory peptides from Atlantic salmon (Salmo salar) skin. Food Res Int 133:109161. https://doi.org/10.1016/j.foodres.2020.109161
Jin T, Y W (2013) Method for preparing dipeptidyl peptidase IV (DPP-IV) inhibitory peptide through using hairtail. (Patent No CN103333940B). https://patents.google.com/patent/CN103333940B/en
Judd N, Griffith WC, Faustman EM (2004) Contribution of PCB exposure from fish consumption to total dioxin-like dietary exposure. Regul Toxicol Pharmacol 40(2):125–135. https://doi.org/10.1016/J.YRTPH.2004.06.004
Karaś M (2019) Influence of physiological and chemical factors on the absorption of bioactive peptides. Int J Food Sci Technol 54:1486–1496. https://doi.org/10.1111/ijfs.14054
Karimi A, Ahmadi Gavlighi H, Amini Sarteshnizi R, Udenigwe CC (2021) Effect of maize germ protein hydrolysate addition on digestion, in vitro antioxidant activity and quality characteristics of bread. J Cereal Sci 97:103148. https://doi.org/10.1016/j.jcs.2020.103148
Ketnawa S, Suwal S, Huang J-Y, Liceaga, AM (2018a) Selective separation and characterisation of dual ACE and DPP-IV inhibitory peptides from rainbow trout (Oncorhynchus mykiss) protein hydrolysates. In J Food Sci Technol 1–12. https://doi.org/10.1111/ijfs.13939
Ketnawa S, Wickramathilaka M, Liceaga AM (2018b) Changes on antioxidant activity of microwave-treated protein hydrolysates after simulated gastrointestinal digestion: purification and identification. Food Chem 254:36–46. https://doi.org/10.1016/J.FOODCHEM.2018.01.133
Kristinsson HG, Rasco BA (2000) Fish protein hydrolysates: production, biochemical, and functional properties. Crit Rev Food Sci Nutr 40:43–81. https://doi.org/10.1080/10408690091189266

Kumar LV, Shakila RJ, Jeyasekaran G (2019) In vitro anti-cancer, anti-diabetic, anti-inflammation and wound healing properties of collagen peptides derived from unicorn leatherjacket (Aluterus monoceros) at different hydrolysis. Turk J Fish Aquat Sci 19(7):551–560. https://doi.org/10.4194/1303-2712-v19_7_02

Kuo-Chiang H, Kit-Pan H, Shih-Li H, Chia-Ling J (2013) Peptide for inhibiting dipeptidyl-peptidase iv. https://patents.google.com/patent/US20140193463

Li R, Zhang P, Barker LE, Chowdhury FM, Zhang X (2010) Cost-effectiveness of interventions to prevent and control diabetes mellitus: a systematic review. Diabetes Care 33(8):1872–1894. https://doi.org/10.2337/dc10-0843

Li Z, Paulson AT, Gill TA (2015) Encapsulation of bioactive salmon protein hydrolysates with chitosan-coated liposomes. J Funct Foods 19:733–743. https://doi.org/10.1016/j.jff.2015.09.058

Li-Chan ECY, Hunag SL, Jao CL, Ho KP, Hsu KC (2012) Peptides derived from Atlantic salmon skin gelatin as dipeptidyl-peptidase IV inhibitors. J Agric Food Chem 60(4):973–978. https://doi.org/10.1021/jf204720q

Lima KO, Pinilla CMB, Alemán A, López-Caballero ME, Gómez-Guillén MC, Montero P, Prentice C (2021) Characterization, bioactivity and application of chitosan-based nanoparticles in a food emulsion model. Polymers 13(19). https://doi.org/10.3390/polym13193331

Liu L, Wang Y, Peng C, Wang J (2013) Optimization of the preparation of fish protein anti-obesity hydrolysates using response surface methodology. Int J Mol Sci 14(2):3124–3139. https://doi.org/10.3390/ijms14023124

López-Sánchez J, Ponce-Alquicira E, Pedroza-Islas R, de la Peña-Díaz A, Soriano-Santos J (2016) Effects of heat and pH treatments and in vitro digestion on the biological activity of protein hydrolysates of Amaranthus hypochondriacus L. grain. J Food Sci Technol 53(12):4298–4307. https://doi.org/10.1007/s13197-016-2428-0

Lozano G, Hardisson A (2003) Fish as food. In: Encyclopedia of food sciences and nutrition, 2nd edn, pp 2417–2423. https://doi.org/10.1016/S0140-6736(01)54974-0

Marcolini E, Babini E, Bordoni A, Di Nunzio M, Laghi L, Maczó A, Picone G, Szerdahelyi E, Valli V, Capozzi F (2015) Bioaccessibility of the bioactive peptide carnosine during in vitro digestion of cured beef meat. J Agric Food Chem 63(20):4973–4978. https://doi.org/10.1021/acs.jafc.5b01157

Marti-Quijal FJ, Remize F, Meca G, Ferrer E, Ruiz MJ, Barba FJ (2020) Fermentation in fish and by-products processing: an overview of current research and future prospects. Curr Opin Food Sci 31:9–16. https://doi.org/10.1016/j.cofs.2019.08.001

Maschmeyer T, Luque R, Selva M (2020) Upgrading of marine (fish and crustaceans) biowaste for high added-value molecules and bio(nano)-materials. Chem Soc Rev 49(13):4527–4563. https://doi.org/10.1039/C9CS00653B

Mata-Cases M, Rodríguez-Sánchez B, Mauricio D, Real J, Vlacho B, Franch-Nadal J, Oliva J (2020) The association between poor glycemic control and health care costs in people with diabetes: a population-based study. Diabetes Care 43(4):751–758. https://doi.org/10.2337/dc19-0573

Matsui T, Oki T, Osajima Y (1999) Isolation and identification of peptidic α-glucosidase inhibitors derived from sardine muscle hydrolyzate. Zeitschrift Fur Naturforschung Sect C J Biosci 54(3–4):259–263

Medenieks L, Vasiljevic T (2008) Underutilised fish as sources of bioactive peptides with potential health benefits. Food Aust 60:581–588

Mora L, González-Rogel D, Heres A, Toldrá F (2020) Iberian dry-cured ham as a potential source of α-glucosidase-inhibitory peptides. J Funct Foods 67:103840. https://doi.org/10.1016/j.jff.2020.103840

Morales-Medina R, García-Moreno PJ, Pérez-Gálvez R, Muñío M, Guadix A, Guadix EM (2015) Seasonal variations in the regiodistribution of oil extracted from small-spotted catshark and bogue. Food Funct 6(8):2646–2652. https://doi.org/10.1039/c5fo00448a

Morales-Medina R, García-Moreno PJ, Pérez-Gálvez R, Muñío MM, Guadix A, Guadix EM (2016a) Nutritional indexes, fatty acids profile, and regiodistribution of oil extracted from four discarded

species of the Alboran Sea: seasonal effects. Eur J Lipid Sci Technol 118(9):1409–1415. https://doi.org/10.1002/ejlt.201500486
Morales-Medina R, Tamm F, Guadix A, Guadix EM, Drusch S (2016b) Functional and antioxidant properties of hydrolysates of sardine (S. pilchardus) and horse mackerel (T. mediterraneus) for the microencapsulation of fish oil by spray-drying. Food Chem 194:1208–1216. https://doi.org/10.1016/j.foodchem.2015.08.122
Natsir H, Dali S, Arif AR, Sartika, Leliani (2019) Activity and kinetics of α-glucosidase inhibition by collagen hydrolysate from thunnus albacares bone. J Phys Conf Ser 1341(3). https://doi.org/10.1088/1742-6596/1341/3/032015
Nelson DL, Cox MM (2008) Lehninger: principles of biochemistry. W H Freeman & Co. https://doi.org/10.1017/CBO9781107415324.004
Neves AC, Harnedy PA, O'Keeffe MB, Alashi MA, Aluko RE, FitzGerald RJ (2017) Peptide identification in a salmon gelatin hydrolysate with antihypertensive, dipeptidyl peptidase IV inhibitory and antioxidant activities. Food Res Int 100:112–120. https://doi.org/10.1016/j.foodres.2017.06.065
Nguyen TH, Wang SL, Nguyen AD, Doan MD, Tran TN, Doan CT, Nguyen VB (2022) Novel α-amylase inhibitor hemi-pyocyanin produced by microbial conversion of chitinous discards. Mar Drugs 20(5):1–18. https://doi.org/10.3390/md20050283
Nirmal NP, Santivarangkna C, Benjakul S, Maqsood S (2022a) Fish protein hydrolysates as a health-promoting ingredient—recent update. Nutr Rev 80(5):1013–1026. https://doi.org/10.1093/nutrit/nuab065
Nirmal NP, Santivarangkna C, Rajput MS, Benjakul S, Maqsood S (2022b) Valorization of fish byproducts: sources to end-product applications of bioactive protein hydrolysate. Compr Rev Food Sci Food Saf 21(2):1803–1842. https://doi.org/10.1111/1541-4337.12917
Nongonierma AB, FitzGerald RJ (2017) Strategies for the discovery and identification of food protein-derived biologically active peptides. Trends Food Sci Technol 69:289–305. https://doi.org/10.1016/J.TIFS.2017.03.003
Nur S, Wierson Y, Sami FJ, Megawati Gani SA, Aisyah AN, Yulia, Marwati (2021) Characterization, antioxidant and α-glucosidase inhibitory activity of collagen hydrolysate from lamuru (caranx ignobilis) fishbone. Sains Malays 50(8):2329–2341. https://doi.org/10.17576/jsm-2021-5008-16
Parthsarathy V, McLaughlin CM, Harnedy PA, Allsopp PJ, Crowe W, McSorley EM, FitzGerald RJ, O'Harte FPMM (2018) Boarfish (Capros aper) protein hydrolysate has potent insulinotropic and GLP-1 secretory activity in vitro and acute glucose lowering effects in mice. Int J Food Sci Technol 54(1):271–281. https://doi.org/10.1111/ijfs.13975
Patil P, Mandal S, Tomar SK, Anand S (2015) Food protein-derived bioactive peptides in management of type 2 diabetes. Eur J Nutr 54(6):863–880. https://doi.org/10.1007/s00394-015-0974-2
Perez-Galvez R, Berge JP (2008) General introduction about by-products: worlwide situation and French focus. In Bergé J-P (ed) Added value to fisheries waste, pp 1–22
Petcharat T, Benjakul S, Karnjanapratum S, Nalinanon S (2021) Ultrasound-assisted extraction of collagen from clown featherback (Chitala ornata) skin: yield and molecular characteristics. J Sci Food Agric 101(2):648–658. https://doi.org/10.1002/jsfa.10677
Remme J, Tveit GM, Toldnes B, Slizyte R, Carvajal AK (2022) Production of protein hydrolysates from cod (Gadus morhua) heads: lab and pilot scale studies. J Aquat Food Prod Technol 31(2):114–127. https://doi.org/10.1080/10498850.2021.2021341
Rivero-Pino F (2021) Obtención de biopéptidos reguladores del índice glucémico para alimentación funcional. Universidad de Granada, Granada. http://hdl.handle.net/10481/68559
Rivero-Pino F (2023) Bioactive food-derived peptides for functional nutrition: effect of fortification, processing and storage on peptide stability and bioactivity within food matrices. Food Chem 406:135046. https://doi.org/10.1016/j.foodchem.2022.135046

Rivero-Pino F, Espejo-Carpio FJ, Guadix EM (2020a) Antidiabetic food-derived peptides for functional feeding: production, functionality and in vivo evidences. Foods 9(8):883. https://doi.org/10.3390/foods9080983

Rivero-Pino F, Espejo-Carpio FJ, Guadix EM (2020b) Bioactive fish hydrolysates resistance to food processing. LWT Food Sci Technol 117(108670). https://doi.org/10.1016/j.lwt.2019.108670

Rivero-Pino F, Espejo-Carpio FJ, Guadix EM (2020c) Evaluation of the bioactive potential of foods fortified with fish protein hydrolysates. Food Res Int 137:109572. https://doi.org/10.1016/j.foodres.2020.109572

Rivero-Pino F, Espejo-Carpio FJ, Guadix EM (2020d) Production and identification of dipeptidyl peptidase IV (DPP-IV) inhibitory peptides from discarded Sardine pilchardus protein. Food Chem 328:127096. https://doi.org/10.1016/j.foodchem.2020.127096

Rivero-Pino F, Espejo-Carpio FJ, Pérez-Gálvez R, Guadix A, Guadix EM (2020e) Effect of ultrasound pretreatment and sequential hydrolysis on the production of Tenebrio molitor antidiabetic peptides. Food Bioprod Process 123:217–224. https://doi.org/10.1016/j.fbp.2020.07.003

Salem et al., 2018 Salem RBS-B, Ktari N, Bkhairia I, Nasri R, Mora L, Kallel R, Hamdi S, Jamoussi K, Boudaouara T, El-Feki A, Toldrá F, Nasri M (2018) In vitro and in vivo anti-diabetic and anti-hyperlipidemic effects of protein hydrolysates from Octopus vulgaris in alloxanic rats. Food Res Int 106:952–963. https://doi.org/10.1016/J.FOODRES.2018.01.068

Shangwu C, Ying Z (2014) Silver carp DPP-IV inhibits polypeptide and its application. (Patent No CN104480176B). https://patents.google.com/patent/CN104480176B/en

Siala R, Khabir A, Lassoued I, Abdelhedi O, Elfeki A, Vallaeys T, Nasri M (2016) Functional and antioxidant properties of protein hydrolysates from grey triggerfish muscle and in vivo evaluation of hypoglycemic and hypolipidemic activities. J Appl Environ Microbiol 4(6):105–119. https://doi.org/10.12691/jaem-4-6-1

Sila A, Alvarez OM, Haddar A, Frikha F, Dhulster P, Nedjar-Arroume N, Bougatef A (2016) Purification, identification and structural modelling of DPP-IV inhibiting peptides from barbel protein hydrolysate. J Chromatogr B 1008:260–269. https://doi.org/10.1016/J.JCHROMB.2015.11.054

Subara D, Jaswir I, Alkhatib MFR, Noorbatcha IA (2018) Synthesis of fish gelatin nanoparticles and their application for the drug delivery based on response surface methodology. Adv Nat Sci Nanosci Nanotechnol 9(4). https://doi.org/10.1088/2043-6254/aae988

Sun X, Acquah C, Aluko RE, Udenigwe CC (2020) Considering food matrix and gastrointestinal effects in enhancing bioactive peptide absorption and bioavailability. J Funct Foods 64:103680. https://doi.org/10.1016/j.jff.2019.103680

Tahrani AA, Bailey CJ, Del Prato S, Barnett AH (2011) Management of type 2 diabetes: new and future developments in treatment. The Lancet 378(9786):182–197. https://doi.org/10.1016/S0140-6736(11)60207-9

Takahashi Y, Kamata A, Konishi T (2021a) Dipeptidyl peptidase-IV inhibitory peptides derived from salmon milt and their effects on postprandial blood glucose level. Fish Sci 87(4):619–626. https://doi.org/10.1007/s12562-021-01530-9

Takahashi Y, Kamata A, Nishimura M, Nishihira J (2021b) Effect of one-week administration of dipeptidyl peptidase-IV inhibitory peptides from chum salmon milt on postprandial blood glucose level: a randomised, placebo-controlled, double-blind, crossover, pilot clinical trial. Food Funct 12(18):8544–8551. https://doi.org/10.1039/d1fo00592h

Takahashi Y, Kamata A, Konishi T (2014) Peptide compound inhibiting dipeptidyl peptidase-iv (dppiv), composition containing same, and method for producing same. (Patent No KR101771867B1). https://patents.google.com/patent/KR101771867B1/en

Tavano OL, Berenguer-Murcia A, Secundo F, Fernandez-Lafuente R (2018) Biotechnological applications of proteases in food technology. Compr Rev Food Sci Food Saf 17:412–436. https://doi.org/10.1111/1541-4337.12326

Theysgeur S, Cudennec B, Deracinois B, Perrin C, Guiller I, Lepoudère A, Flahaut C, Ravallec R (2020) New bioactive peptides identified from a tilapia byproduct hydrolysate exerting effects on DPP-IV activity and intestinal hormones regulation after canine gastrointestinal simulated digestion. Molecules (Basel, Switzerland) 26(1). https://doi.org/10.3390/molecules26010136

Thirukumaran R, Anu Priya VK, Krishnamoorthy S, Ramakrishnan P, Moses JA, Anandharamakrishnan C (2022) Resource recovery from fish waste: prospects and the usage of intensified extraction technologies. Chemosphere 299(March):134361. https://doi.org/10.1016/j.chemosphere.2022.134361

Vázquez JA, Fraguas J, Mirón J, Valcárcel J, Pérez-Martín RI, Antelo LT (2020) Valorisation of fish discards assisted by enzymatic hydrolysis and microbial bioconversion: lab and pilot plant studies and preliminary sustainability evaluation. J Clean Prod 246:119027. https://doi.org/10.1016/J.JCLEPRO.2019.119027

Vildmyren I, Halstensen A, McCann A, Midttun Ø, Ueland PM, Oterhals Å, Gudbrandsen OA (2020) Effect of cod residual protein supplementation on markers of glucose regulation in lean adults: a randomized double-blind study. Nutrients 12(5). https://doi.org/10.3390/nu12051445

Więcaszek B, Sobecka E, Keszka S, Stepanowska K, Dudko S, Biernaczyk M, Wrzecionkowski K (2015) Studies on endangered and rare non-commercial fish species recorded in the Pomeranian Bay (southern Baltic Sea) in 2010–2013. Helgol Mar Res 69(4):411–416. https://doi.org/10.1007/s10152-015-0442-7

Wu H, Abdollahi M, Undeland I (2021) Effect of recovery technique, antioxidant addition and compositional features on lipid oxidation in protein enriched products from cod-salmon and herring backbones. Food Chem 360:129973. https://doi.org/10.1016/j.foodchem.2021.129973

Wu CH, Guo HR, Patel AK, Singhania RR, Chen YA, Kuo JM, Dong CD (2022) Production and characterization of lucrative hypoglycemic collagen-peptide-chromium from tilapia scale. Process Biochem 115:10–18. https://doi.org/10.1016/j.procbio.2022.02.004

Wu S-J, Ho Y-C, Jiang S-Z, Mi F-L (2015) Effect of tannic acid–fish scale gelatin hydrolysate hybrid nanoparticles on intestinal barrier function and α-amylase activity. Food Funct 6. https://doi.org/10.1039/C4FO01015A

Yongkang L (2017) Fish skin protein peptide with DPP-IV inhibition function and preparation method and application thereof. (Patent No CN107164445B). https://patents.google.com/patent/CN107164445B/en

Yunping E (2018) A kind of preparation method of biological source dipeptidyl peptidase-iv inhibitor. (Patent No CN109182435A). https://patents.google.com/patent/CN109182435A/en

Zhang C, Liu H, Chen S, Luo Y (2018) Evaluating the effects of IADHFL on inhibiting DPP-IV activity and expression in Caco-2 cells and contributing to the amount of insulin released from INS-1 cells: in vitro. Food Funct 9(4):2240–2250. https://doi.org/10.1039/c7fo01950e

Zhang C, Gao S, Zhang C, Zhang Y, Liu H, Hong H, Luo Y (2020) Evaluating in vitro dipeptidyl peptidase IV inhibition by peptides from common carp (Cyprinus carpio) roe in cell culture models. Eur Food Res Technol 246(1):179–191. https://doi.org/10.1007/s00217-019-03399-6

Enzymes from Fish Processing Waste Materials and Their Commercial Applications

Sappasith Klomklao, Sakonwat Kuepethkaew, Soottawat Benjakul, Yi Zhang, and Benjamin K. Simpson

Abstract Worldwide population growth as well as the accompanying fast development in urbanization and industrialization, have spurred a massive increase in fisheries production driven mainly by new developments in fishing technologies. Current harvesting and processing practices generate a remarkable amount of fish waste globally on a yearly basis. On average, two-thirds of the whole fish catch is thought to be by-products that are abandoned. Fish by-products and processing waste provide significant disposal challenges for the fishing industry. Thus, waste generated during fish processing must now be disposed of and recycled properly in order to sustain industry and save environment. Fish processing wastes, especially digestive organs offer enormous biotechnological potential as enzyme sources. Fish species' biological variety produces a diverse range of enzymes with distinct characteristics. Despite the wide array of enzymes present in fish, the major ones of economic importance include types commonly used in industrial applications such as proteases, transglutaminases, and lipases. Other major enzymes found in fish viscera include chitinases and collagenases. Fish processing waste-derived enzymes may have distinct features that make them more suitable for industrial uses since they live in a broad range of temperature regimes and have other traits that set them apart from their warm-blooded cousins. Hence, to make the most use of marine resources,

S. Klomklao (✉) · S. Kuepethkaew
Department of Food Science and Technology, Faculty of Agro and Bio Industry, Thaksin University, Phatthalung Campus, Pa-Phayom, Phatthalung 93210, Thailand
e-mail: sappasith@tsu.ac.th

S. Benjakul
Department of Food Technology, Faculty of Agro-Industry, Prince of Songkla University, Hat Yai, Songkhla 90112, Thailand
e-mail: soottawat.b@psu.ac.th

Y. Zhang
Department of Food Science, The Pennsylvania State University, University Park, PA 16802, USA
e-mail: yjz5549@psu.edu

B. K. Simpson
Department of Food Science and Agricultural Chemistry, McGill University, 21111 Lakeshore Road, Ste-Anne-de-Bellevue, Quebec H9X 3V9, Canada
e-mail: benjamin.simpson@mcgill.ca

S. Maqsood et al. (eds.), *Fish Waste to Valuable Products*, Sustainable Materials and Technology, https://doi.org/10.1007/978-981-99-8593-7_8

those enzymes may be recovered from wastes produced during the processing of fish and employed as value-added products or processing aids in a variety of sectors.

1 Introduction

Huge amounts of fish waste are generated by the fish manufacturing industries globally on a yearly basis. Around 70% of the entire weight of some aquatic species is made up of waste products from the processing of fish, including the viscera, head, bones as well as frames (Klomklao 2008). Environmental contamination issues caused by the typical disposal of fish processing waste in landfills or the ocean highlight the importance of properly using and turning fish wastes into valuable products. These abundant wastes are rich in enzymes including proteases, lipases, transglutaminases (TGase), chitinase, and other enzymes which exhibit various biochemical characteristics with their homologs from terrestrial species such as optimum temperature and pH (Venugopal 2016). As inexpensive enzymes might encourage novel industrial uses, the recovery of enzymes from fishing leftovers is crucial. Fish waste enzymes have recently been recovered and characterized, and as a result, several intriguing new uses for these enzymes in a variety of industries, including food processing, have emerged. Owing to the biological diversity of marine animals, a large variety of enzymes with distinctive characteristics may be retrieved and successfully used, which may considerably help to reduce the local pollution issue. This chapter's goal is to give a broad overview of the potential of this plentiful but underused marine resource as a source of enzyme separation for industrial operations.

2 Enzymes Recovery, Characteristics, and Applications from Waste Products of Fish Processing

2.1 *Proteases*

All living organisms rely on proteases for growth and survival. Protease-catalyzed peptide bond hydrolysis is a typical phenomenon in nature. Proteases are mostly generated by the digestive glands of marine animals. Digestive proteases from marine creatures, like those found in plants, animals, and microbes, catalyze the breakdown of proteins into simpler molecules like peptides and amino acids (Klomklao 2008; Klomklao et al. 2012). As compared to their homologous counterparts from land animals, plants, and microbes, marine creatures' proteases have several distinct characteristics because of their adaptations to various environmental situations as well as inter- and intraspecies genetic differences (Simpson 2000; Klomklao et al. 2012). Better catalytic efficiency at low temperatures and decreased thermal stability are some of these distinguishing characteristics (Klomklao et al. 2005).

Proteases from the viscera of marine creatures can be categorized using the same standards as proteases from other animals, plants, or microbes based on how closely they resemble well-known proteases like trypsins, chymotrypsins, chymosins, or cathepsins. They are categorized as acid, neutral, or alkaline proteases based on their pH sensitivity. Moreover, they are classified by common and trade names, preferred specificity, and inhibitor sensitivity. All proteases (peptide hydrolases) fall under subclass 3.4 of the EC system for enzyme nomenclature, which is further classified into exopeptidases (3.4.11-19) and endopeptidases (3.4.21-24), also known as proteinases (Klomklao 2008). Exopeptidases cleave amino acid units from the N or C terminus of polypeptide chains, while endopeptidases hydrolyze peptide bonds at specific locations located throughout the polypeptide chain (Fig. 1). As many exopeptidases, particularly aminopeptidases, are membrane- or intracellular-bound, they are more commonly found in nature but less frequently found as commercial products. Viscera proteases, including those from marine species, are further divided into four groups according to the characteristics of their catalytic sites. These groups include metalloproteases, aspartic proteases, cysteine proteases, and serine proteases (Klomklao 2008). Enzymes in various classes are distinguished by multiple criteria, including the composition of functional groups within their catalytic sites, their specificity toward particular substrates, their sensitivity to inhibitors, and their performance and durability in acidic or alkaline environments (Klomklao et al. 2012).

Numerous species of fish and decapods have been investigated for their viscera proteases. Protease is distributed differently according to species and organs (Table 1). Fish digestive organs are known to contain a variety of proteases, including pepsin, gastricsin, trypsin, chymotrypsin, collagenase, elastase, carboxypeptidase, and carboxyl esterase (Simpson 2000). The two main digestive enzymes present in fish digestive organs are pepsin and trypsin. Trypsin is abundant in the pyloric ceca, pancreas, and intestine whereas pepsin is observed in fish stomachs (Kuepethkaew

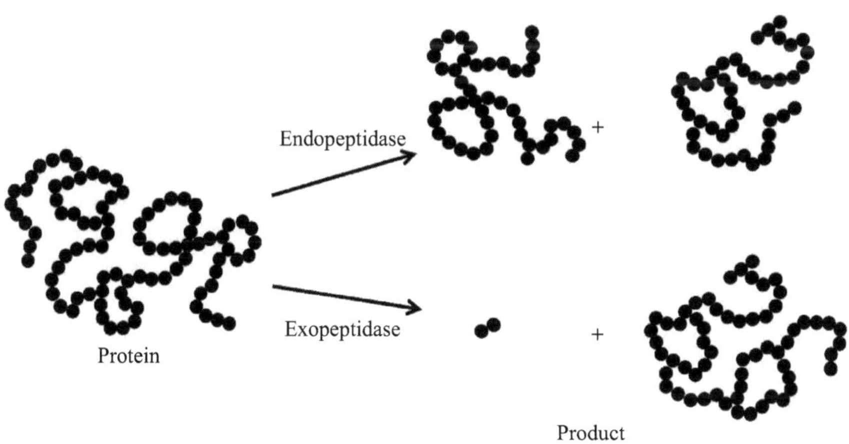

Fig. 1 Action of endopeptidases and exopeptidases on protein structure

et al. 2022; Kishimura et al. 2006). Another digestive enzyme found in the pancreas, intestines, and pyloric cecum of marine creatures is chymotrypsin (Klomklao 2008).

2.1.1 Purification and Characterization of Fish Proteases

– *Pepsin*

The digestive enzymes known as fish pepsins are observed in the gastric fluids of both vertebrate animals such as cod (Zhao et al. 2011) and invertebrate animals such as shrimp (Toomer et al. 2015). They are members of the aspartic endopeptidase family. Compared to their mammalian counterparts, these enzymes appear to have higher specific activity and a less acidic nature (Shahidi and Kamil 2001; Klomklao et al. 2007a). Pepsin is released as a zymogen (pepsinogen; PG), which is then transformed into an active form in the stomach by hydrochloric acid (Klomklao et al. 2007a). Pepsin is activated by the proenzyme being broken down, either in a single step or in a series of stages. Under acidic pH conditions (<5), the removal of prosegments from the active center of enzymes involves the disruption of electrostatic interactions and peptide bonds between the prosegment and the active enzyme, inducing a conformational change in the prosegment and facilitating the formation of the active enzyme (Zhao et al. 2011). The new pepsin thus formed, can proteolytically cleave the prosegments of other pepsinogens, therefore creating more pepsin (Nalinanon et al. 2010a; Zhao et al. 2011). According to Cao et al. (2011), three pepsinogens (PG1, PG2, and PG3) with molecular weights (MW) of 30, 31, and 30 kDa, respectively, were activated in Japanese seabass stomach within 30 min when exposed to acid. Pepsin from bovine has a MW of 35 kDa and is made up of a single 321 amino acid polypeptide chain (Simpson 2000). Pepsins from fish species, nonetheless, have MWs that range from 27 to 42 kDa. Three pepsins (I, II, and III) purified from the stomach of lizardfish were calculated to have MWs of 32, 31 and 30 kDa, respectively (Fig. 2) (Kuepethkaew et al. 2022). By using sodium dodecyl sulfate-polyacrylamide gel electrophoresis, Chen et al. (2009) determined that the molecular weights (MWs) of three pepsins from the stomach of freshwater fish snakehead were, respectively, 32, 33, and 31 kDa. According to SDS-PAGE, the molecular weights of two and four pepsins purified from the stomachs of skipjack tuna and sea bream, respectively, were calculated to be 33.9 and 33.7 kDa and 33 kDa, respectively (Nalinanon et al. 2010b; Zhou et al. 2007). Pepsin from the stomach of albacore tuna was measured to be 36.8 kDa by Nalinanon et al. (2010a). At pH 2.0, pepsins with molecular weights of 37 kDa (pepsin-I and pepsin-III) and 35 kDa (pepsin-II) from rainbow trout stomach were transformed into the corresponding pepsins (Wald et al. 2016). Using SDS-PAGE, the three isolated PGs from the rice field eel's stomach were all assessed to have molecular weights of 36 kDa (Weng et al. 2011). Pepsins A and B purified from the pectoral rattail stomach had MWs of 35 and 31 kDa, respectively, according to Klomklao et al. (2007a). The MWs of three pepsins isolated from the stomach of a European eel were determined to be 37, 35, and 37 kDa, respectively (Wu et al. 2009).

Table 1 Proteases from the viscera of fish and aquatic invertebrates

Enzyme	Fish species	Origin	Optimum		References
			pH	Temperature (°C)	
Pepsin	Pectoral rattail	Stomach	3.0–3.5	45	Klomklao et al. (2007a)
	Mandarin fish	Stomach	3.0–3.5	45–50	Zhou et al. (2008)
	Smooth hound	Stomach	2.0	40	Bougatef et al. (2008)
	Snakehead	Stomach	3.0–3.5	40–45	Chen et al. (2009)
	European eel	Stomach	2.5–3.5	35–40	Wu et al. (2009)
	Albacore tuna	Stomach	2.0	50	Nalinanon et al. (2010a)
	Skipjack tuna	Stomach	2.0–2.5	45–50	Nalinanon et al. (2010b)
	Rice field eel	Stomach	3.0–3.5	35–40	Weng et al. (2011)
	Largemouth bass	Gastric mucosa	1.5–2.0	40–50	Miura et al. (2015)
	Rainbow trout	Stomach	2.5–3.0	30–40	Wald et al. (2016)
	Sábalo	Stomach	2.0	45	Gomez et al. (2018)
	Lizardfish	Stomach	2.0–3.5	40–50	Kuepethkaew et al. (2022)
Trypsin	Bluefish	Pyloric ceca	9.5	55	Klomklao et al. (2007b)
	Atlantic bonito	Pyloric ceca	9.0	65	Klomklao et al. (2007c)
	Walleye pollock	Pyloric ceca	8.0	50	Kishimura et al. (2008)
	Cuttlefish	Hepatopancreas	8.0	70	Balti et al. (2009)
	Skipjack tuna	Intestine	9.0	55–60	Klomklao et al. (2009a)
	Pectoral rattail	Pyloric ceca	8.5	45	Klomklao et al. (2009b)
	Brownstripe red snapper	Pyloric ceca	8.5	60	Khantaphant and Benjakul (2010)
	Silver mojarra	Viscera	8.5	50–55	Silva et al. (2011)

(continued)

Table 1 (continued)

Enzyme	Fish species	Origin	Optimum		References
			pH	Temperature (°C)	
	Zebra blenny	Viscera	9.5	60	Ktari et al. (2012a)
	Albacore tuna	Liver	8.5	55–60	Klomklao and Benjakul (2018)
	Albacore tuna	Spleen	9.0	55	Poonsin et al. (2019a)
	Dolphinfish	Intestinal	8.0	40	Santos et al. (2020)
Chymotrypsin	Monterey sardine	Viscera	8.0	50	Castillo-Yanez et al. (2006)
	Crucian carp	Hepatopancreas	7.5–8.0	40–50	Yang et al. (2009)
	Cuttlefish	Hepatopancreas	8.5	55	Balti et al. (2012)

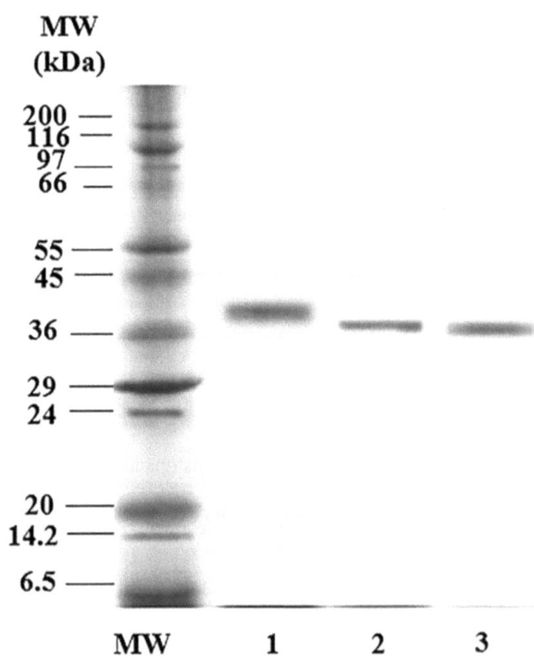

Fig. 2 SDS-PAGE of purified pepsinogens from lizardfish stomach. MW: Molecular weight standards; lane: 1 purified pepsinogen-I; lane: 2 purified pepsinogen-II, lane: 3 purified pepsinogen-III. *Source* Kuepethkaew et al. (2022)

Various techniques for isolating and purifying pepsins from marine creatures have been documented in the literature. Zhou et al. (2008) used ammonium sulfate fractionation (60% saturation), anion exchange (DEAE-Sephacel), and gel filtration (Sephacryl S-200) to isolate four pepsins from the stomach of mandarin fish. Six pepsinogens were isolated from the gastric mucosa of largemouth bass using a combination of DEAE-Sephacel chromatography, Sephadex G-100 gel filtration, and Mono Q fast protein liquid chromatography (Miura et al. 2015). Wu et al. (2009) purified three pepsins from European eel stomach by ammonium sulfate fractionation (20–60% saturation), then a sequence of column chromatographies using DEAE-Sephacel, Sephacryl S-200, and Superdex G-75. Purification of pepsin from rice field eel stomach was accomplished using 20–60% ammonium sulfate fractionation, DEAE-Sephacel anion exchange, and Sephacryl S-200 gel filtration (Weng et al. 2011). A variety of chromatographic methods using DEAE-cellulose, Sephadex G-50, and Sephadex G-75 were used to purify pepsins from the stomach of skipjack tuna (Nalinanon et al. 2010b). Pepsins from the stomach of lizardfish were recently extracted by Kuepethkaew et al. (2022) using ammonium sulfate precipitation, cation exchange (CM-Cellulose), and size exclusion (Sephadex G-75) chromatography.

The activity of pepsin is highly sensitive to variations in pH, temperature, and substrate type. Hemoglobin is the most commonly used substrate for assessing pepsin activity (Klomklao et al. 2007a; Kim and Dewapriya 2014). According to this theory, pepsin proteolyzes acid-denatured hemoglobin (substrate) to produce trichloroacetic acid (TCA) soluble hydrolysis products, mostly tyrosine and phenylalanine (Zhao et al. 2011). At pH 2.0 and 40 °C, smooth hound stomach pepsin had the greatest anti-hemoglobin action (Bougatef et al. 2008). According to Chen et al. (2009), using hemoglobin as a substrate, the three pepsins had maximum activity at pH 3.0, 3.5, and 3.0 at optimal temperatures of 45, 40, and 40 °C, respectively. It was discovered that fish pepsins hydrolyze hemoglobin significantly more quickly than casein (Khaled et al. 2011). The greatest specific activity was shown by three pepsin isoforms from lizardfish stomach against hemoglobin as a substrate, followed by fish skin gelatin, BSA, and bovine skin gelatin (Kuepethkaew et al. 2022). Gildberg and Raa (1983) reported that most marine animals have two or three main pepsins that digest hemoglobin optimally at pH levels between 2 and 4. According to Nalinanon et al. (2010a), albacore tuna pepsin exhibited optimal pH and temperature of 2.0 and 50 °C for pepsin activity. When utilizing hemoglobin as the substrate, Klomklao et al. (2007a) discovered that pepsins A and B from the pectoral rattail stomach were most active at pH 3.0 and pH 3.5, respectively, and at the same optimum temperature of 45 °C. Pepsin 1 and 2 from skipjack tuna's stomach were purified, and the results revealed that they functioned optimally at pH 2.0–2.5 and 45–50 °C (Nalinanon et al. 2010b). When hemoglobin was used as a substrate, pepsins 1, 2, and 3 isolated from the stomach of Japanese seabass had their highest levels of activity at pHs 3.0, 2.5, and 3.0, respectively (Cao et al. 2011). Purified lizardfish stomach pepsins functioned most effectively when temperatures were between 40 and 50 °C and at pH values between 2.0 and 3.5 to hydrolyze hemoglobin (Kuepethkaew et al. 2022). For pH stability, pepsin is relatively stable at a pH range of 2 to approximately 6, but because of denaturation at a pH higher than 6, it rapidly loses activity (Klomklao et al. 2012). Between pH 2 and 5, albacore tuna stomach pepsin was stable, but at pH 6.0, it rapidly lost activity (Nalinanon et al. 2010a). Nalinanon et al. (2010b) revealed that the activity of skipjack tuna pepsin remained stable in the pH range of 2–5. Using a pH range of 3.0–6.0, Castillo-Yanez et al. (2004) discovered that Monterey sardine acidic enzymes were stable. High pH stability of the acid proteases isolated from the digestive organs of the farmed giant catfish was found in the range of 1.0–5.0 (Vannabun et al. 2014). The pH stability of acid proteases from sábalo stomach mucosa was found to be between 2.0 and 5.0 (Gomez et al. 2018). According to Khaled et al. (2011), the aspartic protease from Sardinelle viscera was constant between pH 2.0 and 5.0. Pepsins A and B from pectoral rattails stomach were both found to be stable in the pH range of 2.0–6.0 (Klomklao et al. 2007a). Moreover, Kuepethkaew et al. (2022) noted that the activities of pepsins from the stomach of lizardfish decreased at neutral and alkaline pH, but they were stable at pH ranges between 2.0 and 4.0.

– *Trypsin*

Trypsins (EC 3.4.21.4), belong to a vast family of serine proteases that catalyze the hydrolysis of proteins and peptides, specifically cleaving peptide bonds at the carboxyl side of arginine and lysine residues (Klomklao et al. 2012). Trypsins are essential for many biological functions, such as digestion and the activation of zymogens for other enzymes like chymotrypsin (Cao et al. 2000). Trypsin and trypsin-like proteolytic enzymes have been isolated and characterized from the viscera of various fish species. Despite their resemblance to mammalian trypsins in terms of their biochemical properties, reports suggest that marine trypsin exhibits higher stability under harsh conditions, such as high temperature, pH, and exposure to surfactants or oxidizing agents. The optimal pH range for marine-derived trypsin hydrolysis varies between 6.0 and 11.5, while the optimal temperature range falls between 35 and 65 °C (Klomklao et al. 2006a; Freitas-Júnior et al. 2012). Trypsins derived from marine animals share similarities with mammalian trypsins in terms of their molecular size (22–30 kDa), amino acid composition, and susceptibility to inhibitors.

Many investigations on fish digestive organ trypsins have been performed. Ammonium sulfate (40–60% saturation) fractionation, soybean trypsin inhibitor-Sepharose 4B affinity column, and DEAE-Sephacel column chromatography were used to isolate a 23 kDa MW trypsin from brownstriped red snapper pyloric caeca (Khantaphant and Benjakul 2010). The highest activity of trypsin for hydrolyzing α-*N*-benzoyl-DL-arginine-*p*-nitroanilide (BAPNA) occurred at pH 8.5 and 60 °C. Both soybean trypsin inhibitor and *N*-*p*-tosyl-L-lysine-chloromethylketone (TLCK) significantly inhibited the activity of pure trypsin. Trypsin from the viscera of golden grey mullet with MW of 23 kDa was isolated and characterized by Bkhairia et al. (2016). The maximum activity was discovered at 50 °C, and the best pH for activity was 10. Despite not being inhibited by the metalloprotease inhibitor, 1 mM ethylenediaminetetraacetic acid (EDTA), the isolated enzyme was completely inhibited by soybean trypsin inhibitor (1 mg/ml). In addition, on the basis of MW, the suppression by soybean trypsin inhibitor, phenylmethanesulfonyl fluoride (PMSF), and the N-terminal peptide sequence, trypsin was found to be the predominant form of protease in the viscera of carp fish (Khangembam and Chakrabarti 2015). After being extracted from the pyloric cecum of *Lophiosilurus alexandri*, trypsin was tested, and it was shown to have the highest level of activity at pH 9.0 and 50 °C (Santos et al. 2016). Khandagale et al. (2017) revealed that anion-exchange and soybean trypsin inhibitor (SBTI) affinity chromatography were used to separate the trypsin from the oil sardine viscera after ammonium sulfate precipitation (30–70% saturation). The MW of trypsin was determined to be approximately 24 kDa using SDS-PAGE. The optimal pH and temperature for trypsin-mediated hydrolysis of BAPNA were 8.0 and 60 °C, respectively.

Trypsin from the intestinal of dolphinfish was purified by an ammonium sulfate fractionation (40–60% saturation) and Sephacryl S-100 h (Santos et al. 2020). The dolphinfish trypsin was homogeneously isolated with a MW of 26 kDa on SDS-PAGE and had its highest activity for the hydrolysis of a BAPNA at pH 8.0 and 40 °C. Two distinct trypsin inhibitors—the soybean trypsin inhibitor and benzamidine—as well as one serine-protease inhibitor (PMSF) totally inhibited the enzyme. Klomklao and Benjakul (2018) used Sephacryl S-200, Sephadex G-50, and diethylaminoethyl cellulose to purify two trypsin isoforms (A and B) from liver of albacore tuna. Size exclusion chromatography and SDS-PAGE were used to determine the MWs of trypsins A and B, which were determined to be 21 and 24 kDa, respectively. Trypsins A and B displayed distinct optimal temperatures for activity, with trypsin A showing maximal activity at 60 °C and trypsin B exhibiting maximal activity at 55 °C. However, both enzymes shared the same optimal pH of 8.5 when utilizing Nα-p-tosyl-L-arginine methyl ester hydrochloride (TAME) as a substrate. In addition, an anionic trypsin from the spleen of albacore tuna was isolated and characterized (Poonsin et al. 2019a). Purification was carried out by column chromatographies on Q-Sepharose, Superdex 75, and Arginine sepharose 4B. The MW determined by SDS-PAGE for the purified trypsin was 30 kDa. When Boc-Val-Pro-Arg-MCA was used as the substrate, the enzyme's optimal pH and temperature profiles were 9.0 and 55 °C, respectively (Fig. 3).

The N-terminal amino acid sequences of proteins can serve as valuable tools for identifying specific types of enzymes and may aid in the design of primers for cDNA cloning of these enzymes (Klomklao and Benjakul 2018). In contrast to the Thr residue often present at this location in mammalian pancreatic trypsins, the charged Glu residue at position 6 is a characteristic feature of fish trypsins. Moreover, restricted proteolysis of inactive trypsinogen into its active form results in the N-terminal sequences of all trypsins beginning with IVGG (Klomklao and Benjakul 2018; Klomklao et al. 2007b), and the N-terminal amino acid sequence of the trypsin from the walleye pollock pyloric ceca was reported as IVGGYEC-TKHSQAHQVSLNS (Kishimura et al. 2008). The N-terminal amino acid sequences of trypsins from the pyloric ceca of Pacific cod and saffron cod were reported as IVGGYECTRHSQAHQVSLNS and IVGGYECPRHSQAHQVS-LNS, respectively, (Fuchise et al. 2009). Kishimura et al. (2007) found that N-terminal amino acid sequences from pyloric ceca of trypsins from the jacopever and elkhorn sculpin were IVGGYECKPYSQPHQVSLNS and IVGGYECTPHSQAHQVSLNS, respectively. Cai et al. (2011) reported the N-terminal amino acid sequences of trypsin A

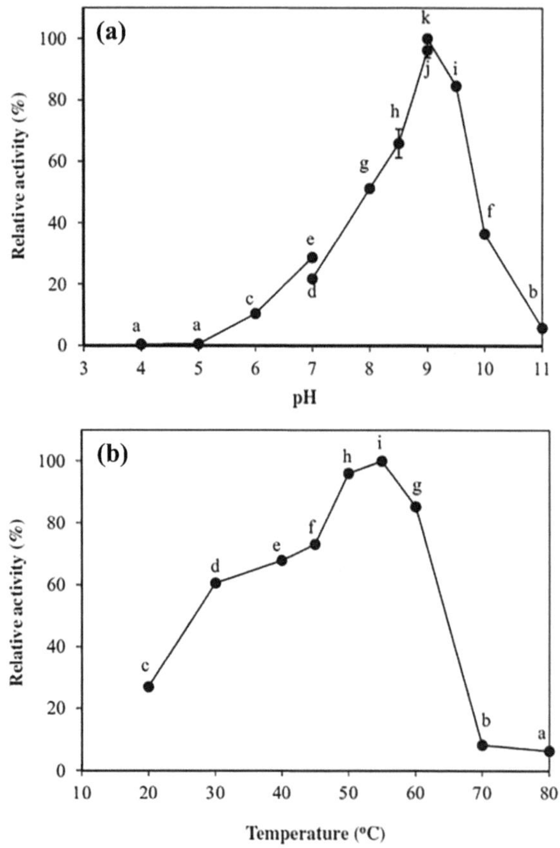

Fig. 3 pH (**a**) and temperature profiles (**b**) of purified anionic trypsin from albacore tuna spleen. *Source* Poonsin et al. (2019a)

and B from the Japanese sea bass hepatopancreas as IVGGYECTVALQAXYLEGGKXYEF and IVGGYECTPYSQPHQVSLNSGYHF, respectively. The N-terminal amino acid sequences of trypsins A and B purified from the liver of albacore tuna were identified by Klomklao and Benjakul (2018) as IVGGYECQAHSQPWQVSLNA and IVGGYECQAHTQPHQVS-LNA, respectively. According to Bkhairia et al. (2016), the 15 N-terminal amino acids from viscera of *Liza aurata* trypsin were found to be IVGGYECTPYSQPHQ, whereas the 20 N-terminal amino acid sequences of trypsin from spleen of albacore tuna were IVGGYECQAHSQPHQVSLNA (Poonsin et al. 2019a).

– *Chymotrypsin*

A digestive endopeptidase enzyme called chymotrypsin is created in the pancreatic tissues of both vertebrates and invertebrates, where it is then released into the duodenum (Zhou and Budge 2011). Chymotrypsin is another member of the serine proteases family of enzymes (Klomklao et al. 2012). Chymotrypsin hydrolyzes amides, esters, and peptide bonds in proteins, with the carboxyl side of the amino acid residue at the P1 position (the first amino acid residue in the N-terminal direction from the cleaved bond). In addition to cleaving these bonds, chymotrypsin has been shown to cleave peptide amide bonds such as tryptophanyl-, tyrosyl-, and phenylalanyl bonds. Furthermore, studies suggest that chymotrypsin can also cleave leucyl and glutamyl bonds (Zhou and Budge 2011). In addition, chymotrypsin targets nonpolar groups such as leucine, however the response time is longer. Furthermore, chymotrypsins A and B exhibit similar specificities in hydrolyzing peptides with phenylalanine, tryptophan, leucine, and tyrosine residues (Hudáky et al. 1999), whereas Raae et al. (1995) discovered differences in specificity between cod chymotrypsins ChT1 and ChT2. Variation in hydrolysis specificity between chymotrypsin isoforms can be exploited to quantify enzyme activity using specific substrates.

In contrast to trypsin, research on chymotrypsins from fish species has been relatively limited, with the earliest report by Overnell dating back to 1973. Extraction, purification, and biochemical properties of chymotrypsin from a variety of marine animals including anchovy (Heu et al. 1995), Atlantic cod (Asgeirsson and Bjarnason 1991), scallop (Le Chevalier et al. 1995), Monterey sardine (Castillo-Yanez et al. 2006), crucian carp (Yang et al. 2009), and cuttlefish (Balti et al. 2012) have been studied. These chymotrypsins typically consist of a single polypeptide chain with molecular weights ranging from 25 to 28 kDa. They exhibit maximal activity within the pH range of 7.5–8.5 and are most stable at a pH of approximately 9.0 (Klomklao 2008). Compared to trypsins, chymotrypsins have a wider range of specificity (Klomklao et al. 2012).

– *Collagenases*

Collagenases are proteases that exhibit high specificity toward collagen, a major component of the extracellular matrix in connective tissue. These enzymes are often considered virulence factors. Collagenases typically target the connective tissue in muscle cells and various organs, leading to tissue destruction and pathology (Yang et al. 2017). Collagenases from viscera of various fish have been purified, isolated, and biochemically characterized. Sovik and Rustad (2006) characterized collagenase in viscera, cut off and liver from cod, saithe, haddock, tusk, and link. At 20 °C in the cut off, 35 °C in the liver, and 50 °C in the viscera, collagenase displayed optimum activity. Internal organ collagenase from filefish was isolated and described (Kim et al. 2002). Ammonium sulfate precipitation (30–80% saturation) was used to purify the material, followed by column chromatographies. The molecular mass of the serine collagenase found in the filefish was estimated to be 27.0 kDa through the use of gel filtration and SDS-PAGE techniques. Upon purification, it was observed that the collagenase demonstrated optimal activity within a pH range of 7.0–8.0 and at a temperature of 55 °C. Byun et al. (2003) used acetone precipitation, Sephadex G-100, DEAE-Sephadex A-50, and Sephadex G-75 to purify collagenase from the tuna (*Thunnus thynnus*) pyloric caeca. At pH 7.5 and temperature of 55 °C were the optimum values for collagenolytic enzyme. Aoki et al. (2003) revealed that purification of three collagenolytic proteases (A1, A2, and

B) from the hepatopancreas of Northern shrimp required a number of chromatographic procedures, including hydroxyapatite column chromatography, gel filtration on Superdex 75, and ion-exchange chromatography on a MonoQ column. The ability of collagenolytic proteases A2 and B to digest native porcine type I collagen was demonstrated at pH 7.5 and 25 °C, whereas protease A1 showed no such activity. Subsequent characterization of these two proteases revealed their optimal pH ranges against DNP-peptide to be 11 for enzyme A2 and 8.5 for enzyme B. Both enzymes exhibited optimal temperature ranges between 40 and 45 °C, although enzyme B demonstrated greater thermal stability than enzyme A2 at pH 7.5. Using a Tris-buffer system, Daboor et al. (2012) extracted collagenase from fish waste (muscles, funs, and boned of the mixture of fish samples of haddock, herring, ground fish, and flounder). After being precipitated the enzyme using ammonium sulfate precipitation (40–80% saturation), the precipitated proteins were purified Sephadex G-100. When insoluble collagen type I was served as the substrate, the temperature of 35 °C and a pH of 7.5 were shown to be the optimal conditions for enzyme activity. A greater activity of the enzyme (72.5 Units/mL) was proven at the optimal pH and temperature of 8.0 and 55 °C, according to Oliveira et al. (2017) who isolated and biochemically characterized the collagenase from smooth weakfish viscera. During 60 min, the enzyme remained stable in a broad pH range (6.5–11.5) and temperature (25–60 °C).

– *Cathepsins*

Cathepsins have been discovered in the viscera of fish and aquatic invertebrates. Among them, cathepsin B, cathepsin L, and cathepsin S are commonly found digestive enzymes in marine animals (Simpson 2000). Digestive cysteine proteases derived from marine animals have been observed to exhibit optimal activity under acidic pH conditions, and are generally inactive under alkaline conditions. Pangkey et al. (2000) reported that in the case of cathepsin S isolated from the hepatopancreas of carp (*Cyprinus carpio*), a four-step purification process was employed, involving ammonium sulfate fractionation, SP-Sepharose, Sephacryl S-200, and Q-Sepharose. The optimal pH and temperature were 7.0 and 37 °C, respectively, when Z-Phe-Arg-MCA was used as the substrate.

Aranishi et al. (1997a) purified cathepsin B from carp hepatopancreas by acid treatment, ammonium sulfate fractionation, and a series of column chromatographies. The optimum pH and temperature of carp cathepsin B were observed at pH 6.0 and 45 °C, respectively. Cathepsin B was also characterized by cod, saithe, haddock, tusk, and link by-products (Sovik and Rustad 2006). In the viscera and liver, cathepsin B activity peaked at 35 °C, however, in the cut off, it peaked at 50 °C. Cathepsin B was purified from sea cucumber gut using ammonium sulfate fractionation (80% saturation) and a series chromatography using DEAE Sepharose CL-6B, Sephadex G-75, and TSK-Gel 3000 SWxl (Sun et al. 2011). The optimum activity of the enzyme was observed at a pH of 5.5 and a temperature of 45 °C. It exhibited notable stability within a pH range of 4.5–6.0, and thermal stability up to 50 °C. Additionally, the cathepsin B purified from sea cucumber gut was found to be strongly inhibited by E-64, iodoacetic acid, and antipain, indicating that it is a cysteine protease that contains sulfhydryl groups.

Purification and biochemical properties of cathepsin L from the hepatopancreas of carp was carried out using ammonium sulfate precipitation (20–80% saturation) and a series of chromatographies, with the enzyme having an affinity for Concanavalin A and Cibacron Blue F3GA (Aranishi et al. 1997b). The purified cathepsin L had the highest activity for carbobenzoxy-L-phenylalanyl-L-arginyl-4-methylcoumaryl-7-amide (Z-Phe-Arg-MCA) at pH 5.5–6.0 and 50 °C and the enzyme was extremely stable at pH 5.0–6.5 and below 40 °C. Cardenas-Lopez and Haard (2009) used a two-step technique using ammonium sulfate precipitation (40–80% saturation) and gel filtration chromatography to purify cathepsin L from the hepatopancreas of giant squid. The MW of the purified cathepsin L was estimated to be 24 kDa using SDS-PAGE and 23.7 kDa via mass spectrometry. In assays utilizing the cathepsin L specific synthetic substrate Z-PAAFC, the enzyme demonstrated an optimal activity at a pH of 4.5 and a temperature of 55 °C.

2.1.2 Fish Protease Applications

Enzymes have become an essential ingredient in various applications such as food, pharmaceutical, and cosmeceutical industries. Among all the enzymes, proteases have been the most extensively studied for industrial bioprocessing. In the fishery industry, proteases serve as processing aids for various products. Enzymes, including those derived from fish, have several advantages over chemical methods, such as superior process control, lower energy requirement, environmental and toxicological safety, and reduced costs. Their potential uses of marine proteases for industrial applications are summarized in Table 2.

– *Carotenoprotein extraction*

Carotenoproteins, which are by-products obtained during shellfish processing, show significant potential as a feed supplement in aquaculture (Shahidi and Kamil 2001). Protein-pigment interactions occurring in nature have been shown to enhance the stability of carotenoids. A complex consisting of a carotenoid and a protein in a stoichiometric ratio is known as a carotenoprotein (Klomklao et al. 2009c). Shrimp and crab shell waste contain approximately 150 mg of carotenoids per kilogram of dry weight. To recover carotenoids from shellfish processing byproducts, several methods have been explored, including fermentation, isolation by organic solvents and vegetable oils, and supercritical fluid extraction (Klomklao et al. 2020). Astaxanthin, the primary carotenoid found in shrimp waste, can be extracted and recovered through trypsin hydrolysis. This extracted carotenoid can be used as an ingredient in aquafeed, as well as to improve the color of various species of salmonids and crustaceans (Venugopal 2016). The pigments responsible for the distinct colors of crustaceans are located within the pigmented layer of the endocuticle. These pigments are primarily carotenoids, which give rise to a range of color appearances, from

Table 2 Potential use of proteases from marine animal sources

Enzyme	Area of application	Application
Pepsin	Pharmaceutical and cosmeceuticals	Collagen extraction, therapeutic agents, fish hydrolysate and bioactive peptide production
	Dairy	Cheese production
	Fishery	Fish silage, caviar production, descaling of fish
	Leather	To remove hair and residual tissue
	Collagen	An effective aid for collagen extraction
Trypsin	Food and fisheries	Production of seafood flavorant, extraction of carotenoprotein, fish sauce
	Dairy	Cheese production
	Paint	Production of pearl essence
	Pharmaceutical	Therapeutic agents, wound healing, veterinary medicine
	Meat	Impact on meat tenderization
Chymotrypsin	Food processing	Meat tenderizing, improving food nutrition value, protein hydrolysate production
	Fisheries	Bone protein removal
	Dairy	Cheese production
	Leather	Dehairing, bating and soaking
	Detergent	Decontaminating agent, cleaning agent

red to orange. Interestingly, different parts of the exoskeleton may contain various carotenoproteins in different proportions, resulting in a complex mixture of colors. (Klomklao et al. 2020). Carotenoproteins can be categorized into two groups based on their composition: true carotenoproteins, which consist solely of carotenoids and proteins in stoichiometric proportions; and carotenolipoproteins, which contain lipids and may not exhibit a stoichiometric relationship between carotenoids and proteins (Klomklao et al. 2009c).

A range of methods have been utilized for the extraction of carotenoproteins from shellfish processing byproducts. Water or dilute salt solutions have been used for the initial extraction of carotenoproteins, followed by fractionation through ammonium sulfate precipitation, and subsequent purification using chromatography on ion exchange celluloses or selective absorption on calcium phosphate or aluminum hydroxide (Klomklao et al. 2020). Shellfish waste contains approximately one-third of protein in its dry matter. An enzymatic process has been developed to increase the extraction efficiency of carotenoprotein from shellfish waste, particularly due to low recovery rates. Using Tunisian barbel trypsin, Sila et al. (2012a) isolated carotenoprotein from the shells of deep-water shrimp. The analysis revealed that the product obtained from the treated shrimp shell had significantly higher levels of protein and fat, and considerably lower levels of chitin and ash compared to the untreated samples. In addition, the efficacy of Tunisian barbel trypsin in recovering total soluble protein and total xanthophylls was found to be comparable to that of bovine trypsin. According to Senphan et al. (2014), the recovery of carotenoproteins was improved when proteases from the hepatopancreas of Pacific white shrimp were added to shrimp shells of the same species. The findings showed that Pacific white shrimp shells treated with hepatopancreas proteases had increased levels of protein, fat, and total essential amino acids. Poonsin et al. (2019b) recently revealed that carotenoprotein was successfully recovered from Pacific white shrimp shells using trypsin extracted from albacore tuna spleen. The carotenoprotein extracted with the aid of the enzyme had significantly higher protein, fat, and pigment contents compared to the untreated shrimp shells. Moreover, it was found to contain astaxanthin and astaxanthin diester as major carotenoids (Table 3). Klomklao et al. (2009c) investigated the recovery of carotenoproteins from black tiger shrimp waste using trypsin extracted from bluefish pyloric ceca. The resulting product was found to have significantly higher protein and pigment content compared to the untreated black tiger shrimp waste, while exhibiting low levels of chitin and ash.

– *Fish sauce*

Fish sauce is a transparent brown liquid with a salty flavor that is typically made by autolyzing fish using endogenous enzymes. The consumption and production of fermented fish sauce is largely concentrated in Southeast Asian provinces, such as Nampla in Thailand, Noucnam in Vietnam, Shotturu and Ishiru in Japan, Patis in the Philippines, Budu in Malaysia, Ketjapikan and Bakasang in Indonesia, Toeuk Trey in Cambodia, Yuilu in China, and Ngapi in Myanmar. Fish sauce provides all essential amino acids as well as other vitamins and minerals (Jiang et al. 2007). Traditional fish sauce is made by fermenting fish combined with salt in a 2:1 or 3:1 ratio and fermenting at room temperature for 12 months or more (Klomklao et al.

Table 3 Carotenoid content and proximate composition of Pacific white shrimp shells and carotenoprotein recovered with or without albacore tuna trypsin

Compositions	Shrimp shell	Carotenoprotein**	
		Control	Albacore tuna trypsin-aided
Total carotenoid content (μg astaxanthin/g sample)	12.05 ± 3.78[a]	33.50 ± 5.66[b]	73.25 ± 4.24[c]
Protein (%)*	38.57 ± 1.18[a]	60.22 ± 0.69[b]	72.37 ± 0.64[c]
Fat (%)*	4.70 ± 0.40[a]	15.98 ± 017[b]	18.79 ± 0.14[a]
Ash (%)*	21.36 ± 0.72[c]	18.09 ± 0.61[b]	7.14 ± 0.06[a]
Chitin (%)*	35.20 ± 0.56[c]	5.61 ± 0.24[b]	1.61 ± 0.03[a]

The different letters in the same row denote the significant differences ($p < 0.05$)
* Dry weight basis
** Values are mean ± standard deviation from triplicate determinations
Source Poonsin et al. (2019b)

2006b; Auttanak et al. 2022). For centuries, the production of fish sauce has been of significant value as a preservation method for fish in Asia, especially in areas with an extensive coastline and high ambient temperature, such as in Southeast Asian countries like Thailand, Cambodia, Malaysia, Philippines, and Indonesia. Fish sauce is a popular condiment used in many Asian cuisines and is a vital source of protein and other essential nutrients (Shahidi and Kamil 2001). Fish sauce is a healthful seasoning for vegetable foods owing to the presence of essential amino acids (Ghaly et al. 2013). For fish sauce preparation, a plethora of raw materials can be employed, and the proteolytic enzymes must be adequate for the solubilization of tissue or protein. Proteins undergo hydrolysis during fermentation, mostly due to the autolysis activity of the fish digestive proteases. In the presence of 20–30% salt, endogenous enzymes continuously break down the fish tissue to produce a transparent liquid with a high concentration of free amino acids and a strong taste and odor. To guarantee that fish sauce solubilizes and develops its taste and color, the fermentation procedure is often carried out for a very long time (Klomklao et al. 2006b).

The traditional method of fish sauce production, which relies solely on autolysis involving fermentation and endogenous enzymes, is a time-consuming process that may take several months up to 3 years for full completion. Fish sauce fermentation takes time, and several ways for speeding up the process have been documented. Various ways for shortening the manufacture time of fish sauce have been presented. Exogenous proteases can expedite the development or production of fish sauce. Plant enzymes were the first to be used. Hence, the application of exogenous enzymes including bromelain from pineapple stems, ficin from figs, or papain from unripe papaya (Benjakul et al. 2010) and the commercial preparations Protamex and and Neutrase, have been exhibited to accelerate the fermentation of fish (Fernandes 2016). However, the flavor attributes of the end product are usually inferior to the traditional product. Additionally, the enzyme extracted from marine animals has been effectively employed as a seafood processing aid, including the expediting of the fermentation of fish sauce. Klomklao et al. (2006b) revealed that addition of skipjack tuna spleen increased the hydrolysis of protein of fish sauce. In comparison to fish sauce made from sardines without skipjack tuna spleen addition, the fish sauce with the addition of skipjack tuna spleen had considerably higher levels of total nitrogen, amino nitrogen (Fig. 4), formaldehyde nitrogen, and ammonia nitrogen. Fish sauce samples that had 20% salt, without or with the inclusion of 10% skipjack tuna spleen, exhibited acceptabilities that were comparable to those of commercial fish sauce (Klomklao et al. 2006b). Male Arctic capelin and Atlantic cod intestines are examples of by-products of fish capture that could be utilized as raw materials to make high-quality fish sauce for human consumption (Gildberg 2001). In this study, it was found that supplementing minced capelin with 5–10% enzyme-rich cod pyloric caeca resulted in a good recovery of fish sauce protein (60%) after 6 months of storage. The addition of hepatopancreas of squid also accelerated the production of capelin sauce (Helgi et al. 2007). The function of several protease enzymes in the manufacturing of fish sauce has been studied by Orejana and Liston (1981) who concluded that the primary enzyme responsible for tissue degradation in fish sauce production appears to be a

Fig. 4 Total amino nitrogen content of fish sauce samples produced from sardine to which different levels of skipjack tuna spleen are added; blends are fermented for 180 days. *Source* Klomklao et al. (2006b)

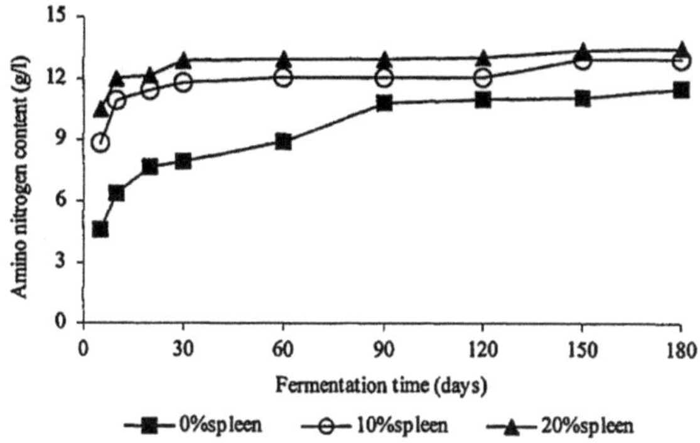

trypsin-like enzyme. Other proteases, such as cathepsin B, have also been found to play a significant role in increasing the amount of soluble protein in fish sauce.

– *Deskinning/descaling*

Deskinning is a critical step in fish processing that involves the removal of fish skin without causing damage to the meat. Currently, deskinning is mainly performed by rough mechanical procedures, which can result in destruction of the flesh and excessive by-products. Some fish species, such as the starry ray (*Raja radiata*), are especially difficult to deskin, and automated mechanical deskinning is virtually impossible. Consequently, manual deskinning must be performed, but this method is labor-intensive and costly, making it economically unfeasible for many underutilized species. Innovative technologies that can overcome the challenges of deskinning and reduce labor and production costs are necessary to fully utilize the potential of these species (Benjakul et al. 2010). An alternate technique would be to utilize enzymes to deskin fish species which are hard to deskin manually or mechanically. The yield of edibles can be increased by enzymatic deskinning. In some cases, the machine-skinning process may not fully remove all skin patches, requiring additional manual removal. To overcome this challenge, enzymatic deskinning has been explored as an alternative approach. Enzymatic deskinning of skate wing, for instance, has been achieved by immersing the fish in an enzyme bath, followed by a rapid warm water treatment. Afterward, water is used to remove the disintegrated skin (Min and Green 2008). Using pepsin extracted from cod, enzymatic deskinning of herring (*Clupea harengus*) has been accomplished (Haard and Simpson 1994). The majority of the proteases found in squid offal extract are collagenolytic enzymes (Leuba and Meyer 1989). Proteases derived from squid intestines have shown potential for removing the double-layered skin of squid. The skin of squid is rich in collagen, with approximately 21.9% collagen per dry tissue and 28.4% per crude protein, whereas the mantle muscle contains approximately 4.6% collagen per dry tissue and 5.4% per crude protein. The skin's collagen content per crude protein is approximately five times that of the muscle (Mizuta et al. 1994). Additionally, various specific uses, including those for manufacturing herring, pollock, squid, skate, shrimp shells, and tuna, have been established and these applications are occasionally accompanied with physical treatment (Rasika et al. 2013). Several methods for processing marine organisms involve the use of enzymes from other marine organisms. For example, acid proteases from cod viscera have been used for processing herring, protease extracts from minced arrowtooth flounder for pollock, and enzymes from squid for processing squid (Fernandes 2016). For the deskinning of catfish nuggets, commercial proteases (Proleather FG-F® and Protease N®) as well as collagenase (CLS1®) were recently evaluated. Proleather FG-F® was successful, and operating settings (enzyme concentration, incubation duration, and temperature) that maximized peritoneal membrane removal were established (Kim et al. 2014). Fish deskinning techniques using enzymes function by dissolving collagenous skin tissue without affecting the muscular tissue. Cold-adapted fish pepsin can be used for enzymatic deskinning at low reaction temperatures (Likhar and Chudasama 2021). Therefore, the milder enzyme method is preferred, especially when mixes of fish proteases capable of operating at low temperatures are utilized (Gildberg et al. 2000).

Biotec-Mackzymal in Norway has developed a complete production line for a similar process of squid enzymatic deskinning, which is now being produced by Carnitech in Denmark. Seatec, a company from Russia and Liechtenstein, is producing a collagenase preparation from crab hepatopancreas that may be used in the deskinning of squid (Díaz-López and García-Carreño 2020).

– *Collagen extraction*

Commercial collagen is mostly derived from bovine and pig skin, cow bones, and other mammalian animal by-products (Pallela et al. 2013). Moreover, collagen is found in marine organisms such as sponges, jellyfish, mollusks (mussels, squid, cuttlefish, and octopus), echinoderms, and fish (Lim et al. 2019). Thus, cheap collagen could be produced from fish by-products. Fish waste for collagen manufacture is mostly composed of bones, skin, scales, fins as well as cartilage that all have high collagen content (Benjakul et al. 2010). The usage of collagen in the cosmetic, food, and beverage industries is driving the expansion of the marine collagen market (Saranya et al. 2020), and fish by-products provide a large and inexpensive supply of collagen for the sector (Jafari et al. 2020). Conventionally, collagen is extracted through an acid-solubilization process (ASP) in which collagen is solubilized in an acid, such as acetic acid, and other non-acid-soluble materials are removed. However, the ASP results in a low yield of collagen. To increase the yield of collagen, pepsin can be used in combination with acid extraction (Table 4). Cross-links at the telopeptide region of collagen molecules are covalent and not easily solubilized by acid. Pepsin, on the other hand, can break down these cross-linkages in the telopeptide regions of collagen while preserving its secondary structure. Therefore, the use of pepsin in the collagen extraction process can significantly improve the yield of collagen (Zhao et al. 2011).

Several investigations have employed fish pepsins to extract collagen. Kaewdang et al (2014) revealed that with the inclusion of pepsin extracted from the stomach of yellowfin tuna increased the yield of collagen from the skin of yellowfin tuna swim bladders up to 12.10% compared with a yield of 1.07% obtained with acid soluble collagen (ASC). According to Kumar and Nazeer (2013), with the addition

Table 4 Pepsins applied for collagen extraction

Raw material	Pepsin sources	Concentration of pepsin used	Digestion time	ASC yield (%)	PSC yield (%)	References
Sharpnose stingray skin	Porcine gastric mucosa	1:40 (w/v)	30 h	20.48	34.84	Ong et al. (2021)
Silver carp scale	Novozymes	3% (w/v)	48 h	5.09	12.06	Wu et al. (2019)
Sturgeon skin	Porcine gastric mucosa	120 Units/g sample	48 h	5.73	10.26	Pei et al. (2019)
Atlantic cod swim bladders	NR	10%	3 days	5.72	11.14	Sousa et al. (2019)
Giant croaker skin	NR	1,389 Units/g	8.67 h	NR	84.85	Yu et al. (2018)
Yellowfin tuna swim bladders	Yellowfin tuna stomach	20 Units/g sample	48 h	1.07	12.10	Kaewdang et al. (2014)
Black carp skin	Bovine gastric mucosa	10% (w/v)	48 h	NR	45.7	Wu et al. (2014)
Horse mackerel skin	NR	1.5% (w/w)	30 h	17.3	22.5	Kumar and Nazeer (2013)
Croaker skin	NR		30 h	21.9	25.7	
Bighead carp fin	Porcine gastric mucosa	0.1% (w/v)	3 days	NR	5.1	Liu et al. (2012)
Bighead carp scale					2.7	
Bighead carp skin					60.3	
Bighead carp bone					2.9	
Bighead carp swim bladder					59.0	
Amiurus nebulosus skin	NR	5,000 Units/g skin	6 h	62.05	97.44	Chen et al. (2011)
Unicorn leatherjacket skin	Albacore stomach	20 Units/g sample	48 h	NR	8.48	Ahmad and Benjakul (2010)
	Yellowfin stomach				8.40	
	Porcine gastric mucosa				7.56	

(continued)

Table 4 (continued)

Raw material	Pepsin sources	Concentration of pepsin used	Digestion time	ASC yield (%)	PSC yield (%)	References
Threadfin bream skin	Albacore tuna stomach	10 Units/g skin	12 h	22.45	74.48	Nalinanon et al. (2008)
	Skipjack tuna stomach		12 h		63.81	
	Tongol tuna stomach		12 h		71.95	
Bigeye snapper skin	Bigeye snapper stomach	20 kUnits/g skin	48 h	5.31	18.70	Nalinanon et al. (2007)

ASC: Acid-solubilized collagen
PSC: Pepsin-solubilized collagen
NR: Not reported

of pepsin, collagen yield extracted from horse mackerel skin was 22.5% (which was higher than the 17.3% obtained with ASC). Pepsin isolated from tuna stomach was demonstrated to be a useful assistance in the extraction of collagen from the skin of fish. According to Ahmad and Benjakul (2010), 8.48 and 8.40% of collagen were recovered from unicorn leatherjackets skin after a 48-h extraction process with pepsins from albacore and yellowfin tuna, respectively. These results were higher than the 7.56% yield achieved with pepsin extraction from porcine gastric mucosa. Nalinanon et al. (2008) demonstrated that after the addition of pepsin from albacore tuna, skipjack tuna, and tongol tuna stomachs, the collagen yields extracted from threadfin bream skin were 74.48, 63.81, and 71.95%, respectively (which were higher than the 22.45% obtained with ASC). The application of bigeye snapper pepsin at a concentration of 20 kUnits/g skin improved the amount of collagen extracted from bigeye snapper skin. Bigeye snapper skin collagen yields were 5.31 and 17.0% (dry basis) after being extracted for 48 h with acid and bigeye snapper pepsin, respectively (Nalinanon et al. 2007).

– *Fish protein hydrolysate*

Fish protein hydrolysate is a valuable product generated by the breakdown of fish proteins into smaller peptides and amino acids. Various techniques are employed to produce fish protein hydrolysate, such as chemical hydrolysis, including both acid and alkaline hydrolysis, autolysis, bacterial fermentation, and enzymatic hydrolysis. The most widely used methods for the production of fish protein hydrolysate are enzymatic and chemical hydrolysis. Acid hydrolysis is a procedure of producing fish protein hydrolysate that involves the use of concentrated hydrochloric acid or occasionally sulfuric acid at high temperatures and pressures. The hydrolysate is then neutralized, resulting in a product that contains large amounts of sodium chloride, which can impair its functionality. Unfortunately, this method also destroys tryptophan, which is a key amino acid. Alkali hydrolysis involves the use of concentrated sodium hydroxide at high temperatures. In addition, various undesired reactions take place throughout the process, which results in the creation of toxic substances and reduces the functionality of the hydrolysate (Kristinsson and Rasco 2000a). The chemical hydrolysis process has the advantages of being low cost, rapid, and resulting in a high protein recovery. However, there is little control over the consistency of the hydrolyzed products, and the free amino acid profile can vary greatly due to the non-specific breakdown of peptide bonds (Celus et al. 2007). The enzymatic approach, while more complex, occurs under mild conditions of temperature, pressure, and pH, and involves the use of proteolytic enzymes which are typically available at low cost. Deleterious reactions are virtually non-existent (Fig. 5).

Fish enzymes, particularly those from the digestive organs, have been demonstrated to be a promising technique for preparing fish protein hydrolysate. Non-commercialized proteases obtained from fish digestive organs, such as pepsin, trypsin, chymotrypsin, gastricsin, and elastase (Benjakul et al. 2014), have been used

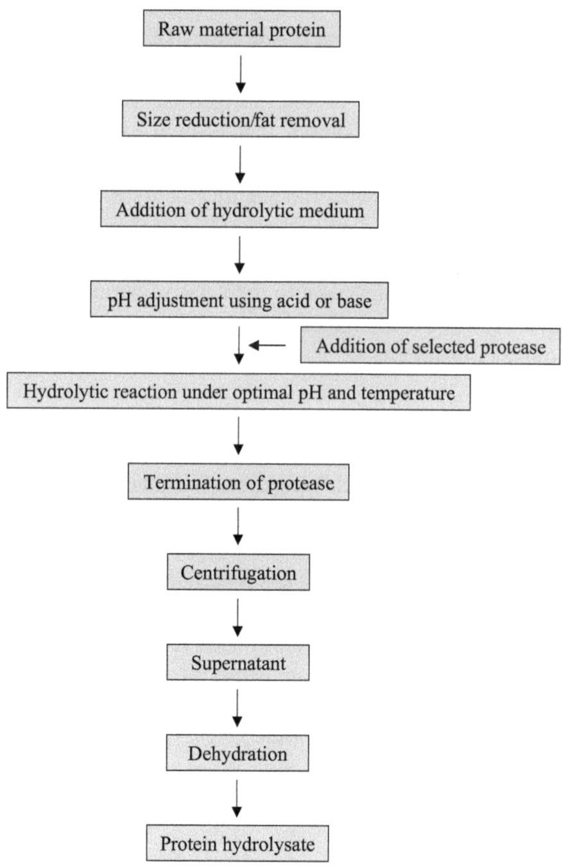

Fig. 5 Flowchart of protein hydrolysate production by an enzymatic process

to generate fish protein hydrolysate (Table 5). Using skipjack tuna pepsin, protein hydrolysate was made from kawakawa (*Euthynnus affinis*) muscle, and it displayed angiotensin I-converting enzyme (ACE) inhibitory effects as well as potential as an antioxidant source (Taheri and Gashti 2019). Proteases from the liver of albacore tuna were applied to prepare protein hydrolysate from the muscle of starry triggerfish. In comparison to starry triggerfish muscle, the obtained protein hydrolysate contained more protein and ash and less moisture and lipid and it also displayed antioxidant activity (Sripokar et al. 2016, 2019). Using Alcalase or Flavourzyme to first hydrolyze proteins from brownstripe red snapper muscle to a degree of 40% hydrolysis (DH), then utilizing pyloric caeca protease to further hydrolyze the proteins, revealed peptides having antioxidant and ACE inhibitory properties (Khantaphant et al. 2011). Zebra blenny muscle protein hydrolysate, which was made using crude alkaline protease extracts from the viscera of the zebra blenny, sardinella, and smooth hound, also contained antioxidant characteristics (Ktari et al. 2012b). Protease from the viscera of hybrid catfish was used to create hydrolysates of the muscle protein from toothed ponyfish with antioxidant activity (Klomklao et al. 2013). Additionally, the skipjack tuna stomach pepsin-prepared hydrolysates of threadfin bream muscle had antioxidant properties (Nalinanon et al. 2011). The current global challenges of limited natural resources and environmental pollution have highlighted the importance of increasing the usage of fish processing by-products. Efficient recovery and utilization of by-products are crucial for maximizing economic benefits. Recent investigations have shown that protein hydrolysates from sardinelle by-products, produced using various proteases, possess antioxidant activity. Among the proteases tested, hydrolysis with crude extract from sardine viscera yielded the hydrolysate with the highest antioxidant activity (Bougatef et al. 2010). Protein physicochemical and functional characteristics are directly impacted by hydrolysis (Kristinsson and Rasco 2000a, b). In comparison to their native counterparts, peptides produced by protein hydrolysis have altered or increased functional characteristics and bioactivities (Kitts and Weiler 2003).

– *Detergent additive*

Detergents are cleaning products that are often applied to eliminate filth. Water softeners and highly advanced surfactants are now observed in detergents. Recently, enzymes are increasingly being used into detergent compositions. For easier removal from textiles, enzymes assist break down complex filth, particularly proteins (such as blood stains), lipids (such as oil spills), and carbohydrates (such as starch) like blood and grass. Proteases are applied to hydrolyze the dirt/stains of proteins. Proteases assist in cleaning up protein-based food stains, whereas amylases clean up starch-based stains. Enzymes are superior than traditional chemicals because they can remove tough stains and decompose prior to contacting waterways, which reduces or eliminates water pollution issues (Naganthran et al. 2017). An enzyme preparation must be compatible with the detergents across a broad range of temperatures in order to be suitable for usage in industrial laundry detergents. The enzyme needs to be sufficiently temperature stable to be effective throughout a variety of washing temperatures, high pHs, as well as the existence of surfactants or oxidizing agents

Table 5 Fish proteases applied for protein hydrolysate production

Protease	Raw materials	References
Skipjack tuna stomach pepsin	*Euthynnus affinis* muscle	Taheri and Gashti (2019)
Farmed giant catfish viscera peptidase	Farmed giant catfish skin	Ketnawa et al. (2017)
Albacore tuna liver protease	Starry triggerfish muscle	Sripokar et al. (2016)
Pacific white shrimp hepatopancreas protease	Splendid squid mantle gelatin	Hamzeh et al. (2016)
Hybrid catfish viscera protease	Toothed ponyfish muscle	Klomklao et al. (2013)
Zebra blenny viscera protease	Zebra blenny muscle	Ktari et al. (2012b)
Sardinella viscera protease		
Smooth hound viscera protease		
Brownstripe red snapper pyloric caeca protease	Brownstripe red snapper muscle	Khantaphant et al. (2011)
Skipjack tuna stomach pepsin	Threadfin bream muscle	Nalinanon et al. (2011)
Bigeye snapper pyloric caeca protease	Bigeye snapper skin	Phanturat et al. (2010)
Sardine viscera protease	Sardinelle heads and viscera	Barkia et al. (2010)
Cuttlefish hepatopancreas protease	Cuttlefish muscle	Balti et al. (2010)
Smooth hound intestine protease	Smooth hound meat	Bougatef et al. (2009)
Mackerel intestine protease	Alaska pollack frame	Je et al. (2005)
Pepsin and mackerel intestine protease	Yellowfin sole frame	Jun et al. (2004)
Tuna pyloric caeca protease	Cod frame protein	Kim et al. (1997)

(Klomklao et al. 2005; Poonsin et al. 2019b). The use of alkaline proteases derived from marine digestive organs, particularly trypsin, has become increasingly popular due to their stability and efficacy under harsh conditions. They can function optimally in environments with high temperatures (50–60 °C), high pH levels, and in the presence of surfactants or oxidizing agents (Poonsin et al. 2019b). Recently, Poonsin et al. (2019b) found that several commercial solid and liquid detergents did not affect the stability of trypsin isolated from the spleen of albacore tuna for 60 min at 40 °C. At a dosage of 7 mg/mL in commercial solid detergent, red scorpionfish viscera protease maintained more than 83% of its activity (Younes et al. 2015). In addition, proteinases from goby were shown to remain 87% of their activity after being treated in commercial solid detergent at 30 °C for a period of 1 h, according to Sila et al. (2012b).

– *Milk coagulation*

Enzymes play a major role in the dairy industry, particularly in the coagulation of milk for cheese production. Milk clotting enzymes that are generally recognized as safe (GRAS) include rennet. The milk coagulation process can be divided into two phases: enzymatic and nonenzymatic. The enzymatic phase begins with the cleavage of the chymosin-sensitive bond of k-caseins (Phe-105-Met-106) at low pH, catalyzed by milk-clotting enzymes such as chymosin. This process results in the formation of para-k-casein and a macropeptide. As a result, k-casein loses its ability to stabilize casein micelles, making them susceptible to coagulation in the presence of calcium. The temperature during the enzymatic phase is kept below 10 °C. The nonenzymatic phase follows the enzymatic phase, where the altered casein micelles undergo aggregation to form a firm gel structure known as curd, while the whey is drained off. The temperature for this phase is between 30 and 39 °C. Both phases of milk clotting overlap (Díaz-López and García-Carreño 2020). Reported sources of rennet and other clotting enzymes include animals, plants, and microbes for cheese-making. However, not all proteolytic enzymes are appropriate for producing high-quality cheese. According to Díaz-López and García-Carreño (2020), excessive general proteolysis can lead to reduced curd yield, excessive protein and fat loss to the whey as well as undesired changes in texture (softening), and flavor (bitter off taste) throughout cheese aging. Chymosin (EC 3.4.23.4), commonly known as rennin, is an acid protease derived from suckling ruminant abomasa. It has a tight substrate specificity and high pH stability at 7.0 (with a range of 5.3–6.3). Chymosin is regarded as the typical milk-clotting enzyme and is employed in order to examine other milk-clotting enzymes. Its capacity to transform colloidal milk casein into curd results in high yields and minimal proteolysis, which assists in the aging of cheese (Manji and Kakuda 1988). Numerous proteases derived from animal, plant, or microbial sources have been considered as rennet replacements. The cheese industry has accepted *Mucor miehei* rennet, a fungal protease preparation, as the most common alternative. However, fungal rennet has some drawbacks, including broad specificity, heat stability, bitterness in cheese, and activity remaining in whey. Proteases from plants have also been used as rennet substitutes, producing a softer curd than calf rennet. Porcine pepsin has been applied as a rennet substitute, but its milk-clotting ability decreases quickly above pH 6.5. By combining this enzyme with bovine rennet and chymosin, a less costly coagulant is created, and low curd yield is avoided (McMahon and Brown 1985). As an alternative to rennet, chicken pepsin has also been exploited, however Cheddar cheese made with this chicken pepsin can have strong off tastes and odors (Gordin and Rosenthal 1978).

Pepsins from fish viscera have several properties that make them suitable for both milk-clotting phases and that avoid the major problems with rennet substitutes pointed out above. Pepsins from Atlantic cod (Haard 1986; Brewer et al. 1984) and Atlantic tuna (*Thunnus obesus*) stomachs (Tavares et al. 1997) have been purified and proposed as milk-coagulating enzymes. Fish pepsin, derived from cold-adapted species such as Atlantic cod, Greenland cod, or Polar cod, has a lower temperature coefficient for milk clotting (1.4–1.7) compared to calf rennet and microbial rennet (2.0–2.8). This allows for clotting to occur with a lower concentration of the enzyme, conserving rennet and reducing the presence of residual curd proteolysis. Additionally, cold-adapted fish pepsins are instability at temperatures over 30 °C, causing milk to clot and then heat-denature rennet during curd formation to inactivate proteases and prevent softening troubles and off-flavors. Because of these properties, cold-adapted fish pepsin is an appropriate rennet alternative.

– *Caviar production*

The term "caviar" refers to cured fish eggs from fish species such as white sturgeon, salmon trout, etc. (Venugopal 2016). The riddling process during caviar preparation can be challenging, which is completed either manually or mechanically. A common issue with the conventional process is the difficulty to remove fish roe from the connective tissues of the roe sac without causing damage to the roe. The process of manually or mechanically separating the roe from the roe sac has a number of drawbacks, including labor-intensiveness, low yield, and limited shelf life owing to mechanical damage to the egg (Shahidi and Kamil 2001). While mechanical processing of roe is possible for various fish species, this is not the case for fatty fish species. The eggs of these fish are firmly attached to the connective tissues of the roe sacs and are too delicate to survive mechanical treatment. In the riddling procedure, many proteases were utilized. Even in cold water, the enzymes utilized were extracted from fish intestine and worked fast to liberate roe particles from connective tissues in the roe sac (Raa 1985). Fish pepsin as well as collagenase from the hepatopancreas of carp, can be utilized to liberate fish roe from the connective tissues of the roe sac (Likhar and Chudasama 2021). In the process of making caviar from the roe of the same species, acidic proteases from the stomach of cold-water species like Atlantic cod (*Gadus morhua*) and orange roughy (*Hoplostethus atlanticus*) have been employed (Xu et al. 1996). The technique has been demonstrated to be successful in increasing caviar production from 70 to 90% (Raa 1990). The use of fish proteases in the manufacture of caviar has several benefits, including improved recovery, reduced egg damage, residue-free cleanliness, reduced labor requirements, and superior product hygiene (Shahidi and Kamil 2001).

2.2 *Lipases*

Lipases (EC 3.1.1.3), also known as triacylglycerol acylhydrolase, are defined as enzymes which catalyze the hydrolysis esters of long-chain aliphatic acids from glycerol at the oil/water interface. Its hydrolysis occurs where the enzyme dispersed, at the interface between the aqueous phase and the insoluble substrate phase (Fig. 6). The most widely accepted kinetic model of lipase activity suggests that the lipase exists in both active and inactive conformations in a solution. The inactive conformation predominates in aqueous solutions, leading to low enzyme activity. While, at the lipid-water interface, the active conformation becomes favored, and thus, water-insoluble substrates are hydrolyzed by the enzyme. The active forms of the enzyme in solution which are adsorbed at the interface may have different conformations (Simpson et al. 2012). Lipases, in addition to hydrolysis, also catalyze synthesis reactions such as esterification, amidation, alcoholysis, acidolysis, and aminolysis

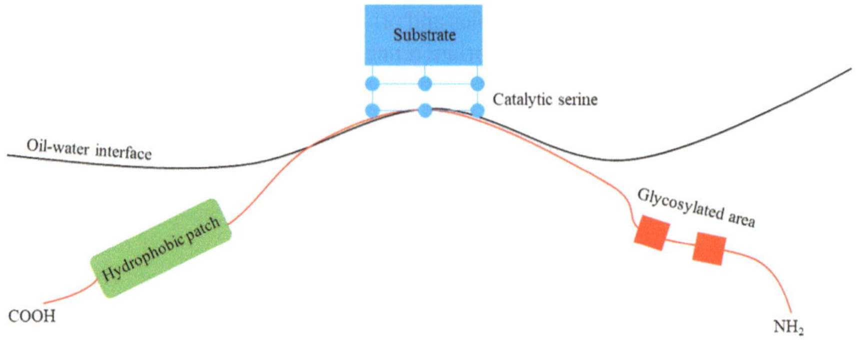

Fig. 6 Hydrolysis of the triglyceride by lipase takes place at the interface of the insoluble substrate and the aqueous phases, the region where enzyme is dissolved

(A) Hydrolysis:

$$RCOOR' + H_2O \longleftrightarrow RCOOH + R'OH$$

(B) Synthesis:

Reactions under this category can be further separated:

(a) Esterification

$$RCOOH + R'OH \longleftrightarrow RCOOR' + H_2O$$

(b) Interesterification

$$RCOOR' + R''COOR\cdot \longleftrightarrow RCOOR\cdot + R''COOR'$$

(c) Alcoholysis

$$RCOOR' + R''OH \longleftrightarrow RCOOR'' + R'OH$$

(d) Acidolysis

$$RCOOR' + R''COOH \longleftrightarrow R''COOR + RCOOH$$

Fig. 7 Lipase reactions

(Casas-Goboy et al. 2012; Lee et al. 2015). Figure 7 illustrates the two major classifications of lipase-catalyzed reactions. The latter three reactions are frequently clustered together under a single word, transesterification.

2.2.1 Purification and Characterization of Fish Lipases

Lipases are commercially obtained from the pancreas and serous glands of ruminants, primarily young calves and pigs. Nevertheless, because of the several benefits of microbial lipases, there has been a trend toward commercial synthesis of lipases from microbial sources such as bacteria, yeasts, and molds (Hasan et al. 2006). Plant lipases can be extracted from oilseeds, oily fruits, and cereal grains. Aquatic animals have gained popularity as a potential source of lipases during the last decade. Investigations on various tissues that might serve as a lipase source revealed that several fish and aquatic creatures exhibit lipase activity and their biochemical properties are summarized in Table 6. Ammonium sulfate precipitation (60% saturation), size exclusion, anion, and cation exchange chromatographies were used to isolate and purify a lipase from the digestive organs of grey mullet (Smichi et al. 2013). Apart from having lipase activity, grey mullet digestive lipase also possesses phospholipase activity, which needs Ca^{2+} and bile salts to be active. When tributyrin, an olive oil emulsion, or phosphatidylcholine were used as substrates and the lipase activity was determined at pH 8.0 and at 55 °C, 64, 55, and 63 Units/mg of specific activity for grey mullet lipase, respectively, were detected (Smichi et al. 2013). Zarai et al. (2012) purified lipase from the hepatopancreas of marine snail by using acid treatment, ammonium sulfate precipitation (60% saturation), and column chromatographies on Sephacryl S-200, mono Q-Sepharose and octyl-Sepharose. SDS-PAGE confirmed the lipase's MW to be 70 kDa. Using tributyrin as a substrate, the optimal pH and temperature of the lipase from marine snail hepatopancreas were 8.5 and 50 °C, respectively. Grey mullet viscera were used to extract lipase, which was then purified by ammonium sulfate fractionation (saturation level: 60%), simultaneous desalting and concentration by ultrafiltration, and affinity chromatography on cholate-EAH-Sepharose 4B (Aryee et al. 2007). The obtained lipase was stable between pH 4.0 and 10.0 and active between pH 7.0 and 10.0, with an optimal pH of 8.0. The optimum temperature for the hydrolysis of *p*-nitrophenyl palmitate (*p*-NPP) by the purified lipase was 50 °C, and the purified enzyme was stable between 10 and 50 °C. Rivera-Pérez et al. (2011) used Superdex 200 size exclusion and Resource Q anionic exchange column chromatography to purify lipase from the midgut gland of whiteleg shrimp. The purified lipase had a MW of 44.8 kDa, and specific activities of 1,787 and 475 Units/mg as determined using triolein and tributyrin as substrates, respectively, at pH 8.0 and 30 °C in the absence of colipase. Natural detergents, like sodium deoxycholate, are effective lipase inhibitors. Kameshwar Sharma et al. (2014) also purified lipase from digestive tract of Indian major carp using ammonium sulfate fractionation (20–80%), dialysis, and anion exchange column chromatography on DEAE-Cellulose. The MW of the purified enzyme was 70 kDa, and when *p*-NPP was used as the substrate, its optimum temperature and pH were 20 and 7.8 °C, respectively.

Recently, recovery and purification of lipase from the viscera of Nile tilapia by thermoseparating aqueous two-phase system (T-ATPS) were carried out (Patchimpet et al. 2021). When using *p*-NPP as a substrate, the fractionated enzyme showed maximum activity at pH 8.5 and 40 °C. At a temperature range of 0–40 °C and a pH range of 8–10, the fractionated lipase from Nile tilapia viscera was stable. It showed high tolerance to both ethanol and methanol (Fig. 8). Moreover, using an aqueous two-phase system (ATPS) and a three-phase partitioning (TPP) system, a lipase was recovered and isolated from the hepatopancreas of Pacific white shrimp (Kuepethkaew et al. 2017a). The isolated lipase was stable at a temperature range of 0–40 °C and a pH range of 7.0–10.0, and it showed its highest activity at pH 8.0 and 55 °C (Fig. 9). Moreover, the isolated lipase demonstrated good tolerance to

Table 6 Characteristics of lipases from digestive organs of fish

Fish species	Origin	Molecular weight (kDa)	Optimum		Substrate	References
			pH	Temperature (°C)		
Nile tilapia	Viscera	~64, 33, 28, 27, 26	8.5	40	*p*-nitrophenyl palmitate	Patchimpet et al. (2021)
Seabass	Liver	60	8.0	50	*p*-nitrophenyl palmitate	Sae-leaw and Benjakul (2018)
Pacific white shrimp	Hepatopancreas	~45, 34, 24	8.0	55	*p*-nitrophenyl palmitate	Kuepethkaew et al. (2017a)
Common stingray	Pancreas	55	ND	ND	Olive oil	Bouchaâla et al. (2015)
Indian major carp	Digestive gut	70	7.8	20	*p*-nitrophenyl palmitate	Kameshwar Sharma et al. (2014)
Golden grey mullet	Viscera	~35	8.0	50	Tributyrin	Smichi et al. (2013)
Marine snail	Hepatopancreas	~70	8.5	50	Tributyrin	Zarai et al. (2012)
Whiteleg shrimp	Midgut gland	44.8	8.0	30	Triolein	Rivera-Pèrez, et al. (2011)
Sardine	Pyloric caeca	~73	9.0	37	Tributyrin	Smichi et al. (2010)
Crab	Hepatopancreas	ND	ND	60	Olive oil	Cherif and Gargouri (2009)
Grey mullet	Viscera	ND	8.0	50	*p*-nitrophenyl palmitate	Aryee et al. (2007)
Crayfish	Digestive gland	~118, 63, 46, 43	8.5	35–40	β-Naphthyl caprylate	López-López et al. (2003)

ND: not determined

both ethanol and methanol, indicating that it might have uses in enzymatic biodiesel manufacturing.

2.2.2 Fish Lipase Applications

Over a decade ago, lipases were widely regarded as having immense potential for applications across numerous industrial fields (Kurtovic et al. 2009). Some progress has since been made, and presently there have been many such applications (Kuepethkaew et al. 2017b, c; Patchimpet et al. 2019, 2020), with many more

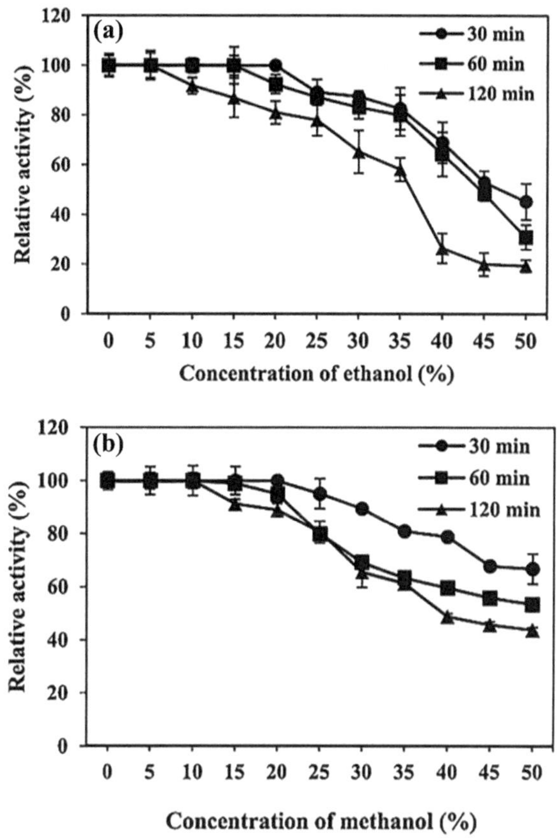

Fig. 8 Ethanol (**a**) and methanol (**b**) stability of fractionated lipase from Nile tilapia viscera. *Source* Patchimpet et al. (2021)

potential applications of lipases still to be discovered. Lipases are highly versatile catalysts due to their broad range of specificities, which makes them potentially valuable in numerous applications.

– *Biodiesel production*

By utilizing lipases as biocatalysts instead of more conventional alkali or acid catalysts, it is convenient to manufacture biodiesel through a transesterification process between triglycerides and short-chain alcohol (Kurtovic et al. 2009). While chemical transesterification using alkali catalysis yields high and rapid conversion levels of triglycerides to their corresponding methyl esters, the process has several drawbacks. It is energy-intensive, recovery of glycerol is challenging, removal of the acidic or alkaline catalyst from the product is required, alkaline wastewater necessitates treatment, and the presence of free fatty acids and water can interfere with the reaction. As demonstrated in Table 7, extracellular and intracellular lipases may efficiently catalyze the transesterification of triglycerides in both aqueous and nonaqueous conditions. It is worth noting that the by-product glycerol can be easily recovered

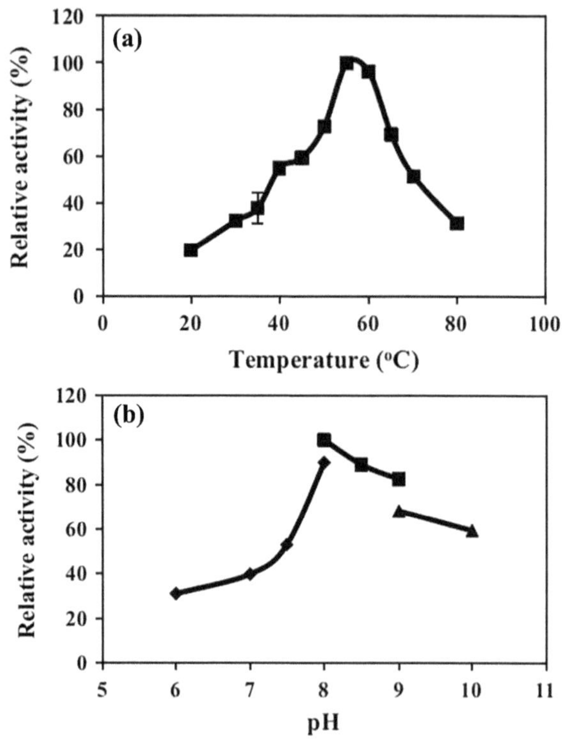

Fig. 9 Temperature (**a**) and pH (**b**) profiles of fractionated lipase from Pacific white shrimp hepatopancreas. *Source* Kuepethkaew et al. (2017a)

without the need for complex processing. Free fatty acids present in waste oils and fats can be fully converted to methyl esters (Fig. 10).

The utilization of lipases for the production of biodiesel fuel has tremendous potential when compared to chemical catalysis (Patchimpet et al. 2019, 2020). However, the use of lipases for commercial biodiesel production is currently costly. To mitigate this cost, the use of inexpensive lipases derived from seafood processing waste, particularly from viscera, may serve as a viable alternative (Patchimpet et al. 2020). A low-cost hepatopancreas lipase was extracted from Pacific white shrimp waste, and has been identified as a promising alternative for use as a biodiesel catalyst. To recover and purify the lipase, a successful process using three-phase partitioning (TPP) and aqueous two-phase system (ATPS) has been developed. It has been demonstrated that the Pacific white shrimp lipase is a useful assistance for producing biodiesel from palm oil (Kuepethkaew et al. 2017b). Optimal conditions for biodiesel preparation were achieved with a 70 kUnit lipase, 4:1 methanol to oil molar ratio, 3% water, 45 °C reaction temperature, and 16-h reaction time, resulting in a maximum yield of 97.01%. The obtained biodiesel characteristics complied with the requirements given by EN 14214 and ASTM D 6751 (Table 8).

– *Detergent additive*

Commercially, lipases are most often used in home detergents, washing powders, and general cleansers. As compared to traditional cleaning agents, detergents with

Table 7 Comparison of chemical and enzymatic methods for biodiesel production

Parameter	Chemical process		Enzymatic process
	Acid process	Alkaline process	
Biodiesel yield	>90%	>96%	>96%
Free fatty acid content in the substrate	Converted to biodiesel	Soap formation	Converted to biodiesel
Water in substrate	Interference with reaction	Interference with reaction	No influence
Purification of methyl esters	Repeated washing	Repeated washing	None
Glycerol recovery	Complex, low-grade glycerol	Complex, low-grade glycerol	Easy, high-grade glycerol
Reaction rate	Slower than for alkaline process	High	Low
Reaction temperature	>100 °C	60–80 °C	20–50 °C
Energy cost	High	Medium	Low
Catalyst reuse	No reusability	Partially lost in post-processing steps	Reusable
Waste water generation	High	High	Low

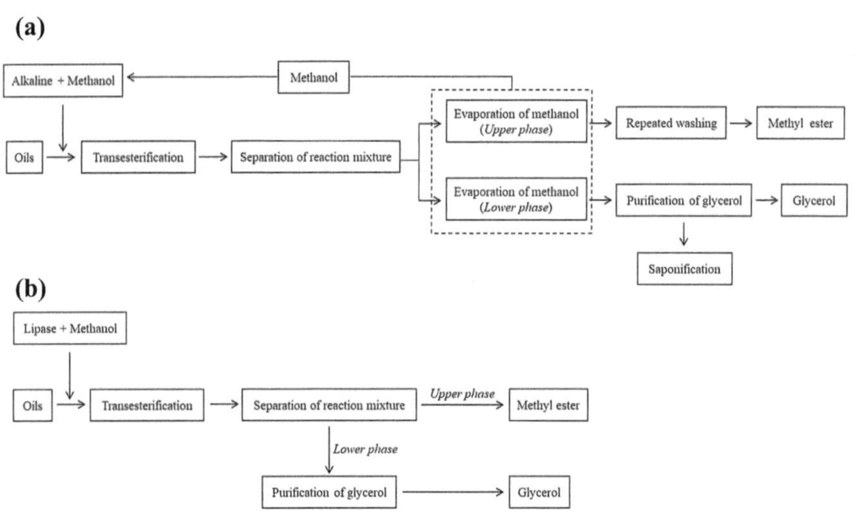

Fig. 10 Flow diagrams comparing biodiesel production using alkali- (**a**) and lipase- (**b**) catalysis processes

Table 8 Some properties of Pacific hepatopancreas lipase-catalyzed biodiesel in comparison with EN 14214 and ASTM D 6751

Properties	Methods	Lipase-catalyzed biodiesel	Biodiesel specification	
			EN 14214	ASTM D 6751
Fatty acid methyl esters content (%, w/w)	EN 14103	97.01 ± 1.89	Min. 96.5	–
Viscosity at 40 °C	ASTM D 445-06	4.04 ± 0.12	3.5–5.0	1.9–6.0
Acid value (mg KOH/g)	ASTM D664-01	0.47 ± 0.03	Max. 0.5	Max. 05
Water content (mg/kg)	EN ISO 12937	320.62 ± 0.08	Max. 500	–
Iodine value (g I_2/100 g)	EN 14111	2.33 ± 0.13	Max. 120	–
Density (15 °C) (kg/m^3)	ASTM D 4052-96	868.52 ± 3.88	860–900	870–890
Pour point (°C)	ASTM D 5950-02	1.00 ± 0.50	–	Report
Cloud point (°C)	ASTM D 2500	8.50 ± 0.50	–	Report

Source Kuepethkaew et al. (2017b)

lipases can save energy, decrease the impact on the environment, and effectively remove lipid stains (Kurtovic et al. 2009). Enzymes, such as lipase, have been incorporated into detergent formulations as additives. This development was driven by the desire to conserve energy by washing clothes at lower temperatures, as well as by environmental concerns regarding the use of non-biodegradable chemicals and the need to reduce pollution (Naganthran et al. 2017). Lipases must be alkalophilic and uninhibited by various ingredients in the detergent formulation in order to be effective (Pandey et al. 1999). The lipase extracted from the hepatopancreas of Pacific white shrimp exhibited robust stability toward 1% non-ionic surfactants, including Tween20, Tween80, and Triton X-100. The residual activity remained above 86.80% (Kuepethkaew et al. 2017c). All of the evaluated commercial liquid and solid detergents including Attack®, Bres®, Omo®, and Pao® had no effect on the Pacific white shrimp lipase. The experiment indicated strong lipase resistance for 30 and 60 min, with maintained activity exceeding 75% (Fig. 11).

2.3 *Transglutaminases*

Transglutaminase (TGase) (EC 2.3.2.13), transferase or protein glutamine γ-glutamyltransferase, is an endogenous fish enzyme that catalyzes the acyl transfer reaction between γ-carboxamide groups of glutamine residues in proteins, polypeptides, and a variety of primary amines. TGase is a sulphydryl enzyme with a conserved

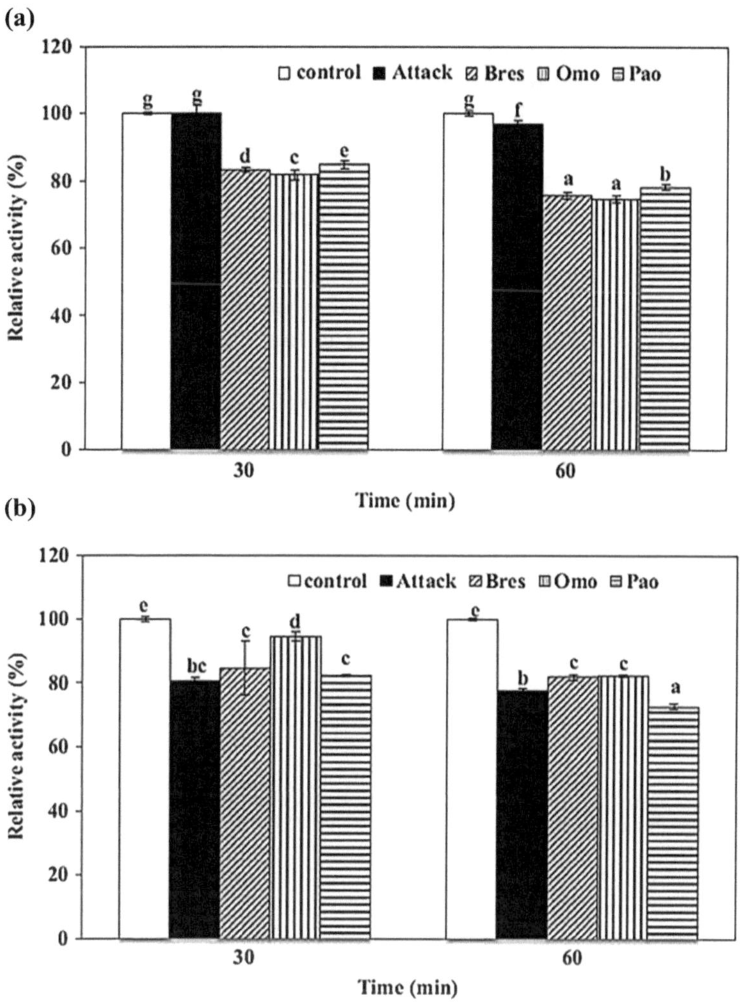

Fig. 11 Stability of lipase from Pacific white shrimp hepatopancreas in presence of various commercial liquid (**a**) and solid (**b**) laundry detergents. *Source* Kuepethkaew et al. (2017c)

pentapeptide active site sequence (Tyr-Gly Gln-Cys-Trp). TGases have been characterized from marine animals and are shown to be monomeric proteins (Likhar and Chudasama 2021). When acceptors are ε-amino groups of lysine residues, the formation of ε-(γ-glutamyl) lysine linkages occur both intra- and intermolecularly. This is accomplished by exchanging the ε-amino groups of the lysine residues for the ammonia at the γ-carboxyamide group of a glutamine residue in a protein molecule (Fig. 12). These linkages are stable and proteolysis-resistant. The sequence and charge of amino acids around the susceptible glutamine residue, as well as local secondary structures, have been found to influence crosslinking capacity (Klomklao et al. 2012). The capacity of TGase to change the physical properties of protein foods is based on the formation of covalent cross-links between proteins (Klomklao et al. 2012). The enzyme-promoted reactions cause dramatic changes in the proteins in food matrices, resulting in enhanced texture and temperature stability, syneresis, emulsifying qualities, gelation, and enhanced water-binding capacity without altering the pH, color, flavor, or nutritional quality of the food, and may even make it more nutritious owing to the potential of incorporating essential amino acids (Gaspar and Góes-Favoni, 2015).

Generally, TGases have been found in the muscle of fish. Fish viscera TGase has been recovered and biochemically characterized in recent years, and as a result, some intriguing novel applications of this TGase in food manufacturing have emerged (Table 9). Hemung and Yongsawatdigul (2008) used ion exchange, gel filtration, and affinity chromatography to recover and purify TGase from the liver of threadfin

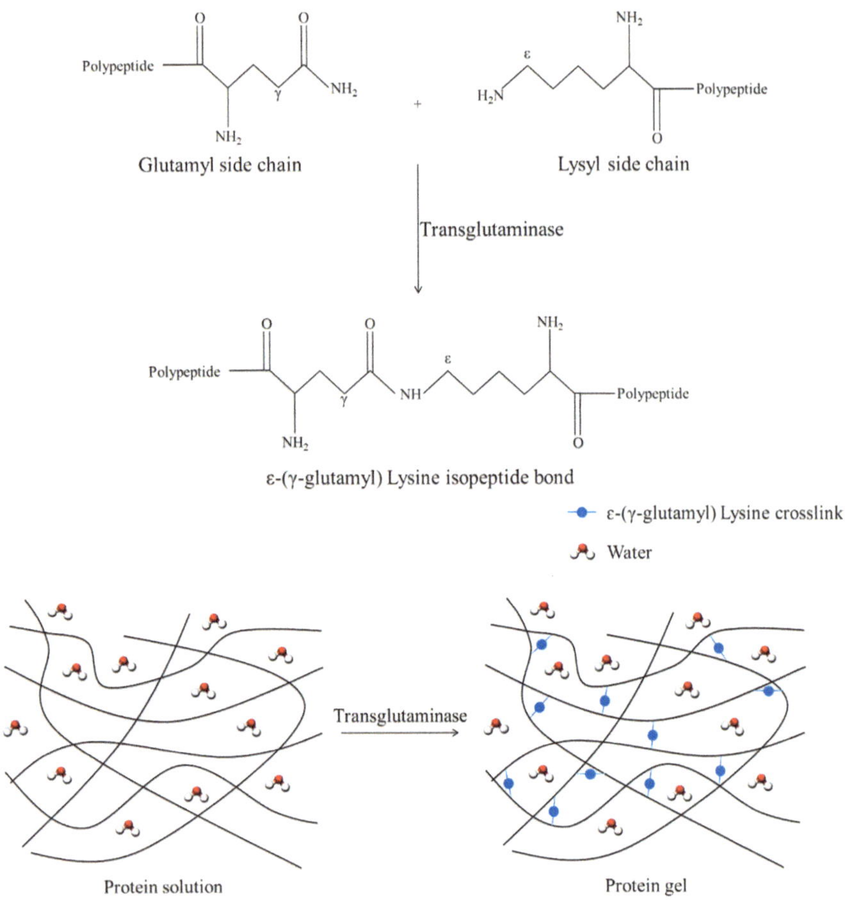

Fig. 12 Cross-linking of proteins by transglutaminase

bream. SDS-PAGE revealed three protein bands with molecular weights of 95, 63, and 43 kDa. pH and temperature were found to be optimal at 8.5–9.0 and 50 °C, respectively. According to Nozawat et al. (1997), fish muscle and viscera TGase was shown to be calcium-dependent. TGase from red sea bream liver, Japanese oyster, scallop, and pollock liver had optimal Ca^{2+} values of 0.5, 25, 10, and 3 mM, respectively (Kumazawa et al. 1996; Yasueda et al. 1994; Nozawa and Seki 2001). It was suggested that the calcium ion caused conformational changes in the enzyme, exposing the cysteine in the active site to the substrate (Ashie and Lanier 2000). According to Noguchi et al. (2001), the calcium ion coupled to a binding site of the red sea bream TGase molecule, causing conformational alterations. Following that, the Tyr covering the catalysis Cys was deleted. The acyl donor then binds to the Cys in the active site, resulting in the formation of an acyl-enzyme intermediate.

Table 9 Enzymatic characteristics of transglutaminase obtained from fish and shellfish

Fish/shellfish source	Molecular weight (kDa)	Activity (Units/mg)	Optimum		References
			pH	Temperature (°C)	
Crayfish	86	ND	8.5	22	Sirikharin et al. (2018)
Antarctic krill	76	53.52	8.0–9.0	0–10	Zhang et al. (2017)
Bigeye snapper	73–95	29.54	ND	37	Binsi and Shamasundar (2012)
Indian oil sardine	73–95	24.85	ND	37	Binsi and Shamasundar (2012)
Tilapia	73–95	22.66	ND	50	Binsi and Shamasundar (2012)
Common carp	73–95	33.01	ND	37	Binsi and Shamasundar (2012)
Threadfin bream liver	95	3,920.35	8.5–9.0	50	Hemung and Yongsawatdigul (2008)
Tropical tilapia	85	196.90	7.5	50	Worratao and Yongsawatdigul (2005)
European sardine	ND	ND	ND	35	Batista et al. (2002)
Squid fill	94	216.3	8.0	20	Nozawa et al. (2001)

ND: not determined

2.3.1 Fish TGase Applications

TGases play a critical role in numerous industries, including food processing, chemical production, pharmaceuticals, biotechnology, textiles, and many more. Foods such as yogurt, cheeses, tofu, and surimi (fish paste), among others with unique physical properties are produced by protein gelation.

TGase, an endogenous enzyme present in fish, is responsible for protein cross-linking and changing the textural characteristics of fish sol during manufacturing, especially in the preparation of surimi. Surimi is a fish mince that is refined through a process of leaching the soluble fraction via repeated washing with water, straining to remove any remaining connective tissue elements, and separating lipids and other undesirable components. Upon heating, surimi forms thermos-irreversible gels that consist of a three-dimensional protein network, primarily composed of actomyosin that is a complex of actin, tropomyosin, troponin, and myosin filaments) (Ohshima et al. 1993; Sikorski et al. 1995). This setting phenomena, known as suwari, is a crucial procedure in the production of surimi-based products. Fish proteins, especially myosin, are cross-linked by endogenous TGase in the temperature range of 5–40 °C, resulting in the network formation known as setting or suwari. Araki and Seki (1993) reported that between 0.10 and 2.41 Units/g of TGase are observed in fish muscle. The function of TGase in setting the surimi from bigeye snapper, threadfin bream, barracuda, and bigeye croaker was validated by Benjakul et al. in 2004. It promoted the development of non-disulfide covalent bonds, particularly when adequate calcium ion was present. The cross-linking of myosin heavy chain was successfully catalyzed by TGase from threadfin bream liver (Hemung and Yongsawatdigul 2008). The TGase from threadfin bream liver exhibited high activity (50%) over a wide pH range (pH 6–10) and relatively greater NaCl level. Recent research found that adding 0.1 Units/mg of TGase from Antarctic krill considerably increased gel strength compared to cold-set gelatin gel without TGase (Zhang et al. 2017).

2.4 *Chitinase*

Chitinolytic enzymes break down chitin and can be classified based on their degradation mechanisms as endo-type chitinolytic enzymes, which yield chitooligosaccharides (GlcNAc)n by randomly degrading chitin internally, and exo-type chitinolytic enzyme, that generate *N*-acetyl-D-glucosamine (GlcNAc) by sequentially degrading chitin from the nonreducing end (Patil et al. 2000; Kakizaki et al. 2015). Based on the amino acid sequence homology within the catalytic domain, the former is known as chitinases (EC 3.2.1.14) and is a member of the glycoside hydrolase (GH) family 18 or 19 (Kakizaki et al. 2015). The latter is known as β-*N*-acetylhexosaminidase (Hex) (EC 3.2.1.52). Chitinases are found in a wide variety of organisms and have been extracted and identified from various tissues of aquatic species (Koga et al.

1999; Fukamizo 2000; Patil et al. 2000). Significant physiological functions of chitinases include morphogenesis, defense, and aggression. Chitinases from fish and mollusk digest chitin present in prey (Gutowska et al. 2004; Matsumiya et al. 1998). Chitinases have been isolated and biochemical characterized from the stomach of several Osteichthyes (Matsumiya et al. 2006; Ikeda et al. 2014), the livers of Japanese common squid (Matsumiya et al. 2002) and golden cuttlefish (Nishino et al. 2014), and the posterior salivary gland of common octopus (Ogino et al. 2014). Chitinases genes were cloned from several Osteichthyes stomach (Ikeda et al. 2014; Kurokawa et al. 2004; Kakizaki et al. 2015), the Chondrichthyes blue shark stomach (Suzuki et al. 2014), the swimming crab hepatopancreas (Fujitani et al. 2014), and several shrimps tissues (Proespraiwong et al. 2010). According to reports, fish chitinases are very active in the stomach (Matsumiya et al. 2006; Ikeda et al. 2009, 2012, 2013). As documented by Matsumiya et al. (2006), Ikeda et al. (2009, 2014), fish stomach chitinases hydrolyzed crystalline α- or β-chitin and crystalline chitin nanofibers with consistent widths of roughly 10–20 nm that were made from crab chitin flakes as gels and solubilized in water.

2.4.1 Fish Chitinase Applications

Chitin is a polysaccharide composed of amino sugars linked by β-1,4 glycosidic bonds, with *N*-acetyl-D-glucosamine (GlcNAc) units as the primary component. It is a renewable biological resource that is widely available worldwide, found in the exoskeletons of arthropods, the cell walls of fungi, and the epidermis of nematodes (Karthik et al. 2014; Khoushab and Yamabhai 2010; Rinaudo 2006). Most naturally occurring chitin is in a rigid, α-crystalline structure that is insoluble in typical solvents, making it hard to employ (Ravi Kumar 2000). Nevertheless, chitin degradation products display a variety of bioactivities that have been related to polymer length and solubility (Liu et al. 2014). They include the enhancement of bifidobacterial proliferation and immunostimulatory impact of chitooligosaccharides ((GlcNAc)n) (Wang et al. 2011; Mei et al. 2013) as well as the improvement of osteoarthritis in GlcNAc (Matsumiya 2004; Kurita 2006; Chen et al. 2010; Kakizaki et al. 2014).

Chitinolytic enzymes are a valuable tool for breaking down chitin into oligomeric units, without relying on chemical depolymerization methods that use concentrated hydrochloric acid and often result in commercial products with inconsistent physicochemical properties. Research on the preparation of the physiological activities of chitin and chitosan oligomers has attracted significant attention in the food and pharmaceutical industries due to their diverse antimicrobial and antitumor properties, immuno-enhancing effects, and protective effects against certain infectious pathogens (Shahidi et al. 1999; Shahidi and Kamil 2001). Affordable chitinolytic enzymes with enhanced catalytic activity are in high demand. Fish chitinase has potential applications in the production of a variety of health products derived from chitin. Thus, marine fishery waste is a promising source of chitinases with potent catalytic activity (Kim and Dewapriya 2014).

3 Conclusions

The genetic diversity observed within the marine environment makes it a tremendously promising source of enzymes. Due to the biological diversity of marine animals, a diverse range of enzymes with distinct properties can be recovered and used successfully. The body of knowledge on the characteristics and mechanisms of action of enzymes from aquatic animals, especially fish manufacturing by-products materials, is quickly expanding, allowing their use in a wide range of industrial applications. Enzymatic approaches serve an important role in mitigating the environmental impact of conventional processes, making them more sustainable and cost-effective.

References

Ahmad M, Benjakul S (2010) Extraction and characterization of pepsin-solubilised collagen from the skin of unicorn leatherjacket (*Aluterus monocerous*). Food Chem 120:817–824

Aoki H, Ahsan N, Matsuo K, Hagiwara T, Watabe S (2003) Purification and characterization of collagenolytic protease from the hepatopancreas of Northern shrimp (*Pandalus eous*). J Agric Food Chem 51:777–783

Araki H, Seki N (1993) Comparison of reactivity of transglutaminase to various fish actomyosins. Nippon Suisan Gakkaishi 95:711–716

Aranishi F, Hara K, Osatomi K, Ishihara T (1997a) Purification and characterization of cathepsin B from hepatopancreas of carp *Cyprinus carpio*. Comp Biochem Physiol B 1117(4):579–587

Aranishi F, Ogata H, Hara K, Ishihara T (1997b) Purification and characterization of cathepsin L from hepatopancreas of carp (*Cyprinus carpio*). Comp Biochem Physiol B 118:531–537

Aryee NA, Simpson BK, Villalonga R (2007) Lipase fraction from the viscera of grey mullet (*Mugil cephalus*): isolation, partial purification and some biochemical characteristics. Enzyme Microb Technol 40:394–402

Asgeirsson B, Bjarnason JB (1991) 'Structural and kinetic properties of chymotrypsin from Atlantic cod (*Gadus morhua*). Comparison with bovine trypsin. Comp Biochem Physiol Part B Biochem Mol Biol 9:327–335

Ashie INA, Lanier TC (2000) Transglutaminase in seafood processing. In: Haard NF, Simpson BK (eds) Seafood enzymes: utilization and influence on postharvest seafood quality. Maecel Dekker, New York, pp 147–166

Auttanak B, Chomnawang P, Huaisan K, Sootsuwan K, Chomnawang C (2022) Influences of salt and koji contents on chemical properties of fish sauce from tilapia frames. Aquac Fish 7:67–73

Balti R, Barkia A, Bougatef A, Ktari N (2009) A heat-stable trypsin from the hepatopancreas of the cuttlefish (*Sepia officinalis*): purification and characterisation. Food Chem 113(1):146–154

Balti R, Bougherra F, Bougatef A, Hayet BK, Nedjar-Arroume N, Dhulster P, Guillochon D, Nasri M (2012) Chymotrypsin from the hepatopancreas of cuttlefish (*Sepia officinalis*) with high activity in the hydrolysis of long chain peptide substrates: purification and biochemical characterisation. Food Chem 130(3):475–484

Balti R, Nedjar-Arroume N, Bougatef A, Guillochon D, Nasri M (2010) Three novel angiotensin I-converting enzyme (ACE) inhibitory peptides from cuttlefish (*Sepia officinalis*) using digestive proteases. 43(4):1136–1143

Barkia A, Bougatef A, Khaled HB, Nasri M (2010) Antioxidant activities of sardinella heads and/or viscera protein hydrolysates prepared by enzymatic treatment. J Food Biochem 34:303–320

Batista I, Salteiro AT, Mateus MJ (2002) Preliminary characterization of European sardine transglutaminase. J Aquat Food Prod Technol 11(3–4):57–64
Benjakul S, Klomklao S, Simpson BK (2010) Enzymes in fish processing. In: Whitehurst RJ, Oort M (eds) Enzymes in food technology, 2nd edn. Blackwell Publishing Ltd., pp 211–235
Benjakul S, Visessanguan W, Pecharat S (2004) Suwari gel properties as affected by transglutaminase activator and inhibitors. Food Chem 85:91–99
Benjakul S, Yarnpakdee S, Senphan T, Halldorsdottir SM, Kristinsson HG (2014) Fish protein hydrolysates: production, bioactivities, and applications. In: Kristinsson HG, Raghavan S (eds) Antioxidants and functional components in aquatic foods, 1st edn. Wiley, pp 237–281
Binsi PK, Shamasundar BA (2012) Purification and characterisation of transglutaminase from four fish species: effect of added transglutaminase on the viscoelastic behaviour of fish mince. Food Chem 132:1922–1929
Bkhairia I, Khaled HB, Ktari N, Miled N, Nasri M, Ghorbel S (2016) Biochemical and molecular characterisation of a new alkaline trypsin from *Liza aurata*: structural features explaining thermal stability. Food Chem 196:1346–1354
Bouchaâla E, Ali MB, Ali YB, Miled N, Gargouri Y, Fendri A (2015) Biochemical characterization and molecular modeling of pancreatic lipase from a cartilaginous fish, the common stingray (*Dasyatis pastinaca*). Biotechnol Appl Biochem 176(1):151–169
Bougatef A, Balti R, Zaied SB, Souissi N, Nasri M (2008) Pepsinogen and pepsin from the stomach of smooth hound (*Mustelus mustelus*): purification, characterization and amino acid terminal sequences. Food Chem 107:777–784
Bougatef A, Hajji M, Balti R, Lassoued I, Triki-Ellouz Y, Nasri M (2009) Antioxidant and free radical-scavenging activities of smooth hound (*Mustelus mustelus*) muscle protein hydrolysates obtained by gastrointestinal proteases. Food Chem 114:1198–1205
Bougatef A, Nedjar-Arroume N, Manni K, Ravallec R, Barkia A, Guillochon D, Nasri M (2010) Purification and identification of novel antioxidant peptides from enzymatic hydrolysates of sardinella (*Sardinella aurita*) by-proteins proteins. Food Chem 118:559–565
Brewer P, Helbig N, Haard NF (1984) Atlantic cod pepsin—characterization and use as a rennet substitute. Can Inst Food Sci Technol J 17:38–43
Byun HG, Park PJ, Sung NJ, Kim SK (2003) Purification and characterization of a serine proteinase from the tuna pyloric caeca. J Food Biochem 26:479–494
Cai QF, Jiang YK, Zhou LG, Sun LC, Liu GM, Osatomi K, Cao MJ (2011) Biochemical characterization of trypsins from the hepatopancreas of Japanese sea bass (*Lateolabrax japonicus*). Comp Biochem Physiol B Biochem Mol Biol 159:183–189
Cao MJ, Chen WQ, Du CH, Yoshida A, Lan WG, Liu GM, Su WJ (2011) Pepsinogens and pepsins from Japanese seabass (*Lateolabrax japonicus*). Comp Biochem Physiol B Biochem Mol Biol 158:259–265
Cao MJ, Osatomi K, Suzuki M, Hara K, Tachibana K, Ishihara T (2000) Purification and characterization of two anionic trypsins from the hepatopancreas of carp. Fish Sci 66:1172–1179
Cardenas-Lopez JL, Haard N (2009) Identification of a cysteine proteinase from jumbo squid (*Dosidicus gigas*) hepatopancreas as cathepsin L. Food Chem 112:442–447
Casas-Goboy L, Duquesne S, Bordes F, Sandoval G, Marty A (2012) Lipases: an overview. Methods Mol Biol 861:3–30
Castillo-Yanez FJ, Pacheco-Aguialr RP, Garcia-Carreno FL, Toro MAND (2004) Characterization of acidic proteolytic enzyme from Montery sardine (*Sardinops sagax caerulea*) viscera. Food Chem 85:343–350
Castillo-Yáñez FJ, Pacheco-Aguilar R, García-Carreño FL, Toro MAND, López MF (2006) Purification and biochemical characterization of chymotrypsin from the viscera of Monterey sardine (*Sardinops sagax caeruleus*). Food Chem 99(2):252–259
Celus I, Brijs K, Delcour JA (2007) Enzymatic hydrolysis of brewers' spent grain proteins and technofunctional properties of the resulting hydrolysates. J Agric Food Chem 55:8703–8710
Chen JK, Shen CR, Liu CL (2010) *N*-Acetylglucosamine: production and applications. Mar Drugs 8:2493–2516

Chen L, Zhao L, Liu H, Wu R (2011) Extraction and partial characterization of pepsin-soluble collagens from the skin of Amiurus nebulosus. Adv Mater Res 236–238:2926–2934
Chen WQ, Cao MJ, Yoshida A, Liu GM, Weng WY, Sun LC, Su WJ (2009) Study on pepsinogens and pepsins from snakehead (*Channa argus*). J Agric Food Chem 57:10972–10978
Cherif S, Gargouri Y (2009) Thermoactivity and effects of organic solvents on digestive lipase from hepatopancreas of the green crab. Food Chem 116:82–86
Daboor SM, Budge SM, Ghaly AE, Brooks MS, Dave D (2012) Isolation and activation of collagenase from fish processing waste. Adv Biosci Biotechnol 3:191–203
Díaz-López M, García-Carreño FL (2020) Application of fish and shellfish enzyme in food and feed products. In: Haard NF, Simpson BK (eds) Seafood enzymes. Marcel Dekker, New York, pp 571–618
Fernandes F (2016) Enzymes in fish and seafood processing. Front Bioeng Biotechnol 59(4):1–14
Freitas-Júnior ACV, Costa HMS, Icimoto MY, Hirata IY, Marcondes M, Carvalho LB, Oliveira V, Bezerra RS (2012) Giant Amazonian fish pirarucu (*Arapaima gigas*): its viscera as a source of thermostable trypsin. Food Chem 133(4):1596–1602
Fuchise T, Kishimura H, Sekizaki H, Nonami Y, Kanno G, Klomklao S, Benjakul S, Chun BS (2009) Purification and characteristics of trypsins from cold-zone fish, Pacific cod (Gadus macrocephalus) and saffron cod (*Eleginus gracilis*). Food Chem 116:611–616
Fujitani N, Hasegawa H, Kakizaki H, Ikeda M, Matsumiya M (2014) Molecular cloning of multiple chitinase genes in swimming crab *Portunus trituberculatus*. J Chitin Chitosan Sci 2:149–156
Fukamizo T (2000) Chitinolytic enzymes: catalysis, substrate binding, and their application. Curr Protein Pept Sci 1:105–124
Gaspar ALC, Góes-Favoni SP (2015) Action of microbial transglutaminase (MTGase) in the modification of food proteins: a review. Food Chem 171:315–322
Ghaly AE, Ramakrishnan VV, Brooks MS, Budge SM, Dave D (2013) Fish processing wastes as a potential source of proteins, amino acids and oils: a critical review. J Microb Biochem Technol 5(4):107–129
Gildberg A (2001) Utilisation of male Arctic capelin and Atlantic cod intestines fish sauce production—evaluation of fermentation conditions. Biores Technol 76:119–123
Gildberg A, Raa J (1983) Purification and characterization of pepsins from the Arctic fish capelin (*Mallotus villosus*). Comp Biochem Physiol A Mol Integr Physiol 75:337–342
Gildberg A, Simpson BK, Haard NF (2000) Uses of enzymes from marine organisms. In: Haard NF, Simpson BK (eds) Seafood enzymes: utilization and influence in postharvest seafood quality. Marcel Dekker, New York, pp 619–640
Gomez AVA, Gomez G, Ghamorro E, Bustillo S, Leiva LG (2018) Digestive aspartic proteases from sábalo (*Prochilodus lineatus*): characterization and application for collagen extraction. Food Chem 269:610–617
Gordin S, Rosenthal I (1978) Efficacy of chicken pepsin as a milk clotting enzyme. J Food Prot 41:684–688
Gutowska MA, Drazen JC, Robison BH (2004) Digestive chitinolytic activity in marine fishes of Monterey Bay, California. Comp Biochem Physiol Part A Mol Integr Physiol 139:351–358
Haard NF (1986) Atlantic cod gastric protease. Characterization with casein and milk substrate and influence of sepharose immobilization on salt activation, temperature characteristics ad milk clotting reaction. J Food Sci 51:313
Haard NF, Simpson BK (1994) Proteases from aquatic organisms and their uses in the seafood industry. In: Martin AM (ed) Fisheries processing: biotechnology applications. Chapman and Hall, UK, pp 132–154
Hamzeh A, Benjakul S, Senphan T (2016) Comparative study on antioxidant activity of hydrolysates from splendid squid (*Loligo formosana*) gelatin and protein isolate prepared using protease from hepatopancreas of Pacific white shrimp (*Litopenaeus vannamei*). J Food Sci Technol 53(9):3615–3623
Hasan F, Shah AA, Hameed A (2006) Influence of culture conditions on lipase production by *Bacillus* sp. FH5. Ind Microbiol 56:247–252

Helgi G, Hjalmarsson J, Park JW, Kristbergsson K (2007) Seasonal effects on the physicochemical characteristics of fish sauce made from capelin (*Mallotus villosus*). Food Chem 103:495–504

Hemung BO, Yongsawatdigul J (2008) Partial purification and characterization of transglutaminase from threadfin bream (*Nemipterus* sp.) liver. J Food Biochem 32:182–200

Heu MS, Kim HR, Pyeun JH (1995) Comparison of trypsin and chymotrypsin from the viscera of anchovy, *Engraulis japonica*. Comp Biochem Physiol B Biochem Mol Biol 112(3):557–567

Hudáky P, Kaslik G, Venekei I, Gráf L (1999) The differential specificity of chymotrypsin A and B is determined by amino acid 226. Eur J Biochem 259:528–533

Ikeda M, Kondo Y, Matsumiya M (2013) Purification, characterization, and molecular cloning of chitinases from the stomach of the threeline grunt *Parapristipoma trilineatum*. Process Biochem 48:1324–1334

Ikeda M, Miyauchi K, Matsumiya M (2012) Purification and characterization of a 56 kDa chitinase isozyme (PaChiB) from the stomach of silver croaker *Pennahia argentatus*. Biosci Biotechnol Biochem 76:971–979

Ikeda M, Miyauchi K, Mochizuki A, Matsumiya M (2009) Purification and characterization of chitinase from the stomach of silver croaker *Pennahia argentatus*. Protein Expr Purif 65:214–222

Ikeda M, Shirase D, Sato T, Ueda M, Hirabayashi S, Matsumiya M (2014) Primary structure and enzymatic properties of chitinase isozymes purified from the stomach of the marbled rockfish *Sebastiscus marmoratus*. J Chitin Chitosan Sci 2:106–116

Jafari H, Lista A, Siekapen MM, Ghaffari-Bohlouli P, Nie L, Alimoradi H, Shavandi A (2020) Fish collagen: extraction, characterization, and applications for biomaterials engineering. Polymers 12(10):2230

Je J, Park PJ, Kim S (2005) Antioxidant activity of peptide isolated from Alaska Pollack (*Theragra chalcogramma*) frame protein hydrolysate. Food Res Int 38(1):45–50

Jiang JJ, Zeng QX, Zhu ZW, Zhang L (2007) Chemical and sensory changes associated with Yu-lu fermentation process—a traditional Chiness fish sauce. Food Chem 104:1629–1634

Jun SY, Park PJ, Jung QK, Kim SK (2004) Purification and characterization of an antioxidative peptide from enzymatic hydrolysate of yellowfin sole (*Limanda aspera*). Eur Food Res Technol 219(1):20–26

Kaewdang O, Benjakul S, Kaewmanee T, Kishimura H (2014) Characteristics of collgens from the swim bladders of yellowfin tuna (*Thunnus albacores*). Food Chem 155:264–270

Kakizaki H, Hamaguchi K, Ikeda M, Matsumiya M (2014) Cloning of a novel chitinase cDNA from the stomach of the Coelacanth *Latimeria chalumnae* (sarcopterygii). J Chitin Chitosan Sci 2:123–129

Kakizaki H, Ikeda M, Fukushima H, Matsumiya M (2015) Distribution of chitinolytic enzymes in the organs and cDNA cloning of chitinase isozymes from the stomach of two species of fish, chub mackerel (*Scomber japonicus*) and silver croaker (*Pennahia argentata*). Open J Mar Sci 5:398–411

Kameshwar Sharma YVR, Boora N, Tyagi P (2014) Isolation, purification and characterization of secondary structure and kinetic study of lipase from Indian major carp, *Catla catla* (catla). Enzyme Eng 3(1):1–8

Karthik N, Akanksha K, Binod P, Pandey A (2014) Production, purification and properties of fungal chitinases—a review. Indian J Exp Biol 52:1025–1035

Ketnawa S, Benjakul S, Martínez-Alvarez O, Rawdkuen S (2017) Fish skin gelatin hydrolysates produced by visceral peptidase and bovine trypsin: bioactivity and stability. Food Chem 215:383–390

Khaled HB, Ghorbel-Bellaaj O, Hmidet N, Jellouli K, Ali NEH, Ghorbel S, Nasri M (2011) A novel aspartic protease from the viscera of Sardinella (*Sardinella aurita*): purification and characterization. Food Chem 128:847–853

Khandagale AS, Mundodi L, Sarojini K (2017) Isolation and characterization of trypsin from fish viscera of oil sardine (*Sardinella longiceps*). Fish Aquat Stud 5(2):33–37

Khangembam BK, Chakrabarti R (2015) Trypsin from the digestive system of carp *Cirrhinus mrigala*: purification, characterization and its potential application. Food Chem 175:386–394

Khantaphant S, Benjakul S (2010) Purification and characterization of trypsin from the pyloric caeca of brownstripe red snapper (*Lutjanus vitta*). Food Chem 120:658–664
Khantaphant S, Benjakul S, Kishimura H (2011) Antioxidative and ACE inhibitory activities of protein hydrolysates from the muscle of brownstripe red snapper prepared using pyloric caeca and commercial proteases. Process Biochem 46:318–327
Khoushab F, Yamabhai M (2010) Chitin research revisited. Mar Drugs 8:1988–2012
Kim SK, Dewapriya P (2014) Enzymes from fish processing waste materials and their commercial applications. In: Kim SK (ed) Seafood processing by-products. Springer, New York, pp 183–196
Kim SK, Jeon YJ, Byeun HG, Kim YT, Lee CK (1997) Enzymatic recovery of cod frame proteins with crude proteinase from tuna pyloric caeca. Fish Sci 63(3):421–427
Kim SK, Park PJ, Kim JB, Shahidi F (2002) Purification and characterization of a collagenolytic protease from the filefish, *Novoden modestrus*. J Biochem Mol Biol 35(2):165–171
Kim T, Silva JL, Parakulsuksatid P, Wang DS (2014) Optimization of enzymatic treatments for deskining of catfish nuggets. J Aquat Food Prod Technol 23:385–393
Kishimura H, Klomklao S, Benjakul S, Chun BS (2008) Characteristics of trypsin from the pyloric ceca of walleye pollock (*Theragra chalcogramma*). Food Chem 106(1):194–199
Kishimura H, Tokuda Y, Klomklao S, Benjakul S, Ando S (2006) Enzymatic characteristics of trypsin from pyloric ceca of spotted mackerel (*Scomber australasicus*). J Food Biochem 30:466–477
Kishimura H, Tokuda Y, Yabe M, Klomklao S, Benjakul S, Ando S (2007) Trypsins from the pyloric ceca of jacopever (*Sebastes schlegelii*) and elkhorn sculpin (*Alcichthys alcicornis*): isolation and characterization. Food Chem 100:1490–1495
Kitts DD, Weiler K (2003) Bioactive proteins and peptides from food sources: applications of bioprocesses used in isolation and recovery. Curr Pharm Des 9:1309–1323
Klomkalo S, Kishimura H, Nonami Y, Benjakul S (2009a) Biochemical properties of two isoforms of trypsin purified from the intestine of skipjack tuna (*Katsuwonus pelamis*). Food Chem 115(1):155–162
Klomklao S, Kishimura H, Benjakul S (2009b) Trypsin from the pyloric ceca of pectoral rattail (*Coryphaenoides pectoralis*): purification and characterization. J Agric Food Chem 57(15):7097–7103
Klomklao S, Benjakul S, Visessanguan W, Kishimura H, Simpson BK (2009c) Extraction of carotenoprotein from black tiger shrimp shell with the aid of bluefish trypsin. J Food Biochem 33:201–217
Klomklao S (2008) Digestive proteinase from marine organisms and their applications. Songklanakarin J Sci Technol 30(7):37–46
Klomklao S, Benjakul S (2018) Two trypsin isoforms from albacore tuna (*Thunnus alalunga*) liver: purification and physicochemical and biochemical characterization. Int J Biol Macromol 107:1864–1870
Klomklao S, Benjakul S, Kishimura H (2013) Functional properties and antioxidative activity of protein hydrolysates from toothed ponyfish muscle treated with viscera extract from hybrid catfish. Int J Food Sci Technol 48:1483–1489
Klomklao S, Benjakul S, Simpson BK (2012) Seafood enzymes: biochemical properties and their impact on quality. In: Simpson BK (ed) Food biochemistry and food processing, 2nd edn. Wiley, UK, pp 263–284
Klomklao S, Benjakul S, Visessanguan W, Kishimura H, Simpson BK (2006a) Purification and characterization of trypsin from the spleen of tongol tuna (*Thunnus tonggol*). J Agric Food Chem 54(15):5617–5622
Klomklao S, Benjakul S, Visessanguan W, Kishimura H, Sompson BK (2006b) Effects of the addition of skipjack tuna (*Katsuwonus pelamis*) on the liquefaction and characteristic of fish sauce made from sardine (*Sardinella gibbosa*). Food Chem 98:440–452
Klomklao S, Kishimura H, Yabe M, Benjakul S (2007a) Purification and characterization of two pepsins from the stomach of pectoral rattail (*Coryphaenoides pectoralis*). Comp Biochem Physiol B Biochem Mol Biol 147:628–689

Klomklao S, Benjakul S, Visessanguan W, Kishimura H, Simpson BK (2007b) Trypsin from the pyloric caeca of bluefish (*Pomatomus saltatrix*). Comp Biochem Physiol B Biochem Mol Biol 148(4):381–389

Klomklao S, Benjakul S, Visessanguan W, Kishimura H, Simpson BK (2007c) 29 kDa Trypsin from the pyloric ceca of Atlantic bonito (*Sarda sarda*): recovery and characterization. J Agric Food Chem 55(11):4548–4553

Klomklao S, Benjakul S, Visessanguan W, Simpson BK, Kishimura H (2005) Partitioning and recovery of proteinase from tuan spleen by aqueous two-phase system. Process Biochem 40:3061–3067

Klomklao S, Poonsin T, Benjakul S, Simpson BK (2020) Byproducts from shellfish harvesting and processing. In Simpson BK, Aryee ANA, Toldra F (eds) Byproducts from agriculture and fisheries: adding value for food, feed, pharma, and fuels. Wiley, pp 219–257

Koga D, Mitsutomi M, Kono M, Matsumiya M (1999) Biochemistry of chitinases. In: Jolles P, Muzzarelli RAA (eds) Chitin and chitinases. Birkhauser Publishing, Basel, pp 111–123

Kristinsson HG, Rasco BA (2000a) Fish protein hydrolysates: production, biochemical, and functional properties. Crit Rev Food Sci Nutr 40:43–81

Kristinsson HG, Rasco BA (2000b) Biochemical and functional properties of Atlantic salmon (*Salmon salar*) muscle proteins hydrolyzed with various alkaline proteases. J Agric Food Chem 48(3):657–666

Ktari N, Jridi M, Bkhairia I, Sayari N, Salah RB, Nasri M (2012a) Functionalities and antioxidant properties of protein hydrolysates from muscle of zebra blenny (*Salaria basilisca*) obtained with different crude protease extracts. Food Res Int 49(2):747–756

Ktari N, Khaled HB, Nasri R, Jellouli K, Ghorbel S, Nasri M (2012b) Trypsin from zebra blenny (*Salaria basilisca*) viscera: purification, characterization and potential application as a detergent additive. Food Chem 130:467–474

Kuepethkaew S, Sangkharak K, Benjakul S, Klomklao S (2017a) Use of TPP and ATPS for partitioning and recovery of lipase from Pacific white shrimp (*Litopenaeus vannamei*) hepatopancreas. J Food Sci Technol 54(12):3880–3891

Kuepethkaew S, Sangkharak K, Benjakul S, Klomklao S (2017b) Optimized synthesis of biodiesel using lipase from Pacific white shrimp (*Litopenaeus vannamei*) hepatopancreas. Renew Energy 104:139–147

Kuepethkaew S, Sangkharak K, Benjakul S, Klomklao S (2017c) Laundry detergent-stable lipase from Pacific white shrimp (*Litopenaeus vannamei*) hepatopancreas: effect of extraction media and biochemical characterization. Int J Food Prop 20(4):769–781

Kuepethkaew S, Zhang Y, Kishimura H, Kumagai Y, Simpson BK, Benjakul S, Damodaran S, Klomklao S (2022) Enzymological characteristics of pepsinogens and pepsins purified from lizardfish (*Saurida micropectoralis*) stomach. Food Chem 366:130532

Kumar NSS, Nazeer RA (2013) Characterization of acid and pepsin soluble collagen from the skin of horse mackerels (*Magalaspis cordyla*) and croaker (*Otolithes ruber*). Int J Food Prop 16:613–621

Kumazawa Y, Nakanishi K, Yasueda H, Motoki M (1996) Purification and characterization of transglutaminase from walleye pollack liver. Fish Sci 62:959–964

Kurita K (2006) Chitin and chitosan: functional biopolymers from marine crustaceans. Mar Biotechnol 8:203–226

Kurokawa T, Uji S, Suzuki T (2004) Molecular cloning of multiple chitinase genes in Japanese flounder, *Paralichthys olivaceus*. Comp Biochem Physiol B Biochem Mol Biol 138:255–264

Kurtovic I, Marshall S, Zhao X, Simpson BK (2009) Lipases from mammals and fishes. Rev Fish Sci 17(1):18–40

Le Chevalier P, Sellos S, Van Wormhoudt A (1995) Purification and partial characterization of chymotrypsin-like proteases from the digestive gland of the scallop *Pecten maximus*. Comp Biochem Physiol B Biochem Mol Biol 110(4):777–784

Lee LP, Karbul HM, Citartan M, Gopinath SGB, Lakshmipriya T, Tang TH (2015) Lipase-secreting *Bacillus* species in an oil-contaminated habitat: promising strains to alleviate oil pollution. Biomed Res Int 5(1):1–9

Leuba JL, Meyer I (1989) Skining of squid. In: Fifth international congress on engineering and food, Cologne, p 240

Likhar V, Chudasama BG (2021) Seafood enzymes and their potential industrial applicatons. J Entomol Zool Stud 9(1):1410–1417

Lim YS, Ok YJ, Hwang SY, Kwak JY, Yoon S (2019) Marine collagen as a promising biomaterial for biomedical applications. Mar Drugs 17:467

Liu CL, Lan CY, Fu CC, Juang RS (2014) Production of hexaoligochitin from colloidal chitin using a chitinase from *Aeromonas schubertii*. Int J Biol Macromol 69:59–63

Liu D, Liang L, Regenstein JM, Zhou P (2012) Extraction and characterisation of pepsin-solubilised collagen from fins, scales, skins, bones and swim bladders of bighead carp (*Hypophthalmichthys nobilis*). Food Chem 133(4):1441–1448

López-López S, Nolasco H, Vega-Villasante F (2003) Characterization of digestive gland esterase–lipase activity of juvenile redclaw crayfish *Cherax quadricarinatus*. Comp Biochem Physiol Part B Biochem Mol Biol 135(2):337–347

Manji B, Kakuda Y (1988) The role of protein denaturation, extent of proteolysis, and storage temperature on the mechanism of age gelation in a model system. J Dairy Sci 71:1455–1463

Matsumiya M (2004) Enzymatic production of *N*-acetyl-D-glucosamine using crude enzyme from the liver of squids. Food Sci Technol Res 10:296–299

Matsumiya M, Arakane Y, Haga A, Mathukrishnan S, Kramer KJ (2006) Substrate specificity of chitinases from two species of fish, greenling, *Hexagrammos otakii*, and common mackerel, *Scomber japonicus* and the insect, tabacco hornworm, *Manducca sexta*. Biosci Biotechnol Biochem 70(4):971–979

Matsumiya M, Miyauchi K, Mochizuki A (1998) Distribution of chitinase and *β-N*-acetylhexosaminidase in the organs of a few squid and a cuttlefish. Fish Sci 64:166–167

Matsumiya M, Miyauchi K, Mochizuki A (2002) Characterization of 38 kDa and 42 kDa chitinase isozymes from the liver of Japanese squid *Todarodes pacificus*. Fish Sci 68:603–609

McMahon DJ, Brown RJ (1985) Effects of enzyme type on milk coagulation. J Dairy Sci 68:628–632

Mei YX, Chen HX, Zhang J, Zhang XD, Liang YX (2013) Protective effect of chitooligosaccharides against cyclophosphamide-induced immunosuppression in mice. Int J Biol Macromol 62:330–335

Min B, Green BW (2008) Use of microbial transglutaminasse and nonmeat proteins to improve functional properties of low NaCl, phosphate-free patties made from channel catfish (*Ictalurus punctatus*) belly flap meat. J Food Sci 73:E218–E226

Miura Y, Kageyama T, Moriyama A (2015) Pepsinogens and pepsins from largemouth bass, *Micropterus salmoides*: purification and characterization with special reference to high proteolytic activities of bass enzymes. Comp Biochem Physiol B Biochem Mol Biol 183:42–48

Mizuta S, Yoshinaka R, Sato M, Sakaguchi M (1994) Isolation and partial characterization of two distinct types of collagen in the squid *Todarodes pacificus*. Fish Sci 60:467–471

Naganthran A, Masomian M, Rahman RNZRA, Ali MSM, Nooh HM (2017) Improving the efficiency of new automatic dishwashing detergent formulation by addition of thermostable lipase, protease and amylase. Molecules 22:1577

Nalinanon S, Benjakul S, Kishimura H (2010a) Biochemical properties of pepsinogen and pepsin from the stomach of albacore tuna (*Thunnus alalunga*). Food Chem 121:49–55

Nalinanon S, Benjakul S, Kishimura H (2010b) Purification and biochemical properties of pepsin from the stomach of skipjack tuna (*Katsuwonus pelamis*). Eur Food Res Technol 231:259–269

Nalinanon S, Benjakul S, Kishimura H, Shahidi F (2011) Functionalities and antioxidant properties of protein hydrolysates from the muscle of ornate threadfin bream treated with pepsin from skipjack tuna. Food Chem 124:1354–1362

Nalinanon S, Benjakul S, Visessanguan W, Kishimura H (2007) Use of pepsin for collagen extraction from the skin of bigeye snapper (*Priacanthus tayenus*). Food Chem 104:593–601

Nalinanon S, Benjakul S, Visessanguan W, Kishimura H (2008) Tuna pepsin: characteristics and its use for collagen extraction from the skin of threadfin bream (*Nemipterus* spp.). J Food Sci 73(5):C413–C419

Nishino R, Suyama A, Ikeda M, Kakizaki H, Matsumiya M (2014) Purification and characterization of a liver chitinase from golden cuttlefish. J Chitin Chitosan Sci 2:238–243

Noguchi K, Ishikawa K, Yokoyama KI, Ohtsuka T, Nio N, Suzuki EI (2001) Crystal structure of red sea bream transglutaminase. J Biol Chem 276:12055–12059

Nozawa H, Seki N (2001) Purification of transglutaminase from scallop striated adductor muscle and NaCl-induced inactivation. Fish Sci 67:493–499

Nozawa H, Cho SY, Seki N (2001) Purification and characterization of transglutaminase from squid gill. Fish Sci 67:912–919

Ogino T, Tabata H, Ikeda M, Kakizaki H, Matsumiya M (2014) Purification of a chitinase from the posterior salivary gland of common octopus *Octopus vulgaris* and its properties. J Chitin Chitosan Sci 2:135–142

Ohshima T, Suzuki T, Koizumi C (1993) New developments in surimi technology. Trends Food Sci Technol 4:157–163

Oliveira VM, Assis CRD, Herculano PN, Cavalcanti MTH, Bezerra RS, Porto ALF (2017) Collagenase from smooth weakfish: extraction, partial purification, characterization and collagen specificity test for industrial application. Boletim do Instituto de Pesca Sao Paulo 43(1):52–64

Ong TY, Shaik MI, Sarbon NM (2021) Isolation and characterization of acid and pepsin soluble collagen extracted from sharpnose stingray (*Dasyatis zugei*) skin. Food Res 5(3):214–224

Orejana FM, Liston FM (1981) Agents of proteolysis and its inhibition in patis (fish sauce) fermentation. J Food Sci 47:198–203

Pallela R, Bhatnagar I, Venkatesan J, Shim Y, Kim S (2013) Application of marine collagen-based scaffolds in bone tissue engineering. In: Kim SE (ed) Marine biomaterials, 1st edn. CRCP Press, USA, pp 519–528

Pandey A, Benkamin S, Soccol CR, Nigam P, Krieger N, Soccol VT (1999) The realm of microbial lipases in biotechnology. Biotechnol Appl Biochem 29(2):119–131

Pangkey H, Hara K, Tachibana K, Cao MJ, Osatomi K, Ishihara T (2000) Purification and characterization of cathepsin S from hepatopancreas of carp *Cyprinus carpio*. Fish Sci 66:1130–1137

Patchimpet J, Sangkharak K, Klomklao S (2019) Lipolytic activity of viscera extract from three freshwater fish species in Phatthalung, Thailand: comparative studies and potential use as dishwashing detergent additive. Biocatal Agric Biotechnol 19:101143

Patchimpet J, Sangkhawak K, Eiad-ua A, Klomklao S (2021) Thermoseparating aqueous two-phase system for lipase recovery and partitioning from Nile tilapia viscera: biochemical properties and effect of ultrasound. J Mol Liq 331:115721

Patchimpet J, Simpson BK, Sangkharak K, Klomklao S (2020) Optimization of process variables for the production of biodiesel by transesterification of used cooking oil using lipase from Nile tilapia viscera. Renew Energy 153:861–869

Patil RS, Ghormade V, Deshpande MV (2000) Chitinolytic enzymes: an exploration. Enzyme Microb Technol 26:473–483

Pei W, Hua Z, Zhenyu S, Dan L, Yanling X (2019) Isolation and characterization of acid-soluble collagen and pepsin-soluble collagen from the skin of hybrid sturgeon. J Wuhan Univ Technol Mater Sci Ed 34(4):950–959

Phanturat P, Benjakul S, Visessanguan W, Roytrakul S (2010) Use of pyloric caeca extract from bigeye snapper (*Priacanthus macracanthus*) for the production of gelatin hydrolysate with antioxidative activity. LWT Food Sci Technol 43:86–97

Poonsin T, Simpson BK, Benjakul S, Visessanguan W, Yoshida A, Osatomi K, Klomklao S (2019a) Anionic trypsin from the spleen of albacore tuna (*Thunnus alalunga*): purification, biochemical properties and its application for proteolytic degradation of fish muscle. Int J Biol Macromol 133:971–979

Poonsin T, Simpson BK, Benjakul S, Visessanguan W, Yoshida A, Klomklao S (2019b) Albacore tuna spleen trypsin: potential application as laundry detergent additive and in carotenoprotein extraction from Pacific white shrimp shells. Biocatal Agric Biotechnol 17:638–646

Proespraiwong P, Tassanakajon A, Rimphanitchayakit V (2010) Chitinases from the black tiger shrimp *Penaeus monodon*: phylogenetics, expression and activities. Comp Biochem Physiol B 156:86–96

Raa J (1985) The use of enzymes in processing of marine food products. In: Reilly A (ed) Spoilage of tropical fish and product development: processing of symposium held in conjunction with the sixth session of the Indo-Pacific Fishery Commission Working Party on fish technology and marketing. Royal Melbourne Institute of Technology, Melbourne, Australia, pp 246–248

Raa J (1990) Biotechnology in aquaculture and the fish processing industry: a success story in Norway. In: Voigt MN, Rotta JR (eds) Advances in fisheries technology for increased profitability. Technomic Publication Company, Lancaster, PA, pp 509–524

Raae AJ, Flengsrudt R, Sletten K (1995) Chymotrypsin isoenzymes in Atlantic cod; differences in kinetics and substrate specificity. Compar Biochem Physiol Part B Biochem Mol Biol 112:393–398

Rasika DMD, Ranadheera CS, Vidanarachchi JK (2013) Applications of marine derived peptides and proteins in the food industry. In: Kim SK (ed) Marine proteins and peptides: biological activities and applications. Wiley, pp 545–587

Ravi Kumar MNV (2000) A review of chitin and chitosan applications. React Funct Polym 46:1–27

Rinaudo M (2006) Chitin and chitosan: properties and applications. Prog Polym Sci 31:603–632

Rivera-Pérez C, del Toro MAN, García-Carreño F (2011) Purification and characterization of an intracellular lipase from pleopods of whiteleg shrimp (*Litopenaeus vannamei*). Compar Biochem Physiol Part B Biochem Mol Biol 158:99–105

Sae-leaw T, Benjakul S (2018) Lipase from liver of seabass (*Lates calcarifer*): characteristics and the use for defatting of fish skin. Food Chem 240:9–15

Santos CWV, Marques MEC, Tenório HA, Miranda EC, Pereira HJV (2016) Purification and characterization of trypsin from Luphiosolurus alexandri pyloric cecum. Biochem Biophys Rep 8:29–33

Santos DMRG, Santos CWV, Souza GB, Albuquerque FS, Oliveira JMS, Pereira HJV (2020) Trypsin purified from *Coryphaena hippurus* (common dolphinfish): purification, characterization, and application in commercial detergents. Biocatal Agric Biotechnol 25:101584

Saranya R, Selvi AT, Jayapriya J, Aravindhan R (2020) Synthesis of fat liquor through fish waste valorization, characterization and applications in tannery industry. Waste Biomass Valor 11:6637–6647

Senphan T, Benjakul S, Kishimura H (2014) Characteristics and antioxidative activity of carotenoprotein from shells of Pacific white shrimp extracted using hepatopancreas proteases. Food Biosci 5:54–63

Shahidi F, Kamil YVAJ (2001) Enzymes from fish and aquatic invertebrates and their application in the food industry. Trends Food Sci Technol 12:435–464

Shahidi F, Janak Kamil YVA, Jeon YJ (1999) Food applications of chitin and chitosans. Trends Food Sci Technol 10:37–51

Sikorski ZE, Gildberg A, Ruiter A (1995) Fish products. In: Ruiter A (ed) Fish and fishery products-composition, nutritive properties and stability. CAB International, Wallingford, pp 315–346

Sila A, Nasri M, Bougatef A (2012a) Isolation and characterization of carotenoprotiens from deep-water pink shrimp processing waste. Int J Biol Macromol 51:953–959

Sila A, Nasri R, Bougatef A, Nasri M (2012b) Digestive alkaline proteases from the goby (*Zosterisessor ophiocephalus*): characterization and potential application as detergent additive and in the deproteinization of shrimp wastes. J Aquat Food Prod Technol 21:118–133

Silva JF, Espósito TS, Marcuschi M, Riverio K, Cavalli RO, Oliveeira V, Bezerra R (2011) Purification and partial characterisation of a trypsin from the processing waste of the silver mojarra (*Diapterus rhombeus*). Food Chem 129:777–782

Simpson BK (2000) Digestive proteinases from marine animals. In: Haard NF, Simpson BK (eds) Seafood enzymes: utilization and influence on postharvest seafood quality. Marcel Dekker, New York, pp 191–214

Simpson BK, Rui X, Klomklao S (2012) Enzymes in food processing. In: Simpson BK (ed) Food biochemistry and food processing, 2nd edn. Wiley, UK, pp 181–206

Sirikharin R, Soderhall I, Soderhall K (2018) Characterization of a cold-active transglutaminase from a crayfish *Pacifastacus Leniusculus*. Fish Shellfish Immunol 80:546–549

Smichi N, Fendri A, Chaâbouni R, Rebah FB, Gargouri Y, Miled N (2010) Purification and biochemical characterization of an acid-stable lipase from pyloric caeca of sardine (*Sardinella aurita*). Biotechnol Appl Biochem 162:1483–1496

Smichi N, Gargouri Y, Miled N, Fendri A (2013) A grey mullet enzyme displaying both lipase and phospholipase activities: purification and characterization. Int J Biol Macromol 58:87–94

Sousa RO, Alves AL, Carvalho DN, Martins E, Oliveira C, Silva TH, Reis RL (2019) Acid and enzymatic extraction of collagen from Atlantic cod (*Gadus morhua*) swim bladders envisaging health-related applications. J Biomater Sci Polym Ed 31(1):20–37

Sovik SL, Rustad T (2006) Effect of season and fishing ground on the activity of cathepsin B and collagenase in by-products from cod species. LWT Food Sci Technol 39:43–53

Sripokar P, Benjakul S, Klomklao S (2019) Antioxidant and functional properties of protein hydrolysates obtained from starry triggerfish muscle using trypsin from albacore tuna liver. Biocatal Agric Biotechnol 17:447–454

Sripokar P, Chaijan M, Benjakul S, Klomklao S (2016) Enzymatic hydrolysis of starry triggerfish (*Abalistes stellaris*) muscle using liver proteinase from albacore tuna (*Thunnus alalunga*). J Food Sci Technol 53(2):1047–1054

Sun LM, Zhu BW, Wu HT, Yu L, Zhou DY, Dong X, Yang JF, Li DM, Ye WX, Murata Y (2011) Purification and characterization of cathepsin B from the gut of the sea cucumber (*Stichopus japonicass*). Food Sci Biotechnol 20(4):919–925

Suzuki T, Kakizaki H, Ikeda M, Matsumiya M (2014) Molecular cloning of a novel chitinase gene from blue shark (*Prionace glauca*; Chondrichthyes) stomach. J Chitin Chitosan Sci 2:143–148

Taheri A, Gashti AB (2019) Antioxidant and ACE inhibitory activities of kawakawa (*Euthynnus affinis*) protein hydrolysate produced by skipjack tuna pepsin. J Aquat Food Prod Technol 24(4):1–19

Tavares JFP, Baptista JAB, Marcone MF (1997) Milk-coagulating enzymes of tuna fish waste as a rennet substitute. Int J Food Sci Nutr 48:169–176

Toomer OT, Do AB, Fu TJ, Williams KM (2015) Digestibility and immunoreactivity of shrimp extracts using an in vitro digestibility model with pepsin and pancreatin. J Food Sci 80(7):T1633–T1639

Vannabun A, Ketnawa S, Phongthai S, Benjakul S, Rawdkuen S (2014) Characterization of acid and alkaline proteases from viscera of farmed giant catfish. Food Biosci 6:9–16

Venugopal V (2016) Enzymes from seafood processing waste and their application in seafood processing. In: Kim SK, Toldrá F (eds) Marine enzymes biotechnology: production and industrial applications, Part I—production of enzymes. Academic Press, Burlington, pp 47–69

Wald M, Rehbein H, Beermann C, Buβmann B, Schwarz K (2016) Purification and characterization of pepsinogen and pepsin from the stomach of rainbow trout (*Oncorhynchus mykiss*). Eur Food Res Technol 242:1925–1935

Wang SL, Liang TW, Yen YH (2011) Bioconversion of chitin containing wastes for the production of enzymes and bioactive materials. Carbohyd Polym 84:732–742

Weng WY, Wu T, Chen WQ, Liu GM, Osatomi K, Su WJ, Cao MJ (2011) Purification and characterization of pepsinogens and pepsins from the stomach of rice field eel (*Monopterus albus Zuiew*). Fish Physiol Biochem 37:543–552

Worratao A, Yongsawatdigul J (2005) Purification and characterization of transglutaminase from Tropical tilapia (*Oreochromis niloticus*). Food Chem 93:651–658

Wu GP, Wang XM, Lin LP, Chen SH, Wu QQ (2014) Isolation and characterization of pepsin-solubilized collagen from the skin of black carp (*Mylopharyngodon piceus*). Adv Biosci Biotechnol 5:642–650

Wu J, Kong L, Zhang J, Chen W (2019) Extraction and properties of acid-soluble collagen and pepsin-soluble collagen from silver carp (*Hypophthalmichthys molitrix*) scales: prerequisite information for fishery processing waste reuse. Pol J Environ Stud 28(4):2923–2930

Wu T, Sun LC, Du CH, Cai QF, Zhang QB, Su WJ, Cao MJ (2009) Identification of pepsinogens and pepsins from the stomach of European eel (*Anguilla anguilla*). Food Chem 115:137–142

Xu RA, Wong RJ, Rogers ML, Fletcher GC (1996) Purification and characterization of acidic proteases from the stomach of the deepwater finfish orange roughy (*Hoplostehus atlanicus*). J Food Biochem 20:31–48

Yang F, Su WJ, Lu BJ, Wu T, Sun LC, Hara K, Cao MJ (2009) Purification and characterization of chymotrypsins from the hepatopancreas of crucian carp (*Carassius auratus*). Food Chem 116(4):860–866

Yang X, Xiao X, Liu D, Wu R, Wu C, Zhang J, Huang J, Liao B, He H (2017) Optimization of collagenase production by *Pseudoalteromonas* sp. SJN2 and application of collagenases in the preparation of antioxidative hydrolysates. Mar Drugs 15:377

Yasueda H, Kumazawa Y, Motoki M (1994) Purification and characterization of a tissue-type transglutaminase from red sea bream (*Pagrus major*). Biosci Biotechnol Biochem 58:2041–2045

Younes I, Nasri R, Bkhairia I, Jellouli K, Nasri M (2015) New proteases extracted from red scorpionfish (Scorpaena scrofa) viscera: characterization and application as a detergent additive and for shrimp waste deproteinization. Food Bioprod Process 94:453–462

Yu F, Zong C, Jin S, Zheng J, Chen N, Huang J, Chen Y, Huang F, Yang Z, Tang Y, Ding G (2018) Optimization of extraction conditions and characterization of pepsin-solubilised collagen from skin of giant croaker (*Nibea japonnica*). Mar Drugs 16(1):29

Zarai Z, Ali MB, Fendri A, Louati H, Mejdoub H, Gargouri Y (2012) Purification and biochemical properties of *Hexaplex trunculus* digestive lipase. Process Biochem 47:2434–2439

Zhang Y, He S, Simpson BK (2017) A cold active transglutaminase from Antarctic krill (*Euphausia superba*): purification, characterization and application in the modification of cold-set gelatin gel. Food Chem 232:155–162

Zhao L, Budge SM, Ghaly AE, Brooks MS, Dave D (2011) Extraction, purification and characterization of fish pepsin: a critical review. Food Process Technol 2:2–6

Zhou L, Budge SM (2011) Extraction, purification and characterization of fish chymotrypsin: a review. Am J Biochem Biotechnol 7(3):104–123

Zhou Q, Fu XP, Zhang LJ, Su WJ, Cao MJ (2007) Purification and characterization of sea bream (*Sparus latus Houttuyn*) pepsinogens and pepsins. Food Chem 103:795–801

Zhou Q, Liu GM, Huang YY, Weng L, Hara K, Su WJ, Cao MJ (2008) Pepsinogens and pepsin from Mandarin fish (*Siniperca chuatsi*). J Agric Food Chem 56:5401–5406

Omega-3 Enriched Fish and Shellfish Oils: Extraction, Preservation, and Health Benefits

Bharathipriya Rajasekaran, Saqib Gulzar, Sugumar Gopalrajan, Masilan Karunanithi, and Soottawat Benjakul

Abstract Fish oil is a rich source of n-3 polyunsaturated fatty acids, especially eicosapentaenoic acid and docosahexaenoic acid, which are crucial for the treatment of several chronic diseases. Fish and shellfish processing discards have been widely utilized as raw materials for oil extraction. Conventional methods including wet reduction, enzymatic-assisted, and solvent-assisted extraction processes result in lower oil recovery, reduced nutritive value, and high amounts of solvent used, respectively. To overcome these limitations, alternative novel extraction technologies have been employed, which increase the yield of oil with minimum loss of nutrients and less usage of solvent. Fish oil is highly susceptible to oxidation that led to the generation of undesirable odor and the formation of toxic compounds. Recent findings for the preservation and stabilization using various techniques are addressed. Some active compounds in fish oil such as astaxanthin and squalene are revisited. Also, the health benefits and viable application of n-3 fatty acids and the selected active components on human health are summarized.

Keywords Antioxidants · Astaxanthin · Encapsulation · Enzymatic hydrolysis · Extraction techniques · Health benefits · n-3 Fatty acids

---

This chapter submitted to "Fish Waste to Valuable Products: Recent Applications and Research Update".

B. Rajasekaran · S. Gulzar · S. Benjakul (✉)
International Center of Excellence in Seafood Science and Innovation, Faculty of Agro-Industry, Prince of Songkla University, Hat Yai, Songkhla 90110, Thailand
e-mail: soottawat.b@psu.ac.th

S. Gopalrajan · M. Karunanithi
Department of Fish Processing Technology, Tamil Nadu Dr. J. Jayalalithaa Fisheries University, Nagapattinam 611002, Tamil Nadu, India
e-mail: sugumar@tnjfu.ac.in

M. Karunanithi
e-mail: masilan@tnfu.ac.in

S. Maqsood et al. (eds.), *Fish Waste to Valuable Products*, Sustainable Materials and Technology, https://doi.org/10.1007/978-981-99-8593-7_9

1 Introduction

Fish and shellfish caught from the sea or harvested from fish farming have been increased for consumption and commercial purposes (Kumar et al. 2018). Recently, demand for food with high nutraceutical properties and bioactive compounds has increased tremendously owing to their enormous beneficial effects (Rajasekaran et al. 2022). Increased consumption of fish and fishery products has created a huge challenge to seafood processing industries in handling and exploiting the wastes in a sustainable manner (Nirmal et al. 2020). In 2018, fish production was 179 million tonnes worldwide, of which 156 million tonnes were consumed by people, and 22 million tonnes were discarded as waste (FAO 2018). In finfish processing, leftovers including head, viscera, skin, scale, bone, etc. are produced, depending on the finished products and processes used. In the case of shrimp processing, inedible parts are removed from the whole shrimp including cephalothorax, hepatopancreas, carapace, and tail. Leftovers account for about 30–35% and 40–50% of the total weight of fish and shrimp, respectively (Kumar et al. 2018; Nirmal et al. 2020). Previously, those wastes were considered residues and dumped into the environment, creating severe pollution problems. Nowadays, byproducts from fish processing industries are better utilized by converting them into valuable products under the concept of 'Zero Waste' (Gulzar et al. 2020). The fish waste consists of around 40–60% oil, and abundance in polyunsaturated fatty acids (PUFA) (Adeoti and Hawboldt 2014). These fatty acids are beneficial for health and are utilized in the treatment and prevention of autoimmune diseases, heart ailments, high blood pressure, arthritis, high cholesterol levels, mental illness, etc. Also, vitamins including A, D, E, and K are abundant in fish oil (Yashodhara et al. 2009). The structure of bioactive compounds from fish and shrimp processing byproducts is documented (Fig. 1). Presently, fishing for their oil is not promoted due to the declining marine fishery resources. Therefore, utilizing leftovers from fish processing industries is a promising alternative approach to produce high-grade fish oil.

In general, fish oil contributes about 2% of the world's consumption of fats. Globally, fish oil production is estimated to be 1.1 million tons, of which only 5% is utilized for human consumption. The remaining is used as an ingredient in animal and aquaculture feed (Ivanovs and Blumberga 2017). Traditionally, fish oil is a major byproduct of the fish meal industry (Benjakul et al. 2019). Oil recovery, fat content, and composition vary based on species, gender, season, and type of byproducts (Gulzar et al. 2020). Conventional processes are widely used for the extraction of fish oil on an industrial scale (Ivanovs and Blumberga 2017). These processes involve either high temperatures or toxic solvents. This results in the deterioration of oil via oxidation, thus reducing its nutritive value (Prameela et al. 2017). In addition, the solvent residue is not safe for human consumption. Moreover, fish oil extracted by conventional methods has a high level of impurities such as free fatty acids, peroxides, dienes, and cholesterol (Ramalhosa et al. 2012). Therefore, several refining processes have been adopted to reduce the impurities, and consequently augment the production cost. The common refining steps employed on an industrial scale

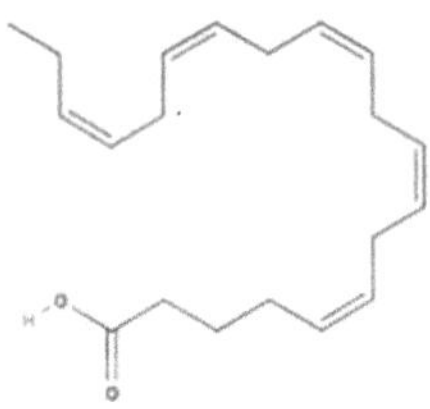

Eicosapentaenoic acid (EPA, 20:5) $C_{20}H_{30}O_2$

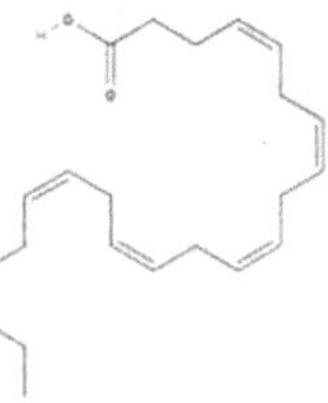

Docosahexaenoic acid (DHA, 22:6) $C_{22}H_{32}O_2$

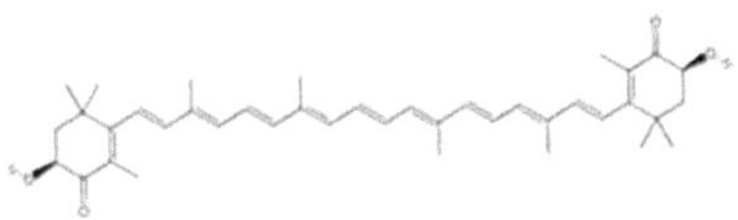

Astaxanthin $C_{40}H_{52}O_4$

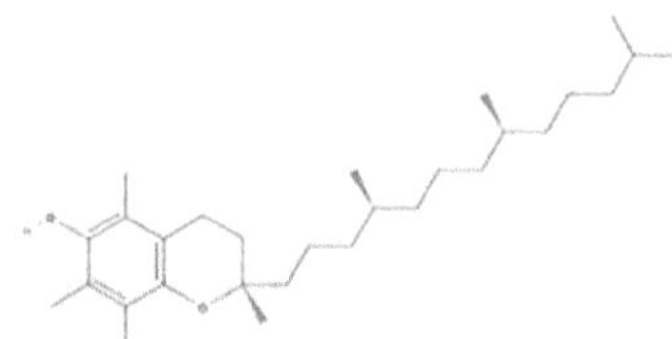

Alpha-Tocopherol $C_{29}H_{50}O_2$

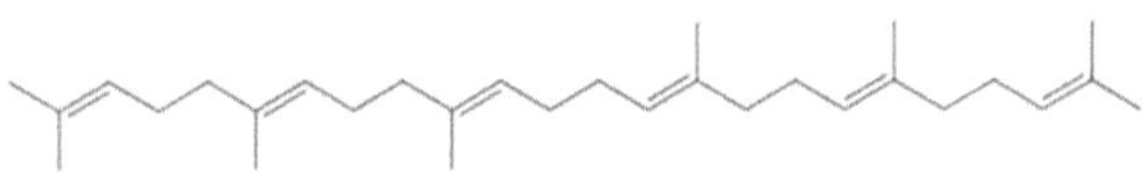

Squalene $C_{30}H_{50}$

Fig. 1 Structure of bioactive compounds from fish and shrimp processing discards

include neutralization, degumming, bleaching, and deodorization processes (Rizliya and Mendis 2014). The quantity and quality of oil are mainly influenced by the extraction methods used. Therefore, the appropriate extraction method needs to be selected to produce oil with negligible impurities and minimize the need for further refining (Gulzar et al. 2020). Another major challenge related to the usage of fish oil is the rapid oxidative deterioration of PUFA which reduces consumer acceptance. This could be overcome by the addition of antioxidants and encapsulation, etc. (Rajasekaran et al. 2022). The application of an appropriate dose of antioxidants and potential encapsulation significantly maintains the stability, solubility, dispersibility, and bioavailability of active compounds in oil (Khoshnoudi-Nia et al. 2022). In this chapter, conventional and novel methods of oil extraction with recent developments were reviewed and challenges were also highlighted. Furthermore, recent techniques to preserve fish oil and its active component such as astaxanthin were addressed. Finally, the potential advantages of bioactive substances and their biological mechanisms were revisited.

2 Extraction of Oil from Fish and shellfish Processing Discards

Oil has been extracted from fish processing discards by several conventional and novel methods. The oil recovery, composition, and quality are greatly influenced by selected methods for oil extraction (Gulzar et al. 2020). Moreover, each method has its own merits and demerits, thus affecting the quality and safety of the resulting oil. Different methods for oil extraction from fish and shellfish processing leftovers are demonstrated in Table 1.

2.1 *Conventional Methods of Oil Extraction*

2.1.1 Wet Reduction

A common technique to extract unrefined fish oil is wet reduction (Chantachum et al. 2000). It includes four stages involving cooking, pressing, and decanting, followed by centrifugation. Initially, the raw material is heated via steam cooking at a high temperature to rupture and coagulate the proteins. As a consequence, bound oil can be released easily followed by hydraulic pressing, mainly a twin-screw press, and centrifugation to remove liquid containing oil from solid residues (Bonilla-Méndez and Hoyos-Concha 2018). The wet reduction process is economically profitable when byproducts from fatty fish such as sardine, herring, salmon, and tuna are used for oil extraction. However, it is not feasible for lean fish, in which oil content is too low (Rubio-Rodríguez et al. 2012). Bako et al. (2014) postulated that the yield of oil from mackerel discards was 18.7% and oil quality was within the acceptable standards. Oil from *Sardinella longiceps* extracted by heating had lower primary and secondary oxidative products than solvent-assisted extraction (Chakraborty and Joseph 2015). The processing time and temperature directly influence the quality of the extracted oil. Longer exposure time in the heat exchanger accelerates oxidation, whereas a higher yield could be obtained at a high temperature (85 °C). Heating could minimize the microbial load and inactivate the endogenous enzymes (Ivanovs and Blumberga 2017). However, a lower yield was reported at 95 °C due to protein denaturation, in which a tightly packed structure is formed, thus preventing oil release. In contrast, crude oil was extracted effectively from herring byproducts at 95 °C for 8 min (Bonilla-Méndez and Hoyos-Concha 2018). Protein normally undergoes irreversible denaturation at a temperature range 90–100 °C. The coagulation temperature of fish protein is around 60–75 °C, but typical cooking is performed at 95–100 °C. Therefore, optimum temperature and process time should be standardized for raw materials from different sources to attain the highest extraction efficacy of the oil. Chantachum et al. (2000) found that precooking of tuna head (85 °C; 30 min) yielded oil with higher oxidation, while oil from a non-precooked sample showed lower lipid oxidation. In another study, the oil separation from herring by the wet reduction process at

Table 1 Different methods for oil extraction from fish and shellfish processing leftovers

Methods	Process	Main factor	Merits	Demerits	References
Wet reduction	Use high temperature and pressure to coagulate protein and facilitate the release of oil from the solid matrix	Temperature, pressure, and extraction time	Widely used on an industrial scale; higher extraction yield; economical and profitable; simple method	Extract crude oil needs several refining processes; releases a high amount of heat; emulsion formation, loss of active compounds, enhanced oxidative reactions	Ivanovs and Blumberga (2017)
Enzymatic hydrolysis	Use exogenous proteolytic enzymes to digest the solid matrix for extraction of oil	Type of enzyme, time, temperature, and enzyme-substrate ratio	Higher stability of bioactive compounds (astaxanthin); good quality oil; commercial low-cost protease can be used for the hydrolysis process; organic solvent is not required	Not suitable for large scale production; expensive method; need to maintain optimum conditions for the hydrolysis process	Qi-Yuan et al. (2016)
Solvent-assisted extraction	Use non-polar organic solvent to extract oil	Type of solvent, polarity, and toxicity	Widely used method; economical; high extraction yield of carotenoids; sophisticated equipment is not required	Extraction yield is low; generation of a huge amount of solvent waste; difficult to remove solvents from the final product	Dalei and Shaoo (2015)
Ultrasound-assisted extraction	Use ultrasound waves to produce the cavitation effect, which disrupts the cell for easy and deeper penetration of solvents	Frequency, extraction time, medium, and power supply	Widely used to extract active compounds; extraction yield is high; extraction time and solvent consumption are reduced	Mechanical effect of ultrasound induces oxidation and hydrolysis of PUFA; high power consumption; difficult to adopt for large scale	Ali et al. (2019)

(continued)

Table 1 (continued)

Methods	Process	Main factor	Merits	Demerits	References
Supercritical fluid extraction	Use supercritical fluid ($SC\text{-}CO_2$) to extract the oil due to its high diffusivity	Temperature, pressure, type of extraction, extraction time, and CO_2 flow	Efficient extraction of bioactive compounds; specificity and purity of extracted compound; trace solvents can be removed from the final product; low operating temperatures (40–80°C)	High capital and operational cost; require special equipment; high power consumption; need a constant supply of clean $SC\text{-}CO_2$	Sánchez-Camargo et al. (2012)
Pulsed electric field extraction	Use high-voltage electric current to pass through the solid matrix and to create minute pores facilitating solvent penetration to extract oil	Electric field strength, the conductivity of medium and transmembrane potential of matrix	Non-thermal extraction technique; efficient extraction of thermolabile compounds; inactivate deteriorate enzymes by electroporation; Extraction time is short	High capital cost; complex process; not applicable for dry substances due to low conductivity	Li et al. (2016)

70 °C had lower lipid oxidation with higher stability than enzymatic hydrolysis using Alcalase (Carvajal et al. 2015). Menegazzo et al. (2014) postulated that crude oil extraction from Nile tilapia processing discards was done at 40 °C for 3 h. The yield of fish oil was greater when heating the visceral depot fat of seabass under the vacuum than the solvent extraction method (Sae-leaw and Benjakul 2017). Heat applied can favor the liberation of oil and simultaneously inactivates the lipolytic enzyme which causes hydrolysis of the lipid (production of free fatty acids). However, a high temperature could partially alter the PUFA due to deteriorative reactions such as oxidation and hydrolysis (Bonilla-Méndez and Hoyos-Concha 2018). The alterations of oil quality depend upon various factors including temperature, heating time, degree of unsaturation of oil, etc. In general, the wet reduction process may harm the environment due to the release of a high amount of heat (Ivanovs and Blumberga 2017).

2.1.2 Enzymatic Hydrolysis Process

In general, oil has been extracted from fish processing discards via enzymatic hydrolysis using proteases. The enzyme will hydrolyze the proteins, thereby facilitating the release of oil without using high temperatures or solvents. Liaset et al. (2003) postulated that around 80% of lipids were recovered from salmon frames via enzymatic hydrolysis using the commercial protease, Protamex. Linder et al. (2005) investigated different proteases such as Flavourzyme, Neutrase, and Alcalase for hydrolyzing the salmon head. The highest amount of oil released was achieved when Alcalase was used. Extraction of oil using enzymatic hydrolysis had a similar or higher yield as compared to the wet reduction process (Benjakul et al. 2019). Moreover, good quality oil can be obtained using enzymatic hydrolysis with the application of mild processing conditions. In addition, peptides in hydrolysate can also be utilized as beneficial functional components (Rustad et al. 2011). Oil extracted from the tuna head through enzymatic hydrolysis using Alcalase had the lowest acid value and oxidation products as well as the highest level of PUFA retained than the oils obtained from wet reduction and solvent-assisted extraction processes (de Oliveira et al. 2017). Indian major carp viscera were subjected to enzymatic hydrolysis using different proteases such as Alcalase, Protex 7L, Neutrase, and Protease-P-Amano. The highest recovery of oil was obtained when using Protease-P-Amano (74.9%), followed by Alcalase (61.7%) (Hathwar et al. 2011). Qi-Yuan et al. (2016) applied response surface methodology to optimize the factors influencing oil from mackerel viscera via enzymatic hydrolysis using Neutrase, pH, temperature, agitation, incubation time, availability of substrate, mode of action, and concentration of enzyme which markedly influenced the yield. The yield obtained with the optimized condition was 78.66%. This method is suitable for oil extraction from lean fish processing byproducts having low-fat content. Recently, extraction of astaxanthin via enzymatic hydrolysis from shrimp byproducts is gaining popularity (Ahmadkelayeh and Hawboldt 2020). Astaxanthin extracted via enzymatic hydrolysis had higher stability and bioavailability than solvent-assisted extraction due to the presence of protein in the form of carotenoproteins. De Holando and Netto (2006) reported a high yield of astaxanthin around 64.6% through enzymatic hydrolysis using Alcalase. The main drawback is the difficulty of recovering the oil phase after hydrolysis associated with the emulsion formed (Benjakul et al. 2019). Moreover, a suitable pH should be maintained for the maximal activity of the enzymes during hydrolysis. Therefore, adjusting the pH using acid or alkali could result in the generation of unwanted inorganic substances, especially sodium chloride in resulting hydrolysate (Benjakul et al. 2019).

2.1.3 Solvent-Assisted Extraction Process

Solvent-assisted extraction was frequently used for oil extraction from fish processing discards with the aid of solvents. Types of solvents used for the extraction directly influence the quantity and quality of the oil. The common criteria considered for the selection of solvent are toxicity, polarity, safe handling, easy removal of solvent, etc. (Gulzar et al. 2020). Presently, organic solvents such as ethyl acetate, acetone, ethanol, methyl ethyl ketone, hexane, and isopropanol are permitted to be used in food industries for oil extraction with the permitted residual level in the final products (Gulzar et al. 2020). Nevertheless, their solvents such as chloroform, dichloromethane, and dimethyl sulfoxide are restricted due to their toxicity (FDA 2010). Gulzar and Benjakul (2018) investigated the efficiency of different solvents to extract oil from cephalothorax, a shrimp processing byproduct, in which the highest yield was obtained using isopropanol as compared with acetone, chloroform, and n-hexane. However, oil extracted using n-hexane had greater carotenoid attributed to the high solubility of carotenoid in n-hexane or n-heptane (Sánchez-Camargo et al. 2011). In general, the oil consists of both polar (phospholipids) and non-polar (carotenoid) compounds. Thus, effective extraction can be obtained by the combination of polar and non-polar solvents at an appropriate ratio (Sachindra et al. 2006). The common procedure to extract oil for lab scale is the Bligh and Dyer method, in which solvent mixtures including chloroform, methanol, and water are used. Extraction yield is higher in the Bligh and Dryer method than Soxhlet extraction method (Macías-Sánchez et al. 2010). Gulzar and Benjakul (2018) reported that the mixture of hexane and isopropanol (1:1) had the highest lipid extraction yield and carotenoid content from shrimp cephalothorax. A significant amount of astaxanthin (53 μg/g) was obtained when extracted from byproducts of Paulo shrimp by the solvent mixture of n-hexane: isopropyl alcohol at a ratio of 60:40 (Sánchez-Camargo et al. 2011). Acetone was used for oil extraction from deep-sea shrimp, and the yield was 43.44 μg/g (Dalei and Shaoo 2015). Therefore, the yield and astaxanthin from crustacean oil vary based on the type of solvent, species, and source of raw materials used. Moreover, the selection of inappropriate solvents results in the consumption of excess solvents and a lower yield (Ivanovs and Blumberga 2017). The main drawback of the solvent extraction process is the disposal issue, though some solvents can be recycled for further uses. Moreover, the solvent can cause environmental pollution if not recycled or refined properly (Gulzar et al. 2020). The difficulties in the removal of residual solvents still require the development of potential novel extraction processes to ensure the safety of the resulting oil.

2.2 Novel Extraction Processes

2.2.1 Ultrasonic Assisted Extraction

To overcome the limitations of conventional extraction processes such as low extraction yield and excess solvent usage, ultrasonic-assisted extraction (UAE) is combined with the existing solvent extraction process to remarkably increase the quantity of lipids and reduce the solvent volumes (Mason et al. 1996). UAE is a physical process that utilizes high-energy sound waves with frequencies higher than the human hearing level (20 kHz), reaching up to 10 MHz (Gallo et al. 2018). During propagation through a medium, these high-energy waves disrupt the cell–matrix of the substrate to facilitate the deeper penetration of solvent for subsequent extraction of target compounds, particularly oil. The ultrasound waves produce a cavitation effect in the liquid solvent medium, in which microbubbles are created owing to the constant compression and rarefaction of the longitudinal waves (Povey and Mason 1998). These microbubbles are unstable and burst shortly after formation. However, the bursting of these microbubbles creates a strong localized effect, in which high pressure and temperature are generated which can reach up to 400 MPa (Flint and Suslick 1991). The high localized pressure disrupts the microstructure of substrates in the vicinity, thus destroying the tissue integrity and causing release of target compounds (Flint and Suslick 1991). The formation of thousands of microbubbles and their explosion literally tears down the substance, thus favoring the complete solvent penetration and extraction of lipids (Povey and Mason 1998). UAE also reduces the extraction time of lipids, but it can negatively impact bioactive compounds, mainly associated with enhanced oxidation.

UAE is commonly utilized in food industry, especially for the extraction and homogenization of bioactive substance (Gallo et al. 2018). In the seafood industry, UAE is employed for oil extraction from fish and shrimp processing discards (Al Khawli et al. 2019; Gulzar et al. 2020). Oil from Asian swamp eel fillet trimmings was extracted with enhanced yield using UAE (Abdullah et al. 2010). UAE pre-treatment resulted in the higher extraction yield of oil from the head of *Labeo rohita* by 8.74%, compared to the untreated counterpart (Bruno et al. 2019). Application of sonication for the squalene extraction from fish livers increased yield, reduced extraction time, and lowered solvents consumed (Ali et al. 2019). Oil from shrimp cephalothorax was obtained with enhanced yield by employing UAE (80% amplitude for 25 min) (Gulzar and Benjakul 2018). The quantity of shrimp oil extracted from the cephalothorax increased by almost 1.5 times by the combination of a heat pre-treatment followed by UAE (Gulzar and Benjakul 2019). Nevertheless, there are some disadvantages of the UAE process, since it can accelerate the oxidation and hydrolysis caused by the cavitation effect (Gulzar and Benjakul 2018). Therefore, some additional protective means have been implemented to maintain the quality of extracted oils. Those include the incorporation of natural antioxidants, extraction of oil under the inert atmosphere, and controlling the temperature rise during ultrasonication (Gulzar and Benjakul 2020a).

2.2.2 Supercritical Fluid Extraction

Supercritical fluid extraction (SFE) is an emerging technique widely used to overcome the limitations of conventional extraction methods. Supercritical fluids (SCF) are substances that exist at the temperature and pressure beyond their critical point, where it is difficult to differentiate gas and liquid phases (Carlès 2010). The physical properties of SCF lie between a liquid and a gas. For example, the viscosity of the SCF is very low and the diffusivity is high, whereas the SCF possesses solvating properties like a liquid (Morrell 2012). In SFE, SCF is used to extract low polarity lipids in an oxygen-free and moderate temperature environment, which results in high-quality lipids, which are free from oxidative deterioration (Rubio-Rodríguez et al. 2012). Moreover, SFE is fast and effective due to the high diffusivity of SCF into the substrate and rapid separation of solvent after extraction (Del Valle and Aguilera 1999). In SFE, CO_2 is mostly used as SCF owing to its low-cost, non-flammable, and non-toxic nature, and it is considered as GRAS (Sahena et al. 2009). The ease of separation after extraction is another major advantage of using CO_2 as SCF, since it is a gas at room temperature. Due to these advantages, SFE is a green technology to obtain bioactive compounds (Ivanovs and Blumberga 2017). In the seafood processing industry, SFE has been extensively used owing to the presence of highly unstable compounds such as PUFAs, carotenoids, vitamins, and some indigenous antioxidants which are prone to oxidation (Rubio-Rodríguez et al. 2012; Sánchez-Camargo et al. 2012).

Oil from fish processing byproducts can be obtained by SFE with minimum damage to n-3 fatty acids. Moreover, due to selective extraction capabilities of SFE, it prevents the co-extraction of polar contaminants, such as certain inorganic derivates containing heavy metals (Rubio-Rodríguez et al. 2012). Several refining processes such as the de-acidification of fish oil can be easily achieved by modifying the density and solubility of SCF by adjusting its temperature and pressure (Kawashima et al. 2006). Extraction of lipids from shrimp processing discards using SFE led to an excellent astaxanthin recovery. Sánchez-Camargo et al. (2012) documented an upsurge in astaxanthin recovery from Brazilian redspotted shrimp (*Farfantepenaeus paulensis*) waste by 57.59% with the aid of SFE. The low temperature and oxygen-deficient environment of extraction by SFE has also proven to be favorable for enhanced recovery of n-3 PUFA (Amiguet et al. 2012). Krill oil extracted by SFE showed enhanced extraction yield, better astaxanthin retention, and augmented oxidative stability (Ali-Nehari et al. 2012). Nevertheless, the complex equipment, high cost of operation, and selective solubility of SCFs have made SFE unsustainable for commercial applications (Bin et al. 2020).

2.2.3 Pulsed Electric Field Extraction

Pulsed electric field (PEF) is a non-thermal processing technique, in which electroporation plays a profound role in extracting bioactive compounds with higher yield and good quality (López-Pedrouso et al. 2019). Electroporation is a phenomenon of creating minute pores in the tissues or intracellular matrix by placing the material between two electrodes and applying high-voltage electric pulses for a short time (Redondo et al. 2018). These pores facilitate solvent penetration and eventually enhance the extraction yield. Since the electric field is applied only for a short time, temperature increases barely in the product, which prevents the degradation of heat-labile compounds (Toepfl et al. 2006). The potency of PEF to facilitate extraction is dependent on physical and chemical properties of the material, the strength of the electric field, the electrical conductivity of the medium, and transmembrane potential (TMP) of the material (Gulzar et al. 2020). TMP typically acts as an ionic gradient across the cell membrane. By passing high-voltage pulses for a short duration, polarity is induced in the cell and produces a strong dipole moment, proportional to the applied electric field, causing irreversible pore formation in the cells (Zbinden et al. 2013). The cell organelles are therefore leaked into the solvent through these pores (Baiano and Del Nobile 2016).

PEF has a huge potential to replace the current conventional extraction process, either as a pre-treatment or in combination with other novel techniques (Donsì et al. 2010). In the seafood industry, PEF has been used modestly to valorize the processing discards, enhance the extraction yield, and preserve the quality of bioactive substances (Xi et al. 2021). Antioxidants from sea bass and sea bream waste (gills, bones, and heads) were extracted using PEF at a high yield (Franco et al. 2020). Pre-treatment of shrimp cephalothorax by PEF prior to UAE led to increased quantity of shrimp oil by 7.13% (Gulzar and Benjakul 2019). Moreover, the quality of extracted shrimp oil was superior to that of oil extracted by UAE alone, indicating that PEF could prevent lipid oxidation and hydrolysis to a large extent. Apart from oil extraction, PEF has been used for protein extraction from marine sources. A higher extraction yield of abalone viscera protein was obtained when PEF was used in combination with an enzymatic-assisted process (Li et al. 2016). Several advantages including high-quality extracts, low-operational temperature, and high disruption of biological cells in fish processing byproducts can be achieved using PEF without inducing detrimental effects (Puertolas et al. 2016). However, the utilization of PEF to extract oil from fish or shellfish processing leftover on a commercial scale is still not fully developed due to some challenges and obstacles involving the high cost, difficulty in handling equipment, and complex process involved (Gulzar et al. 2020).

2.2.4 Other Novel Extraction Processes

Among other novel extraction processes, high hydrostatic pressure (HHP) processing is noteworthy to separate bioactive compounds by applying high pressures (100–1000 MPa) at low temperatures (5–30 °C) (Gulzar et al. 2020). Application of high pressure disrupts the structure of material by breaking weak bonds including hydrogen, hydrophobic and electrostatic bonds in cell membranes, and destroying salt bridges, which ultimately improves solvent permeability (Bermúdez-Aguirre and Barbosa-Cánovas 2011). HHP has been substantially used in the seafood industry for the extraction of n-3 fatty acids (Ali et al. 2021), for obtaining bioactive peptides from fish byproducts (Yu et al. 2018) and for extracting astaxanthin from shrimp byproducts (Li et al. 2017). High pressure extraction of PUFA from liquid effluents generated by fish canning industry produced higher quantity of oil at the lowest cost (Monteiro et al. 2018). High pressure extraction contributed to increased yield of astaxanthin in a shorter time from dried shrimp waste with enhanced antioxidant activity (Li et al. 2017). The intrinsic factors including pH, composition as well as processing temperature and pressure levels influence the quantity and quality of the extracted products. To improve the extraction efficiency, further research is required for HHP process parameter optimization for the extraction of oil and other bioactive compounds on a commercial scale. Merits and demerits of conventional and novel approaches for the oil extraction and active components are summarized in Table 1.

3 Preservation and Stabilization of Oil Using Different Techniques

3.1 Encapsulation

Despite numerous beneficial effects of fatty acids present in fish or shrimp oils, the practical application is still limited mainly due to the rapid oxidation and hydrolysis of PUFA. Those reactions produce toxic and volatile compounds (off-odor). Furthermore, these changes negatively affect the nutritional value of oil (Rajasekaran et al. 2022). Encapsulation is one of the promising techniques for the preservation and stabilization of sensitive compounds like PUFA, astaxanthin, etc. (Gulzar et al. 2020). Encapsulation improves the dispersibility, stability, and bioavailability of active components without influencing the sensory attributes of the final product (Venugopalan et al. 2021). Encapsulation also masks fishy odor and delays oxidation and hydrolysis by preventing direct contact with oxygen and water during storage. Fortification of encapsulated fish oil in a food matrix shows high oxidative stability and better sensory attributes than those incorporated with unencapsulated fish oil (Venugopalan et al. 2021). There are several ways to encapsulate fish oil, and bioactive compounds as presented in Fig. 2. Micro- and nanoencapsulation are widely used for the protection of sensitive bioactive compounds and to ensure the target delivery of active compounds. Recently, novel techniques such as microfluidization and ultrasonication are widely applied to encapsulated EPA and DHA (Venugopalan et al. 2021). Ultrasound provided the microcapsules loaded with shrimp oil having high encapsulation efficiency (93.64%) than microfluidization (75.18%). Encapsulated oil had higher oxidative stability than unencapsulated oil over 8 weeks of

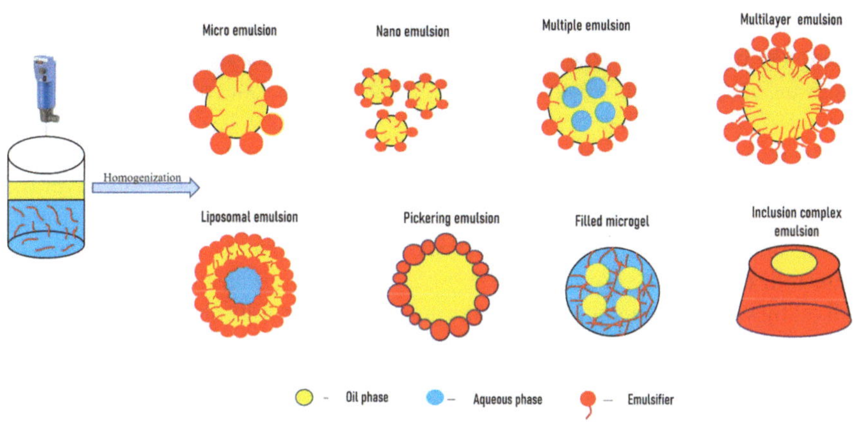

Fig. 2 Encapsulation techniques used to encapsulate fish oil and bioactive compounds

storage (Gulzar and Benjakul 2020b). Oxidative stability of oil in the microcapsule is determined by wall materials used to prevent the oxidation of PUFA. Takeungwongtrakul et al. (2015) reported that microencapsulated shrimp oil using sodium caseinate, fish gelatin, and glucose syrup at a ratio of 1:1:4 as wall materials up to 3% improved the loaf volume and nutritive value of bread. Moreover, biscuits incorporated with encapsulated shrimp oil were stable toward lipid oxidation during storage over 12 days, while maintaining good sensory acceptability (Takeungwongtrakul and Benjakul 2017). Microencapsulated fish oil using chitosan and modified starch (1:9) was used to fortify bread up to 5% without adverse effects on sensory attributes (Hasani et al. 2020).

Nanoemulsion can be used to stabilize active compounds, depending upon the size of the droplet. Nanoemulsion with droplet diameter less than 100 nm possess higher bioavailability and physical stability (Venugopalan et al. 2021). The incorporation of nanoencapsulated fish oil using 25% gum Arabic and 4% tween 80 as emulsifiers did not show any adverse effects on consumer acceptability in fermented low-fat milk (Moghadam et al. 2019). Recently, Pickering emulsion is gaining more interest than conventional surface-active emulsifiers due to its low cost, environment-friendliness, controlled release of active components, and excellent physical and oxidative stability (Ding et al. 2020). In Pickering emulsion, food grade colloidal particles are generally used to form a thin film over oil droplets to stabilize the emulsion (Liu et al. 2021). Pickering emulsion of fish oil had a smaller droplet size than conventional emulsion (Ding et al. 2020). The smaller droplet size reveals greater stability against coalescence. Hosseini and Rajaei (2020) postulated that mayonnaise enriched with fish oil stabilized by chitosan-steric acid nanogel improved the stability of mayonnaise more effectively than non-emulsified fish oil added to mayonnaise. Multilayer emulsion is known as double emulsion, and has a complex structure as compared to the conventional micro and nanoemulsion (Jamshidi et al. 2018). There are two types of multilayer emulsion. They are water–oil–water (W/O/W) emulsion and oil–water–oil emulsion (O/W/O), in which W/O/W is more common and effective for the encapsulation of fish oil. For W/O/W emulsion formation, smaller water droplets disperse inside the larger droplets which themselves disperse in the water phase (Fig. 2) (Venugopalan et al. 2021). The multilayer emulsion has a broad application in different sectors including cholesterol reduction, control, and/or target release of the bioactive component such as vitamin D and PUFA (Venugopalan et al. 2021). Multilayer fish oil emulsion stabilized using fucoidan-whey protein concentrate delayed the formation of oxidative products, thereby increasing the stability of encapsulated fish oil (Jamshidi et al. 2018). Yu et al. (2017) postulated the encapsulation efficiency of fish oil by various techniques including spray-dried microcapsules, complex coacervation, and double-layer emulsion using gelatin-gum Arabic as emulsifiers. Among all the samples, the double emulsion method yielded higher encapsulation efficiency (93.7%) owing to the double layers over the surface oil, thus enhancing its protective activity toward the environment.

Another novel encapsulation method, utilized in the pharmaceutical, cosmetics, and food industries, is liposomal technology. Liposomes are phospholipids that contain both polar and non-polar regions organized in a bilayer structure (Fig. 2) (Gulzar and Benjakul 2020b). Recently, liposomal technology is widely used to preserve and deliver bioactive compounds (Venugopalan et al. 2021). Liposomal technology ensures high encapsulation efficiency, bioavailability, and target release of bioactive substances as well as provides better oxidative stability of shrimp oil (Gulzar and Benjakul 2020b). Nanoliposomes ranging from 50 to 200 nm showed higher oxidative stability than encapsulated fish oil (>200 nm). Liposomal encapsulation can be done by various techniques including ethanol injection and thin film formation, etc. (Ajeeshkumar et al. 2021). In the injection process, phosphatidyl choline and targeted bioactive compound are dissolved in ethanol, followed by injection into water to form liposome (Ajeeshkumar et al. 2021). Ethanolic coconut husk extract-loaded liposome produced by the injection method using soybean phosphatidylcholine and cholesterol (4:1) had high encapsulation efficiency with the more uniform distribution of liposomes (Olatunde et al. 2020). In the thin film formation process, phosphatidyl choline, cholesterol and targeted bioactive compounds are dissolved in the appropriate solvent and hydration with suitable buffer, e.g., phosphate buffer and water to form liposomes. Hydrolyzed collagen-loaded liposomes were prepared using thin film formation by soybean phosphatidyl choline added with cholesterol or glycerol dissolved in ethanol, followed by hydration. The obtained hydrolyzed collagen liposomes had high encapsulation efficiency of 74.54 and 85.42%, respectively (Chotphruethipong et al. 2020). Gulzar and Benjakul (2020b) documented that the liposomal technique improved the oxidative stability of encapsulated shrimp oil during storage of 8 weeks and could mask the fishy odor. Shrimp oil nanoliposomes using an injection method and enriched up to 10% in skim milk showed acceptability and maintained stability under refrigerated condition (Gulzar and Benjakul 2020c). In an in vitro digestion study, controlled release of lipid phase was noticeable in liposome-loaded fish oil, thus ensuring the bioavailability of PUFA (Wang et al. 2015). Other encapsulation techniques include microgel encapsulation, nanofibers, and inclusion complexes (Venugopalan et al. 2021). Edible microgels are made from food-grade protein and polysaccharide that contains physical network and chemical crosslinking. Entrapment of the lipid phase inside the microgel can be done by various techniques such as injection, coacervation, and molding methods (Yan et al. 2020). For the injection method, microgel with oil trapped inside it was prepared by injecting oil droplets and gelling biopolymer into gelling solution (Fig. 2) (Venugopalan et al. 2021). In the case of the coacervation method, microgel encapsulation can be done by adding solutions of two oppositely charged biopolymers with oil (positively charged protein and negatively charged polysaccharide), also known as the phase separation method (Venugopalan et al. 2021). Fish oil encapsulated via a coacervation approach with a different combination of biopolymers (hydroxypropyl methylcellulose-maltodextrin and whey protein-gum Arabic) was prepared (Chang and Nickerson 2018). Nanofibers are long and thin fibrous substances made from protein or polysaccharides. These nanofibers can be used to encapsulate sensitive hydrophobic compounds like PUFA and vitamin D. Busolo et al. (2019) reported that DHA encapsulation using zein nanofibers via the electro-spraying method enhanced its oxidative stability. An inclusion complex method is an approach to trap lipid phase into a substance containing hydrophobic cavity, e.g., cyclodextrin (Fig. 2). Choi et al. (2010) postulated that the higher encapsulation efficiency of fish oil was obtained using cyclodextrin via inclusion complexes.

3.2 *Addition of Antioxidants*

Effective antioxidants must be added to prevent the generation of free radicals and to slow down oxidative processes (Rajasekaran et al. 2022). Synthetic antioxidants including butylated hydroxytoluene, propyl gallate, butylated hydroxyanisole (BHA), and tert-butylhydroquinone (TBHQ) may be harmful to human health (Hrebień-Filisińska 2021). Nowadays, clean labeled products (without synthetic additives) are gaining popularity among consumers. Therefore, antioxidants from natural sources have been paid more interest (Buamard and Benjakul 2017). Apart from antioxidant activity, natural antioxidants are known for various beneficial effects such as antimicrobial, antimutagenic, and antitumor properties (Buamard and Benjakul 2017). The antioxidant activity depends upon the polarity and solubility of the compounds. Hydrophilic antioxidants are generally effective in bulk oil, whereas hydrophobic compounds prevent oxidation in the emulsion system (Hrebień-Filisińska 2021). The α-tocopherol is used as a natural antioxidant to stabilize fish oil and/or oil-in-water emulsion. Addition of α-tocopherol at a low level (<100 ppm) is effective to scavenge free radicals, while a higher concentration leads to hydroperoxide accumulation (Zuta et al. 2007). Budilarto and Kamal-Eldin (2015) reported that tocopherol (1200–2000 ppm) and ascorbyl palmate (100–900 ppm) are not effective at higher concentrations, whereas phosphatidylcholine (1000–9000 ppm) and L-lysine (1000–9000 ppm) were found effective over a wide range of concentrations against oxidation of cod liver oil. Tocopherol donates hydrogen to peroxyl radicals, leading to the generation of tocopherol reactive radicals during storage

(Rajasekaran et al. 2022). These radicals act as 'pro-oxidants' that abstract hydrogen from the lipid phase and accelerate oxidation. Pro-oxidant activity of tocopherol at later stages of storage was reported in several studies (Takeungwongtrakul and Benjakul 2013; Rajasekaran et al. 2022). Pro-oxidant activity of α-tocopherol is concentration-dependent. With excessive concentration, it may induce oxidation (Takeungwongtrakul and Benjakul 2013). This drawback can be overcome by the usage of multi-component antioxidants in the system, and it is a promising approach for the stabilization of shrimp oil against oxidation during storage (Rajasekaran et al. 2022). Incorporation of tocopherol in combination with ascorbic palmitate in the fish oil emulsion using lecithin inhibited the formation of primary oxidative products during 30 days of storage (Drusch et al. 2007). Moen et al. (2017) developed mixed antioxidants including tocopherol (19.1%), ascorbyl palmate (13.3%), rosemary extract (53.2%), and green tea catechin (0.54%) for the stabilization of EPA and DHA. Mixed antioxidants have a synergistic effect and show superior protection to tocopherol. A high antioxidant effect was found in shrimp oil-in-water emulsion with the inclusion of tocopherol and tannic acid (Takeungwongtrakul and Benjakul 2013). Chitoligosaccharides obtained from shrimp shells had a synergistic effect with α-tocopherol in shrimp oil-in-water emulsion using bovine serum albumin-chitosan complex (Rajasekaran et al. 2022). Free radicals generated during autoxidation of tocopherol can be scavenged by typical hydrogen donation from other antioxidants in the system. Thus, it prolonged the induction period of oxidation, thereby preventing the lipid phase from radical attack (Takeungwongtrakul and Benjakul 2013). Bakır et al. (2013) documented that ascorbic acid with tocopherol lengthened the inhibition period of linoleic acid peroxidation induced by either copper or copper-ascorbic acid system.

3.3 Addition of Phenolic Compounds

The secondary metabolism of plants produces phenolic compounds, which act as bioactive substances with one or more phenolic rings (Maqsood et al. 2014). Antioxidant activity of phenolic compounds varies based on their structure and the presence of the hydroxyl group (OH–). Phenolic compounds are efficient in both bulk oil and emulsion (Maqsood et al. 2014). Sánchez-Alonso et al. (2011) documented that caffeic acid effectively retarded oxidation in cod liver oil owing to its higher partitioning coefficient in the oil. The higher antioxidant effect of caffeic acid as compared to BHA was observed when incorporated into cod liver oil (Leonardis and Macciola 2003). Maqsood and Benjakul (2010) documented the preventive effect of various phenolic compounds (catechin, caffeic acid, ferulic acid, gallic acid, and tannic acid) in a menhaden oil-in-water emulsion, in which tannic acid showed the highest effectiveness in preventing oxidation. Among the tested concentrations of caffeic acid (0.02–0.1%), the concentration of 0.06% was found to be effective in stabilizing EPA and DHA (Navarro-García et al. 2016). Flavonoids are effective in stabilizing fish oil. Quercetin prevented oxidation in fish oil-in-water emulsion stabilized using whey protein isolate (Azizi et al. 2019) and was more effective than BHT at high temperatures (70 °C) (Huber et al. 2009). However, the application of flavonoids is restricted owing to their poor solubility in oil. Hydroxytyrosol, a byproduct obtained from the production of olive, is used to stabilize PUFA in cod liver oil (Pazos et al. 2008), and it was effective even at low concentrations (0.0025–0.04%) in pure triacylglycerols from fish oils (Mahdavianmehr et al. 2016). Plant extracts such as tea, rosemary, oregano, mustard seed, and clove oils were used for the stabilization of liver oil from cod and Alaska pollock, in which catechin-rich tea extract exhibited superior antioxidant activity to tocopherol and BHT (O'Sullivan et al. 2005). Grape seed extract prevented the oxidation of PUFA in fish oil (Luther et al. 2007). Topuz et al (2015) documented that the extract of pomegranate peel, a byproduct of pomegranate juice industries, showed higher antioxidant activity in anchovy oil at high concentrations (500 and 1000 ppm) than those with a low concentration (100 ppm) at 60 °C. Apple peel extract inhibited the oxidation of bulk fish oil induced by heat and UV radiation (Rupasinghe et al. 2010). Extract from dry apple peel followed by removal of sugar and organic acid had increased antioxidant properties in bulk fish oil (Rupasinghe et al. 2010). Microencapsulated fish oil added with ginger essential oil and oregano extract showed enhanced oxidative stability (Jeyakumari et al. 2018). Bay leaf extract enhanced the oxidative stability of microencapsulated fish oil at various temperatures (23, 40, and 60 °C) (Yesilsu 2019). Ethanolic extract from coconut husk exhibited good antioxidant properties by reducing the thiobarbituric acid reactive substances generated in a shrimp oil-in-water emulsion (Buamard and Benjakul 2017). Rosemary extract showed a synergistic effect in combination with tocopherol, ascorbyl palmate, and lecithin (Drusch et al. 2007). The greater antioxidant effect of rosemary extract is owing to the carnosic acid and its derivates, which act as cascade antioxidants (Tounekti and Munné-Bosch 2012). The addition of antioxidants at an appropriate dose therefore enhances the stability of fish oil during extended storage.

4 Active Components in Fish and Shrimp Oil

4.1 Astaxanthin

Discards from crustacean processing industries are rich sources of carotenoids. Lipid-soluble pigments are present in different forms (Gulzar et al. 2020). The most valuable and prevalent (65–98%) carotenoid from the marine source is astaxanthin, by considering its anti-cardiovascular, anti-inflammatory, and anti-aging properties (Ahmadkelayeh and Hawboldt. 2020). Astaxanthin is a reddish-orange colored pigment, also known as a marine carotenoid (Fig. 1) (Ambati et al. 2014). The enzymes β-carotene hydroxylase and β-carotene ketolase produce astaxanthin from β-carotene or zeaxanthin, respectively. It is mostly produced from byproducts of crustacean, particularly shrimp and krill (Šimat et al. 2022). It is commonly used in the food, feed, pharmaceutical, and nutraceutical industries. The global astaxanthin market was 1371.24

million USD in 2020, and it is expected to upsurge in the upcoming years (Šimat et al. 2022). The human body cannot produce astaxanthin; it must be acquired through food sources and dietary supplements. By considering its potential benefits, USFDA approved and recommended the addition of astaxanthin in foodstuff as a natural coloring agent (E161) (Dalei and Sahoo 2015). Astaxanthin from the natural source occurs in two forms, namely esterified (stable) and unesterified (unstable). Ambati et al. (2014) reported that Antarctic krill has a more esterified form of astaxanthin than Atlantic salmon. Astaxanthin has antioxidant properties and is reported to have stronger antioxidant activity as compared to other compounds such as vitamin C and β-carotene (Šimat et al. 2022). Astaxanthin is an amphiphilic molecule that exerts 500 times greater antioxidant activity than α-tocopherol and 10 times than other commercial antioxidants (Gulzar et al. 2020). It is grouped in ketocarotenoid under the xanthophylls group, which has OH– (hydroxyl) and CO– (keto) groups at the terminals that function as strong electron donors, and scavenge the free radicals. This typical structure of astaxanthin is highly responsible for its antioxidant activity (Ahmadkelayeh and Hawboldt. 2020). Also, astaxanthin increased the activity of some antioxidant enzymes including manganese superoxide dismutase, glutathione transferase, etc. (Takahashi et al. 2002). Astaxanthin contents from the shrimp oil extracted from cephalothorax and hepatopancreas were 1.2 mg/g oil and 1.9 mg/g oil, respectively (Raju et al. 2021). Astaxanthin recovery from various shrimp processing leftovers using different extraction methods is given in Table 2. The astaxanthin in shrimp oil extracted from cephalothorax by UAE and vacuum-microwave heat pre-treatment under an inert atmosphere was found to be higher (1.36 mg/g oil) than that of the oil extracted using the solvent extraction process (0.98 mg/g oil) (Gulzar and Benjakul 2020a).

4.2 Squalene

Squalene ($C_{30}H_{50}$) is a natural polyunsaturated hydrocarbon (Fig. 1) present in shark liver oil. The deep-sea shark family squalidae is the most abundant source of squalene (Kim and Karadeniz 2012). It has been documented that shark (*Centrophorus artomarginatus*) liver oil has more quantity of squalene (80–85%) with 97.5% purity. Squalene from shark liver oil is referred to as cure-all, miracle oil or gift from the sea (Gopakumar 2012). It possesses potential benefits such as a cytoprotective agent. It also has antioxidant activity, cholesterol-lowering ability, and lowers high blood pressure (Gopakumar 2012). Squalene is a key intermediate for the synthesis of cholesterol, vitamin D, and steroid hormones in humans (Suresh et al. 2018). Considering its rapid transdermal absorption, non-greasiness, and antibacterial properties, it draws more attention from the anti-aging and skincare cosmetics manufacturing industries. Dietary intake of squalene enhanced high-density lipoprotein and lowered low-density lipoprotein, and triglycerides (Kim and Karadeniz 2012). Squalene showed a promising action against lung and colon cancer. Moreover, it enhances the immune system and improves the function of liver and kidney (Gopakumar 2012). Bindu

Table 2 Different methods for oil extraction from fish and shellfish processing leftovers

Species	Extraction method	Astaxanthin yield (μg/g waste)	References
Penaeus indicus	Solvent extraction	Acetone (40.6 μg/g); methanol (29.0 μg/g); ethyl, methyl ketone (36.8); ethyl acetate (36.9 μg/g); ethanol (31.9 μg/g); petroleum ether (12.1 μg/g); hexane (13.1), acetone: hexane (38.5 μg/g); isopropanol: hexane (43.9 μg/g)	Sachindra et al. (2006)
Penaeus monodon	Enzymatic hydrolysis	Trypsin (87.91 μg/g); alcalase (47.0 μg/g); pancreatin (57.0 μg/g)	De Holando and Netto (2006)
Farfantapenaeus paulensis	Supercritical fluid extraction	CO_2, ethanol, 50°C, and 30 MPa (1325 μg/g); CO_2, methanol, 60°C, and 20 MPa (53.17 μg/g); CO_2, water, 60°C, and 20 MPa (2.70 μg/g)	Sánchez-Camargo et al. (2012)
Pandalus borealis	Ultrasound-assisted extraction	Ethanol at 40 kHz frequency (50.32 μg/g); deep eutectic solvent at 20 kHz frequency (218 μg/g head, 146 μg/g shell)	Bin et al. (2020)
Aristeus antennatus	Microwave-assisted extraction	Acetone: ethanol: n-hexane (1:1:2) at 30 W for min (67.3 μg/g)	Ahmadkelayeh and Hawboldt (2020)
Penaeus monodon	High-pressure extraction	Acetone: methanol (7:3) at 210 MPa for 10 min (59.97 μg/g); ethanol with 0.1% acetic acid at 60 bar for 15 min (24 μg/g)	Gulzar et al. (2020)

et al. (2015) reported that squalene exhibits antimicrobial activity against foodborne pathogens. Squalene from spot-tail shark liver (*Carcharhinus sorrah*) was extracted and concentrated via fractional crystallization, followed by column purification. The yield of squalene is 6.8 g/100 g liver with a purity of ≥94% (Ali et al. 2019). Squalene extracted from the liver of two deep-sea sharks (*Carcharhinus plunketi* and *Somniosus pacificus*) was up to 50% (Bakes and Nichols 1995). The amount of squalene in shark liver is greatly influenced by habit and species (Achouri et al. 2018). Moreover, extraction using the green extraction process via ultrasound-assisted direct in situ saponification efficiently extracted squalene with a high yield (0.13–6.86 g/100 g) than conventional solvent extraction (0.10–5.52 g/100 g) using methanol (Ali et al. 2019).

4.3 Vitamins

Fish oil contains fat-soluble vitamins such as vitamin A (retinal), D (cholecalciferol), and E (α-tocopherol), which play key roles in maintaining human health (Suresh et al. 2018). Vitamin E (Fig. 1) has been known as a potent natural antioxidant preventing oxidative stress (Merdzhanova et al. 2014). Vitamin content in fishery byproducts varies, based upon the type, species, age, sex, season, type of byproduct, and extraction method (Gulzar et al. 2020). Vitamin A is known to improve vision and develop the immune system. Vitamin E acts as an antioxidant and develops the muscular and reproductive systems (Afonso et al. 2016). Vitamin D is crucial for calcium regulation. It acts on target organs such as bone, kidney, and intestinal mucosa for the regulation calcium and phosphate metabolisms (Chatterjea and Shinde 2011). The level of vitamins was higher in lipid storage organs such as the liver (fish) and hepatopancreas (shrimp) than that found in meat. Vitamin A and D are higher in halibut, cod, and tuna liver oil (Venugopal 2005). Suresh and Prabhu (2013) reported that sardine oil contained an average of 125 μg/g of vitamin A and D. Also, vitamin E levels in various fishes such as 30 μg/g in menhaden oil, over 60 μg/g in anchovy oil, and 25 μg/g in capelin oil were documented. López-Cervantes et al. (2006) reported that oil from shrimp processing wastes contained vitamin A (0.9–1.6 mg/100 g) and vitamin E (26.2–49.0 mg/100 g), respectively. Mathur et al. (2015) documented the antioxidant effect of tocopherol in the inhibition of lipid peroxidation, cardiovascular disease, and atherosclerosis. Vitamin E was reported to be 7.73 and 50.5 mg/100 g of meat from brown shrimp and fermented shrimp waste (Merdzhanova et al. 2014).

5 Health Benefits of n-3 Fatty Acid-Enriched Oil and Active Components

Increased usage of vegetable oils increased the consumption of n-6 fatty acids, which are pro-inflammatory and prothrombotic (Lange 2020). Fish oil and shrimp oil have been known as the excellent source of n-3 fatty acids (Fig. 1). Various health benefits of n-3 fatty acids are shown in Fig. 3. The dietary ratio of n-6 to n-3 fatty acids has drastically shifted from 1:1 to approximately 10:1 in the modern diet (Covington 2004). The n-3 fatty acids are essential fatty acids that are not synthesized in the body and are obtained only from the diet (Gulzar et al. 2020). In humans, α-linolenic acid is converted into EPA and DHA, which are the precursors of eicosanoids such as prostaglandins, thromboxane, and leukotrienes. These compounds exert several beneficial effects such as anti-inflammatory, antithrombotic, and antiarrhythmic (Simopoulos 2001). However, conversion rate is approximately less than 1 percent. This insufficient enzymatic conversion also increases the dietary ratio of n-6 to n-3 fatty acids (Lange 2020). Therefore, dietary intake of fish oil is required and can be

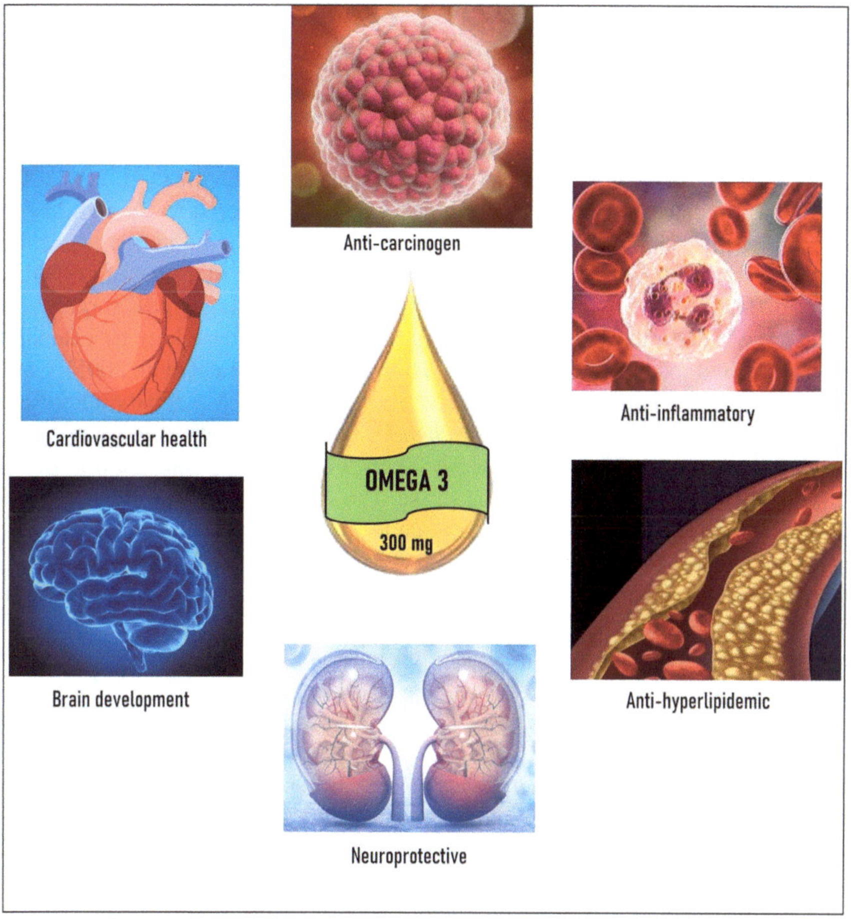

Fig. 3 Health benefits of omega-3 fatty acids

directly absorbed in the body and acts as a precursor of eicosanoids without enzymatic conversion (Bunn and Keith 2002). Also, astaxanthin confers health benefits in preventing several diseases.

5.1 Cardiovascular Disease Treatment

Cardiovascular disease is one of the serious concerns worldwide causing sudden death due to heart-related ailments (Tørris et al. 2018). EPA and DHA are the main components in fish oil and show promising effects on the prevention and treatment of coronary heart disease via different mechanisms (Chen et al. 2022). Consumption of n-3 fatty acids lowered the production of atherosclerotic plaques, thus decreasing the risk of thrombus formation in blood vessels (Von Schacky et al. 1999). Also, the n-3 fatty acids impart a protective effect against cardiovascular disease by reducing platelet aggregation, increasing high-density lipoprotein, decreasing low-density lipoprotein, lowering blood pressure, and improving vasodilation (Yashodhara et al. 2009). Dietary intake of fish oil prevents arrhythmias, related to the abnormal rhythm of the heartbeat that causes sudden cardiac arrests (Dallongeville et al. 2003). The risk of ventricular arrhythmias and cardiac death was lowered to 38 and 28% in myocardial infarction patients (Leaf et al. 2005). Myocardial infarction was effectively lowered with the intake of n-3 fatty acids at a dosage of 2 g per day (Gulzar et al. 2020). The n-3 fatty acids prevent atrial fibrillation during bypass surgery. The reduction of heart rate is due to the ability to alter electrophysiology by extending PR interval and reducing QT interval in electrocardiogram (Leaf et al. 2005). Coronary heart disease patients showed preventive effects against coronary atherosclerosis, stenosis, and restenosis after coronary angioplasty when consuming a fish oil-incorporated diet (Yashodhara et al. 2009). Furthermore, consumption of n-3 fatty acids lowers the risk of ischemic stroke. The study conducted with 4775 adults showed the incidence of stroke was lowered to 27 to 30% by the intake of n-3 fatty acid-enriched diets (Mozaffarian et al. 2005). Moreover, the incidence of congestive heart failure was reduced significantly in a cohort study using 2735 adult patients when supplemented with EPA and DHA (Mozaffarian et al. 2005). A study with apoE-mice showed that fish oil treatment resulted in reduced atherosclerotic lesions, thus preventing the thickening of arteries wall (Wang et al. 2004). Regular dietary intake of n-3 fatty acid at a dose of 1 g per day belated the incidence of heart-related aliments (Lee et al. 2008). Bonaccio et al. (2017) documented that intake of fish oil capsules showed benefits in terms of cardioprotection and stroke prevention. The myocardial infarction and secondary cardiovascular complications were lowered by 59% due to the intake of fish and shellfish in a regular diet (Gulzar et al. 2020). Furthermore, administration of synthetic astaxanthin (CDX-085) to C57BL/6 mice over 6–8 weeks augmented blood flow in basal arteries and attenuated the formation of occlusive thrombosis (Khan et al. 2010). Sprague–Dawley rats fed with the astaxanthin-incorporated diet showed a depleted level of cardiovascular disease markers (Xu et al. 2014).

5.2 *Antioxidant and Anti-inflammatory Effects*

Free radicals and reactive oxygen species (ROS) generated by the cell-regulating process cause oxidative stress, resulting in cell injuries/death (Oppedisano et al. 2020). These compounds are majorly responsible for several diseases such as Alzheimer's, steatohepatitis, aging, cardiovascular, and diabetes (Chen et al. 2022). Free radicals (peroxynitrite) also increase the synthesis of pro-inflammatory cytokines by regulating the cyclooxygenase pathway (COX) (Rodrigo et al. 2013). Mollace et al. (2013) reported that n-3 fatty acids showed a protective effect on endothelial cells and cardiomyocytes against oxidative stress and cell death. Moreover, it exhibited substantial antioxidant activity by restoring imbalanced endogenous antioxidant moieties and regulating the antioxidant signaling pathway (Oppedisano et al. 2020). In general, DHA in the mitochondria is crucial for the synthesis of adenosine triphosphate (ATP) by oxidative phosphorylation. Fish oil rich in DHA is involved in various pathways to decrease oxidative stress by increasing cytochrome c oxidase activity and manganese-dependent superoxide dismutase (Mn-SOD) activity in the mitochondrial membrane (Oppedisano et al. 2020). Supplementation of fish oil inhibited membrane peroxidation and improved the antioxidant superoxide dismutase (SOD) activity in the rat model (Li et al. 2019). Another possible mechanism related to antioxidant activity is the preventive effect against ROS-induced ischemia–reperfusion injury and increased glutathione peroxidase (GSH-Px) levels (Rodrigo et al. 2013). Herrera et al. (2015) documented a lowered oxidative stress marker such as F2-isoprostane in human urine due to supplementation of EPA and DHA in the diet. The n-3 fatty acid can regulate inflammatory pathways by inhibiting IkB phosphorylation process and reducing signaling pathways such as TLR- and TNF-α signaling pathways (Alzoubi and Al-Domi 2017). The administration of fish oil as a supplement increased the activity of antioxidant enzymes in apoE-mice (Wang et al. 2004). An in vivo rat study revealed a higher antioxidant effect in the plasma after the administration of n-3 fatty acids (Erdogan et al. 2004). Fish oil is used for the treatment of rheumatoid arthritis, an autoimmune disease, causing severe pain and inflammation in joints. PUFA has the ability to convert into natural anti-inflammatory compounds in the human body, e.g., prostaglandins and leukotrienes (Chen et al. 2022). Also, n-3 fatty acids showed promising effects in the prevention of inflammatory bowel disease and kidney disease (Chen et al. 2022). Bhattacharya et al. (2007) documented a higher anti-inflammatory effect in C57BL/6 mice by retarding pro-inflammatory cytokines such as TNF-α, IL-6, and IL-1β. Also, antioxidant enzymes (catalase and glutathione transferase) were higher in the kidneys of mice with the intake of fish oil. In addition, astaxanthin was reported to prevent oxidative stress and inflammation in several studies. Astaxanthin can directly inhibit ROS generation and hydrogen peroxide-induced NF-kB activation (Lee and Bai 2003). Furthermore, the addition of astaxanthin reduced the expression of inflammatory markers due to its high anti-inflammatory activity (Fassett and Coombes 2011).

5.3 Mental Health and Brain Development

The role of PUFA has been well recognized in brain development and neuroprotection. DHA is the major component contributing to the formation of brain membrane phospholipids (Swanson et al. 2012). Brain development and its function are directly proportional to the intake of n-3 fatty acids (Bazinet and Layé 2014). In humans, EPA and DHA are obtained from α-linolenic acid in several steps such as desaturation, elongation, and β-oxidation (Lange 2020). However, biological conversion is inadequate and a slow process. Thus, n-3 fatty acid dietary supplements improve the function of the brain by changing the biophysical characteristics of cell membranes which includes activating receptors, cell signaling pathways, and controlling the endocannabinoid system (Bazinet and Layé 2014). The n-3 fatty acids modulate several brain functions including gene transcription, neurotransmission, and neuroinflammation (Bazinet and Layé 2014). The intake of fish oil capsule prevents depression, dementia, schizophrenia, bipolar disorder, and attention-deficit hyperactivity disorder (Lange 2020). Also, EPA and DHA showed preventive effects against Alzheimer's and Parkinson's diseases (Chen et al. 2022). Intake of n-3 fatty acids has a beneficial impact on dopamine level and lipid profile by improving the gene expression of hydroxytryptamine 1A (HT1A) in the Wistar rat in vivo study (Chen et al. 2022). Rotenone rats fed with a fish oil-added diet showed enhanced neuroprotection and could be used for the treatment of Parkinson's disease (Denny-Joseph 2015). Recently, a detailed report on the role of n-3 fatty acids in brain development, mental health, and psychiatric disorders was reviewed by Lange (2020).

5.4 *Other Medical Conditions*

Intake of fish oils normalizes glucose metabolism and imparts protective effects against diabetes mellitus (Covington 2004). Also, it showed protection against various carcinomas in humans. The antitumor effects of n-3 fatty acids are owing to the suppression of cell proliferation and controlled synthesis of prostaglandins (Wang et al. 2014). The intake of n-3 fatty acids reduces the risk of hyperlipemia by decreasing serum triglycerides levels to 25–30% and increasing the high-density lipoprotein to 1–3%, respectively. Salama et al. (2013) documented that hepatoprotective effect of cod oil in rat against sodium nitrite induced hepatotoxicity. In another report, consumption of fish oil reduced the acetaminophen-induced hepatic injury in rats (Kalra et al. 2012). Basheer et al. (2017) reported that administration of EPA and DHA exhibited protective action in the liver by lowering isoniazid-rifampin-induced hepatotoxicity in rats. Apart from the above-mentioned bioactivities, n-3 fatty acid consumption provides numerous health benefits, such as wound healing, gut microbiota modulation, and skincare (Gulzar et al. 2020). The wound-healing property of astaxanthin was also investigated in mice. Mice fed with astaxanthin showed a high level of healing markers, fibroblast growth factor, and collagen type I α 1 (Col1A1) production (Meephansan et al. 2017). In the acetone-induced dry skin rat model, oral administration of EPA and DHA reduced dryness, pruritus, trans-epidermal water loss, and improved skin hydration (Barcelos et al. 2015). Davinelli et al. (2018) documented the preventive effect of astaxanthin against skin cancer and aging.

5.5 Dietary Recommendations of Fish Oil

The American Heart Association recommends a person with no heart ailments can consume fish twice a week, and once a week for a person with coronary heart disease (Covington 2004). The commercially available fish oil capsule normally contains 1 g, comprising 180 mg of EPA and 120 mg of DHA. Thus, the recommended dose of n-3 fatty acids is 0.9 g per day which can be achieved by taking three 1 g capsules of fish oil per day (USFDA 2004). The daily recommendation of n-3 fatty acids is 1 g per day for cardioprotection. A higher dosage of approximately 2–4 g per day is recommended to reduce triglycerides and joint pain caused by rheumatoid arthritis (Covington 2004). The Environmental Protection Agency (EPA) and Food and Drug Administration (FDA) recommended children and pregnant women to avoid the consumption of oil or meat from shark, king mackerel, and tilefish that may contain a high level of mercury (Covington 2004). FDA provided generally regarded as safe (GRAS) status for dietary intake of marine n-3 fatty acids up to 3 g per day (USFDA 2004). The recommended intakes of EPA and DHA by various organizations are presented in Table 3. Astaxanthin extracted from shrimp processing discards are considered as safe for consumption in its free or ester form (Šimat et al. 2022). Its bioavailability depends on dosage, form (mono or diester), and food intake (before or after food). In food supplements, the recommended level of astaxanthin is 8 mg/day and the acceptable daily intake for adults is 0.034–0.2 mg/kg of body weight (Šimat et al. 2022).

6 Future Prospects

Fish oil or shrimp oil is an excellent source of PUFA, carotenoid, and other active components. Mostly, conventional methods are adopted to extract oil on an industrial scale using solvents or at high temperatures. Recent studies using UAE, SFE, and PEF show a promising result with higher yields and reduced consumption of solvents. Despite its benefits, a few methods affect the quality of the extracted oil. Therefore, some novel pre-treatment techniques are needed to overcome the drawbacks. The economic viability and scalability of the novel extraction methods need to be elucidated for commercialization. To improve the stability of oil, encapsulation techniques must be optimized or standardized for the specific food product without affecting its sensory attributes. Development of fish oil or shrimp oil as functional foods and beverages is still required, in which marketable products can be available. As a consequence, the consumption of fish oil or shrimp oil will increase among the consumers, especially after the removal of cholesterol. Furthermore, clinical studies are required to ensure the health benefit of fish oil or shrimp oil for reducing the risk of heart-related diseases or other diseases.

Table 3 Different methods for oil extraction from fish and shellfish processing leftovers

Organization	Recommended level	Astaxanthin yield (μg/g waste)	References
United States Food and Drug Administration (USFDA)	3 g DHA and EPA/day	Acetone (40.6 μg/g); methanol (29.0 μg/g); ethyl, methyl ketone (36.8); ethyl acetate (36.9 μg/g); ethanol (31.9 μg/g); petroleum ether (12.1 μg/g); hexane (13.1), acetone: hexane (38.5 μg/g); isopropanol: hexane (43.9 μg/g)	Sachindra et al. (2006)
International Society for the Study of Fatty Acids and Lipids (ISSFAL)	0.22 g DHA and EPA/day	Trypsin (87.91 μg/g); alcalase (47.0 μg/g); pancreatin (57.0 μg/g)	De Holando and Netto (2006)
British Nutrition Foundation (BNF)	1.1–1.4 g DHA and EPA/day	CO_2, ethanol, 50°C, and 30 MPa (1325 μg/g); CO_2, methanol, 60°C, and 20 MPa (53.17 μg/g); CO_2, water, 60°C, and 20 MPa (2.70 μg/g)	Sánchez-Camargo et al. (2012)
Health and Wealth Canada	1.0–1.8 g DHA and EPA/day	Ethanol at 40 kHz frequency (50.32 μg/g); deep eutectic solvent at 20 kHz frequency (218 μg/g head, 146 μg/g shell)	Bin et al. (2020)
Institute of Medicine (IOM)	0.5 g DPA and EPA/day	Acetone: ethanol: n-hexane (1:1:2) at 30 W for min (67.3 μg/g)	Ahmadkelayeh and Hawboldt (2020)

Source Rizliya and Mendis (2014)

7 Conclusion

The demand for oil extracted from fish or shrimp processing wastes has been increasing globally for their wide range of health benefits. This chapter summarizes the oil extraction methods and stabilization techniques to preserve active compounds. Discards from fish/crustacean processing can be used to extract oil thus improving waste management and ensuring sustainability of marine resources. Also, shrimp oil containing carotenoid pigments (astaxanthin) can be used as natural colorants in oil-based foods such as mayonnaise, soups, butter, and confectionery products, while providing health benefits. The consumption of functional food fortified with fish oil exerts potential benefits on human health.

Acknowledgements Chair professor grant (Grant No: P-20-52297) are gratefully acknowledged. This research was also supported by National Science, Research and Innovation Fund (NSRF) and Prince of Songkla University (PSU) (Grant No: AGR6602079N). PSU president scholarship for Bharathipriya Rajasekaran, prachayacharn grant from PSU (Grant No: AGR6502111N).

Conflict of Interest The authors have declared no conflicts of interest.

References

Abdullah S, Mudalip SA, Shaarani SM, Pi NC (2010) Ultrasonic extraction of oil from *Monopterus albus*: effects of different ultrasonic power, solvent volume and sonication time. J Appl Sci (Faisalabad) 10(21):2713–2716

Achouri N, Smichi N, Kharrat N, Rmili F, Gargouri Y, Miled N, Fendri A (2018) Characterization of liver oils from three species of sharks collected in Tunisian coasts: *in vitro* digestibility by pancreatic lipase. J Food Biochem 42(1):e12453

Adeoti IA, Hawboldt K (2014) A review of lipid extraction from fish processing by-product for use as a biofuel. Biomass Bioenerg 63:330–340

Afonso C, Bandarra NM, Nunes L, Cardoso C (2016) Tocopherols in seafood and aquaculture products. Crit Rev Food Sci Nutr 56(1):128–140

Ahmadkelayeh S, Hawboldt K (2020) Extraction of lipids and astaxanthin from crustacean by-products: a review on supercritical CO_2 extraction. Trends Food Sci Technol 103:94–108

Ajeeshkumar KK, Aneesh PA, Raju N, Suseela M, Ravishankar CN, Benjakul S (2021) Advancements in liposome technology: preparation techniques and applications in food, functional foods, and bioactive delivery: a review. Compr Rev Food Sci Food Saf 20(2):1280–1306

Al Khawli F, Pateiro M, Domínguez R, Lorenzo JM, Gullón P, Kousoulaki K, Ferrer E, Berrada H, Barba FJ (2019) Innovative green technologies of intensification for valorization of seafood and their by-products. Mar Drugs 17(12):689

Ali AMM, Bavisetty SCB, Prodpran T, Benjakul S (2019) Squalene from fish livers extracted by ultrasound-assisted direct in situ saponification: purification and molecular characteristics. J Am Oil Chem Soc 96(9):1059–1071

Ali A, Wei S, Liu Z, Fan X, Sun Q, Xia Q, Liu S, Hao J, Deng C (2021) Non-thermal processing technologies for the recovery of bioactive compounds from marine by-products. LWT 147:111549

Ali-Nehari A, Kim SB, Lee YB, Lee HY, Chun BS (2012) Characterization of oil including astaxanthin extracted from krill (*Euphausia superba*) using supercritical carbon dioxide and organic solvent as comparative method. Korean J Chem Eng 29(3):329–336

Alzoubi MR, Al-Domi HA (2017) Could omega-3 fatty acids a therapeutic treatment of the immune-metabolic consequence of intermittent hypoxia in obstructive sleep apnea? Diabetes Metab Syndr 11(4):297–304

Ambati RR, Phang SM, Ravi S, Aswathanarayana RG (2014) Astaxanthin: sources, extraction, stability, biological activities and its commercial applications—a review. Mar Drugs 12(1):128–152

Amiguet VT, Kramp KL, Mao J, McRae C, Goulah A, Kimpe LE, Blais JM, Arnason JT (2012) Supercritical carbon dioxide extraction of polyunsaturated fatty acids from Northern shrimp (*Pandalus borealis* Kreyer) processing by-products. Food Chem 130(4):853–858

Azizi M, Li Y, Kaul N, Abbaspourrad A (2019) Study of the physicochemical properties of fish oil solid lipid nanoparticle in the presence of palmitic acid and quercetin. J Agric Food Chem 67(2):671–679

Baiano A, Del Nobile MA (2016) Antioxidant compounds from vegetable matrices: biosynthesis, occurrence, and extraction systems. Crit Rev Food Sci Nutr 56(12):2053–2068

Bakes MJ, Nichols PD (1995) Lipid, fatty acid and squalene composition of liver oil from six species of deep-sea sharks collected in southern Australian waters. Comp Biochem Physiol B Biochem Mol Biol 110(1):267–275

Bakır T, Yıldoğan Beker B, Sönmezoğlu İ, İmer F, Apak R (2013) Antioxidant and prooxidant effects of α-tocopherol in a linoleic acid-copper (II)-ascorbate system. Eur J Lipid Sci Technol 115(3):372–376

Bako T, Umogbai VI, Obetta SE (2014) Extraction and characterization of Mackery (*Scomber scombrus*) oil for industrial use. Researcher 6(8):80–85

Barcelos RC, de Mello-Sampayo C, Antoniazzi CT, Segat HJ, Silva H, Veit JC, Piccolo J, Emanuelli T, Bürger ME, Silva-Lima B, Rodrigues LM (2015) Oral supplementation with fish oil reduces dryness and pruritus in the acetone-induced dry skin rat model. J Dermatol Sci 79(3):298–304
Basheer AS, Siddiqui A, Paudel YN, Hassan MQ, Imran M, Najmi AK, Akhtar M (2017) Hepatoprotective and antioxidant effects of fish oil on isoniazid-rifampin induced hepatotoxicity in rats. PharmaNutrition 5(1):29–33
Bazinet RP, Layé S (2014) Polyunsaturated fatty acids and their metabolites in brain function and disease. Nat Rev Neurosci 15(12):771–785
Benjakul S, Sae-leaw T, Simpson BK (2019) Byproducts from fish harvesting and processing. In: Byproducts from agriculture and fisheries: adding value for food, feed, pharma, and fuels, pp 179–217
Bermúdez-Aguirre D, Barbosa-Cánovas GV (2011) An update on high hydrostatic pressure, from the laboratory to industrial applications. Food Eng Rev 3(1):44–61
Bhattacharya A, Sun D, Rahman M, Fernandes G (2007) Different ratios of eicosapentaenoic and docosahexaenoic omega-3 fatty acids in commercial fish oils differentially alter proinflammatory cytokines in peritoneal macrophages from C57BL/6 female mice. J Nutr Biochem 18(1):23–30
Bin LK, Janakiraman AK, Abd Razak FS, Uddin AH, Sarker MZI, Ming LC, Goh BH (2020) Supercritical fluid technology and its pharmaceutical applications: a revisit with two decades of progress. Indian J Pharm Educ Res 54(2):1–11
Bindu BSC, Mishra DP, Narayan B (2015) Inhibition of virulence of Staphylococcus aureus—a food borne pathogen—by squalene, a functional lipid. J Funct Foods 18:224–234
Bonaccio M, Ruggiero E, Di Castelnuovo A, Costanzo S, Persichillo M, De Curtis A, Cerletti C, Donati MB, De Gaetano G, Iacoviello L, Vermylen J (2017) Fish intake is associated with lower cardiovascular risk in a Mediterranean population: prospective results from the Moli-sani study. Nutr Metab Cardiovasc Dis 27(10):865–873
Bonilla-Méndez JR, Hoyos-Concha JL (2018) Methods of extraction refining and concentration of fish oil as a source of omega-3 fatty acids. Ciencia y Tecnología Agropecuaria 19(3):645–668
Bruno SF, Kudre TG, Bhaskar N (2019) Impact of pretreatment-assisted enzymatic extraction on recovery, physicochemical and rheological properties of oil from Labeo rohita head. J Food Process Eng 42(3):e12990
Buamard N, Benjakul S (2017) Ethanolic coconut husk extract: in vitro antioxidative activity and effect on oxidative stability of shrimp oil emulsion. Eur J Lipid Sci Technol 119(11):1700131
Budilarto ES, Kamal-Eldin A (2015) Stabilization of cod liver oil with a quaternary combination of α-tocopherol and synergists: method of assessment. Eur J Lipid Sci Technol 117(10):1598–1606
Bunn PA Jr, Keith RL (2002) The future of cyclooxygenase-2 inhibitors and other inhibitors of the eicosanoid signal pathway in the prevention and therapy of lung cancer. Clin Lung Cancer 3(4):271–277
Busolo MA, Torres-Giner S, Prieto C, Lagaron JM (2019) Electrospraying assisted by pressurized gas as an innovative high-throughput process for the microencapsulation and stabilization of docosahexaenoic acid-enriched fish oil in zein prolamine. Innov Food Sci Emerg Technol 51:12–19
Carlès P (2010) A brief review of the thermophysical properties of supercritical fluids. J Supercritical Fluids 53(1–3):2–11
Carvajal A, Slizyte R, Storrø I, Aursand M (2015) Production of high quality fish oil by thermal treatment and enzymatic protein hydrolysis from fresh Norwegian spring spawning herring by-products. J Aquat Food Prod Technol 24(8):807–823
Chakraborty K, Joseph D (2015) Cooking and pressing is an effective and eco-friendly technique for obtaining high quality oil from *Sardinella longiceps*. Eur J Lipid Sci Technol 117(6):837–850
Chang C, Nickerson MT (2018) Encapsulation of omega 3-6-9 fatty acids-rich oils using protein-based emulsions with spray drying. J Food Sci Technol 55(8):2850–2861
Chantachum S, Benjakul S, Sriwirat N (2000) Separation and quality of fish oil from precooked and non-precooked tuna heads. Food Chem 69(3):289–294

Chatterjea MN, Shinde R (2011) Textbook of medical biochemistry. Wife Goes On
Chen J, Jayachandran M, Bai W, Xu B (2022) A critical review on the health benefits of fish consumption and its bioactive constituents. Food Chem 369:130874
Choi MJ, Ruktanonchai U, Min SG, Chun JY, Soottitantawat A (2010) Physical characteristics of fish oil encapsulated by β-cyclodextrin using an aggregation method or polycaprolactone using an emulsion–diffusion method. Food Chem 119(4):1694–1703
Chotphruethipong L, Battino M, Benjakul S (2020) Effect of stabilizing agents on characteristics, antioxidant activities and stability of liposome loaded with hydrolyzed collagen from defatted Asian sea bass skin. Food Chem 328:127127
Covington M (2004) Omega-3 fatty acids. Am Fam Physician 70(1):133–140
Dalei J, Sahoo D (2015) Extraction and characterization of astaxanthin from crustacean shell waste from shrimp processing industries. Int J Pharm Sci Res 6(6):2532–2537
Dallongeville J, Yarnell J, Ducimetière P, Arveiler D, Ferrières J, Montaye M, Luc G, Evans A, Bingham A, Hass B, Ruidavets JB (2003) Fish consumption is associated with lower heart rates. Circulation 108(7):820–825
De Holanda HD, Netto FM (2006) Recovery of components from shrimp (*Xiphopenaeus kroyeri*) processing waste by enzymatic hydrolysis. J Food Sci 71(5):C298–C303
de Oliveira DA, Licodiedoff S, Furigo A Jr, Ninow JL, Bork JA, Podestá R, Block JM, Waszczynskyj N (2017) Enzymatic extraction of oil from yellowfin tuna (*Thunnus albacares*) by-products: a comparison with other extraction methods. Int J Food Sci Technol 52(3):699–705
Del Valle JM, Aguilera JM (1999) Revision: Extracción con CO_2 a alta presión. Fundamentos y aplicaciones en la industria de alimentos/Review: high pressure CO_2 extraction. Fundamentals and applications in the food industry. Food Sci Technol Int 5(1):1–24
Denny Joseph KM (2015) Combined oral supplementation of fish oil and quercetin enhances neuroprotection in a chronic rotenone rat model: relevance to Parkinson's disease. Neurochem Res 40(5):894–905
Ding M, Zhang T, Zhang H, Tao N, Wang X, Zhong J (2020) Gelatin molecular structures affect behaviors of fish oil-loaded traditional and Pickering emulsions. Food Chem 309:125642
Donsì F, Ferrari G, Pataro G (2010) Applications of pulsed electric field treatments for the enhancement of mass transfer from vegetable tissue. Food Eng Rev 2(2):109–130
Drusch S, Serfert Y, Scampicchio M, Schmidt-Hansberg B, Schwarz K (2007) Impact of physicochemical characteristics on the oxidative stability of fish oil microencapsulated by spray-drying. J Agric Food Chem 55(26):11044–11051
Erdogan H, Fadillioglu E, Ozgocmen S, Sogut S, Ozyurt B, Akyol O, Ardicoglu O (2004) Effect of fish oil supplementation on plasma oxidant/antioxidant status in rats. Prostaglandins Leukot Essent Fatty Acids 71(3):149–152
FAO (2018) The state of world fisheries and aquaculture 2018—Meeting the sustainable development goals. FAO, Rome, Italy. ISBN 978-92-5-130562-1
Fassett RG, Coombes JS (2011) Astaxanthin: a potential therapeutic agent in cardiovascular disease. Mar Drugs 9(3):447–465
FDA (2010) Food and drug administration. In: FDA (ed) Status (L.O.F.A., ed) Food and drug administration, MD, USA
Flint EB, Suslick KS (1991) The temperature of cavitation. Science 253(5026):1397–1399
Food US (2004) Drug administration centre for food safety and applied nutrition. Agency response letter. GRAS notice no. GRN 000105, October 15, 2002
Franco D, Munekata PE, Agregán R, Bermúdez R, López-Pedrouso M, Pateiro M, Lorenzo JM (2020) Application of pulsed electric fields for obtaining antioxidant extracts from fish residues. Antioxidants 9(2):90
Gallo M, Ferrara L, Naviglio D (2018) Application of ultrasound in food science and technology: a perspective. Foods 7(10):164
Gopakumar K (2012) Therapeutic applications of Squalene-A review

Gulzar S, Benjakul S (2018) Ultrasound waves increase the yield and carotenoid content of lipid extracted from cephalothorax of Pacific white shrimp (*Litopenaeus vannamei*). Eur J Lipid Sci Technol 120(5):1700495
Gulzar S, Benjakul S (2019) Effect of pre-treatments on yield and properties of lipid extracted from cephalothorax of Pacific white shrimp (*Litopenaeus vannamei*) by ultrasonic assisted process. LWT 100:106–113
Gulzar S, Benjakul S (2020a) Impact of pretreatment and atmosphere on quality of lipids extracted from cephalothorax of Pacific white shrimp by ultrasonic assisted process. Food Chem 309:125732
Gulzar S, Benjakul S (2020b) Characteristics and storage stability of nanoliposomes loaded with shrimp oil as affected by ultrasonication and microfluidization. Food Chem 310:125916
Gulzar S, Benjakul S (2020c) Fortification of skim milk with nanoliposomes loaded with shrimp oil: properties and storage stability. J Am Oil Chem Soc 97(8):929–940
Gulzar S, Raju N, Nagarajarao RC, Benjakul S (2020) Oil and pigments from shrimp processing by-products: extraction, composition, bioactivities and its application—a review. Trends Food Sci Technol 100:307–319
Hasani S, Ojagh SM, Ghorbani M, Hasani M (2020) Nano-encapsulation of lemon essential oil approach to reducing the oxidation process in fish burger during refrigerated storage. J Food Biosci Technol 10(1):35–46
Hathwar SC, Bijinu B, Rai AK, Narayan B (2011) Simultaneous recovery of lipids and proteins by enzymatic hydrolysis of fish industry waste using different commercial proteases. Appl Biochem Biotechnol 164(1):115–124
Herrera EA, Farías JG, González-Candia A, Short SE, Carrasco-Pozo C, Castillo RL (2015) Ω3 Supplementation and intermittent hypobaric hypoxia induce cardioprotection enhancing antioxidant mechanisms in adult rats. Mar Drugs 13(2):838–860
Hosseini RS, Rajaei A (2020) Potential Pickering emulsion stabilized with chitosan-stearic acid nanogels incorporating clove essential oil to produce fish-oil-enriched mayonnaise. Carbohyd Polym 241:116340
Hrebień-Filisińska A (2021) Application of natural antioxidants in the oxidative stabilization of fish oils: a mini-review. J Food Process Preserv 45(4):e15342
Huber GM, Rupasinghe HV, Shahidi F (2009) Inhibition of oxidation of omega-3 polyunsaturated fatty acids and fish oil by quercetin glycosides. Food Chem 117(2):290–295
Ivanovs K, Blumberga D (2017) Extraction of fish oil using green extraction methods: a short review. Energy Procedia 128:477–483
Jamshidi A, Shabanpour B, Pourashouri P, Raeisi M (2018) Using WPC-inulin-fucoidan complexes for encapsulation of fish protein hydrolysate and fish oil in W1/O/W2 emulsion: characterization and nutritional quality. Food Res Int 114:240–250
Jeyakumari A, Zynudheen AA, Parvathy U, Binsi PK (2018) Impact of chitosan and oregano extract on the physicochemical properties of microencapsulated fish oil stored at different temperature. Int J Food Prop 21(1):943–956
Kalra J, Ali B, Kalra S, Pant KK (2012) Fish oil and its role in acetaminophen induced hepatic injury. Asian J Exp Biol Sci 3(4):826–829
Kawashima A, Iwakiri R, Honda K (2006) Experimental study on the removal of dioxins and coplanar polychlorinated biphenyls (PCBs) from fish oil. J Agric Food Chem 54(26):10294–10299
Khan SK, Malinski T, Mason RP, Kubant R, Jacob RF, Fujioka K, Denstaedt SJ, King TJ, Jackson HL, Hieber AD, Lockwood SF (2010) Novel astaxanthin prodrug (CDX-085) attenuates thrombosis in a mouse model. Thromb Res 126(4):299–305
Khoshnoudi-Nia S, Forghani Z, Jafari SM (2022) A systematic review and meta-analysis of fish oil encapsulation within different micro/nanocarriers. Crit Rev Food Sci Nutr 62(8):2061–2082
Kim SK, Karadeniz F (2012) Biological importance and applications of squalene and squalane. Adv Food Nutr Res 65:223–233

Kumar V, Muzaddadi AU, Mann S, Balakrishnan R, Bembem K, Kalnar Y (2018) Utilization of fish processing waste: a waste to wealth approach

Lange KW (2020) Omega-3 fatty acids and mental health. Glob Health J 4(1):18–30

Leaf A, Albert CM, Josephson M, Steinhaus D, Kluger J, Kang JX, Cox B, Zhang H, Schoenfeld D (2005) Prevention of fatal arrhythmias in high-risk subjects by fish oil n-3 fatty acid intake. Circulation 112(18):2762–2768

Lee SJ, Bai SK, Lee KS, Namkng S, Na HJ, Ha KS et al (2003) Astaxanthin inhibits nitric oxide production and inflammatory gene expression by suppressing I (kappa) B kinase-dependent NF-(kappa) B activation. Mol Cells 16(1):97–105

Lee JH, O'Keefe JH, Lavie CJ, Marchioli R, Harris WS (2008) Omega-3 fatty acids for cardioprotection. In: Mayo clinic proceedings, vol 83, no 3. Elsevier, pp 324–332

Leonardis AD, Macciola V (2003) Effectiveness of caffeic acid as an anti-oxidant for cod liver oil. Int J Food Sci Technol 38(4):475–480

Li M, Lin JIE, Chen J, Fang T (2016) Pulsed electric field-assisted enzymatic extraction of protein from abalone (*Haliotis discus* hannai Ino) viscera. J Food Process Eng 39(6):702–710

Li J, Sun W, Ramaswamy HS, Yu Y, Zhu S, Wang J, Li H (2017) High pressure extraction of astaxanthin from shrimp waste (*Penaeus Vannamei* Boone): effect on yield and antioxidant activity. J Food Process Eng 40(2):e12353

Li J, Li H, Li H, Guo W, An Z, Zeng X, Li W, Li H, Song J, Wu W (2019) Amelioration of PM2. 5-induced lung toxicity in rats by nutritional supplementation with fish oil and Vitamin E. Respir Res 20(1):1–9

Liaset B, Julshamn K, Espe M (2003) Chemical composition and theoretical nutritional evaluation of the produced fractions from enzymic hydrolysis of salmon frames with Protamex™. Process Biochem 38(12):1747–1759

Linder M, Fanni J, Parmentier M (2005) Proteolytic extraction of salmon oil and PUFA concentration by lipases. Mar Biotechnol 7(1):70–76

Liu C, Fan L, Yang Y, Jiang Q, Xu Y, Xia W (2021) Characterization of surimi particles stabilized novel pickering emulsions: Effect of particles concentration, pH and NaCl levels. Food Hydrocolloids 117:106731

Lopez-Cervantes J, Sanchez-Machado DI, Rios-Vazquez NJ (2006) High-performance liquid chromatography method for the simultaneous quantification of retinol, α-tocopherol, and cholesterol in shrimp waste hydrolysate. J Chromatogr A 1105(1–2):135–139

López-Pedrouso M, Lorenzo JM, Zapata C, Franco D (2019) 5—Proteins and amino acids. In: Barba FJ, Saraiva JMA, Cravotto G, Lorenzo JM (eds) Bioaccessibility and bioavailability of nutrients and bioactive compounds. Woodhead publishing series in food science, technology and nutrition. Woodhead Publishing, pp 139–169

Luther M, Parry J, Moore J, Meng J, Zhang Y, Cheng Z, Yu LL (2007) Inhibitory effect of Chardonnay and black raspberry seed extracts on lipid oxidation in fish oil and their radical scavenging and antimicrobial properties. Food Chem 104(3):1065–1073

Macías-Sánchez MD, Fernandez-Sevilla JM, Fernández FA, García MC, Grima EM (2010) Supercritical fluid extraction of carotenoids from *Scenedesmus almeriensis*. Food Chem 123(3):928–935

Mahdavianmehr H, Farhoosh R, Sharif A (2016) Mechanism of the inhibitory effect of hydroxytyrosol on lipid oxidation in different bulk oil systems. J Am Oil Chem Soc 93(9):1233–1242

Maqsood S, Benjakul S (2010) Comparative studies of four different phenolic compounds on in vitro antioxidative activity and the preventive effect on lipid oxidation of fish oil emulsion and fish mince. Food Chem 119(1):123–132

Maqsood S, Benjakul S, Abushelaibi A, Alam A (2014) Phenolic compounds and plant phenolic extracts as natural antioxidants in prevention of lipid oxidation in seafood: a detailed review. Compr Rev Food Sci Food Saf 13(6):1125–1140

Mason TJ, Paniwnyk L, Lorimer JP (1996) The uses of ultrasound in food technology. Ultrason Sonochem 3(3):S253–S260

Mathur P, Ding Z, Saldeen T, Mehta JL (2015) Tocopherols in the prevention and treatment of atherosclerosis and related cardiovascular disease. Clin Cardiol 38(9):570–576
Meephansan J, Rungjang A, Yingmema W, Deenonpoe R, Ponnikorn S (2017) Effect of astaxanthin on cutaneous wound healing. Clin Cosmet Investig Dermatol 10:259
Menegazzo ML, Petenuci ME, Fonseca GG (2014) Production and characterization of crude and refined oils obtained from the co-products of Nile tilapia and hybrid sorubim processing. Food Chem 157:100–104
Merdzhanova A, Dobreva DA, Stancheva M, Makedonski L (2014) Fat soluble vitamins and fatty acid composition of wild Black sea mussel, rapana and shrimp. Ovidius Univ Ann Chem 25(1):15–23
Moen V, Stoknes I, Breivik H (2017) Antioxidant efficacy of a new synergistic, multicomponent formulation for fish oil omega-3 concentrates. J Am Oil Chem Soc 94(7):947–957
Moghadam FV, Pourahmad R, Mortazavi A, Davoodi D, Azizinezhad R (2019) Use of fish oil nanoencapsulated with gum arabic carrier in low fat probiotic fermented milk. Food Sci Anim Resour 39(2):309
Mollace V, Gliozzi M, Carresi C, Musolino V, Oppedisano F (2013) Re-assessing the mechanism of action of n−3 PUFAs. Int J Cardiol 170(2):S8–S11
Monteiro A, Paquincha D, Martins F, Queirós RP, Saraiva JA, Švarc-Gajić J, Nastić N, Delerue-Matos C, Carvalho AP (2018) Liquid by-products from fish canning industry as sustainable sources of ω3 lipids. J Environ Manage 219:9–17
Morrell JJ (2012) 14-Protection of wood-based materials. In: Handbook of environmental degradation of materials, 2nd edn. Elsevier Inc., pp 407–439
Mozaffarian D, Longstreth WT, Lemaitre RN, Manolio TA, Kuller LH, Burke GL, Siscovick DS (2005) Fish consumption and stroke risk in elderly individuals: the cardiovascular health study. Arch Intern Med 165(2):200–206
Navarro-García G, Gámez-Meza N, Medina-Juárez LA, Ortega-García J, Cota-Quiñones E, Ramirez-Suarez JC (2016) Natural antioxidants in the stability of ray liver oil. Ciência Rural 47
Nirmal NP, Santivarangkna C, Rajput MS, Benjakul S (2020) Trends in shrimp processing waste utilization: an industrial prospective. Trends Food Sci Technol 103:20–35
Olatunde OO, Benjakul S, Vongkamjan K, Amnuaikit T (2020) Influence of stabilising agents on the properties of liposomal encapsulated ethanolic coconut husk extract. Int J Food Sci Technol 55(2):702–711
Oppedisano F, Macrì R, Gliozzi M, Musolino V, Carresi C, Maiuolo J, Bosco F, Nucera S, Caterina Zito M, Guarnieri L, Scarano F (2020) The anti-inflammatory and antioxidant properties of n-3 PUFAs: their role in cardiovascular protection. Biomedicines 8(9):306
O'Sullivan A, Mayr A, Shaw NB, Murphy SC, Kerry JP (2005) Use of natural antioxidants to stabilize fish oil systems. J Aquat Food Prod Technol 14(3):75–94
Pazos M, Alonso A, Sánchez I, Medina I (2008) Hydroxytyrosol prevents oxidative deterioration in foodstuffs rich in fish lipids. J Agric Food Chem 56(9):3334–3340
Povey MJW, Mason TJ (eds) (1998) Ultrasound in food processing. Blackie Academic & Professional, London, UK
Prameela K, Venkatesh K, Immandi SB, Kasturi APK, Krishna CR, Mohan CM (2017) Next generation nutraceutical from shrimp waste: the convergence of applications with extraction methods. Food Chem 237:121–132
Puertolas E, Koubaa M, Barba FJ (2016) An overview of the impact of electrotechnologies for the recovery of oil and high-value compounds from vegetable oil industry: energy and economic cost implications. Food Res Int 80:19–26
Qi-Yuan L, Jun-Qing Q, Xiao-Ge W (2016) Optimization of enzymatic fish oil extraction from mackerel viscera by response surface methodology. Int Food Res J 23(3):992
Rajasekaran B, Singh A, Nagarajan M, Benjakul S (2022) Effect of chitooligosaccharide and α-tocopherol on physical properties and oxidative stability of shrimp oil-in-water emulsion stabilized by bovine serum albumin-chitosan complex. Food Control 137:108899

Raju N, Gulzar S, Buamard N, Ma L, Ying X, Zhang B, Benjakul S (2021) Comparative study of astaxanthin, cholesterol, fatty acid profiles, and quality indices between shrimp oil extracted from hepatopancreas and cephalothorax. Front Nutr 8

Ramalhosa MJ, Paíga P, Morais S, Alves MR, Delerue-Matos C, Oliveira MBPP (2012) Lipid content of frozen fish: comparison of different extraction methods and variability during freezing storage. Food Chem 131(1):328–336

Redondo D, Venturini ME, Luengo E, Raso J, Arias E (2018) Pulsed electric fields as a green technology for the extraction of bioactive compounds from thinned peach by-products. Innov Food Sci Emerg Technol 45:335–343

Rizliya V, Mendis E (2014) Biological, physical, and chemical properties of fish oil and industrial applications. In: Seafood processing by-products. Springer, New York, NY, pp 285–313

Rodrigo R, Prieto JC, Castillo R (2013) Cardioprotection against ischaemia/reperfusion by vitamins C and E plus n−3 fatty acids: molecular mechanisms and potential clinical applications. Clin Sci 124(1):1–15

Rubio-Rodríguez N, Sara M, Beltrán S, Jaime I, Sanz MT, Rovira J (2012) Supercritical fluid extraction of fish oil from fish by-products: a comparison with other extraction methods. J Food Eng 109(2):238–248

Rupasinghe HV, Erkan N, Yasmin A (2010) Antioxidant protection of eicosapentaenoic acid and fish oil oxidation by polyphenolic-enriched apple skin extract. J Agric Food Chem 58(2):1233–1239

Rustad T, Storrø I, Slizyte R (2011) Possibilities for the utilisation of marine by-products. Int J Food Sci Technol 46(10):2001–2014

Sachindra NM, Bhaskar N, Mahendrakar NS (2006) Recovery of carotenoids from shrimp waste in organic solvents. Waste Manage 26(10):1092–1098

Sae-leaw T, Benjakul S (2017) Lipids from visceral depot fat of Asian seabass (*Lates calcarifer*): compositions and storage stability as affected by extraction methods. Eur J Lipid Sci Technol 119(11):1700198

Sahena F, Zaidul ISM, Jinap S, Saari N, Jahurul HA, Abbas KA, Norulaini NA (2009) PUFAs in fish: extraction, fractionation, importance in health. Compr Rev Food Sci Food Saf 8(2):59–74

Salama MF, Abbas A, Darweish MM, El-Hawwary AA, Al-Gayyar MM (2013) Hepatoprotective effects of cod liver oil against sodium nitrite toxicity in rats. Pharm Biol 51(11):1435–1443

Sánchez-Alonso I, Careche M, Moreno P, González MJ, Medina I (2011) Testing caffeic acid as a natural antioxidant in functional fish-fibre restructured products. LWT Food Sci Technol 44(4):1149–1155

Sánchez-Camargo AP, Martinez-Correa HA, Paviani LC, Cabral FA (2011) Supercritical CO_2 extraction of lipids and astaxanthin from Brazilian redspotted shrimp waste (*Farfantepenaeus paulensis*). J Supercrit Fluids 56(2):164–173

Sánchez-Camargo AP, Meireles MÂA, Ferreira AL, Saito E, Cabral FA (2012) Extraction of ω-3 fatty acids and astaxanthin from Brazilian redspotted shrimp waste using supercritical CO_2+ ethanol mixtures. J Supercrit Fluids 61:71–77

Šimat V, Rathod NB, Čagalj M, Hamed I, Generalić Mekinić I (2022) Astaxanthin from crustaceans and their byproducts: a bioactive metabolite candidate for therapeutic application. Mar Drugs 20(3):206

Simopoulos AP (2001) N−3 fatty acids and human health: defining strategies for public policy. Lipids 36(S1):S83–S89

Suresh PV, Prabhu GN (2013) Seafood. In: Valorization of food processing by-products. Taylor & Francis, New York, pp 685–736

Suresh PV, Kudre TG, Johny LC (2018) Sustainable valorization of seafood processing by-product/discard. In: Waste to wealth. Springer, Singapore, pp 111–139

Swanson D, Block R, Mousa SA (2012) Omega-3 fatty acids EPA and DHA: health benefits throughout life. Adv Nutr 3(1):1–7

Takahashi M, Tsuboyama Kasaoka N, Nakatani T, Ishii M, Tsutsumi S, Aburatani H, Ezaki O (2002) Fish oil feeding alters liver gene expressions to defend against PPARα activation and ROS production. Am J Physiol Gastrointest Liver Physiol 282(2):G338–G348

Takeungwongtrakul S, Benjakul S (2013) Oxidative stability of shrimp oil-in-water emulsions as affected by antioxidant incorporation. Int Aquat Res 5(1):1–12

Takeungwongtrakul S, Benjakul S (2017) Biscuits fortified with micro-encapsulated shrimp oil: characteristics and storage stability. J Food Sci Technol 54(5):1126–1136

Takeungwongtrakul S, Benjakul S, Aran H (2015) Characteristics and oxidative stability of bread fortified with encapsulated shrimp oil. Ital J Food Sci 27:476–486

Toepfl S, Heinz V, Knorr D (2006) Applications of pulsed electric fields technology for the food industry. Pulsed electric fields technology for the food industry. Springer, Boston, MA, pp 197–221

Topuz OK, Yerlikaya P, Uçak İ, Gümüş B, Büyükbenli HA, Gökoğlu N (2015) Influence of pomegranate peel (*Punica granatum*) extract on lipid oxidation in anchovy fish oil under heat accelerated conditions. J Food Sci Technol 52(1):625–632

Tørris C, Småstuen MC, Molin M (2018) Nutrients in fish and possible associations with cardiovascular disease risk factors in metabolic syndrome. Nutrients 10(7):952

Tounekti T, Munné-Bosch S (2012) Enhanced phenolic diterpenes antioxidant levels through non-transgenic approaches. Crit Rev Plant Sci 31(6):505–519

Venugopal V (2005) Seafood processing: adding value through quick freezing, retortable packaging and cook-chilling. CRC Press

Venugopalan VK, Gopakumar LR, Kumaran AK, Chatterjee NS, Soman V, Peeralil S, Mathew S, McClements DJ, Nagarajarao RC (2021) Encapsulation and protection of omega-3-rich fish oils using food-grade delivery systems. Foods 10(7):1566

Von Schacky C, Angerer P, Kothny W, Theisen K, Mudra H (1999) The effect of dietary Ω-3 fatty acids on coronary atherosclerosis: a randomized, double-blind, placebo-controlled trial. Ann Intern Med 130(7):554–562

Wang HH, Hung TM, Wei J, Chiang AN (2004) Fish oil increases antioxidant enzyme activities in macrophages and reduces atherosclerotic lesions in apoE-knockout mice. Cardiovasc Res 61(1):169–176

Wang W, Zhu J, Lyu F, Panigrahy D, Ferrara KW, Hammock B, Zhang G (2014) ω-3 polyunsaturated fatty acids-derived lipid metabolites on angiogenesis, inflammation and cancer. Prostaglandins Other Lipid Mediat 113:13–20

Wang Q, Lv S, Lu J, Jiang S, Lin L (2015) Characterization, stability, and in vitro release evaluation of carboxymethyl chitosan coated liposomes containing fish oil. J Food Sci 80(7):C1460–C1467

Xi J, Li Z, Fan Y (2021) Recent advances in continuous extraction of bioactive ingredients from food-processing wastes by pulsed electric fields. Crit Rev Food Sci Nutr 61(10):1738–1750

Xu J, Long KZ, Wu F, Xue B, Li YX, Cao Y (2014) Efficient synthesis of dimethyl carbonate via transesterification of ethylene carbonate over a new mesoporous ceria catalyst. Appl Catal A 484:1–7

Yan X, Ma C, Cui F, McClements DJ, Liu X, Liu F (2020) Protein-stabilized Pickering emulsions: formation, stability, properties, and applications in foods. Trends Food Sci Technol 103:293–303

Yashodhara BM, Umakanth S, Pappachan JM, Bhat SK, Kamath R, Choo BH (2009) Omega-3 fatty acids: a comprehensive review of their role in health and disease. Postgrad Med J 85(1000):84–90

Yeşilsu AF, Özyurt G (2019) Oxidative stability of microencapsulated fish oil with rosemary, thyme and laurel extracts: a kinetic assessment. J Food Eng 240:171–182

Yu F, Li Z, Zhang T, Wei Y, Xue Y, Xue C (2017) Influence of encapsulation techniques on the structure, physical properties, and thermal stability of fish oil microcapsules by spray drying. J Food Process Eng 40(6):e12576

Yu C, Cha Y, Wu F, Fan W, Xu X, Du M (2018) Effects of ball-milling treatment on mussel (*Mytilus edulis*) protein: structure, functional properties and *in vitro* digestibility. Int J Food Sci Technol 53(3):683–691

Zbinden MDA, Sturm BS, Nord RD, Carey WJ, Moore D, Shinogle H, Stagg-Williams SM (2013) Pulsed electric field (PEF) as an intensification pretreatment for greener solvent lipid extraction from microalgae. Biotechnol Bioeng 110(6):1605–1615

Zuta PC, Simpson BK, Zhao X, Leclerc L (2007) The effect of α-tocopherol on the oxidation of mackerel oil. Food Chem 100(2):800–807

Fish Waste and By-Product as a Source of Calcium

Tao Yin, Ibrahim Khalifa, Juan You, Lin Peng, and Ramy M. Khoder

Abstract Fish head, bones, and scales are the main solid fish waste or by-product, which are rich in calcium. The calcium obtained from these fish by-products can be used in feed, food, clinical, and biological materials after proper processing. This chapter first takes a glace of the fish by-products that can be used as calcium sources, with the focus on fish bones. The chemical composition, physicochemical characteristics, and bioavailability of the calcium sources are summarized. The main methods of extracting and processing the calcium source are introduced and compared by classifying chemical, enzymatic, and physical methods. The products derived from the calcium source are divided into fish bone foods, powder/paste/slurry, calcium supplements, and hydroxyapatite products. Finally, production lines for the calcium products made from fish by-products are depicted. This book chapter provides readers with a comprehensive knowledge of the calcium source made from fish by-product.

Keywords Bioavailability · Calcium · Extraction · Feed · Fish bone · Fish waste · Fish by-product · Hydroxyapatite · Powder · Paste · Milling

---

T. Yin (✉) · J. You · L. Peng · R. M. Khoder
College of Food Science and Technology, Key Laboratory of Environment Correlative Dietology, Ministry of Education, Huazhong Agricultural University, Wuhan, Hubei Province 430070, P. R. China
e-mail: yintao@mail.hzau.edu.cn

J. You
e-mail: juanyou@mail.hzau.edu.cn

L. Peng
e-mail: pengling@webmail.hzau.edu.cn

R. M. Khoder
e-mail: ramy_khodir@fagr.bu.edu.eg

T. Yin · J. You · L. Peng
National R & D Branch Center for Conventional Freshwater Fish Processing, Wuhan 430070, Hubei, China

I. Khalifa · R. M. Khoder
Food Technology Department, Faculty of Agriculture, Benha University, Moshtohor, Toukh 13736, Egypt
e-mail: ibrahiem.khalifa@fagr.bu.edu.eg

S. Maqsood et al. (eds.), *Fish Waste to Valuable Products*, Sustainable Materials and Technology, https://doi.org/10.1007/978-981-99-8593-7_10

1 Introduction

Thousands of tons of fish waste and by-product are produced during fish processing, including fish head, fish bones, fish scale, fish viscera, fish skin, etc. However, these by-product has not been effectively utilized, which not only causes environmental pollution, but also wastes a lot of valuable resources. Depending on the species, the head, bone, and scale account for about 30% of the total body weight of the fish. Fish head, bone, and scale can be used as a source of calcium. However, at present, most fish head and bone are combined in the solid waste stream and enter the fishmeal processing plant. In the fishmeal plant, all solid waste is heated/cooked and dried. Oil is separated before drying. Then the large bone is separated and ground as fish meal. Calcium in fish by-product is natural and has high bioavailability (Malde et al. 2010). Therefore, it has been the main direction and hot spot of scientific research to use calcium from fish by-product for the enhancement of food nutrition or to develop calcium health food to improve its economic added value (Nemati et al. 2016; Benjakul et al. 2017). Because of its good biocompatibility with human body, hydroxyapatite extracted from the fish by-product can be widely used as implant materials in plastic surgery, dentistry, and other fields (Boutinguiza et al. 2012). In addition, the hydroxyapatite can also be used as specific adsorption materials in the chromatographic columns (Ashokan et al. 2021).

The utilization of calcium from fish by-product usually goes through three stages: recovery, extraction, and product development. In the recovery stage of by-product that can be used as calcium source, different methods should be adopted by comprehensively considering the sanitary condition, recovery rate, and economic cost required for the development of different products. Most of the calcium in fish by-product exists in the form of hydroxyapatite. After hydroxyapatite is deposited on the surface of collagen fibers, the mineralized collagen fiber is formed, which is the basic unit of the hierarchical structure of fish bone and scale. In the development and utilization of calcium in fish by-product, two main schemes can be adopted. The first is to extract calcium ion or hydroxyapatite and then further process it; the other is to use it after crushing without extraction. Due to the complex hierarchical structure of fish bones and the extremely low solubility of hydroxyapatite, a combination of various processes is often used in the extraction of calcium ions and calcium compounds to improve the extraction rate. Fish bone is hard and tough, which makes it difficult to be crushed into desired size directly. Before crushing, pretreatments such as enzymatic hydrolysis and high-temperature cooking are often used to soften the bone to reduce the difficulty of crushing (Jafarpour et al. 2020). With the development of modern food processing technology, the ways of fish bone crushing have become diverse (Wu et al. 2012; Cui et al. 2021; Boutinguiza et al. 2012).

This chapter first reviews the recovery and characteristics of fish waste and by-product which possess the potential to be utilized as calcium source, and chemical composition, properties, and bioavailability of the recovered calcium compounds. Then, extraction and processing method, and applications of the calcium compounds from fish waste are summarized. Finally, production lines of several calcium products made from fish by-product are depicted.

2 Raw Materials as Calcium Source

2.1 *Fish Bone*

2.1.1 Backbone

The proportion of fish bones to total body weight is about 9%. The proportion of fish backbone in fish bones is the largest, about 80%. The fish backbone is composed of a number of vertebrae, which are connected one by one from the back of the head to the base of the tail to form a segmented column (Fig. 1). The basic function of fish backbone is to support the body and protect the spinal cord, internal organs, and main blood vessels. The backbone, ribs, skull, and scale of fish have multi-level hierarchical structure, which is formed by the mineralized collagen fibers of basic units through different arrangements (e.g., parallel, stagger, interweave, and enclosed arrangements). The mineralized collagen fibers are composed of hexagonal hydroxyapatite crystals and triple helix collagen (Fig. 1). The thermal and mechanical properties of bone in different parts are different, which may be mainly related to the mineralized form of collagen fibers and the arrangement structure of mineralized collagen fibers. The fish backbone is a typical trabecula, which is mainly composed of lamellar bundles. In the process of surimi production, the fish carcass is transported into the de-boning machine through the plastic belt. Under the extrusion of the belt and the stainless steel drum, the fish meat enters the drum through the pores of the drum and is separated from the fish backbone and other components (Shi et al. 2021). The fish backbone is easy to recover. In the processing of fish fillet, the backbone can be separated from the fish meat by a filleting machine. Generally, a residual part of fish meat is adhered to the fish backbone which can be removed by cooking, alkali, or enzymatic hydrolysis in the subsequent processing (Nemati et al. 2016).

2.1.2 Rib Bone

Fish rib bone accounts for about 15% weight of fish bones. The ribs are physiologically connected with the backbone (Fig. 1), and their main function is to protect visceral tissues. Fish ribs are mainly composed of densely arranged fibrous lamellar bone along the long axis (Jiao et al. 2020). During the production of fish fillets such as salmon, the ribs can be pulled out by a stinger and then collected.

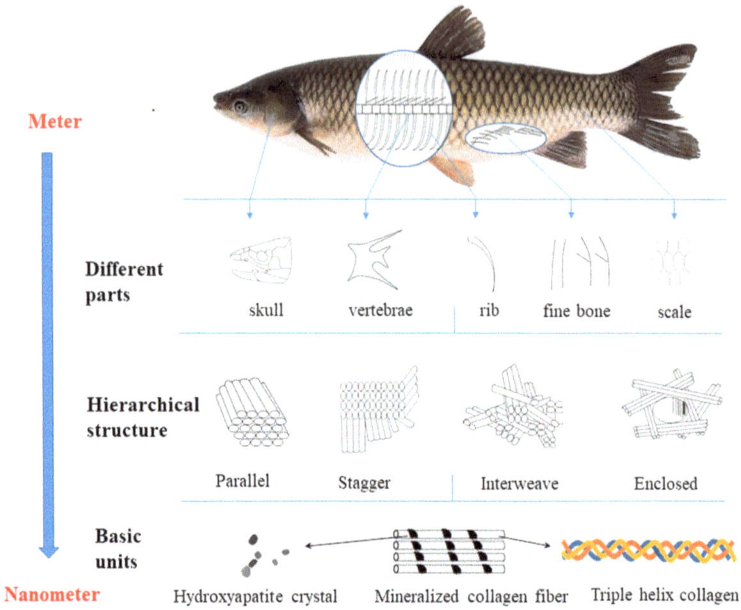

Fig. 1 Schematic diagram of fish bone from macro to micro structure

2.1.3 Fine Bone

Fine bone accounts for about 5% of fish bones. Fine bones are generally distributed in the trunk muscles on both sides of the fish (Fig. 1). The trunk muscles on each side are divided into supra-axial and infra-axial muscles by the horizontal septum of connective tissue from the spine, namely, the dorsal and abdominal muscles. The fine bones are mainly located in the diaphragms of the supra-axial and infra-axial muscles (after the abdominal cavity). One end is close to the backbone, and the other end extends outward along the diaphragms. They are free and have no connection with the backbone. The number and arrangement of fine bones are related to the species of fish. Fine bones are commonly found in lower bony fishes, such as herring and carp fishes. The fine bones are generally difficult to be recovered alone.

2.2 Fish Head

Fish head accounts for about 16% of the total weight of the fish. The fish skull is the main part of the fish head (Fig. 1). The fish skull can be divided into two parts: the brain and the pharyngeal skull. The brain is located in the upper part of the whole skull, which is used to contain the brain, olfactory, visual, auditory, and other sensory organs. The pharyngeal skull is located in the lower part of the whole skull, which is arranged in an arc and surrounds both sides of the oropharyngeal cavity and the front of the esophagus. The fish skull is mainly composed of lamellar bones. The skull is mainly composed of a series of planar mineralized collagen fiber layers, including thick aligned fibrils and thin-layer oriented disordered fibrils. In addition, most collagen fibrils orthogonal to the lamellar plane are low mineralized (Atkins et al. 2015). In the processing, it can be removed by a decapitation machine. The fish head can be easily recovered and the integrity of raw materials can be maintained. The content of fat and connective tissue in fish head is high, so it is necessary to remove these components before extracting calcium.

2.3 Fish Scale

Fish scales are deformations of the dermis of fish skin (Fig. 1), accounting for 2%-5% of the weight of fish. Fish scales act as a protective barrier to prevent fish from being invaded by water microorganisms. The fish scale can be divided into two layers: the upper layer is the bone layer, the main component is hydroxyapatite, and some collagen fibers are scattered; The lower layer is a fibrous layer. The collagen fibers with a diameter of 70–80 nm are closely arranged in parallel in the same thin layer, and form an angle of 90° with the collagen fibers in the adjacent thin layers, thus forming an orthogonal splint structure composed of multiple thin layers (Feng et al. 2020). The method of obtaining fish scales generally adopts the mechanical descaling method, which applies a series of mechanical actions such as rubbing and stripping to separate the fish scales from the fish body. There is a large amount of mucus on the surface of fish scales, which should be cleaned with clean water before processing. At present, fish scales are mainly used as raw materials for processing collagen. The by-product of collagen production can be further used for calcium extraction and processing.

2.4 Fish Waste Mixture

In surimi processing, the fish carcass with bones is used for meat collection after the fish scales and heads are removed respectively. After the separation of bone and meat, the by-product which is the mixture of fish bone, fish skin, and other parts are obtained. In the process of surimi refine, the fish meat reaches the outside of the fine filter through the sieve hole, and the by-products of mixed connective tissue and fish fine bones are separated (Park et al. 2014). In the processing of small size fish, fish protein is extracted and separated by enzymatic hydrolysis, pH-shift, and other methods, and the mixed by-product containing fish bone and scale can be obtained.

3 Chemical Composition, Properties, and Bioavailability

3.1 *Chemical Composition*

3.1.1 Calcium Content

The calcium content in fish head, fish bone, fish scale, and mixture of fish processing by-product from different sources is shown in Table 1. The calcium content in these primary processing raw materials varies significantly. The calcium content in fish bone is the highest, which is 255.07–294.98 mg/g, and the calcium content in fish head is the lowest, which is 16.6–22.5 mg/g. The difference of calcium content in by-products from different sources is mainly related to the composition of mineral salt, collagen, and fat in raw materials. The mineral salt in fish head is mainly distributed in fish skull. However, the calcium content in other parts of fish head, such as fish eyes, fish brain, and fish lips, is very low. Therefore, the whole calcium content of fish head is the lowest. The calcium content of different fish species from the same source is also slightly different. Taking the calcium content in the backbone as an example, the calcium content in the backbone of cod is relatively high, while that of grass carp is relatively low, which is mainly related to the difference in the fat content in the backbone of fish.

Fish bone can be processed into fish bone powder and fish bone paste after crushing. The content of calcium in the fish bone powder prepared from the backbone of silver carp was 236.90 mg/ g, while the content of calcium in the fish bone paste was 391.09 mg/g. The difference of calcium content between the two products may be related to the different pretreatment process before crushing. After high-energy ball milling, the particle size of fish bone powder and fish bone paste can even reach nanometer level. According to Yin et al. (2015), the calcium content in nano fish bone was 236 mg/g.

3.1.2 Form and Crystalline Structure

Hydroxyapatite and calcium carbonate are the dominate calcium in fish by-product. And the content of calcium carbonate is generally less than 1%. Calcium in fish by-product raw materials and processed products mainly exists in the following forms: inorganic calcium (hydroxyapatite, calcium carbonate, calcium chloride, calcium citrate, etc.), organic calcium (peptide chelated calcium) (Fig. 2). In order to improve the bioavailability of calcium, calcium chloride and calcium citrate with high solubility are often prepared by adding hydrochloric acid and citric acid (Fig. 1). In recent years, peptide chelated calcium has become a research hotspot because of its high bioavailability. After identification, one of the peptide sequences with high calcium chelating capacity is Val-Leu-ser-Gly-Gly-thr-thr met-ala-met-Tyr-thr-Leu-Val (Peng et al. 2017). For the peptide-chelated calcium, the amino, carboxyl, and side chain groups of the peptide react with calcium ions. The carbonyl and imino

Table 1 Calcium contents in primary raw materials and semi-processed products of fish waste (dry base)

		Fish species	Calcium (mg/g)	References
Primary raw materials	Fish Head	Eel	22.50	(Zeng et al. 2002)
		Silver carp	16.60	(Huang et al. 2008)
	Fish Backbone	Cod	294.98	(Kołodziejska et al. 2008)
		Silver carps	274.50	(Li et al. 2017)
		Grass carps	255.07	(Li et al. 2017)
	Fish Scale	Croaker	180.31	(Wang et al. 2019)
		Silver carp	28.20	(Zhang et al. 2010)
		Grass carp	51.90	(Liu et al. 2000)
	Deboning discharge	Threadfin bream	146.95	(Wiriyaphan et al. 2012)
	Refining discharge	Threadfin bream	50.39	(Wiriyaphan et al. 2012)
Semi processed products	Fish bone powder	Silver carp	236.90	(Yin et al. 2016a)
	Fish bone paste	Silver carp	391.09	(Yin et al. 2016b)
	Nano fish bone	Silver carp	236.00	(Yin et al. 2015)

groups in the peptide chain may also participate in the coordination of calcium ions, so it has high coordination rate and stability.

Hydroxyapatite (HAP), also known as hydroxyapatite, basic calcium phosphate, with the molecular formula of ($Ca_{10}(PO_4)_6(OH)_2$). The crystal structure is hexagonal system (Fig. 1), which belongs to L_6PC symmetric type and P_{63}/M space group. Its cell parameters are a = b = 0.9418 nm, c = 0.6884 nm, and theoretical ca/p = 1.67. Calcium carbonate is a white fine crystalline powder, tasteless and odorless. There are amorphous and crystalline forms. The crystalline forms can be divided into orthorhombic and hexagonal systems (anhydrous calcium carbonate is a colorless orthorhombic crystal, and hexahydrate calcium carbonate is a colorless monoclinic crystal, which is columnar or rhombic.

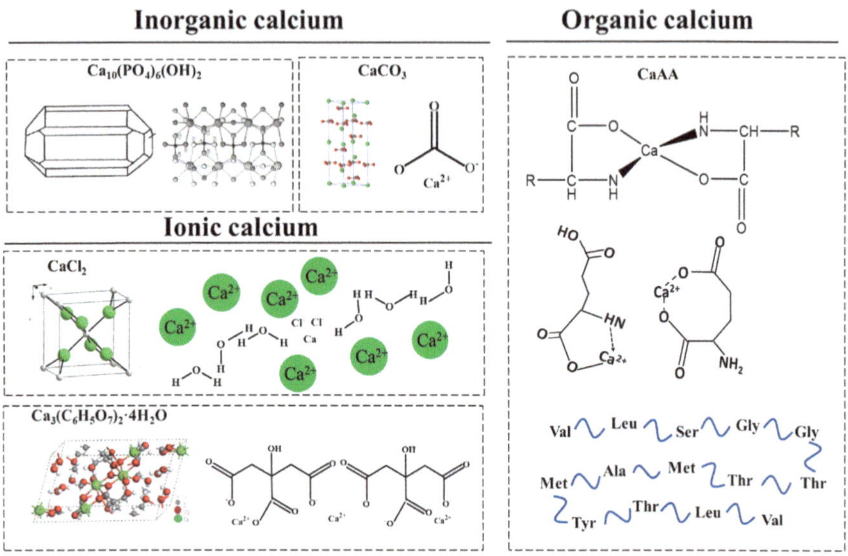

Fig. 2 Form and crystalline structure of calcium compounds derived from fish by-product. *Note*: CaAA represents peptide-chelated calcium

3.2 Physiochemical Properties

3.2.1 Solubility

Calcium compounds (mainly hydroxyapatite) in fish by-product are difficult to dissolve in water, acid, and alkaline solutions at room temperature, mainly because calcium in fish by-product is mainly deposited in the network structure formed by collagen fibers in the form of hydroxyapatite. The solubility of hydroxyapatite is 0.4 ppm, which is most stable at 25 °C and pH > 4.2. Its dissolution characteristics are affected by many factors, including the weight percentage of hydroxyapatite and water, the specific surface of hydroxyapatite, the pH value of the solution, and the grain defects and impurity content of hydroxyapatite.

3.2.2 Particle Size

The particle size of calcium containing particles in fish by-product is related to the way how they are crushed. Through coarsely crushing by a bone paste mill, the particle size range is 74 ~ 833 μm. The particle size ranges from 2 to 20 μm after dry ball milling. Yin et al. (2015) reported that the average particle size range of fish bone crushed by wet high-energy ball milling was 100 ~ 700 nm. The particle size determines its bioavailability and application range.

3.2.3 Potential

The zeta potential of calcium containing particles from fish by-product is negatively charged, and its value is related to the particle size. When the absolute value of zeta potential is greater than 30, the particles can be dispersed stably. It is reported that the absolute value of zeta potential in nano fish bone increases significantly from 7 to 14 mV with the decrease of particle size from 481 to 111 nm, which may be related to the release of positively charged calcium ions in fish bone particles during ball milling (Yin et al. 2015). Yin et al. (2018) prepared pure nano-hydroxyapatite particles by hydrothermal synthesis, with an absolute potential of 23.2 MV.

3.2.4 Affinity

Calcium in fish by-products mainly exists in the form of hydroxyapatite. Because of the regular hexagonal structure of the hydroxyapatite crystal plane, it can distinguish small differences in the geometric arrangement of atoms on the surface of adsorbed molecules. According to its absorptivity, it can be used as the filling material of chromatographic column to separate a variety of proteins. There are two different adsorption sites C-site and P-site on the surface of hydroxyapatite crystal. The adsorption mechanism is mainly as follows: when the OH- near the Ca ion on the crystal surface falls off instantaneously, the local area is positively charged, forming a C-site; when the local Ca ion on the surface is vacant at a certain moment, the Ca position connected with six O atoms forms a P-site. When hydroxyapatite interacts with proteins, acidic proteins are mainly adsorbed at c-sites, and basic proteins are mainly adsorbed at p-sites (Ashokan et al. 2021).

3.3 *Calcium Bioavailability*

3.3.1 In Vitro Test

Bioavailability of calcium products made from different fish by-products and evaluated by various methods are summarized in Table 2. Nemati et al. (2016) reported that the in vitro bioavailability of calcium in tuna bone powder (TBP) fortified biscuit and tricalcium phosphate (TCP) fortified biscuit were 38.9% and 39.5% respectively. Benjakul et al. (2017) found that the bioavailability of calcium in powder extracted from pre-cooked tuna bones decreased from 8.57% to 4.63% after calcining at 900 °C.

3.3.2 Cell Test

Idowu et al. (2019) reported that the solubility of salmon frame powders (Bio-cal-H, Bio-cal-L) and $CaCO_3$ in the gastrointestinal tract were 7.65%, 8.41%, and 0.62%

Table 2 Evaluation of calcium bioavailability of products derived from fish waste

	Product category	Outcomes	References
In vitro test	Tuna bone powder (TBP) and Tricalcium phosphate (TCP) fortified cookies	Availability of calcium in TBP-fortified cookies or TCP-fortified cookies was comparable at 38.9% and 39.5%, respectively	(Nemati et al. 2016)
	Tricalcium phosphate (TCP), fish bone (FB), nano-fish bone (NFB), algae calcium (AC)	Bioaccessibility for TCP, FB, NFB, and AC was 1.2%, 0.69%, 1.8% and 4%, respectively	(Lee et al. 2020)
	Skipjack tuna biocalcium without (BC) and with calcination (CB)	Bioavailability for BC and CB was 8.57% and 4.63%, respectively	(Benjakul et al. 2017)
Cell test	Salmon frame powders (Bio-cal-H、Bio-cal-L)	The bioavailability of Bio-cal-H, Bio-cal-L and $CaCO_3$ was 43.02%, 38.18%, 21.48%	(Idowu et al. 2019)
	Tricalcium phosphate (TCP) Fish bone (FB) Nano-fish bone (NFB) Algae calcium (AC)	Caco-2 cells uptake of calcium in TCP, FB, NFB, and AC handling groups was 9.5, 13.5, 15.5, and 13.5 μg/mg protein, respectively; intestinal transport at 30.5, 33.5, 31.5, and 30.5 μg, respectively	(Lee et al. 2020)
	Nano-hydroxyapatite(n-HAP, Sardinella longiceps)	The cell proliferation of 10, 50, 100, and 250 μg/mL n-HAP was 105%, 124%, 141.3%, and 125.3%, respectively. The calcium deposition of control, 50, 100 μg/mL n-HAP was 0.24, 0.54, 0.99, respectively	(Surya et al. 2021)
Animal test	Haddock calcium powder	Compared with $CaCO_3$, the calcium absorption and retention rate of Haddock calcium powder increased by 10% and 12%, respectively	(Huo et al. 2010)

(continued)

Table 2 (continued)

	Product category	Outcomes	References
	Microwave Tuna bone powder	The calcium absorption rate and retention rate were 59.62% and 58.31%, respectively, in the microwave Tuna bone powder group	(Yang et al. 2020)
	Pacific cod bone peptide chelated calcium (BCP)	Apparent calcium absorption rate of 80.0%, and calcium retention rate of 79.65%	(Peng et al. 2017)
	Silver carp bone powder-fish protein hydrolysate mixture	The mixture with average size < 74 μm had an apparent calcium absorption rate of 47.84%, a calcium retention rate of 44.94% and a bone calcium content of 24.18%	(Xie et al. 2014)
Clinic test	Cod bones, salmon bones and $CaCO_3$	The Ca absorption of cod bones, salmon bones, $CaCO_3$ was 21.9%, 22.5%, 27.4%, respectively	(Malde et al. 2010)

respectively. The calcium transport capacity (calcium bioavailability) through CaCo-2 monolayer was 43.02%, 38.18%, and 21.48% respectively. The results showed that salmon skeleton biological calcium could be used as a potential source of calcium supplement. Surya et al. (2021) determined different concentrations (10, 50, 100, 250 μg/ml) of n-HAP on the activity of human osteoblast (MG-63) cells. The cell proliferation rates at each concentration were 105%, 124%, 141.3%, and 125.3%, respectively. And they found n-HAP with concentrations at 50 and 100 μg/ ml obviously promoted the mineralization of osteoblast MG-63 cells.

3.3.3 Animal Test

Peng et al. (2017) studied the bioavailability of Pacific cod bone peptide-chelated calcium in rats. It was found that the apparent calcium absorption rate, calcium retention rate, and femoral calcium content were significantly higher than those of the model group and $CaCO_3$ group, while the serum ALP was significantly lower than that of the model group. Yang et al. (2020) prepared tuna bone calcium powder by microwave method. The study showed that the calcium absorption rate of mice in the microwave fish bone powder group was 59.62%, and the calcium retention rate 58.31%, which was significantly higher than that of common fish bone meal group, calcium carbonate group, and low calcium control group. Xie et al. (2014) prepared fish bone powder and fish protease hydrolysate mixtures with low, medium, and high particle sizes. According to their results, the evaluation indexes of all calcium bioavailability of the silver carp bone powder-fish protein hydrolysate mixture were significantly higher than those of other experimental groups and calcium carbonate control and low calcium control groups. And the apparent calcium absorption rate, calcium retention rate, and bone calcium content of the mixture with the smallest particle size were the highest (47.84%, 44.94%, and 24.18%, respectively). The above results showed that the bioavailability of calcium products made from fish by-products is high. And the calcium bioavailability dependent on processing method, particle size, and composition, etc.

3.3.4 Clinic Test

Malde et al. (2010) evaluated the calcium bioavailability of cod bone, salmon bone, and $CaCO_3$ in young healthy men by ^{47}Ca whole-body counting method. The bioavailability of cod bone, salmon bone, and $CaCO_3$ was 21.9%, 22.5%, and 27.4%, respectively (Table 2). The results showed that the bones of Atlantic salmon and Atlantic cod were suitable as natural calcium sources, which can be utilized as functional foods or supplements.

4 Extraction and Processing Method

4.1 Chemical Method

4.1.1 Acid Solution

The extraction and processing methods of calcium in fish by-products are summarized in Table 3. At present, acid extraction is the most widely used method, which can dissolve calcium from fish by-products. Generally, food grade acids are used, including hydrochloric acid, lactic acid, acetic acid, citric acid, malic acid, etc. Sriuttha et al. (2014) extracted calcium from tilapia bone powder using acetic acid, and obtained an extract with a concentration of 2,376 mg/ L. The factors that affect the extraction of calcium from fish by-products include the type of acid, acid concentration, temperature, time, and the ratio of by-product to acid. In addition, the size of by-product particles is also a key factor affecting the extraction efficiency. Before acid dissolution, reducing the particle size can increase the extraction efficiency. The main reason may be that the crushing process destroys the structure of collagen fiber network and hydroxyapatite crystal, and increases the specific surface area of fish bone powder, which is conducive to the release of calcium ions (Li et al. 2020).

Table 3 Extraction and processing of calcium in fish by-products

	Source	Method used	Suitable conditions	References
Chemical method	Tilapia frame	Acetic acid	Added acetic acid (0.25 M) at a ratio of 1:50 (weight per volume) and stirred for 48 h at 28°C	(Sriuttha et al. 2014)
	Tuna frame	Sodium hydroxide	Boiled in 2% KOH solution for 30 min at a ratio of 1 part tuna frame to 3 parts NaOH by weight over volume (w/v)	(Nemati et al. 2016)
Enzymatic method	Cod frames	Alcalase and Neutrase	Under 50 °C and enzyme to solid ratio of 1.5%, subjected to neutrase for 3 h and then alcalase was added and incubated for the next 3 h	(Jafarpour et al. 2020)
Physical method	Sword fish and tuna bone	Calcination	Calcined at 600 or 950 °C and maintained isothermally for 12 h	(Boutinguiza et al. 2012)
	Grass carp backbone	Millstone milling	Subjected to colloid mill for 5 min	(Nawaz et al. 2019)
	Silver carp bones	Grinding	Feeding pressure of 0.85 MPa, grinding pressure of 1.0 MPa, feeding rate of 0.0325 g/s, feeding particle size of 150–300 lm and one pass through the grinder	(Wu et al. 2012)
	Silver carp backbone	Dry media milling	Rotation speed of 400 revolutions per minute, ratio of media to fish bone powder at 4:1, media diameter of 2mm and 2.5 h of milling time	(Yin et al. 2016a)
	Silver carp backbone	Wet media milling	0.5 mm diameter bead, agitation speed at 3000 rpm, a filling ratio of 85%, and 6 h of milling time	(Yin et al. 2015)
	Tuna bones	Steam explosion	Steam pressure of 0.6 MPa, reaction duration of 5 min	(Cui et al. 2021)

4.1.2 Alkaline Solution

The use of an alkaline solution to remove organic molecules, particularly protein and lipids, is a useful procedure. The ash/protein ratio is an essential criterion for determining the purity of bone powder. As a result, alkaline aid solubilization may be the most practicable way for producing a high-purity fish bone powder (Hemung 2013). Nemati et al. (2016) extracted Tuna bone powder (TBP) using an alkaline process and found that the TBP was tiny in particle size, white in color, and odorless. These characteristics are ideal for use in the food fortification process.

4.2 Enzyme Method

Enzymatic hydrolysis is the preferred method over chemical hydrolysis for various reasons, including moderate reaction conditions, fewer unwanted products, and good product quality and yield. Alcalase has been discovered to be a very effective enzyme in the hydrolysis of fish proteins, because it may achieve a high degree of hydrolysis in a short amount of time under moderate circumstances. A few investigations on the enzymatic hydrolysis of fish by-products utilizing alcalase enzyme, such as pacific whiting solid waste, catfish frames, and sardinella by-product have been undertaken (Table 3). In addition to alkaline protease, papain, flavor protease, and neutral protease are also used alone or in combination to process calcium products from fish by-products. According to Jafarpour et al. (2020), enzymatic hydrolysis of cod frame (MCF) was performed using alcalase and neutrase, either singly or consecutively, to generate phosphorus and calcium-rich bone powder. After enzymatic hydrolysis, significant levels of phosphorus and calcium (330 and 583 g/kg, respectively) were recovered from the cod frame.

4.3 Physical Method

4.3.1 Calcination

Compared with chemical method and enzymatic method, calcination method is more environmentally friendly, simpler, and cheaper to process. At high temperature, inorganic substances in by-products, mainly hydroxyapatite, remain after calcination. The calcination conditions have an effect on the crystal form of hydroxyapatite. Boutinguiza et al. (2012) reported that material obtained at 600 °C is a B type hydroxyapatite. At 950 °C a biphasic material was found: biological hydroxyapatite/beta-TCP in a 87/13 ratio.

4.3.2 Grinding and Milling

Milling and grinding are common methods to prepare calcium containing products by processing fish by-products. In the process of milling and grinding, raw materials are crushed under the force of collision, extrusion, and friction. The crushing efficiency depends on the amount of energy transferred to the material per unit time. The smaller the particle size of fish bone particles obtained by crushing, the higher the calcium release rate, the smaller the adverse sensory effects, and the wider the application range (Yin et al. 2017). In the physical crushing of fish bones, there must be a limit value of the actual particle size of raw materials, which mainly depends on the tendency of product particles to re-accumulate and the dynamic balance established between accumulation and crushing. Therefore, after the particle size of raw material particles reaches the limit value, increasing the crushing time can not further reduce the particle size, but can only increase energy consumption. With the gradual decrease of the particle size of the material, the *van der Waals* force increases significantly, the small particles gather among the powders, and the particle size of the powders even increases with the extension of the processing time (Yin et al. 2016a). The ultimate particle size of fish bone particles is affected by the crushing mode. As so far, the particle size of fish bone crushed by wet ball milling is the smallest. Recently, Yin et al. (2015) described a method for producing Nano fish bone (NFB) slurry with an average particle size of 110 nm utilizing high-energy wet medium milling.

4.3.3 Steam Explosion

Steam explosion is an environmentally friendly and revolutionary technology. It enables high-temperature, high-pressure reactions in which steam penetrates the interior space of raw materials, followed by rapid decompression for physical treatment. In a recent study, a steam explosion pretreatment technique was developed to prepare tuna bone powder (Cui et al. 2021). Their results indicated that the median particle size (D_{50}) of steam explosion pretreated tuna bone powder was much lower than that of normal biological calcium tuna bone powder.

5 Calcium Products and Application

5.1 Fish Bone Foods

In Southeast Asia, it is one of the important ways for people to supplement calcium source by eating small whole fish with bones, such as deep fried juvenile sardine. The backbone of larger fish (such as Pseudosciaena crocea) is cut into sections, pickled with seasoning, wrapped with bread bran and fried, which can be prepared into golden and crisp snack. These products maintain the natural form of raw materials and do not need deep processing. As a result, increasing the value of fish bones by turning them into nutritious foods is a plausible method.

5.2 Powder/Paste/Slurry

Fish by-products are dried and crushed into powder, then mixed with other main and auxiliary materials and granulated into fish or animal feed, which is the main way of processing at present. After the clean fish backbone is ground by dry method and wet method, the edible fish bone powder or fish bone paste can be obtained. Furthermore, fish bone powder or paste can be prepared into nano-scale fish bone slurry by high-energy wet ball milling (Yin et al. 2015). Fish bone powder, paste, and slurry are intermediate products of fish by-products, which can be added to fish sausage, baked foods, snack foods, tofu, and other foods. In addition to supplementing calcium in food, calcium ion released from fish bone intermediate products can also improve the texture of food by forming ionic bonds and activating endogenous transglutaminase activity (Yin and Park 2014; Khoder et al. 2020).

5.3 Calcium Supplements

After deep processing, fish by-products can be prepared into chewable tablets, soft capsules, and other calcium supplements. Their core components are generally acid soluble calcium or peptide chelated calcium. Calcium chelated with peptides is thought to be a good choice for improving calcium absorption. Several calcium-chelating peptides have recently been discovered in hoki frame, tilapia scale, and other sources. Salmon ossein oligopeptides (SOOP) were made from the bones of Atlantic salmon (*Salmo salar* L.). According to the report by Liu et al. (2015) that SOOP with 22 peptide sequences has a calcium-chelating capability of 52.47%.

5.4 Hydroxyapatite Products

The main component of mineral salt in fish bone is hydroxyapatite (HAP). Because of its excellent characteristics, such as biocompatibility with human tissue, bone conductivity, bone induction, no inflammation and immune response, HAP is widely used as an implant material in plastic surgery, dentistry, and other fields (Boutinguiza et al. 2012). Hydroxyapatite prepared from fish by-products can also be used to make chromatographic column filler or additive because of its porous structure and special surface properties, which can nonspecifically adsorb bioactive macromolecules (Ashokan et al. 2021).

6 Production Lines of Calcium Products

6.1 Calcium Supplements

6.1.1 Calcium Capsule

The production process flow diagram of calcium soft capsule is depicted in Fig. 3. Fresh or thawed fish bones are washed to remove impurities such as lipids and blood. The fish bones are cut into about 3 cm long sections, immersed in water, and transferred to a high-pressure cooking pot. It is thermally treated for 1 h at 120 °C and 0.1 MPa. After that, the fish bones are rinsed with tap water 3 times to remove the adhered flesh and floating lipids. The fish bones are coarsely crushed using a grinder with a sieve plate aperture of 2 mm. The obtained bone particles are milled the first time under the conditions of plate gap 1 mm and rotating speed of 1500 rpm, and then the second time under the conditions of plate gap of 0.1 mm and rotating speed of 2000 rpm (Yin et al. 2016b).

After adjusting the water content to 90%, the fish bone paste is crushed to an average particle size of 110 nm using a high-energy media mill. The particle size of fish bone particles in the obtained slurry can be adjusted by controlling the mixing speed, media particle diameter, media filling rate and other parameters (Yin et al. 2015). Ball mill is the core equipment for processing nano fishbone. The nano fish bone slurry was mixed with other ingredients and sterilized under ultra-high pressure. The calcium soft capsule is obtained by aseptic filling with a capsule filling machine.

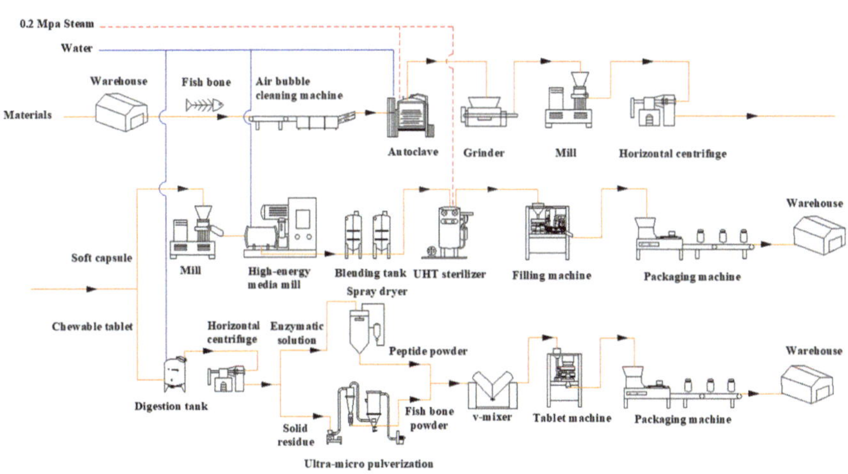

Fig. 3 Production line of calcium soft capsule and table made from fish by-products

6.1.2 Calcium Tablet

The production process flow diagram of calcium tablet is depicted in Fig. 3. The same as above, fish bone paste is obtained after being treated by a grinder and a millstone miller. Then the paste is added with 3 times weight of water. The paste is hydrolyzed with alkaline protease (2500 U/g fish bone) at 45 °C for 3 h, followed by flavor protease (1000 U/g fish bone) at 40 °C for 3 h. The solid residue and hydrolysate are separated by centrifugation. After drying, the solid residue was superfine crushed to obtain fish bone powder with an average particle size of less than 300 nm; The hydrolysate is concentrated under vacuum pressure, freeze-dried, and superfine crushed to obtain peptide powder. The fish bone powder and peptide powder are thoroughly mixed with casein phosphopeptide powder, vitamin D_3, dextrin, mannitol, sodium carboxymethyl cellulose, etc. After that, the mixture is pressed to tablet and package in plastic bottle.

6.2 *HAP Products*

6.2.1 HAP for Chromatographic Column

The production process flow diagram of HAP for chromatographic column is depicted in Fig. 4. First, the preparation of fish bone powder is required. Fish bones are put into a continuous steam cooker and heat at 100 °C for 15 ~ 30 min to remove the residual meat, oil, blood, etc. The fish bones are separated and crushed into small fish bone particles, and then immersed in a soaking tank with a temperature of 50 °C and a concentration of 1 M Na_2CO_3 solution for 5 h to completely remove the residual oil and protein in the bones. The soaked fish bone particles are transferred to the bubble cleaner through the elevator, fully cleaned to neutral pH, and then dried in an oven at 45 ~65 °C for 6 ~8 h. Finally, the dried fish bone particles are subjected to grinding to below 100 mesh sieve to obtain fish bone powder.

The prepared fish bone powder and water are evenly mixed in a stirring tank, and then pumped into the spray dryer through the feed pump to prepare a particle size of 0 ~15 μM hydroxyapatite microspheres. In spray drying, the particle size of the product can be controlled by adjusting the spray air pressure and the solid content of the slurry. Hydroxyapatite microspheres less than 2 μm are deposited at the bottom of a sedimentation tank to avoid plugging during subsequent column filling. Finally, the high-pressure homogenization column filling machine is used to fill the hydroxyapatite microsphere homogenate into the chromatographic column through the high-pressure pump.

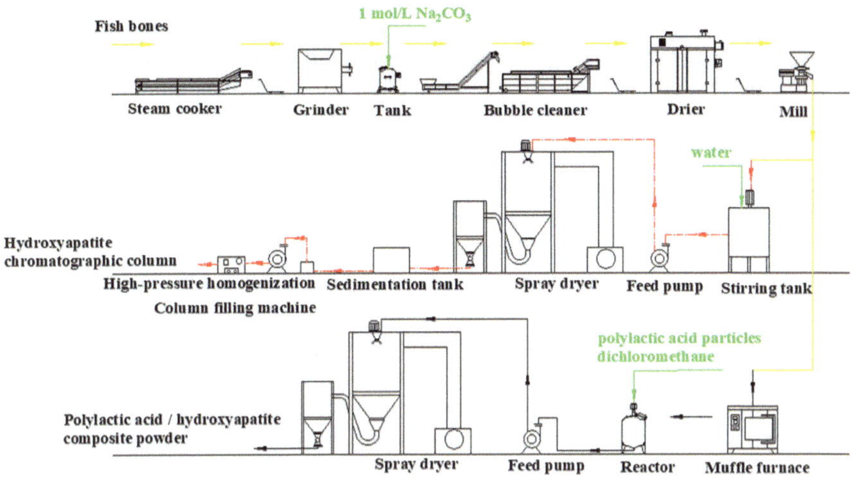

Fig. 4 Production line of HAP products made from fish by-products

6.2.2 HAP for Medical Treatment

The preparation of polylactic acid/hydroxyapatite composite powder, which is used for bone tissue scaffold, is depicted in Fig. 4. The fish bone powder prepared the same as above is calcined in a muffle furnace. It is first quickly heated up to 300 °C for 60 min, up to 450 °C for 60 min, then heat up to 700 °C for 60 min, and then cooled down to room temperature to obtain hydroxyapatite particles (Boutinguiza et al. 2012). The purpose of segmented calcination is to gradually remove water and organic matter from fish bone powder and increase its specific surface area. Then, polylactic acid particles and methylene chloride are added into the reactor. After the solid is dissolved, the hydroxyapatite particles are added and fully stirred, and pumped into the spray dryer by the feed pump to obtain 5 ~ 11 μm porous polylactic acid / hydroxyapatite composite microspheres.

7 Conclusion and Prospect

Fish by-products that can be used as calcium sources include fish bones, fish heads, fish frame, fish scale, and mixtures. The size, proportion, composition, recovery difficulty, and hygiene level of these by-products vary significantly, which determines their suitable uses and types of processed products. There are many reports on the content, physicochemical properties, and bioavailability of calcium compounds in fish head and fish bone, which provide a scientific basis for its application. Reducing the particle size of fish bone or calcium compounds to the nanometer level can improve their functional properties, which has become a research hotspot in recent years. The safety of these nanoparticles has gradually attracted the attention of scientists and government regulators. However, there are few related studies.

Most fish by-products exist in the form of mixtures, such as mixture containing fish bones, fish skin which is generated during the de-boning process of surimi production. However, there is little research on these mixtures. In addition, it is necessary to carry out a comparative study on the characteristics of calcium in fish by-products and calcium from other sources.

At present, the calcium compounds in fish by-products are mainly used for feed processing, which can not effectively improve the added value. Fish bone food, calcium supplements, medical and biochemical materials are the main high value-added products processed with calcium compounds in fish by-products. In the future, it is necessary to develop more high value-added products with calcium compounds in fish by-products as raw materials and realize commercialization, which would make our fisheries environmentally more sustainable and profitable.

References

Ashokan A, Rajendran V, Kumar T, Jayaraman G (2021) Eggshell derived hydroxyapatite microspheres for chromatographic applications by a novel dissolution - precipitation method. Ceram Int 47:18575–18583

Atkins A, Reznikov N, Ofer L, Masic A, Weiner S, Shahar RJAB (2015) The three-dimensional structure of anosteocytic lamellated bone of fish, 13, 311–323

Benjakul S, Mad-Ali S, Senphan T, Sookchoo PJJoFB (2017) Biocalcium powder from precooked skipjack tuna bone: Production and its characteristics', e12412

Boutinguiza M, Pou J, Comesaa R, Lusquios F, Science BLJM, CE (2012) Biological hydroxyapatite obtained from fish bone, 32(3), 478–486

Cui Y, Yang L, Lu W, Yang H, Zhang Y, Zhou X, Ma Y, Feng J, Shen Q (2021) Effect of steam explosion pretreatment on the production of microscale tuna bone power by ultra-speed pulverization. Food Chem 347:129011

Feng H, Li X, Deng X, Li X, Guo J, Ma K, Jiang BJRA (2020) The lamellar structure and biomimetic properties of a fish scale matrix. RSC Advance 10:875–885

Hemung B-O (2013) Properties of tilapia bone powder and its calcium bioavailability based on transglutaminase assay. Int J Biosci, Biochem Bioinform 3:306–309

Huang CH, Zeng BP, Dong JB (2008) Comparison of nutrients in the head of black carp, grass carp, silver carp and bighead carp. J Hunan Univ Arts Sci (Nat Sci Ed), 20(3), 46–48,57

Huo J, Deng S, Tong G (2010) Studies on preparation of haddock calcium tablet and its biological utilization. J Fish China 34(3):382–387

Idowu AT, Benjakul S, Sae-Leaw T, Sookchoo P, Kishimura H, Suzuki N, Kitani Y (2019) Amino acid composition, volatile compounds and bioavailability of biocalcium powders from salmon frame as affected by pretreatment. J Aquat Food Prod Technol 28(1):1–9

Jafarpour A, Gomes RM, Gregersen S, Sloth JJ, Srensen A (2020) Characterization of Cod (Gadus morhua) frame composition and its valorization by enzymatic hydrolysis. J Food Compos Anal 89:103469

Jiao YY, Okada M, Hara ES, Xie SC, Nagaoka N, Nakano T, Matsumoto T (2020) Micro-architectural investigation of teleost fish rib Inducing pliant mechanical property. Materials 13:5099

Khoder RM, Yin T, Liu R, Xiong S, You J, Hu Y, Huang Q (2020) Effects of nano fish bone on gelling properties of tofu gel coagulated by citric acid. Food Chem 332:127401

Kołodziejska I, Skierka E, Sadowska M, Kołodziejski W, Niecikowska C (2008) Effect of extracting time and temperature on yield of gelatin from different fish offal. Food Chem 107(2):700–706
Lee D-Y, Oh J-H, Uhm J-T, Kim I-H, Park M-J, Moon S-H (2020) Impact of acidity regulator and excipient nutrients on digestive solubility and intestinal transport of calcium from calcium phosphate and carbonate. Food Funct 11:10655–10664
Li J, Yin T, Xiong S, Huang Q, You J, Hu Y, Liu R, Li YJ (2020) Mechanism on releasing and solubilizing of fish bone calcium during nano-milling. J Food Process Eng 43:13354
Li Y, Xiong S, Yin T, You J, Hu Y (2017) Comparative study on the physicochemical properties of six types of nano-sized fish bones. Mod Food Sci Technol 33(7):125–133
Liu Q, Liu C, Wang C (2000) Scale nitriernt composition and effect on serum lipids in experimental hypercholesterolemic rats. J J Fish Sci China 7(4):56–59
Liu W-Y, Lu J, Gao F, Gu R-Z, Lin F, Ren D-F, Cai M-Y (2015) Preparation, characterization and identification of calcium-chelating Atlantic salmon (Salmo salar L.) ossein oligopeptides. Eur Food Res Technol 241(6):851–860
Malde MK, Bügel S, Kristensen M, Malde K, Graff IE, Pedersen JI (2010) Calcium from salmon and cod bone is well absorbed in young healthy men: a double-blinded randomised crossover design. Nutr Metablsm 7(1):61
Nawaz A, Xiong Z, Xiong H, Chen L, Wang PK, Ahmad I, Hu C, Irshad S, Ali SW (2019) The effects of fish meat and fish bone addition on nutritional value, texture and microstructure of optimised fried snacks. Int J Food Sci Technol 54(4):1045–1053
Nemati M, Kamilah H, Huda N, Ariffin FJIJoFS, Nutrition (2016) In vitro calcium availability in bakery products fortified with tuna bone powder as a natural calcium source, 67(5), 535–540
Park JW, Graves D, Draves R, Yongsawatdigul J (2014) Manufacture of surimi: harvest to frozen block. In: Park JW (ed) Surimi and surimi seafood, 3rd edn. CRC Press, Boca Raton, FL, pp 55–100
Peng Z, Hou H, Zhang K, Li BJFC (2017) Effect of calcium-binding peptide from Pacific cod (Gadus macrocephalus) bone on calcium bioavailability in rats, 221, 373–378
Shi L, Yin T, Huang QL, You J, Hu Y, Jia D, Xiong SB (2021) Effects of filleting methods on composition, gelling properties and aroma profile of grass carp surimi. Food Sci Human Wellness 10:307–314
Sriuttha M, Chantarangsee M, Hemung BO (2014) Characteristics of tilapia bone powder and calcium xxtraction by acid solution. Asian J Chem, 26(Supp.), S103-S106
Surya P, Nithin A, Sundaramanickam A, Sathish M (2021) Synthesis and characterization of nano-hydroxyapatite from Sardinella longiceps fish bone and its effects on human osteoblast bone cells. J Mech Behav Biomed Mater 119(2021):1–9
Wang X, Wang X, Huang Z, Liu Z, Huang T (2019) Element distribution analysis of crucian carp, silver carp and snakehead. China Brewing 38(8):163–167
Wiriyaphan C, Chitsomboon B, Yongsawadigul J (2012) Antioxidant activity of protein hydrolysates derived from threadfin bream surimi byproducts. Food Chem 132(1):104–111
Wu G, Zhang M, Wang Y, Mothibe KJ, Chen W (2012) Production of silver carp bone powder using superfine grinding technology: Suitable production parameters and its properties. J Food Eng 109(4):730–735
Xie WW, Yin T, Zhang J, Liu R, Zhao SM, Xiong SB (2014) Effects of fish bone powder particle size on calcium bioavailability of fish bone powder-fish protein hydrolysate mixture. Food Sci (In Chin) 35(7):211–216
Yang L, Wang H, Zhou X, Ma Y, Wang J, Shen Q (2020) Microwave-assisted preparation and calcium bioavailability evaluation of fishbone flour of tuna (Thunnus albacares). Food Sci 41(4):235–242
Yin ML, Zhao SK, Zheng Y, Shi SD, Sun YQ (2018) Preparation, characterization and calcium supplement efficiency of hydroxyapatite nanocrystals. Chemical Reagents 40(3):212–216
Yin T, Du H, Zhang J, Xiong S (2016a) Preparation and characterization of ultrafine fish bone powder. J Aquat Food Prod Technol 25(7):1045–1055
Yin T, Park JW (2014) Effects of nano-scaled fish bone on the gelation properties of Alaska pollock surimi. Food Chem 150:463–468

Yin T, Park JW, Xiong S (2015) Physicochemical properties of nano fish bone prepared by wet media milling. LWT Food Sci Technol 64(1):367–373
Yin T, Park JW, Xiong S (2017) Effects of micron fish bone with different particle size on the properties of silver carp (*Hypophthalmichthys molitrix*) surimi gels. J Food Qual 2017:1–8
Yin T, Shi L, Zhang J, Xiong SB, You J, Hu Y (2016b) Effects of processing methods on nutrition of the micro sized fish bone pastes. J Huazhong Agric Univ 35(6):124–128
Zeng SK, Zhang CH, Jiang ZH (2002) Study on the comparation of the food nutrient contents between the muscle and head of Muraenesox Cinereus. Mar Sci 26(5):13–15
Zhang J, Duan R, Chao YE, Konno K (2010) Isolation and characterzation of collagens from scale of solver carp (Hypophthalmichthys molitrix). J Food Biochem 34(6):1343–1354

Chitin, Chitosan, and their Derivatives from Seafood Waste and Processing Byproducts

Ajay Mittal, Avtar Singh, Martin Xavier, C. N. Ravishankar, and Soottawat Benjakul

Abstract Processing of crustaceans such as shrimp, crab, etc., or molluscs, e.g., squid generates large amounts of byproducts, especially shells and squid pens. Improper disposal of those byproducts may cause deleterious implications on environment and human health. The valorization of shells or squid pens via conversion to several high-value bioactive compounds could pave a way for sustainable utilization of marine resources. Crustacean shells and squid pens serve as a potential source of chitin, which can further be deacetylated into chitosan (CS), a positively charged polysaccharide that exhibits biocompatibility and biodegradability. Derivatization or degradation of CS can enhance bioactivities including antioxidant, antimicrobial, antitumor activities, etc. The present chapter covers the established methods of chitin and CS extraction from seafood processing waste. Moreover, the methods for preparation of various CS derivatives, e.g., chitooligosaccharides or chemically modified CS are also revisited. Recent trends in applications of CS derivatives in the food sector are also explored. Furthermore, CS derivatives for biomedical uses are addressed.

Keywords Chitin · Chitosan derivatives · Chitooligosaccharides · Food applications · Biomedical uses

A. Mittal · A. Singh · S. Benjakul (✉)
International Center of Excellence in Seafood Science and Innovation, Faculty of Agro-Industry, Prince of Songkla University, Hat Yai, Songkhla 90110, Thailand
e-mail: soottawat.b@psu.ac.th

M. Xavier · C. N. Ravishankar
Fisheries Resources, Harvest and Post-Harvest Division, ICAR-Central Institute of Fisheries Education, Mumbai 400061, Maharashtra, India

S. Benjakul
Department of Food and Nutrition, Kyung Hee University, Seoul 02447, Republic of Korea

S. Maqsood et al. (eds.), *Fish Waste to Valuable Products*, Sustainable Materials and Technology, https://doi.org/10.1007/978-981-99-8593-7_11

1 Introduction

As the population and the economy have grown in both developed and developing countries, consumers have demanded a better lifestyle including high quality, healthy, and safer foods or food ingredients. Thus, non-conventional ingredients or additives, particularly from seafood processing byproducts, have drawn progressive attention to meet the requirements of consumers. In general, seafood processing byproducts are promising sources of proteins, essential amino acids, unsaturated fatty acids, glycosaminoglycans, vitamins, and minerals, thus serving as the diet of consumers for a complete food (Nanda et al. 2021). Among the edible aquatic species, crustaceans including shrimp, crab, and lobster have great commercial importance due to their delicacy and high market price. Globally, every year, crustacean processing generates approximately 6 to 8 million tons of various byproducts, in which shells and head (cephalothorax) from shrimp impart at high portion (Pinheiro et al. 2021). These byproducts could cause environmental pollution, if not properly managed or discharged.

Generally, crustacean shells contain proteins, minerals, and chitin in the range of 30–40, 30–50, and 20–30 wt%, respectively along with pigments (Ahmadkelayeh and Hawboldt 2020). The squid pen contained approximately 1% ash, 2.3% lipids, 64% protein, and 31% chitin (Singh et al. 2019c). The recovery of bioactive molecules, especially chitin via demineralization followed by deproteinization, respectively from the crustacean shell or squid pen would be beneficial for the economy of seafood processing industries to gain the value-added products instead of discharge into the environment. Shrimp, lobster, and crab shells consist of α-chitin, while squid pen possesses β-chitin (Singh et al. 2021c). Due to chitin insolubility, deacetylation with strong alkali or enzymes has been conducted to render acid-soluble chitosan (CS). Low toxicity, bio-compatibility, and higher biodegradability of CS increase its popularity for biomedical applications (Pillai et al. 2009). Cancer therapy, anti-aging, medication delivery, and tissue engineering are the other benefits of CS (Singh et al. 2019c; Mittal et al. 2021a). Moreover, CS can be used for food packaging as well as a natural antioxidant and antimicrobial agent. Nevertheless, water insolubility of CS due to compact structure associated with high intermolecular and intramolecular hydrogen bonding limits its applications in foods and other fields. Functional groups, i.e., $-NH_2$ at C2 and –OH at C3 or C6 on the glucosamine unit of the CS chain allowed the chemical modification of CS, in which water solubility and bioactivities can be improved (Singh et al. 2021c). Various chemical modifications including acylation (Zhang et al. 2017), carboxylation (Jimtaisong and Saewan 2014), and quaternization (Andreica et al. 2020) at the C-2, C-3, and C-6 of CS have been done. Moreover, hydrolysis of CS via either chemical-assisted or enzyme-assisted processes has been performed to improve solubility and augment biological activities (Mittal et al. 2022a).

This book chapter aims to address the updated information on the extraction of chitin and CS from seafood processing byproducts. Moreover, different methods for preparation of CS and its structural characterization are revisited. In addition, bioactivities, biomedical or nutraceutical properties, and food applications of CS derivatives are discussed comprehensively.

2 Chitin

2.1 Chitin Structure and Physicochemical Properties

Professor Henri Braconnot firstly discovered chitin in 1811. Odier discovered a similar substance in insects and plants in 1823 and called it "chitine" (Muzzarelli and Muzzarelli 2009). Lassaigne found nitrogen in chitin in the 1840s (Domard et al. 1997) and Henri Braconnot called it "fungine." Chitin is the world's second most abundant biopolymer, which possesses both nitrogen and carbon (C:N = 8:1) and is generated almost as much as cellulose per year (Struszczyk 2006). Chitin is a linear amino-polysaccharide, primarily composed of β-(1 → 4)-linkage between two 2-acetamido-2-deoxy-β-D-glucopyranose units (Fig. 1). The α and β chitins are classified as the anti-parallel and parallel polymer chain arrangements, respectively (Singh et al. 2021c). The α-chitin is the most stable due to its high crystallinity index (80%). Chitin possesses a strong inter- and intramolecular hydrogen bonding in its network, which contributes to its insoluble character in various common solvents (Zargar et al. 2015). It can be dissolved in inorganic acids such as hydrochloric, sulfuric, and phosphoric acids at high concentration (Casadidio et al. 2019). Additionally, other solvents including dichloro and trichloroacetic acids, *N,N*-dimethylacetamide, lithium thiocyanate, *N*-methyl-2-pyrrolidone, hexafluoroisopropyl alcohol, and hexafluoraceton can dissolve chitin (Casadidio et al. 2019). However, those solvents are poisonous, corrosive, hazardous, or only partially biodegradable. Due to the insolubility of chitin, its uses are limited in the food processing and biomedical industries (Park and Kim 2010).

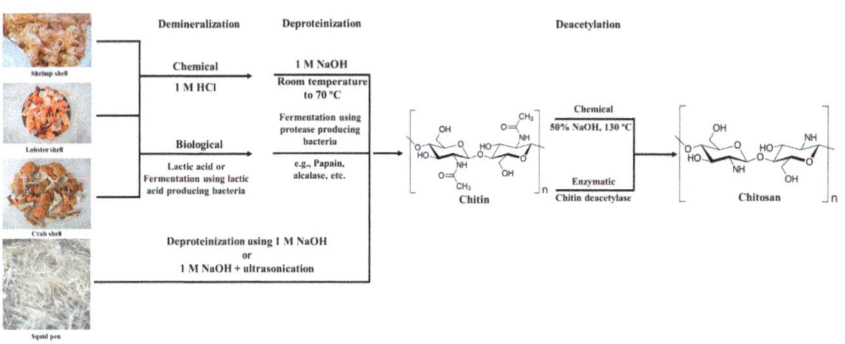

Fig. 1 Preparation scheme for chitin and chitosan from different seafood processing by-products

2.2 *Extraction of Chitin*

2.2.1 Chemical Extraction

In general, the chitin extraction process consists of two main steps: demineralization and deproteinization (Fig. 1). Furthermore, decolorization, an obligatory step, is conducted to abolish pigments including β-carotene or astaxanthin with the aid of different solvents such as hydrogen peroxide, acetone, sodium hypochlorite, etc. (El Knidri et al. 2018). *Penaeus monodon* shells were used for chitin extraction, in which shells were subjected to demineralization using HCl (1 M) and deproteinization using NaOH (3 M), at room temperature, in which 75 min were used for each process (Srinivasan et al. 2018). Chitin was obtained from washed and dried shrimp shell powder via deproteinization implemented for 24 h using 1 M NaOH (Teli and Sheikh 2012). The powder was subsequently demineralized and discolored with the help of 1 M HCl and $KMnO_4$/oxalic acid mixture, respectively. Demineralization of *Metapenaeus stebbingi* shells was performed using 1.7 N HCl for 6 h at 25 °C and those shells were further deproteinized with aid of 2.5 N NaOH for 6 h at 65 °C to obtain chitin (Kucukgulmez et al. 2011). Moreover, chitin was isolated from Pacific white shrimp shell using HCl (1 M; 1:20, w/v) for 2 h and NaOH (1 M; 1:30, w/v) for 80 min at 70 °C for demineralization and deproteinization, respectively (Mittal et al. 2021a). On the other hand, squid pen contains a minimal quantity of minerals and pigments. Sometimes, demineralization and decolorization can be omitted during the chitin extraction process. Chitin was extracted from squid (*Loligo formosana*) pen through an ultrasonication-assisted process, in which the lowest remaining protein but the highest yield (34%) were achieved when the amplitude of 69% for 41.46 min was implemented using NaOH (1 M; 1:18, w/v) for deproteinization (Singh et al.2019c). Similarly, β-chitin from squid (*Loligo chenisis*) was obtained via deproteinized and demineralized using NaOH (4%, w/v) for 10 h at 80 °C. It had molecular weight (MW) and protein content of 8.5×10^3 kDa and 0.63%, respectively along with negligible content of minerals (Cuong et al. 2016). The conventional extraction of chitin demands ample quantity of hazardous chemicals, e.g., HCl and NaOH for demineralization and deproteinization, respectively, which are subsequently released into the environment and may cause a negative impact.

2.2.2 Biological Extraction

Biological extraction of chitin employs microorganisms, which produce proteolytic enzymes and organic acids, providing a cleaner and greener process. Chitin from *Litopenaeus vannamei* shell waste was extracted using protease-producing bacterium and lactic acid-producing bacterium, which were identified as *Alcaligenes faecalis* and *Bacillus coagulans*, respectively (Rakshit et al. 2021). Lactic acid fermentation at various times (0–60 h) with the aid of *Lactobacillus plantarum* was performed to extract chitin from crab (*Allopetrolisthes punctatus*) shell (Castro et al. 2018).

High-quality chitin with a maximum yield of 7% was attained after fermentation of 60 h followed by purification using 0.4 M NaOH and 0.5 M HCl, thus lowering the amount of chemicals required. Moreover, chitin was produced from lobster shell biomass via the successive treatment of *Serratia marcescens* db11 and *Lactobacillus plantarum*, which results in the removal of protein (87.19%) and $CaCO_3$ (89.59%) (Chakravarty et al. 2018). Recently, chitin was produced from shrimp shell powder through microbial fermentation for 7 days using protease-producing bacterium, *Bacillus cereus* HMRSC30 (Cahyaningtyas et al. 2022). The microbial fermentation produced protease (198 U/mL), which could achieve deproteinization and demineralization of 97 and 54%, respectively. In addition, chitin was extracted from fresh and cooked fermented *Litopenaeus vannamei* shell. The fermentation was performed for 16 days using *Halobacterium salinarum* and *Halococcus dombrowskii* (Dayakar et al. 2021). Among both haloarchaea, cooked shrimp shell inoculated with *Halococcus dombrowskii* showed highest removal of protein and carotenoids. Fermentation using protease and lactic acid generating bacteria is a green approach for fully utilizing shrimp shell waste, while minimizing waste disposal throughout the established process.

Enzyme-assisted extraction of chitin is gaining popularity because of its environmentally friendly nature. Deproteinization using protease from *Bacillus cereus* SV1 using an enzyme/substrate ratio of 1:20 at 40 °C for 3 h, followed by demineralization using HCl (1.25 M) at room temperature for 6 h to obtain chitin from shrimp processing waste (Manni et al. 2010). The aforementioned treatment rendered 90% recovery and the extensive elimination of protein and minerals from waste was achieved. Moreover, various commercially available food grade enzymes including lysozyme, trypsin VI, papain, alcalase, and flavorzyme were used to hydrolyze proteins when dried shrimp heads were subjected to deproteinization for chitin extraction. Among all treatments, alcalase at pH 8.0 was the most effective commercial enzyme for shrimp head deproteinization (Valdez-Peña et al. 2010). Thereafter, demineralization of aforementioned deproteinized shrimp head was performed using lactic acid in combination with microwave irradiation (400W) at 121 °C for 30 min. Thus, amalgamation of enzymatic and physical methods enhanced chitin recovery.

2.3 *Chitin Derivatives*

Chitin has micro- and nano-fibril structure that is intensively organized and incorporates at both amorphous and crystalline regions. Chitin microfibrils isolated from crustacean shells or squid pen as nanocrystals having 6–60 and 100–800 nm in width and length, respectively or nanofibres with 10–100 nm and several micrometres in width and length, respectively. The physical characteristics of chitin microfibrils are strongly affected with origin and extraction conditions (Ngasotter et al. 2022). Box-Behnken design using response surface methodology (RSM) was used for the optimization of nanochitin extraction using microwave irradiation, in which nanochitin was obtained from yellow lobster shells when treated with HCl (1 M) for 10 min at 124.75 W. In addition, shrimp shells were treated with 1 M HCl at 50.21 W for 14.34 min to obtain nanocrystals and for squid pen nanofiber, squid pen were treated at 54.08 W for 29.08 min in the presence of 1 M HCl (Fernández-Marín et al. 2021). The obtained nanochitins from shrimp shell, lobster shell, and squid pen had length of 386, 315, and >900 nm, respectively.

3 Chitosan Structure and Physicochemical Properties

Chitosan (CS) is a partially deacetylated chitin, containing 2-amino-2-deoxy-glucose (deacetylated moieties) and 2-acetamido-deoxy-D-glucose (acetylated moieties) connected with the aid of β-(1 → 4) linkages (Fig. 1) (Islam et al. 2017; Park et al. 2010). Degree of acetylation (DA) is primarily used to distinguish chitin from CS. When DA is higher and lower than 50%, the polymer is called chitin and CS, respectively (Croisier and Jérôme 2013; Gonil and Sajomsang 2012). Under acidic condition, CS acts as a polyelectrolyte, and its solubility is improved due to protonation of $-NH_2$ group present at C-2 of the glucosamine unit (Rinaudo 2006). CS has a plethora of peculiar properties including non-toxic, and biodegradable properties. Moreover, CS has been intensively documented for various bioactivities (De Queiroz Antonino et al. 2017).

3.1 Preparation of CS

3.1.1 Chemical Methods

MW, degree of deacetylation (DDA), and acetyl group distribution in the CS chain strongly affect the behavior of CS solution (Huang and Tsai 2020; Singh et al. 2019c). Thus, CS with desired MW and DDA has the enhanced solubility, thus widening applications in various fields. The preparation of CS from chitin mainly involves chemical and enzymatic methods (Fig. 1) (Younes and Rinaudo 2015). Between both methods of extraction, chemical-assisted extraction has been widely used for CS manufacturing due to less cost and ease at a commercial level (Abdou et al. 2008; Al-Sagheer and Muslim 2010; Benhabiles et al. 2012, Tokatlı et al. 2018). Deacetylation is the process of removing acetyl groups using alkali at high concentration and high temperature. The alkaline treatment eliminates the acetyl groups, yielding D-glucosamine units with free $-NH_2$ groups instead of N-acetyl-D-glucosamine units. Deacetylation of chitin can be achieved in two ways namely (1) heterogeneous and (2) homogenous processes (Sannan et al. 1976; Chang et al. 1997). During heterogenous deacetylation, chitin is heated in NaOH solution for several hours, rendering a DDA of 85–99%. In homogeneous process, firstly chitin is dispersed in the concentrated NaOH (3 g chitin in 30 g NaOH/45 g H_2O) for 3 h at 25 °C and then cooled at 0 °C. This method produces soluble CS having

DDA between 48 and 55% (Younes and Rinaudo 2015). However, post-reaction separation of soluble CS from NaOH solution is complex. Therefore, this process is tedious to scale up on the industrial level. Furthermore, the non-uniform distribution of acetyl glucosamine and glucosamine residues along with blockage of acetyl group distribution in CS chains determines different physicochemical properties during homogenous deacetylation (Yadav et al. 2019). Due to aforementioned disadvantages of homogenous deacetylation, heterogenous deacetylation is prioritized by industries due to the ease in the separation of insoluble CS from NaOH residues and other components such as pigments and protein.

Chitin extracted from *Penaeus monodon* shells was deacetylated using NaOH (50%; 1:50, w/v) at 90 °C for 50 min (Srinivasan et al. 2018). Also, chitin from *Metapenaeus stebbingi* shells was deacetylated by 50% NaOH at 120 °C (Kucukgulmez et al. 2011). CS was prepared from tail shells of lobster (*Homarus americanus*) chitin using NaOH (50%; 1:10, w/v) for 20 h at 65 °C. MW and DDA of CS are primarily governed by alkali concentration, time and temperature of reaction, and repetition of process (Younes and Rinaudo 2015). DDA of CS was basically augmented at the initial reaction and subsequently became slower during course of time (Tsaih and Chen 2003). The high temperature and alkali concentration aided in the diffusion of NaOH into chitin matrix (Chang et al. 1997). DDA of CS from different sources (*Penaeus aztecus*, *Penaeus durarum*, squid pen, cuttlefish pen, crab shell, crayfish shell) increased up to 80–90% with augmenting NaOH concentrations from 10 to 40% (Abdou et al. 2008). CS's DDA was 92 and 95% after alkaline deacetylation at 99 and 140 °C, respectively. Shrimp (*Litopenaeus vannamei*) shell chitin was deacetylated using 50% NaOH at 130 °C for 4 h, yielding CS with DDA of 85.28% (Mittal et al. 2021a). Moreover, the deacetylation condition was optimized using several approaches, e.g., RSM, in which the influence and function of each factor on DDA and MW were taken into account (Singh et al. 2019c). RSM was employed to optimize deacetylation of chitin from shrimp waste (130 °C for 90 min), in which CS (MW:150 kDa and DDA: 90%) was produced (Weska et al. 2007). Hwang et al. (2002) documented MW (100–1100 kDa) and DDA (67 to 96%) of CS when RSM was adopted for deacetylation. The MW was reduced with augmentation in DDA of CS, especially when higher reaction temperature, longer reaction time, and higher NaOH concentration were employed (Mittal et al. 2021a). CS with desired MW and DDA has enhanced solubility and wide applications.

Due to looser confirmation of β-chitin than α-chitin, the former shows higher reactivity toward different solvents (Singh et al. 2019b, c). Squid pen CS was prepared with the aid of NaOH (50%, w/v) at varying temperatures (60–130 °C) (Singh et al. 2019c; Elieh-Ali-Komi and Hamblin 2016). CS was prepared from squid (*Loligo formosana*) pen chitin using NaOH (50%, w/v) at different temperatures (110–130 °C) and times (2–8 h), in which highest DDA (87%) was gained when chitin was treated in NaOH solution at 130 °C for 2 h (Singh et al. 2019c). The yields of 60 and 54% were achieved when chitin was deacetylated at 110 and 130 °C, respectively for 2 h. Regardless of temperature, the lowest yield was observed with increasing extraction time. This was associated with CS degradation or loss of fragmented molecules during the neutralization or washing processes. Furthermore, as temperature and time increased, DDA of CS rose. Nevertheless, no variation in DDA was noticeable when the deacetylation was performed for >2 h at 130 °C (Singh et al. 2019c). To evaluate the effect of DDA on CS properties, chitin deacetylation using 50% NaOH was performed for various times (2–8 h) at 100 °C (Chandumpai et al. 2004). The viscosity and average MW consistently decreased from 10,240 to 9,110 kDa as time of deacetylation was augmented from 2 to 8 h. As a result, deacetylation time and temperature significantly influenced the DDA and yield of CS, which could further determine its functional properties. However, alkali treatment can pollute the environment, requires enormous energy, and yields poor quality CS. Thus, biological processes including microbial fermentation or enzyme-assisted deacetylation could be used to overcome those limitations.

3.1.2 Biological Methods

Deacetylation with the aid of an enzyme can convert chitin to CS using chitin deacetylase. It has been known to save energy and protect the environment. However, the various drawbacks such as extraction, breeding, selection, and cultivation of enzyme-producing bacteria limits the usage of enzymatic method (Wang et al. 2020b). Therefore, the selection of appropriate bacterial strains producing chitin deacetylase is crucial for efficacy in deacetylation. Moreover, chitin deacetylase-producing bacteria have been investigated as potential replacements for current fungal strains due to their easy and faster growth than fungi in large fermentation systems (Wang et al. 2020b). Thus, bacterial strains may have greater potential. Shrimp chitin was converted to CS through bacterial fermentation process using *Alcaligenes faecalis* Alca F2018 having chitin deacetylase activity (Amer et al. 2022).

4 Derivatives of Chitosan: Process and Structural Characteristics

CS constituted functional groups including acetylamino and glycosidic bonds along with $-NH_2$/acetamido, –OH groups at C-2, C-3, and C-6, respectively (Jeon et al. 2001; Kuroiwa et al. 2002). Among all functional groups, the acetylamino and glycosidic bonds are not easy to break due to high stability for further modification of the CS. Moreover, the –OH group at C-3 could not react spontaneously due to restricted rotation and steric hindrance. On the other hand, $-NH_2$ group has critical role in hydrogen bonding (Zhang et al. 2010). Protonation of the $-NH_2$ group in the aqueous acidic renders cationic CS, which has increased solubility, thereby promoting CS as a polyelectrolyte complex with other molecules (Hu and Luo 2016; Luo and Wang 2014). In general, chemical modification of CS has been performed to improve its

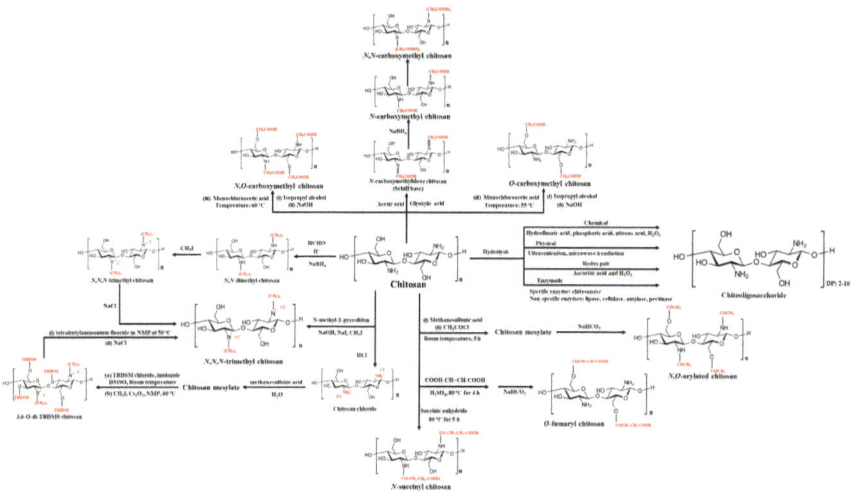

Fig. 2 Chemical reactions involved in preparation of different chitosan derivatives

physicochemical properties, in which applications of CS derivatives in food or relevant fields can be widened. The scheme for chemical reactions of CS modification via different processes is given in Fig. 2.

4.1 Acylated Chitosan

The introduction of aliphatic or aromatic acyl groups to CS chains using organic acids and their derivatives such as anhydride and acyl chloride refers to "acylation" of CS. Acylation can be done on either $–NH_2$ group or two -OH groups present on the glucosamine unit of CS. When the acylation reaction occurs at $–NH_2$ group at C2 of CS chain through an amide linkage, it is called "*N*-acylation." Various anhydrides including acetic, propionic, and hexanoic acids were introduced to CS (4%, w/v) via acylation under the homogenous condition to synthesize *N*-acetyl CS, *N*-propionyl CS, and *N*-hexanoyl CS, respectively (Hu et al. 2007). All *N*-acylated CSs prepared using each aforementioned anhydride at a molar ratio of 0.5/1 showed similar DDA of about 40% (less than DDA of CS) as determined by pH titration method, indicating successful substitution of anhydride on the CS chain. Moreover, *N*-succinyl-CS with 67% degree of substitution (DS) was obtained by conjugation of succinyl group using HCl and ethanol on the *N*-terminal of alkaline-treated CS (Tang et al. 2016). The reaction was conducted for 5 h at 80 °C. The resultant derivative was water-soluble and solubilized in a wide pH range (2–12), compared to CS, which was only soluble from pH 2 to 6 and insoluble in water. Moreover, acylation can be performed at the –OH group either at C3 or C6 of CS chain through the formation of ester linkage, known as *O*-acylation. Nevertheless, it is comparatively complicated due to the protection and deprotection of the -NH_2 group during the reaction (Sahariah et al. 2019). Therefore, very few studies were available on the *O*-acyl CS derivatives. *O*-fumaryl-CS was prepared via partial acylation of CS and fumaric acid using H_2SO_4 as protector to -NH_2 group at 80 °C for 4 h at different molar ratios (1:1 to 5:1; fumaric acid to CS) (Feng and Xia 2011). DS of *O*-fumaryl-CS was increased from 0.07 to 0.48 with increasing molar ratio of fumaric acid to CS. *O*-acylation of CS nanofibers was performed using different fatty acid anhydrates as acylation agents such as short (C2-C6) or long (C8 and C12) chain and pyridine as a catalyst, while trifluoracetic acid was the protecting agent (Zhang et al. 2017). *O*-acylated nanofibers exhibit both hydrophobic and hydrophilic characteristics and have varying chain length of the substituted acyl groups (Garg and Jana 2011). The observed characteristic bands in the FTIR spectra of acylated CS derivatives were found as follows: (a) stretching and bending of C–H due to –CH_3 group between 3000 and 2840 and 1435 cm^{-1}, respectively; (b) C=O stretching due to amide I at 1640 cm^{-1}, representing *N*-acylation; (c) band at 1250 and 1198 cm^{-1} for C–O stretching, and (d) ester C=O stretching at 1740 cm^{-1}, confirming *O*-acylation (Piegat et al. 2020). Reaction conditions and coupling system used directly determine band intensity and wavenumbers of resulting derivative.

4.2 Carboxylated CS

Carboxylation is a process, which introduces the carboxyl (–COOH) group to the –OH and –NH_2 group at C6 and C2, respectively of CS chains with the aid of glyoxylic acid or chloroalkanoic acid. Carboxymethylation of CS has been employed frequently to enhance water solubility (Shariatinia 2018; Upadhyaya et al. 2013; Zhou et al. 2021). Various types of carboxymethyl CS (CMCS) can be obtained by direct alkylation of CS. However, degree of substitution (DS) and site-specificity including –NH_2 and –OH group denoted as *N* and *O* at C2 and C6, respectively, are affected by reaction conditions (Bukzem et al. 2016). *N*-CMCS was prepared via reaction of free –NH_2 group of CS with glyoxylic acid and then reducing it with $NaBH_4$.

When the DS is $\geq$1, carboxymethyl substitution happens at –OH and –NH_2 at C6 and C2, respectively to form *N,O*-CMCS. *O*-CMCS possessed a carboxymethyl group at -OH site of CS, while *N,O*-CMCS had a carboxymethyl group at both –NH_2 and –OH sites of CS (Wang et al. 2020a). *O*-CMCS and *N,O*-CMCS were produced by adding CS flakes in isopropanol followed by the addition of NaOH solution (40–60%, w/v) and monochloroacetic acid with mild heating (50–60 °C) for 1–24 h or freezing (Sahariah et al. 2019). Kusuma et al. (2015) prepared *N,O*-CMCS by adding CS (5 g) in 50 mL of isopropyl alcohol followed by the addition of 10 N NaOH (13 mL) and monochloroacetic acid (30 g). The resulting mixture was incubated at 60 °C and stirred for 3 h. After the reaction, the solid was washed with methanol till pH 7 was obtained, followed by drying at 60 °C. In another study, *N,O*-CMCS was synthesized by adding CS (10 g) in isopropyl alcohol (100 mL) and 10 N NaOH (25 mL). Afterward, monochloroacetic acid (12 g) was mixed with resultant solution and stirred continuously for 3 h at 60 °C. The reaction mixture (pH 7) was then filtered to obtain *N,O*-CMCS. The retentate was rinsed several times with methanol or ethanol. The *N,O*-CMCS was filtered again and then dispersed in absolute methanol or ethanol and subjected to drying under vacuum or in an oven at 60 °C to attain *N,O*-CMCS powder (Nadira et al. 2022).

Based on the FTIR spectrum of *N*-CMCS, wavenumber a at 1600 and 1411 cm^{-1} corresponded to the both asymmetric and symmetric stretch vibration of COONa, respectively. The peak at 1600 cm^{-1} indicates that carboxymethylation successfully occurred at the $-NH_2$ or the OH of CS. ^{1}H-NMR of *N,O*-CMCS exhibited characteristic peaks at 3.2 ppm (C2 of $-N-CH_2-COOH$) and 4.0 ppm (C6 of $-O-CH_2-COOH$) along with peaks associated with CS. This verified the replacement of carboxymethyl at –OH and $-NH_2$ on glucosamine unit (Xu et al. 2021). Moreover, the absence of a peak from 3.19–3.40 ppm in the ^{1}H-NMR spectrum indicated the synthesis of *O*-CMCS.

4.3 *Quaternary Ammonium CS*

The tailoring of CS by altering the primary $-NH_2$ groups at C2 to quaternary ammonium salts derivative with a stable positive charge could improve the water solubility of CS (Fabiano et al. 2020). Also, this can induce the disruption of the hydrogen bonding, thus unfolding CS chains. Generally, quaternary ammonium CS is synthesized via either transformation of the primary $-NH_2$ group of CS into quaternary salt or CS linked with quaternary salt by a lateral spacer (Andreica et al. 2020). *N,N,N*-trimethyl CS (TMCS) iodide was first quaternized CS derivative. Successive alternative steps of *N*-alkylation, then reduction and methylation were performed, respectively using formaldehyde in the acidic media, sodium borohydride, and methyl iodide (CH_3I) (Muzzarelli and Tanfani 1985). However, *N*-monosubstituted or *N,N*-disubstituted secondary products were formed during reaction, in which the obtained CS derivative was insoluble in water. Despite multistep reactions of the aforementioned method, the methylation brings about water-soluble quaternized CS derivatives. Water-soluble TMCS possessing average DS between 10 and 45% was prepared using NaOH (15 wt.%), CH_3I (5.7 mL), *N*-methyl-2-pyrrolidone (15 mL as a solvent), and sodium iodide (2.4 g as a catalyst) (Curti et al. 2003). The trimethyl group was augmented with the increasing amount of NaOH (20–40 wt.%) and CH_3I (10 mL). However, yield and water-insoluble fraction decreased and increased, respectively, with increasing concentration of the aforementioned reactants (Curti et al. 2003). This was plausibly due to undesirable *O*-methylation as a side reaction under strongly alkaline conditions and high temperatures (90–95 °C) (Xu et al. 2010). To decrease the impact of *O*-methylation, a pre-alkylation of CS was performed using various aldehydes except for formaldehyde to obtain a Schiff-base intermediate followed by quaternization of CS using CH_3I (Wu et al. 2016a). The electron density was increased at the nitrogen atom following introduction of $-CH_3$ group, which further facilitated the quaternary ammonium salt formation (Andreica et al. 2020). This method could prevent chain depolymerization, deacetylation, and *O*-methylation. In another method, *N,N*-dimethyl CS (DMC) was first prepared through Eschweiler-Clarke reaction using formaldehyde and formic acid, followed by the methylation with CH_3I in NMP, leading to TMCS without imination reaction (Verheul et al. 2008).

However, the aforementioned methods are not fully efficient, and *O*-methylation occurred during the reaction. CS reacts with methanesulfonic acid to produce CS metasylate with subsequent insertion of tert-butyldimethylsilyl (TBDMS) groups on it to obtain 3,6-di-O-tert-butyldimethylsilyl-CS. The aforementioned compound was highly soluble in several common organic solvents (Rúnarsson et al. 2008). Fully quaternized TMCS was produced by treating the *O* protected di-tert-butyldimethylsilyl-3,6-*O*-CS precursor (TBDMS-CS) with CH_3I and cesium carbonate in NMP (Benediktsdóttir et al. 2011). Finally, the -OH groups can be deprotected using tetrabutylammonium fluoride solution in NMP.

FTIR spectra of TMCS showed characteristics band as follows (Xu et al. 2010): (a) asymmetric stretching vibration of C–H in methyl groups at ~1475 cm^{-1}; (b) C=O stretching of amide I band between 1630–1660 cm^{-1}; (c) the amide II band (N–H bending vibrations) and C–N stretching at 1600 cm^{-1}. It disappeared in TMCS due to the substitution of the protons on the $-NH_2$ group of CS by the $-CH_3$ groups. For ^{1}H-NMR of TMCS, proton signals at 2.80–3.45 ppm and 3.30–3.72 ppm were associated with $-N(CH_3)_2$ and $-N(CH_3)_3^+$ group, respectively, depending upon the pH of the NMR solvent. In addition, if *O*-methylation occurred, the peaks associated with it appeared at 3.5 ppm, while protons of the acetyl group of CS appeared at around 2 ppm. In the ^{13}C-NMR spectrum, the signals for $N(CH_3)_3^+$ and $N(CH_3)_2$ appeared at 55 and 44 ppm, respectively (Wu et al. 2016a).

4.4 *Hydrolysis of CS*

Chitooligosaccharide (COS) is a depolymerized CS product possessing varied degree of polymerization (DP) (Mittal et al. 2022a; Benjakul et al. 2022). CS having DP $\leq$ 20 and an average MW $\leq$ 3.9 kDa is named COS (Singh et al. 2021c). Generally, hydrolysis of CS was performed via enzymatic and non-enzymatic processes (Liang et al. 2018). Physical and chemical methods are categorized as non-enzymatic hydrolysis process, whereas enzymatic hydrolysis is performed with the aid of specific and non-specific enzymes (Singh et al. 2021c).

Physical methods including microwave, lambda radiation, and ultrasonication at high intensity, which are generally unable to cleave CS efficiently (Singh et al. 2021c). Physical processes are based on the energy absorbed by CS moieties, which could break the hydrogen bond. Nevertheless, physical techniques for depolymerization of CS have not been widely used because of high energy consumption as well as production of COS with DP higher than 20.

Chemical hydrolysis of CS was performed with the aid of hydrochloric acid or mixture of acids as well as electrolytes such as hydrofluoric, phosphoric, and nitrous acids (Mourya et al. 2011). In addition, CS hydrolysis was done through oxidative reductive methods using H_2O_2 or persulfate (Mourya et al. 2011). Organic acids like formic, lactic, acetic, and trichloroacetic acids have been used for hydrolysis (Mourya et al. 2011). The acid hydrolysis mechanism was based on the formation of conjugated acid via addition of proton to oxygen atom of glycosidic linkage between two glucosamine monomers. Following that, cyclic carbonium-oxonium ion is generated through heterolysis of the exocyclic O-5 to C-1 bond, which further produces COS when reacted with water (Singh et al. 2021c). However, some reactions are difficult to eliminate the hazardous secondary compounds generated during hydrolysis (Lodhi et al. 2014).

Hydroxyl radicals were formed when ascorbic acid (AsA) reacts with hydrogen peroxide (H_2O_2), known as redox pair. Those radicals effectively hydrolyze CS (Zou et al. 2020). Commercially available CS (MW and DDA: 210 kDa and 92%, respectively) was depolymerized using AsA/H_2O_2 redox pair reaction in combination with ultrasonication (Wu et al. 2016b). Similarly, AsA and H_2O_2 at a molar ratio of 0.05/0.1 were used to hydrolyze CS (MW: ~2100 kDa and DDA: 82%) (Mittal et al. 2022a). The resulting COS had MW, DDA, and DP of 0.7 kDa, 91%, and 2–8, respectively. When H_2O_2 is used alone for CS hydrolysis to produce COS, the resulting COS showed the lower yield and lower antioxidant and antimicrobial activities as compared to COS prepared using redox pair (Mittal et al. 2022a).

Owing to environmental concerns, enzymatic hydrolysis is preferred over other methods for COS production from CS with required MW and DP. CS was hydrolyzed using specific (chitosanases) or non-specific (proteases, amylase, lipase, carbohydrase, etc.) enzymes (Lodhi et al. 2014). Ismail (2019) prepared COS from medium MW CS at 50 °C for different hydrolysis times (0.5–6 h) using thermostable chitosanase extracted from *Bacillus cereus* strain SSW1 through solid-state fermentation of shrimp by-products. Similarly, COS with DP of 2–6 was prepared when CS was hydrolyzed for various times (1–24 h) at 37 °C using chitosanase, extracted from *Purpureocillium lilacinum* CFRNT12 (Nidheesh et al. 2015). COS with DP of 2–5 was attained when CS (MW: 120 kDa and DDA: 87%) was hydrolyzed using chitosanase from *Penicillium oxalicum* M2 at 50 °C for 20–180 min. Generally, chitosanases acquired from various sources for COS preparation or chitooligomers are based on endo-degrading action or endo-hydrolysis (Cao et al. 2022). However, low affordability and availability of chitosanases limits their application on commercial scale. Therefore, non-specific enzymes have been extensively used for COS production with different ranges of MW and DP (Lodhi et al. 2014). Generally, CS hydrolysis by non-specific enzymes is carried out via hydrolyzing of glycosidic bond, resulting in the formation of smaller units (Singh et al. 2021c). CS consists of four types of glycosidic bond (GlcN–GlcNAc–, –GlcN–GlcN–, –GlcNAc–GlcN–, and –GlcNAc–GlcNAc–) (Singh et al. 2019a), which can be cleaved depending on the specificity of enzyme, pH, and DDA (Lee et al. 2008). COS with various degree of depolymerization (DDP) from squid pen CS was prepared. Different enzymes (lipase, amylase, and pepsin) at pH 5 were used for varying hydrolysis time (0–72 h) at 37 °C (for pepsin) and 50 °C (for lipase and amylase) (Singh et al. 2019a). Similarly, COS with DP of 2–6 was obtained when lipase was employed for CS hydrolysis at 55 °C and pH 4.2–5 (Lee et al. 2008). Likewise, carbohydrase (α-amylase and β-amylase at pH 4 and 5, respectively) hydrolyzed commercial CS at temperatures of 90 and 50 °C, respectively (Rokhati et al. 2013). Various enzymes of distinct specificities and origins were implemented for the hydrolysis of CS (DDA of 93%) (Roncal et al. 2007). Non-specific enzymes including lipase A, cellulase, and pepsin could hydrolyze CS, similar to chitosanase. Among the aforementioned enzymes, CS oligosaccharides (DP: 16) prepared using pepsin (1:100, w/w) at 40 °C and pH 4 had a higher yield (52%). Similarly, Renuka et al. (2021) produced COS from *Parapeneopsis stylifera* shell CS with aid of different non-specific enzymes (α-amylase, papain, pepsin, and β-amylase), in which COS prepared using pepsin provided higher yield (65%), DDA (84%), and water solubility (69%) along with lower MW (5 kDa) as compared to other treatments. Pepsin-treated CS produced reducing sugars quickly, thus the enzyme exhibited chitosanolytic activity while operating in an exo-splitting mode (Xu et al. 2020). As a result, various enzymes may produce COS with varying DP/DDP and MW. However, using non-specific enzymes to hydrolyze high viscosity CS solutions (MW > 2000 kDa), resulted in a low degree of hydrolysis. Moreover, higher amount of enzyme was required, resulting in the reduced production efficiency (Sanchez et al. 2017).

COS prepared from shrimp shells showed characteristic bands in the FTIR spectra as following: (a) the bands from 3500 to 3250 cm^{-1} representing the stretching of –OH and –NH groups, respectively along with hydrogen bonding; (b) $-CH_3$ asymmetrical and $-CH_2$ asymmetrical stretching vibrations assigned at 2881 and 2930 cm^{-1}, respectively; (c) the amide-I band with lower intensity at 1647 cm^{-1} for C=O stretching of acetyl group; d) the bands between 1590 and 1570 cm^{-1} corresponding amide-II, NH symmetric bending vibration, and C–N stretching; (e) bending and deformation of methylene and methyl groups at 1424 cm^{-1}; (f) symmetrical CH_3 deformation at 1375 and 1370 cm^{-1}; (g) C–O–C bridge asymmetric stretching (glycosidic bond) near 1150 cm^{-1}; (h) the C–O asymmetrical stretching assigned to primary and secondary alcohol at 1030 cm^{-1}; and glucosamine ring stretch appearing between 890 and 900 cm^{-1} (Mittal et al. 2022a; Singh et al. 2019a). Moreover, in ^{1}H-NMR spectra of COS, protons attributed to an acetyl group, H1 of GlcN, H3-H6 of pyranose ring, H2 of GlcN, and acetyl protons of AcOH at C6 appeared at $\delta = 1.94$, 4.92, 3.79–3.60, 3.06, and 2.09 ppm, respectively. In ^{13}C-NMR, COS showed signals for C-2, C-6, C-5, C-3, C-4, and C-1 around at 56, 60, 70, 77, 75, and 98 ppm, respectively (Mittal et al. 2022a).

5 Bioactivities and Applications of CS and Its Derivatives

CS-based derivatives and its native form have potential for applications in various sectors. Owing to different bioactivities, they have been introduced into food sector to augment the quality as well as shelf-life of different food matrices, especially fresh produces.

5.1 *Chitosan (CS)*

With antioxidant and antimicrobial activities of CS film, it can be applied as coating material for different fresh produces including fruits/vegetables, meat, and seafood because of its superior film-forming characteristics (Singh et al. 2021c). CS and epigallocatechin gallate grafted CS-based composite films possessed antioxidant activities and inhibited the growth of *Pseudomonas aeruginosa* and *Listeria monocytogenes* (Mittal et al. 2021c). Pumpkin slices were cut freshly and packed in vacuum conditions, then coated with CS and kept for 21 days at 4 °C. The slices had microbial load <6 log CFU/g throughout the storage. Higher β-carotene and better textural parameters were obtained as compared to those found in uncoated samples (Yüksel et al. 2022). The ber (*Ziziphus mauritiana*) fruits were soaked in CS solution at different concentrations (0, 0.5, and 1%) and kept at 5 and 25 °C for 28 days. The ber coated with CS (1%) and stored at 5 °C had improved quality and prolonged shelf-life as compared to ber stored at 25 °C (14 days) (Hesami et al. 2021). When chicken breast flesh was soaked in pomegranate juice, then coated with CS along with 2% *Zataria multiflora* essential oil and kept at 4 °C, the shelf-life was increased to 20 days, whereas the control could be kept for 5 days (Bazargani-Gilani et al. 2015). Another study found that the CS-gelatin-based coating (gelatin: 3 and 6%; CS: 0.5 and 1.0%; glycerol: 6%) could extend the shelf-life of beef steak and retained its color when kept for 8–14 days at 4 °C (Cardoso et al. 2016). Moreover, composite film from CS/CS-epigallocatechin gallate conjugate was able to extend the shelf-life of the Asian seabass with sensorial acceptability for 18 days of refrigerated storage when packed under a vacuum (Mittal et al. 2021b). Similarly, when red sea bream fillets were wrapped with 1% cinnamon-perilla essential oil Pickering emulsion/CS nanoparticles/anthocyanidin film, fillets could be stored for 14 days while the control had shelf-life only for 6 days (Zhao et al. 2022).

5.2 *Acylated CS*

In general, the addition of hydrophobic branches confers novel physicochemical properties, such as mucoadhesiveness, antimicrobial activity, drug delivery, etc. *O*-palmitoyl CS incorporated with iron-loaded liposome showed enhanced absorption of iron in Caco-2 cells by 1.5-fold as compared to CS (Zariwala et al. 2018). Acyl-CS was added to the cefixime-self-nano emulsion drug delivery system (CFX-SNEDDS) to prepare acyl CS-CFX-SNEDDS (Saifullah et al. 2021). Various studies revealed higher controlled release of cefixime and increased drug plasma concentration in rabbits for acyl CS-CFX-SNEDDs as compared to CFX-SNEDDs and control CFX. Acyl CS-CFX-SNEDDs with improved physicochemical properties enhanced the oral functioning of cefixime. The increased absorption of the drug is due to the incorporation of acyl CS mainly owing to its mucoadhesive property.

N-succinyl CS at 8 mg/mL was able to inhibit 90% of total bacterial growth and showed sustained bacteriostatic action against *Escherichia coli* and *Staphylococcus aureus* when incubated for 24 h due to the active functional groups (–NH_2, –COOH, and –OH groups), which might chelate various essential ions or metals required for bacterial growth. This could inhibit the uptake of trace elements, and eventually suppress bacterial proliferation (Tang et al. 2016). Moreover, *N*-succinyl CS showed a higher wound healing effect relative to CS, indicating the ability of granulation tissue formation and epithelialization of the former (Tang et al. 2016). *N,O*-acylated CS derivative with linoleic acid exhibited higher bacterial inhibition activity against *Staphylococcus aureus, Escherichia coli,* and *Helicobacter pylori* than CS due to the increased hydrophobic-hydrophobic interactions with mucin driven by linoleic acid (Piegat et al. 2020).

5.3 Carboxymethyl CS

Carboxymethyl CS (CMCS) has various applications such as moisture-retention, antimicrobial and antioxidant agents, delivery systems, and emulsion stabilizers (Jimtaisong and Saewan 2014). It is well known that CS and its derivatives contain positive electric charges and relatively large molecular weights, allowing for skin adherence and subsequent skin moisturizing (Shariatinia 2018). CMCS derivative, 6-*O*-CMCS, had higher moisture absorption capability as compared to other derivatives. The DDA, MW, and substitution sites strongly affected the moisture retention capacity (Chaiwong et al. 2020). The moisturizing capacity of mangosteen extract deodorant cream was enhanced via the incorporation of CMCS (1%, w/v) (Chaiwong et al. 2022). The aforementioned formulation possessed deodorizing properties and also showed antioxidant and antimicrobial activities. Owing to moisture retention capacity, CMCS was applied as a wound-healing agent. Cheng et al. (2019) prepared CMCS grafted with collagen peptides (CMCS-CP), which showed significant effects on scratch closure, epidermis regeneration, and deposition of collagen fiber on 7th day along with the full recovery of the epidermis on 14th day and burn wounds on 21st day of the treatment.

Recently, CMCS oligosaccharide (CMCO; DS: 1.2) was studied for its cryoprotective effect on frozen surimi from silver carp. CMCO at 0.6% was mixed with surimi and stored at −18 °C for 60 days. Freezing-induced denaturation of myofibrillar protein was lowered as compared to control (4% sucrose, 4% sorbitol) (Zhu et al. 2022). Moreover, gel strength, water-holding capacity, and whiteness of surimi added with CMCO were improved. CMCS also used for shelf-life enhancement of fresh produces. *N,O*-CMCS (1%, w/v) in combination with essential oil from oregano (1%) extended the shelf-life to 14 days of chicken meat fillets when inoculated with *Listeria monocytogenes* and packed in air at 4 ºC (Khanjari et al. 2013). While control samples and samples treated with oregano essential oil exceeded the permissible microbial count (7 log cfu/g) on 6th and 10th day of storage, respectively.

5.4 Quaternized CS

Antimicrobial activity is facilitated by electrostatic interactions between positively charged quaternized CS and negatively charged bacterial cell membrane. Moreover, the hydrophobic alkyl group on quaternized CS potentially interacted with the lipoidal cell membrane or at the inner surface of cell wall of bacteria, thus causing bacterial death (Pathak et al., 2021). Generally, the antimicrobial activity of quaternized CS is affected by degree of quaternization (DQ), MW, and location of positive charge. TMC chloride inhibited *E. coli* and *S. aureus* growth, however, sample with enhanced DQ, i.e., 64% showed higher bacterial inhibition relative to other counterparts, irrespective of the bacteria tested (Sajomsang et al. 2009). CS derivatives possessing a high density of quaternary amino groups had more potential in inhibiting bacteria. At the same DQ (50%), antimicrobial efficacy of TMCS and *N*-diethylmethyl CS nanoparticles was compared (Sadeghi et al. 2008). Higher antimicrobial efficacy of the former than the latter was found against *Staphylococcus aureus*, which was plausibly due to smaller alkyl groups causing easy penetration into the bacteria. Thus, alkyl chain length immensely affects antimicrobial activity of quaternized CS derivatives, irrespective of DQ. Moreover, hydrolyzed TMCS (degree of trimethylation: 93%) showed higher inhibition toward *Staphylococcus aureus* growth when MW increased up to 20 kDa (Sahariah et al. 2019). In general, low MW quaternized CS was easily soluble in water and was able to penetrate bacteria effectively, intercalate with genetic material, and suppress the transcription process, hence causing bacterial death. Conversely, high MW quaternized CS derivatives such as *N,N*-di-ethyl-*N*-methyl quaternized CS adsorbed at the surface of bacteria aided in the electrostatic interactions, obstruct transportation of essential nutrients, and ultimately cause alteration in cell membrane permeability or form polyelectrolyte complexes between quaternized CS and peptidoglycan layer of the bacteria (Pathak et al. 2021; Avadi et al. 2004). Moreover, composite coating based on hydrophilic quaternary ammonium CS and poly (vinyl alcohol) inhibited *Escherichia coli*, *Botrytis cinerea*, and *Staphylococcus aureus* growth up to 99% and also prolonged shelf-life of strawberries from 3 to 5 days at 25 °C (Min et al. 2020).

5.5 Chitooligosaccharide

Chitooligosaccharide (COS) has been known to have a wide range of bioactivities, which can be exploited in several fields, such as foods, health promotion, etc. COS can provide H-ion to fatty acids via abstraction of H-atom from the -NH_2 group at the glucosamine unit. COS has been known to inhibit free radicals, thus acting as a natural antioxidant (Laokuldilok et al. 2017). COS with high MW has poorer antioxidant activity than lower MW COS (Kyung et al. 2007). This is related to greater exposure of the free –NH_2 and –OH group of the latter in radical scavenging and hydrogen donation. Shrimp shell COS with lower MW (0.7 kDa) showed higher

DPPH and ABTS radical scavenging activities, reducing power and metal chelating activity (MCA) than COS with higher MW (1.2 kDa) (Mittal et al. 2022a). Moreover, Singh et al. (2019a) prepared squid pen COS using lipase. COS exhibited various antioxidant activities.

Generally, antimicrobial activity of COS is related with the change in membrane integrity due to interaction with charged COS, which further causes leakage of intracellular components and cell death (Kumirska et al. 2011). In addition, invasion of COS into the cells causes interference in the central dogma process, which includes the disruption of translation process in bacteria and interaction with microbial DNA (Singh et al. 2021c). Microbial inactivation might also be caused by chelation of vital nutrients and metal ions (No et al. 2007). COS prepared from shrimp shell CS via ascorbic acid/H_2O_2 redox pair hydrolysis inhibited different spoilage and pathogenic bacteria (Mittal et al. 2022a). Similarly, squid pen COS prepared via enzymatic hydrolysis using pancreatic lipase was able to inhibit the aforementioned bacteria along with *Salmonella enterica* (Singh et al. 2019a). Apart from increasing gel strength of sardine surimi gel, squid pen COS could extend shelf-life when gel was stored for 10 days at 4 °C (Singh et al. 2019a, b). Singh (2020a) reported that treatment of COS from squid pen in conjugation with high voltage atmospheric plasma (HV-CAP) suppressed the growth of *Pseudmonas aeruginosa* when inoculated in Asian sea bass slices. Also, COS from squid pen could impede the lipid and protein oxidation of Asian seabass slices induced by HV-CAP. COS also lowered the proliferation of pathogenic and spoilage bacteria of HV-CAP treated slices when stored for 15 days at 4 °C (Singh 2020b). In addition, COS inhibited the production of metmyoglobin formation and the discoloration of yellowfin tuna slices during refrigerated storage at 4 °C for 9 days. COS with varying concentrations (200–800 ppm) could retard the aforementioned deterioration due to its reducing capacity in any atmospheric packaging (Singh et al. 2021a, b). Therefore, it can be used to sustain the meat color containing red pigment, such as beef, tuna, etc. Furthermore, COS along with α-tocopherol (1:1) at 0.4 g/L effectively reduced lipid oxidation in shrimp oil based oil-in-water emulsion when stored at 25 °C for 30 days, thus maintaining the oxidation stability and retaining polyunsaturated fatty acids at higher extent (Rajasekaran et al. 2022).

Ability of COS to reduce body weight, triglyceride/cholesterol levels in blood, or lipid build up in the liver and adipose tissues has been documented (Benjakul et al. 2022). Cho et al. (2008) reported differentiation inhibition of adipocyte cell, lower lipid accumulation, and decrease in the adipogenic markers expression when 3T3-L1 cells were treated with COS. COS has also been shown to decrease the desire to eat by decreasing ghrelin (the hunger hormone), thus reducing fat content in the body (Kao et al. 2012). COS-fed obese rats had high-density lipoproteins (HDL) or cardioprotective lipid-containing particles at higher levels than in the control (Huang et al. 2015; Kang et al. 2012). Those proteins are able to remove tissue cholesterol and then transferred to the liver. Moreover, COS can protect β-cells from extreme high glucose levels by enhancing proliferation of pancreatic cell, which results in increased production of insulin to reduce glucose levels (Karadeniz et al. 2010). Additionally, type 2 diabetes mellitus may be treated by suppressing different digestive enzymes including α-amylase and α-glucosidase, which might result in a decrease in glucose-derived caloric intake. COS and its various polyphenol conjugates inhibit α-amylase and α-glucosidase activities, which can further cease the hydrolysis of dietary carbohydrates into monosaccharides such as glucose via digestive enzymes (Mittal et al. 2022b).

6 Conclusion

Seafood waste, especially from crustacean leftover or squid pen, can be effectively converted into chitin and CS via chemical and biological methods. Although extraction of CS using bacteria or enzymes is eco-friendly, it is not effective on a commercial scale. Thus, chemical methods have still been adopted for CS preparation. Due to the insolubility of CS in the aqueous medium, various CS derivatives including acylated CS, carboxymethyl CS, quaternary ammonium CS, and COS, etc., have been prepared extensively to enhance the bioactivities of CS. Also, CS derivatives can be applied in biomedical and food industries owing to their biocompatibility. Acylated CS is effectively used as a drug carrier, while carboxymethyl CS has potential application in cosmetics as moisturizer. Moreover, quaternized ammonium CS shows enhanced antimicrobial activity. Additionally, COS showed digestive enzyme inhibitory agents, which lead to the lowering of diabetes and obesity. Also, it can extend shelf-life of various fresh produces. CS derivatives have excellent potential to act as nutraceutical and food additives. Thus the related industries could be encouraged to commercialize chitin or CS, and their derivatives for numerous applications.

7 Future Prospects

CS and its derivatives have had significant progress in terms of preparation and improved biological activities in the recent years. Nevertheless, many unresolved issues and challenges remain unaddressed. Excessive and high concentration of chemicals used for CS extraction causes deleterious impacts on environment. Therefore, optimization of green technology on a commercial scale for the CS preparation still required to end up with promising technology. Moreover, CS derivatives have great potential in food and biomedical industries due to their various bioactivities, however their safety for human consumption or utilization needs to be elucidated. In addition, the information on the relationship between structure–activity is still limited and action mechanism needs further intensive research. It can be achieved with the aid of computational simulation and in silico modeling. Hence, solution for these challenges and encouragement from industries toward application of CS derivatives could help in efficient seafood waste utilization, leading to improved economy of allied sectors.

Acknowledgment Authors acknowledged the support from Prachayacharn grant by Prince of Songkla University (Grant No: AGR6602079N).

References

Abdou ES, Nagy KS, Elsabee MZ (2008) Extraction and characterization of chitin and chitosan from local sources. Biores Technol 99:1359–1367

Ahmadkelayeh S, Hawboldt K (2020) Extraction of lipids and astaxanthin from crustacean by-products: a review on supercritical CO_2 extraction. Trends Food Sci Technol 103:94–108

Al-Sagheer F, Muslim S (2010) Thermal and mechanical properties of chitosan/hybrid composites. J Nanomater. Article ID 490679. https://doi.org/10.1155/2010/490679

Amer OA, Ali SS, Azab M, El-Shouny WA, Sun J, Mahmoud Y.-G (2022) Exploring new marine bacterial species, Alcaligenes faecalis Alca F2018 valued for bioconversion of shrimp chitin to chitosan for concomitant biotechnological applications. Int J Biol Macromol 196:35–45

Andreica B-I, Cheng X, Marin L (2020) Quaternary ammonium salts of chitosan. A critical overview on the synthesis and properties generated by quaternization. Eur Polym J 139:110016

Avadi M, Sadeghi A, Tahzibi A, Bayati K, Pouladzadeh M, Zohuriaan-Mehr M, Rafiee-Tehrani M (2004) Diethylmethyl chitosan as an antimicrobial agent: synthesis, characterization and antibacterial effects. Eur Polymer J 40:1355–1361

Bazargani-Gilani B, Aliakbarlu J, Tajik H (2015) Effect of pomegranate juice dipping and chitosan coating enriched with *Zataria multiflora* Boiss essential oil on the shelf-life of chicken meat during refrigerated storage. Innov Food Sci Emerg Technol 29:280–287

Benediktsdóttir BE, Gaware VS, Rúnarsson ÖV, Jónsdóttir S, Jensen KJ, Másson M (2011) Synthesis of *N, N, N*-trimethyl chitosan homopolymer and highly substituted N-alkyl-N, N-dimethyl chitosan derivatives with the aid of di-tert-butyldimethylsilyl chitosan. Carbohyd Polym 86:1451–1460

Benhabiles MS, Salah R, Lounici H, Drouiche N, Goosen MFA, Mameri N (2012) Antibacterial activity of chitin, chitosan and its oligomers prepared from shrimp shell waste. Food Hydrocolloids 29:48–56

Benjakul S, Singh A, Mittal A (2022) Chitooligosaccharides: preparation and applications in food and nutraceuticals. In: Kim SK (ed) Chitooligosaccharides, pp 203–221. Springer, Cham. https://doi.org/10.1007/978-3-030-92806-3_13

Bukzem AL, Signini R, Dos Santos DM, Lião LM, Ascheri DPR (2016) Optimization of carboxymethyl chitosan synthesis using response surface methodology and desirability function. Int J Biol Macromol 85:615–624

Cahyaningtyas HAA, Suyotha W, Cheirsilp B, Prihanto AA, Yano S, Wakayama M (2022) Optimization of protease production by *Bacillus cereus* HMRSC30 for simultaneous extraction of chitin from shrimp shell with value-added recovered products. Environ Sci Pollut Res 29:22163–22178

Cao S, Gao P, Xia W, Liu S, Wang B (2022) A novel chitosanase from *Penicillium oxalicum* M2 for chitooligosaccharide production: purification, identification and characterization. Mol Biotechnol

Cardoso GP, Dutra MP, Fontes PR, Ramos ADS, Gomide LAD, Ramos EM (2016) Selection of a chitosan gelatin-based edible coating for color preservation of beef in retail display. Meat Sci 114:85–94

Casadidio C, Peregrina DV, Gigliobianco MR, Deng S, Censi R, Di Martino P (2019) Chitin and chitosans: Characteristics, eco-friendly processes, and applications in cosmetic science. Mar Drugs 17:369

Castro R, Guerrero-Legarreta I, Bórquez R (2018) Chitin extraction from Allopetrolisthes punctatus crab using lactic fermentation. Biotechnol Rep 20:e00287

Chaiwong N, Leelapornpisid P, Jantanasakulwong K, Rachtanapun P, Seesuriyachan P, Sakdatorn V, Leksawasdi N, Phimolsiripol Y (2020) Antioxidant and moisturizing properties of carboxymethyl chitosan with different molecular weights. Polymers 12:1445

Chaiwong N, Phimolsiripol Y, Leelapornpisid P, Ruksiriwanich W, Jantanasakulwong K, Rachtanapun P, Seesuriyachan P, Sommano SR, Leksawasdi N, Simirgiotis MJ (2022) Synergistics of carboxymethyl chitosan and mangosteen extract as enhancing moisturizing, antioxidant, antibacterial, and deodorizing properties in emulsion cream. Polymers 14:178

Chakravarty J, Yang C-L, Palmer J, Brigham CJ (2018) Chitin extraction from lobster shell waste using microbial culture-based methods. Appl Food Biotechnol 5:141–154

Chandumpai A, Singhpibulporn N, Faroongsarng D, Sornprasit P (2004) Preparation and physicochemical characterization of chitin and chitosan from the pens of the squid species, *Loligo lessoniana* and *Loligo formosana*. Carbohyd Polym 58:467–474

Chang KLB, Tsai G, Lee J, Fu W-R (1997) Heterogeneous N-deacetylation of chitin in alkaline solution. Carbohyd Res 303:327–332

Cheng Y, Hu Z, Zhao Y, Zou Z, Lu S, Zhang B, Li S (2019) Sponges of carboxymethyl chitosan grafted with collagen peptides for wound healing. Int J Mol Sci 20:3890

Cho E-J, Rahman A, Kim S-W, Baek Y-M, Hwang H-J, Oh J-Y, Hwang H-S, Lee S-H, Yun J-W (2008) Chitosan oligosaccharides inhibit adipogenesis in 3T3-L1 adipocytes. J Microbiol Biotechnol 18:80–87

Croisier F, Jérôme C (2013) Chitosan-based biomaterials for tissue engineering. Eur Polymer J 49:780–792

Cuong HN, Minh NC, Van Hoa N, Trung TS (2016) Preparation and characterization of high purity β-chitin from squid pens (*Loligo chenisis*). Int J Biol Macromol 93:442–447

Curti E, De Britto D, Campana-Filho SP (2003) Methylation of chitosan with iodomethane: effect of reaction conditions on chemoselectivity and degree of substitution. Macromol Biosci 3:571–576

Dayakar B, Xavier KM, Das O, Porayil L, Balange AK, Nayak BB (2021) Application of extreme halophilic archaea as biocatalyst for chitin isolation from shrimp shell waste. Carbohydr Polym Technol Appl 2:100093

De Queiroz Antonino RSCM, Lia Fook BRP, De Oliveira Lima VA, De Farias Rached RÍ, Lima EPN, Da Silva Lima RJ, Peniche Covas CA, Lia Fook MV (2017) Preparation and characterization of chitosan obtained from shells of shrimp (*Litopenaeus vannamei* Boone). Mar Drugs 15:141

Domard A, Roberts GA, Varum KMJDA (1997) Advances in chitin science, pp 410

El Knidri H, Belaabed R, Addaou A, Laajeb A, Lahsini A (2018) Extraction, chemical modification and characterization of chitin and chitosan. Int J Biol Macromol 120:1181–1189

Elieh-Ali-Komi D, Hamblin MR (2016) Chitin and chitosan: production and application of versatile biomedical nanomaterials. Int J Adv Res 4:411

Fabiano A, Beconcini D, Migone C, Piras AM, Zambito Y (2020) Quaternary ammonium chitosans: the importance of the positive fixed charge of the drug delivery systems. Int J Mol Sci 21:6617

Feng Y, Xia W (2011) Preparation, characterization and antibacterial activity of water-soluble O-fumaryl-chitosan. Carbohyd Polym 83:1169–1173

Fernández-Marín R, Hernández-Ramos F, Salaberria AM, Andrés MÁ, Labidi J, Fernandes SC (2021) Eco-friendly isolation and characterization of nanochitin from different origins by microwave irradiation: optimization using response surface methodology. Int J Biol Macromol 186:218–226

Garg S, Jana AK (2011) Characterization and evaluation of acylated starch with different acyl groups and degrees of substitution. Carbohyd Polym 83:1623–1630

Gonil P, Sajomsang W (2012) Applications of magnetic resonance spectroscopy to chitin from insect cuticles. Int J Biol Macromol 51:514–522

Hesami A, Kavoosi S, Khademi R, Sarikhani S (2021) Effect of chitosan coating and storage temperature on shelf-life and fruit quality of *Ziziphus mauritiana*. Int J Fruit Sci 21:509–518

Hu Q, Luo Y (2016) Polyphenol-chitosan conjugates: Synthesis, characterization, and applications. Carbohyd Polym 151:624–639

Hu Y, Du Y, Yang J, Tang Y, Li J, Wang X (2007) Self-aggregation and antibacterial activity of *N*-acylated chitosan. Polymer 48:3098–3106

Huang L, Chen J, Cao P, Pan H, Ding C, Xiao T, Zhang P, Guo J, Su Z (2015) Anti-obese effect of glucosamine and chitosan oligosaccharide in high-fat diet-induced obese rats. Mar Drugs 13:2732–2756

Huang Y-L, Tsai Y-H (2020) Extraction of chitosan from squid pen waste by high hydrostatic pressure: Effects on physicochemical properties and antioxidant activities of chitosan. Int J Biol Macromol 160:677–687

Hwang KT, Jung ST, Lee GD, Chinnan MS, Park YS, Park HJ (2002) Controlling molecular weight and degree of deacetylation of chitosan by response surface methodology. J Agric Food Chem 50:1876–1882

Islam S, Bhuiyan MR, Islam MN (2017) Chitin and chitosan: structure, properties and applications in biomedical engineering. J Polym Environ 25:854–866

Ismail SA (2019) Microbial valorization of shrimp byproducts via the production of thermostable chitosanase and antioxidant chitooligosaccharides. Biocatal Agric Biotechnol 20:101269

Jeon Y-J, Park P-J, Kim S-K (2001) Antimicrobial effect of chitooligosaccharides produced by bioreactor. Carbohyd Polym 44:71–76

Jimtaisong A, Saewan N (2014) Utilization of carboxymethyl chitosan in cosmetics. Int J Cosmet Sci 36:12–21

Kang N-H, Lee WK, Yi B-R, Park M-A, Lee H-R, Park S-K, Hwang K-A, Park HK, Choi K-C (2012) Modulation of lipid metabolism by mixtures of protamine and chitooligosaccharide through pancreatic lipase inhibitory activity in a rat model. Lab Anim Res 28:31

Kao C-H, Hsiang C-Y, Ho T-Y (2012) Assessment of chitosan-affected metabolic response by peroxisome proliferator-activated receptor bioluminescent imaging-guided transcriptomic analysis. PLoS ONE 7:e34969

Karadeniz F, Artan M, Kong C-S, Kim S-K (2010) Chitooligosaccharides protect pancreatic β-cells from hydrogen peroxide-induced deterioration. Carbohyd Polym 82:143–147

Khanjari A, Karabagias I, Kontominas M (2013) Combined effect of N,O-carboxymethyl chitosan and oregano essential oil to extend shelf life and control Listeria monocytogenes in raw chicken meat fillets. LWT-Food Sci Technol 53:94–99

Kucukgulmez A, Celik M, Yanar Y, Sen D, Polat H, Kadak AEJFC (2011) Physicochemical characterization of chitosan extracted from *Metapenaeus stebbingi* shells 126:1144–1148

Kumirska J, Weinhold MX, Thöming J, Stepnowski P (2011) Biomedical activity of chitin/chitosan based materials influence of physicochemical properties apart from molecular weight and degree of *N*-acetylation. Polymers 3:1875–1901

Kuroiwa T, Ichikawa S, Hiruta O, Sato S, Mukataka S (2002) Factors affecting the composition of oligosaccharides produced in chitosan hydrolysis using immobilized chitosanases. Biotechnol Prog 18:969–974

Kusuma HA, Farikhin Al-Sa'bani A, Darmokoesoemo H (2015) N,O-carboxymethyl chitosan: an innovation in new natural preservative from shrimp shell waste with a nutritional value and health orientation. Procedia Food Sci 3:35–51

Kyung W, Kim RL, Thomas (2007) Antioxidative activity of chitosans with varying molecular weights. Food Chem 101(1):308–313. https://doi.org/10.1016/j.foodchem.2006.01.038

Laokuldilok T, Potivas T, Kanha N, Surawang S, Seesuriyachan P, Wangtueai S, Phimolsiripol Y, Regenstein JM (2017) Physicochemical, antioxidant, and antimicrobial properties of chitooligosaccharides produced using three different enzyme treatments. Food Biosci 18:28–33

Lee D-X, Xia W-S, Zhang J-L (2008) Enzymatic preparation of chitooligosaccharides by commercial lipase. Food Chem 111:291–295

Lodhi G, Kim Y-S, Hwang J-W, Kim S-K, Jeon Y-J, Je J-Y, Ahn C-B, Moon S-H, Jeon B-T, Park P-J (2014) Chitooligosaccharide and its derivatives: preparation and biological applications. Biomed Res Int 2014:1–14

Luo Y, Wang Q (2014) Recent development of chitosan-based polyelectrolyte complexes with natural polysaccharides for drug delivery. Int J Biol Macromol 64:353–367

Manni L, Ghorbel-Bellaaj O, Jellouli K, Younes I, Nasri M (2010) Extraction and characterization of chitin, chitosan, and protein hydrolysates prepared from shrimp waste by treatment with crude protease from *Bacillus cereus* SV1. Appl Biochem Biotechnol 162:345–357

Min T, Zhu Z, Sun X, Yuan Z, Zha J, Wen Y (2020) Highly efficient antifogging and antibacterial food packaging film fabricated by novel quaternary ammonium chitosan composite. Food Chem 308:125682

Mittal A, Singh A, Aluko RE, Benjakul S (2021a) Pacific white shrimp (*Litopenaeus vannamei*) shell chitosan and the conjugate with epigallocatechin gallate: antioxidative and antimicrobial activities. J Food Biochem 45:e13569

Mittal A, Singh A, Benjakul S (2021b) The impact of chitosan film or chitosan/chitosan–epigallocatechin gallate conjugate composite film on the quality changes of Asian sea bass (*Lates calcarifer*) slices stored in air or under vacuum packaging. Int J Food Sci Technol 56:6025–6038

Mittal A, Singh A, Benjakul S, Prodpran T, Nilsuwan K, Huda N, de la Caba K (2021c) Composite films based on chitosan and epigallocatechin gallate grafted chitosan: characterization, antioxidant and antimicrobial activities. Food Hydrocolloids 111:106384

Mittal A, Singh A, Hong H, Benjakul S (2022a) Chitooligosaccharides from shrimp shell chitosan prepared using H_2O_2 or ascorbic acid/H_2O_2 redox pair hydrolysis: characteristics, antioxidant and antimicrobial activities. Int J Food Sci Technol. https://doi.org/10.1111/ijfs.15696

Mittal A, Singh A, Zhang B, Visessanguan W, Benjakul S (2022b) Chitooligosaccharide conjugates prepared using several phenolic compounds via ascorbic acid/H_2O_2 free radical grafting: characteristics, antioxidant, antidiabetic, and antimicrobial activities. Foods 11:920

Mourya V, Inamdar N, Choudhari YM (2011) Chitooligosaccharides: synthesis, characterization and applications. Polym Sci, Ser A 53:583–612

Muzzarelli R, Muzzarelli C (2009) Chitin and chitosan hydrogels. In: Phillips GO, Williams PA (eds) Handbook of hydrocolloids, pp 849–888. Woodhead Publishing Elsevier, Sawston, United Kingdom. https://doi.org/10.1533/9781845695873.849

Muzzarelli RA, Tanfani F (1985) The *N*-permethylation of chitosan and the preparation of *N*-trimethyl chitosan iodide. Carbohyd Polym 5:297–307

Nadira P, Mujeeb VA, Rahman PM, Muraleedharan K (2022) Effects of cashew leaf extract on physicochemical, antioxidant, and antimicrobial properties of *N, O*–carboxymethyl chitosan films. Carbohydr Polym Technol Appl 3:100191

Nanda PK, Das AK, Dandapat P, Dhar P, Bandyopadhyay S, Dib AL, Lorenzo JM, Gagaoua M (2021) Nutritional aspects, flavour profile and health benefits of crab meat based novel food products and valorisation of processing waste to wealth: a review. Trends Food Sci Technol 112:252–267

Ngasotter S, Sampath L, Xavier KM (2022) Nanochitin: an update review on advances in preparation methods and food applications. Carbohyd Polym 291:119627

Nidheesh T, Pal GK, Suresh P (2015) Chitooligomers preparation by chitosanase produced under solid state fermentation using shrimp by-products as substrate. Carbohyd Polym 121:1–9

No H, Meyers S, Prinyawiwatkul W, Xu Z (2007) Applications of chitosan for improvement of quality and shelf life of foods: a review. J Food Sci 72:R87–R100

Park BK, Kim M-M (2010) Applications of chitin and its derivatives in biological medicine. Int J Mol Sci 11:5152–5164

Park JH, Saravanakumar G, Kim K, Kwon IC (2010) Targeted delivery of low molecular drugs using chitosan and its derivatives. Adv Drug Deliv Rev 62:28–41

Pathak K, Misra SK, Sehgal A, Singh S, Bungau S, Najda A, Gruszecki R, Behl T (2021) Biomedical applications of quaternized chitosan. Polymers 13:2514

Piegat A, Żywicka A, Niemczyk A, Goszczyńska A (2020) Antibacterial activity of *N,O*-acylated chitosan derivative. Polymers 13:107

Pillai CKS, Paul W, Sharma CP (2009) Chitin and chitosan polymers: chemistry, solubility and fiber formation. Prog Polym Sci 34:641–678

Pinheiro ACDAS, Martí-Quijal FJ, Barba FJ, Tappi S, Rocculi P (2021) Innovative non-thermal technologies for recovery and valorization of value-added products from crustacean processing by-products—an opportunity for a circular economy approach. Foods 10:2030

Rajasekaran B, Singh A, Nagarajan M, Benjakul S (2022) Effect of chitooligosaccharide and α-tocopherol on physical properties and oxidative stability of shrimp oil-in-water emulsion stabilized by bovine serum albumin-chitosan complex. Food Control 137:108899

Rakshit S, Mondal S, Pal K, Jana A, Soren JP, Barman P, Mondal KC, Halder SK (2021) Extraction of chitin from *Litopenaeus vannamei* shell and its subsequent characterization: an approach of waste valorization through microbial bioprocessing. Bioprocess Biosyst Eng 44:1943–1956

Renuka V, Ravishankar CNR, Krishnamoorthy E, Zynudheen AA (2021)Comparison of depolymerization of chitosan into chitooligosaccharides from the shrimp shell waste of Parapeneopsis stylifera by non-specific enzymes. Res Square

Rinaudo M (2006) Chitin and chitosan: properties and applications. Prog Polym Sci 31:603–632

Rokhati N, Widjajanti P, Pramudono B, Susanto H (2013) Performance comparison of α- and β-amylases on chitosan hydrolysis. ISRN Chem Eng 2013:186159

Roncal T, Oviedo A, De Armentia IL, Fernández L, Villarán MC (2007) High yield production of monomer-free chitosan oligosaccharides by pepsin catalyzed hydrolysis of a high deacetylation degree chitosan. Carbohyd Res 342:2750–2756

Rúnarsson ÖV, Holappa J, Malainer C, Steinsson H, Hjálmarsdóttir M, Nevalainen T, Másson M (2010) Antibacterial activity of *N*-quaternary chitosan derivatives: synthesis, characterization and structure activity relationship (SAR) investigations. Eur Polym J 46:1251–1267

Rúnarsson ÖV, Holappa J, Nevalainen T, Hjálmarsdóttir M, Järvinen T, Loftsson T, Einarsson JM, Jónsdóttir S, Valdimarsdóttir M, Másson M (2007) Antibacterial activity of methylated chitosan and chitooligomer derivatives: synthesis and structure activity relationships. Eur Polym J 43:2660–2671

Rúnarsson ÖV, Malainer C, Holappa J, Sigurdsson ST, Másson M (2008) Tert-Butyldimethylsilyl O-protected chitosan and chitooligosaccharides: useful precursors for *N*-modifications in common organic solvents. Carbohyd Res 343:2576–2582

Sadeghi A, Dorkoosh F, Avadi M, Saadat P, Rafiee-Tehrani M, Junginger H (2008) Preparation, characterization and antibacterial activities of chitosan, *N*-trimethyl chitosan (TMC) and *N*-diethylmethyl chitosan (DEMC) nanoparticles loaded with insulin using both the ionotropic gelation and polyelectrolyte complexation methods. Int J Pharm 355:299–306

Sahariah P, Cibor D, Zielińska D, Hjálmarsdóttir MÁ, Stawski D, Másson M (2019) The effect of molecular weight on the antibacterial activity of *N, N, N*-trimethyl chitosan (TMC). Int J Mol Sci 20:1743

Saifullah S, Kanwal T, Ullah S, Kawish M, Habib SM, Ali I, Munir A, Imran M, Shah MR (2021) Design and development of lipid modified chitosan containing muco-adhesive self-emulsifying drug delivery systems for cefixime oral delivery. Chem Phys Lipid 235:105052

Sajomsang W, Gonil P, Saesoo S (2009) Synthesis and antibacterial activity of methylated *N*-(4-*N*, *N*-dimethylaminocinnamyl) chitosan chloride. Eur Polymer J 45:2319–2328

Sanchez A, Mengibar M, Rivera-Rodriguez G, Moerchbacher B, Acosta N, Heras A (2017) The effect of preparation processes on the physicochemical characteristics and antibacterial activity of chitooligosaccharides. Carbohyd Polym 157:251–257

Sannan T, Kurita K, Iwakura Y (1976) Studies on chitin, 2. Effect of deacetylation on solubility. Die Makromolekulare Chemie: Macromol Chem Phys 177:3589–3600

Shariatinia Z (2018) Carboxymethyl chitosan: Properties and biomedical applications. Int J Biol Macromol 120:1406–1419

Singh A, Benjakul S, Peng Z, Zhang B, Deng S (2021a) Effect of squid pen chitooligosaccharide and epigallocatechin gallate on discoloration and shelf-life of yellowfin tuna slices during refrigerated storage. Food Chem 351:129296

Singh A, Benjakul S, Zhang B, Deng S, Mittal A (2021b) Effect of squid pen chitooligosaccharide on discoloration and shelf-life of yellowfin tuna slices packed under different modified atmospheric packaging during refrigerated storage. Food Control 125:108013

Singh A, Benjakul S, Prodpran T (2019a) Chitooligosaccharides from squid pen prepared using different enzymes: characteristics and the effect on quality of surimi gel during refrigerated storage. Food Prod, Process Nutr 1:1–10

Singh A, Benjakul S, Prodpran T (2019b) Effect of chitooligosaccharide from squid pen on gel properties of sardine surimi gel and its stability during refrigerated storage. Int J Food Sci Technol 54:2831–2838

Singh A, Benjakul S, Prodpran T (2019c) Ultrasound-assisted extraction of chitosan from squid pen: molecular characterization and fat binding capacity. J Food Sci 84:224–234

Singh A, Benjakul S, Olatunde OO, Xun C, Wu P (2020a) The combined effect of squid pen chitooligosaccharide and high voltage cold atmospheric plasma on the quality of Asian sea bass slices inoculated with *Pseudomonas aeruginosa*. Turk J Fish Aquat Sci 21:41–50

Singh A, Benjakul S, Xun C, Wu P (2020b) The combined effect of squid pen chitooligosaccharides and high voltage cold atmospheric plasma on the shelf-life extension of Asian sea bass slices stored at 4 °C. Innov Food Sci Emerg Technol 64:102339

Singh A, Mittal A, Benjakul S (2021c)Chitosan, chitooligosaccharides and their polyphenol conjugates: preparation, bioactivities, functionalities and applications in food systems. Food Rev Intl

Singh A, Mittal A, Benjakul S (2020c) Full utilization of squid meat and its processing by-products: revisit. Food Rev Intl 38:455–479

Srinivasan H, Kanayairam V, Ravichandran R (2018) Chitin and chitosan preparation from shrimp shells *Penaeus monodon* and its human ovarian cancer cell line, PA-1. Int J Biol Macromol 107:662–667

Struszczyk MH (2006) Global requirements for medical applications of chitin and its derivatives. Polish Chitin Society, Monograph XI; Polish Chitin Society: Łódz, Poland, pp 95–102

Sun T, Yao Q, Zhou D, Mao F (2008) Antioxidant activity of *N*-carboxymethyl chitosan oligosaccharides. Bioorg Med Chem Lett 18:5774–5776

Tang F, Lv L, Lu F, Rong B, Li Z, Lu B, Yu K, Liu J, Dai F, Wu D (2016) Preparation and characterization of N-chitosan as a wound healing accelerator. Int J Biol Macromol 93:1295–1303

Teli M, Sheikh J (2012) Extraction of chitosan from shrimp shells waste and application in antibacterial finishing of bamboo rayon. Int J Biol Macromol 50:1195–1200

Tokatlı K, Demirdöven A (2018) Optimization of chitin and chitosan production from shrimp wastes and characterization. J Food Process Preserv 42:e13494

Tsaih ML, Chen RH (2003) The effect of reaction time and temperature during heterogenous alkali deacetylation on degree of deacetylation and molecular weight of resulting chitosan. J Appl Polym Sci 88:2917–2923

Upadhyaya L, Singh J, Agarwal V, Tewari RP (2013) Biomedical applications of carboxymethyl chitosans. Carbohyd Polym 91:452–466

Valdez-Peña AU, Espinoza-Perez JD, Sandoval-Fabian GC, Balagurusamy N, Hernandez-Rivera A, De-La-Garza-Rodriguez IM, Contreras-Esquivel JC (2010) Screening of industrial enzymes for deproteinization of shrimp head for chitin recovery. Food Sci Biotechnol 19:553–557

Verheul RJ, Amidi M, Van Der Wal S, Van Riet E, Jiskoot W, Hennink WE (2008) Synthesis, characterization and in vitro biological properties of O-methyl free N,N,N-trimethylated chitosan. Biomaterials 29:3642–3649

Wang W, Meng Q, Li Q, Liu J, Zhou M, Jin Z, Zhao K (2020a) Chitosan derivatives and their application in biomedicine. Int J Mol Sci 21:487

Wang W, Xue C, Mao X (2020b) Chitosan: structural modification, biological activity and application. Int Joynrla Biol Macromol 164:4532–4546

Weska R, Moura JMD, Batista LDM, Rizzi J, Pinto LDA (2007) Optimization of deacetylation in the production of chitosan from shrimp wastes: Use of response surface methodology. J Food Eng 80:749 753

Wu M, Long Z, Xiao H, Dong C (2016a) Recent research progress on preparation and application of *N, N, N*-trimethyl chitosan. Carbohyd Res 434:27–32

Wu TT, Wu CH, Xiang YC, Huang JQ, Luan LL, Chen SG, Hu YQ (2016b) Kinetics and mechanism of degradation of chitosan by combining sonolysis with H_2O_2/ascorbic acid. RSC Adv 6:76280–76287
Xu C, Guan S, Xu J, Gong W, Liu T, Ma X, Sun C (2021) Preparation, characterization and antioxidant activity of protocatechuic acid grafted carboxymethyl chitosan and its hydrogel. Carbohyd Polym 252:117210
Xu T, Xin M, Li M, Huang H, Zhou S (2010) Synthesis, characteristic and antibacterial activity of N, N, N-trimethyl chitosan and its carboxymethyl derivatives. Carbohyd Polym 81:931–936
Xu W, Mohan A, Pitts NL, Udenigwe C, Mason B (2020) Bile acid-binding capacity of lobster shell-derived chitin, chitosan and chitooligosaccharides. Food Biosci 33:100476
Yadav M, Goswami P, Paritosh K, Kumar M, Pareek N, Vivekanand V (2019) Seafood waste: a source for preparation of commercially employable chitin/chitosan materials. Bioresour Bioprocess 6:1–20
Younes I, Rinaudo M (2015) Chitin and chitosan preparation from marine sources. Structure, properties and applications. Mar Drugs 13:1133–1174
Yüksel Ç, Atalay D, Erge HS (2022) The effects of chitosan coating and vacuum packaging on quality of fresh-cut pumpkin slices during storage. J Food Process Preserv 46:e16365
Zargar V, Asghari M, Dashti A (2015) A review on chitin and chitosan polymers: structure, chemistry, solubility, derivatives, and applications. ChemBioEng Rev 2:204–226
Zariwala MG, Bendre H, Markiv A, Farnaud S, Renshaw D, Taylor KM, Somavarapu S (2018) Hydrophobically modified chitosan nanoliposomes for intestinal drug delivery. Int J Nanomed 13:5837
Zhang J, Xia W, Liu P, Cheng Q, Tahi T, Gu W, Li B (2010) Chitosan modification and pharmaceutical/biomedical applications. Mar Drugs 8:1962–1987
Zhang Z, Jin F, Wu Z, Jin J, Li F, Wang Y, Wang Z, Tang S, Wu C, Wang Y (2017) O-acylation of chitosan nanofibers by short-chain and long-chain fatty acids. Carbohyd Polym 177:203–209
Zhao R, Guan W, Zheng P, Tian F, Zhang Z, Sun Z, Cai L (2022) Development of edible composite film based on chitosan nanoparticles and their application in packaging of fresh red sea bream fillets. Food Control 132:108545
Zhou W, He Y, Liu F, Liao L, Huang X, Li R, Zou Y, Zhou L, Zou L, Liu Y (2021) Carboxymethyl chitosan-pullulan edible films enriched with galangal essential oil: characterization and application in mango preservation. Carbohyd Polym 256:117579
Zhu X, Zhu M, He D, Li X, Shi L, Wang L, Xu J, Zheng Y, Yin T (2022) Cryoprotective roles of carboxymethyl chitosan during the frozen storage of surimi: protein structures, gel behaviors and edible qualities. Foods 11:356
Zou MY, Nie SP, Yin JY, Xie MY (2020) Ascorbic acid induced degradation of polysaccharide from natural products: a review. Int J Biol Macromol 151:483–491

Biomedical Applications of Chitin, Chitosan, Their Derivatives, and Processing By-Products from Fish Waste

Subhasis Roy, Shubhamitra Chaudhuri, Prasenjit Mukherjee, and Samit Kumar Nandi

Abstract The use of natural resources has increased by 254% from 1970 to 2017 globally. Similarly, the consumption of fish and fish products has also gained popularity over the last few years. As per available records and data, it has been projected that about 75% of the fish biomass is discarded as biological waste which can be explored, recycled, and processed for further validation. Fish waste as a rich source of enzymes, bioactive peptides, polymers, and many other bioactive compounds has been studied by several researchers in the last decades. In this chapter, we would focus only on chitin, chitosan, its derivatives, and processing by-products in advanced applications dimensions. They have a huge potential for use in the field of biomedical engineering, and in this article, we will concentrate on their usage in the areas of growth factor delivery, cancer diagnostics, cartilage and tendon repair, dentistry, drug administration, gene delivery, and bone tissue creation. Besides, they have also got used in food and cosmetic industries, nutraceuticals, and bioremediation.

Keywords Chitin · Chitosan · Fish waste · Biomedical applications

1 Introduction

Recent significant increases in several diseases concluding cardiovascular system, the change in lifestyle and food habits are the major advised prescription. As a part of this, the consumption of fish and fish products have been increased dramatically (Coppola et al. 2021; FAO 2018). As a result of it, the fish byproduct and biomass have also increased which environmentally, socially, and economically can best be utilized

S. Roy · S. Chaudhuri · P. Mukherjee
Department of Veterinary Clinical Complex, West Bengal University of Animal & Fishery Sciences, Kolkata, India

S. K. Nandi (✉)
Department of Veterinary Surgery & Radiology, West Bengal University of Animal & Fishery Sciences, Kolkata, India
e-mail: samitnandi1967@gmail.com

S. Maqsood et al. (eds.), *Fish Waste to Valuable Products*, Sustainable Materials and Technology, https://doi.org/10.1007/978-981-99-8593-7_12

with effective fish waste management strategies (Ferraro et al. 2010; Mo et al. 2018). Among the by-products of the fish industry, different important bioactive and valuable components have been reviewed for use in the biomedical and pharmaceutical fields like enzymes, chitin, polyunsaturated fatty acids, minerals, collagen, peptides, etc. (Shahidi et al. 2019; Shavandi et al. 2019).

In this chapter, we would exclusively discuss only on chitin, chitosan, and their derivatives in advanced application dimensions. Because they are natural biopolymers, chitin and chitosan are non-toxic, biocompatible, and biodegradable. Chitosan is a chitin derivative produced by deacetylating chitin through enzymatic hydrolysis. They can be used to make a variety of forms, such as membranes, gels, nanoparticles, microparticles, nanofibers, beads, scaffolds, and sponges.

Chitin, which is the second most prevalent polymer after cellulose and is a white, nonelastic, rigid, nitrogenous polymer of natural origin, was first discovered in 1884. It is a part of the exoskeleton of arthropods and a component of yeast and fungi's cell walls. Chitosan, which is produced by enzymatic hydrolysis or deacetylation, is the primary chitin derivative. Although it cannot be dissolved in water or organic solvents, it can be dissolved in a variety of acids, such as acetic, hydrochloric, nitric, and perchloric (Rinaudo 2006; Sankararamakrishnan and Sanghi 2006; Kurita 2006). Chitosan has an electrically positive charge and is hence able to adhere to negatively charged surfaces. As the cell surface is anionic, chitosan is believed to adhere, due to electrostatic interactions (Dash et al. 2011). Three different group types in chitosan make it possible to produce copolymerized components specifically for use in tissue engineering. Different modified chitosan can be created with various targeted effects depending on the level of deacetylation (DDA) and molecular weight (MW). Researchers have created a variety of chitin and chitosan composites that have shown to be future attractive candidates in the biomedical fields.

2 Chitosan Processing

The following steps (Fig. 1) are maintained for yielding chitosan (Rinaudo 2006; Roberts 1992).

Based on the source the extraction process varies. The traditional extraction process involves steps like demineralization, deproteination, bleaching, and deacetylation.

Fig. 1 Processing of Chitosan

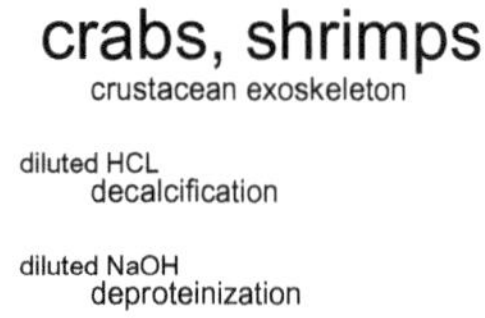

3 Chitin/Chitosan from Aquatic Ecosystem

Chitin may be isolated from shrimps, crabs, lobsters, and crayfish in the form of granules, sheets, and powders (Khor and Lim 2003; Khor 2014; Ehrlich et al. xxxx). Chitin has also been isolated from sponges and found to be effective in biomedical engineering (Ehrlich et al. 2007; Brunner et al. 2009a, 2009b). Chitin is been documented to be obtained from protozoa or alga, foraminifera (Jeuniaux and Voss-Foucart 1991). Diatoms, hydroids, coelenterates, brachiopods, polychaetes, pogonophorans, mollusks, and crustaceans have also been documented to yield chitin (Kurita 2006; Brunner et al. 2009a). The shell of all crabs and shrimps has been enlisted to provide the α-chitin (Stepnowski et al. 2004).

4 Different Composite of Chitin and Chitosan

Base	Composite and type	Reference
Chitin	The sponge-like 3D chitin	Kim et al. 2008; Rinaudo 2008; Abe et al. 2004)
Chitin	Electrospun water-soluble carboxymethyl	Menon et al. xxxx)
Chitin	Fiber	Mikhailov et al. 2001)
Chitin	Electrospun transparent nano mats	Shamshina et al. 2018)
Chitosan	Heparin-like composite fibrous membranes	Li et al. 2018)
Chitosan	Nanofibers with carbon nanotubes, Fe_3O_4, and TiO_2	Bahmani et al. 2020)
Chitosan	Cellulose chitosan multifilament fiber	Zhu et al. 2019)
Chitosan	Films and membranes	Hu et al. 2016; Youssef et al. 2016)
Chitin	Cellulose hydrogels	Shamshina et al. 2014)
Chitin	MCC aerogels	Shen et al. 2016)
Chitosan	CMC and Glycerol 2 phosphate hydrogels	Azadi et al. 2018)
Chitosan	Hydroxypropylmethylcellulose and glycerol thermosensitive hydrogel	Wang et al. 2016)
Chitosan	Cellulose hydrogel	Kabir et al. 2018)

Biomedical applications in the area of.

1. *Bone Tissue Engineering*

For effective bone tissue engineering, any implant/scaffold must be biocompatible, biodegradable, and bioactive. Chitin and chitosan along with these properties also possess flexibility and porosity, but they lack mechanical strength and are unstable too (Mathur and Narang 1990). Hence different composites of chitin and chitosan have evolved to mitigate the disadvantages and have been used in bone tissue engineering. Enhancement of mechanical strength with different composites like alginate for resisting compressive force (Li et al. 2005) and poly(lactic-co-glycolic)acid(PLAGA) (Jiang et al. 2006) composite was investigated with promising results. Further in one study the efficacy of heparin-modified chitosan-PLGA scaffold and recombinant human bone morphogenic protein (rhBMP-2)-heparin chitosan-PLGA scaffold was investigated for osteointegration, osteoblastic proliferation, and differentiation, and rapid bone formation with a promising result (Jiang et al. 2010). Additionally, different composites have also been examined in bone tissue engineering like arginine–glycine–aspartic acid conjugated UV-cross-linked chitosan (Tsai et al. 2012), poly-caprolactone-chitosan scaffolds (Wu et al. 2010; Thuaksuban et al. 2011), growth factors releasing porous poly-caprolactone-chitosan scaffolds (Im et al. 2003), bone morphogenic protein-2 and 7 incorporated chitosan scaffold (Yilgor et al. 2009), chitosan-collagen scaffold incorporated with rhBMP-2-PLGA (Shi et al. 2009), chitosan scaffold functionalized with heparin (Gümüşderelioğlu and Aday 2011). In one study poly L Lactic acid-chitosan scaffold was soaked in Ca^{2+} and PO_4^{3-} solutions for increasing the osteoconductive properties of this scaffold and the same was investigated to facilitate bone regeneration (Prabaharan et al. 2007). All of the above studies yield better results in terms of maintaining scaffold porosity, increasing osteoconductivity, improving mechanical attributes, reducing the rate of degradation of materials, and finally augmenting bone formation.

Apart from this, nano-hydroxyapatite (nHAp), being osteoinductive and conductive, also have been investigated by incorporating chitin (Chang et al. 2013). Similarly, the incorporation of nHAp with α-chitin and β-chitin, chitosan hydrogels for bone tissue engineering is investigated (Kumar et al. 2011; Sudheesh Kumar et al. 2011; Madhumathi et al. 2009). Studies on HAp-incorporated chitin and chitosan, different polymeric and ceramic-like silk fibroin, carboxymethyl cellulose, gelatin, and carbon nanotube-incorporated chitin and chitosan have been reviewed and found to have improved mechanical strength with better and uniform mineralization (Fig. 2) (Li et al. 2006; Frohbergh et al. 2012). Osteoblast obtained from mesenchymal stem cells was seeded into a HAp-chitin scaffold and was studied for osteogenic properties (Ge et al. 2004). Similarly, goat mesenchymal stem cells were also studied in chitin composite for new bone tissue regeneration with good results (Liu et al. 2010). Reports on bioactive glass (BGC)-chitin/chitosan composite for use in load-bearing areas including nanocomposite, nano-silvers, solvent cast bioactive glass, the hybrid scaffold of Bioactive glass, surface modified BGC is recorded for better mechanical strength, biocompatibility in the bone and dentistry (EBSCOhost xxxx; Hench et al. 1971; Jones 2013). Hybrid Chitin/chitosan-silica/titania and zirconia composite for bone formation and regeneration is also demonstrated by different researchers (Toskas et al. 2013; Jongwattanapisan et al. 2011).

2. *Cartilage Repair*

Most of the time cartilage damage demands surgical interventions especially replacement owing to the fact of degenerative changes that may occur due to disease, trauma, or genetic irregularities. Conventional surgeries do not guarantee all the time

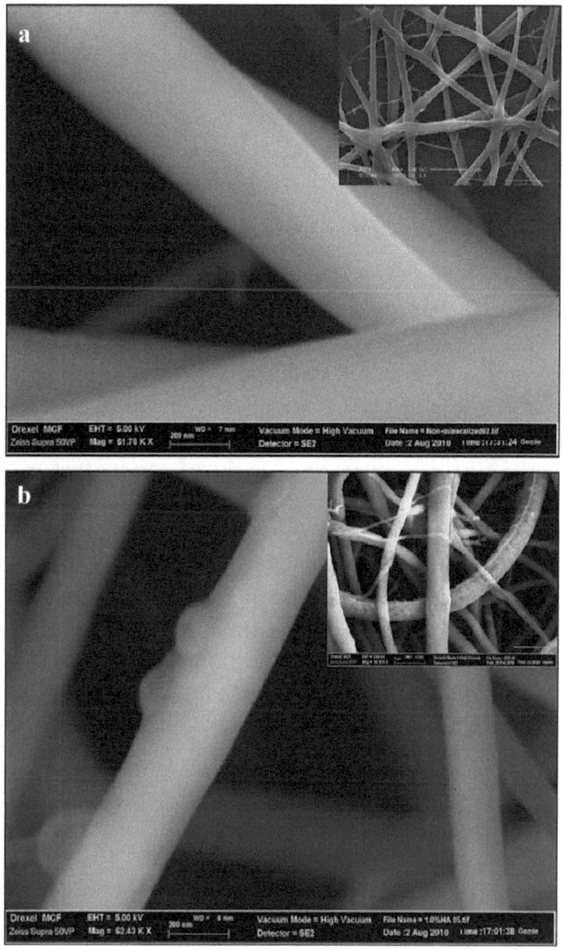

Fig. 2 Examination of electrospun scaffolds morphology by atomic force and scanning electron microscopy. SEM images of chitosan nanofibers with 7% HA content and 0.1% genipin cross-linked CTS-GP **a** and 0.1% genipin cross-linked 1.0% GP **b** show distinctive nanofiber shapes at low magnification. (**Reproduced from Biomaterials Journal, Frohbergh, et al., 2012, with permission**)

a scarless operative site and hence suffer from loss of mobility and reduced functionality. Here lies the need for a cell scaffold composite that can stimulate, regenerate, and remain active in the scaffold environment. Chitosan has been found to have chondrogenesis properties. Porcine chondrocyte-seeded chitosan scaffold was studied with a positive outcome for chondrogenesis (Use and of Chitosan as a Cell Scaffold Material for Cartilage Tissue Engineering xxxx; Griffon et al. 2006). Chitosan microspheres incorporated with transforming growth factor β showed improved chondrocyte growth (Kim et al. 2003; Lee et al. 2004). Chitosan-collagen-genipin scaffold demonstrated better viability of chondrocytes in rabbits (Yan et al. 2010). Utilizing the hemostatic characteristic of chitosan composite in cartilage regeneration resulted in the development of hyaline cartilage and the variation of pluripotent cells into chondrocytes (Hoemann et al. 2007; Hao et al. 2010). Chitin and chitosan hydrogels showed significant achievement in terms of regeneration of damaged cartilage and wound healing (Xi et al. 1999; Jin et al. 2009; Tan et al. 2009). Different hydrogel composite of chitosan is demonstrated to provide better chondrocyte survival (Park et al. 2013). The chitosan-based fibrous scaffold has been studied with better deposition of the extracellular matrix (Subramanian et al. 2004). The hybrid scaffold of chitosan was evaluated for the mechanical integrity of cartilage formation (Neves et al. 2011).

3. *Tendon and Ligament Repair*

As with cartilage, tendon and ligament repair also have drawbacks of scar tissue formation. For any technology to be used in the repair process must meet high tensile strength along with regenerative capacity. A polyelectrolyte complex of alginate-chitosan gave better adhesion attributes (Majima et al. 2005). Similar findings along with enhanced strength were found in the hyaluronic acid chitosan complex (Funakoshi et al. 2005a). Other studies using hyaluronic acid chitosan complexes with osteoblast seeds also produced results that were more favorable in terms of increased mechanical strength (Fig. 3) (Funakoshi et al. 2005b; Irie et al. 2011). As chitin is degraded first, hence alone chitin fabric sometimes gave poor regeneration of non-healed ligament with evidence of scar tissue, but if a composite is formed with PCL the outcome was found to be encouraging (Funakoshi et al. 2006; Sato et al. 2000). The role of chitosan in treating ligament injury lies in favoring the deposition of collagen type I (Tendon healing in vivo and in vitro: chitosan improves range of motion after flexor tendon repair 2013). Similar studies were also carried out by other scientists indicating the outcome at per (Shao et al. 2010a, 2010b).

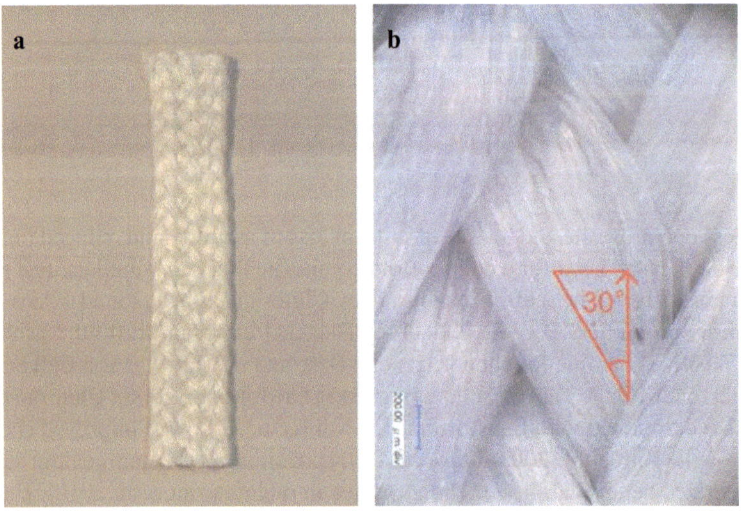

Fig. 3 **a** A braiding machine was used to produce a three-dimensional scaffold from 13 braids that measured 40 mm long, 7 mm broad, and 0.7 mm thick. **b** The braided fibers and longitudinal axis formed a 30-degree angle. (**Reproduced from J. Biomed. Mater. Res. A Irie et al., 2010 with permission)**

4. *Skin Regeneration*

As already discussed the hemostatic properties of chitin and chitosan, may be well employed for the healing of varied types of dermatological conditions (Taravel and Domard 1995, 1996; Ma et al. 2001). Cotton fiber type chitosan with rapid healing and presence of polymorphonuclear cells (Ueno et al. 1999), growth factor incorporated chitosan with better healing (Mizuno et al. 2003), chitosan-alginate polyelectrolyte complex with better wound stability (Yan et al. 2000; Wang et al. 2002), chitosan-collagen scaffold providing more fibroblast infiltration (Ma et al. 2003), chitosan acetate bandaging with improved antibacterial and anti-inflammatory property (Burkatovskaya et al. 2008), growth factor incorporated collagen-chitosan complex in burn wound with enhanced neo-angiogenesis (Guo et al. 2011), atelocollagen-chitosan in excisional wound healing198, chitosan hydrogels for skin tissue healing (Kumar et al. 2013) have been investigated and documented for ease of ready to hand reference.

5. *Liver Regeneration*

As the liver's extracellular matrix contains glycosaminoglycans, hence chitin and chitosan are used in different hepatic engineering and regeneration because these glycosaminoglycans are also a part of chitin and chitosan (Li et al. 2003a) and have been recorded to modulate the actions of vascular endothelial cells (Chupa et al. 2000). Different composites along with their applications have been listed below.

Composites	Application	References
Chitosan-collagen	Increased hepatocyte compatibility	Wang et al. 2003)
Chitosan-collagen-heparin	Artificial liver with more blood compatibility	Wang et al. 2005)
Chitosan-galactose	Increased hepatocyte attachment	Park et al. 2003)
Chitosan-fructose	Increased liver cell metabolic activities	Li et al. 2003a; Li et al. 2003b)
Chitosan-fibroin	Increased hepatocyte attachment	She et al. 2009)
Chitosan-galactosylated hydroxyapatite	More albumin secretion and elimination of urea	Fan et al. 2010)
Chitosan-poly ether-ether ketone	The proliferation of progenitor cells	Piscioneri et al. 2011)

6. *Nerve Regeneration*

Peripheral nerve regeneration and repair frequently involve chitin and chitosan. Schwann cell, one of the key cells in nerve tissue engineering showed better migration, adhesion, and proliferation with chitosan fiber (Yuan et al. 2004). The effectiveness of the chitosan complex in additional research on cell adhesion, proliferation, and migration has been evaluated (Cao et al. 2005; Mingyu et al. 2004; Cheng et al. 2003; Chiono et al. 2008). The viability and proliferation of different neural cells were studied in chitosan-collagen composite (Yang et al. 2010) and laminin-coated chitosan membrane (Guo et al. 2012).

5 Application in the Field of the Delivery System

Any kind of regeneration, repair, and healing is always associated with growth factors and can be influenced by different drugs and gene therapy. In these domains, chitin and chitosan are crucial. Here we will briefly review the effect of these in tabular form.

Different drug delivery implant has been prepared like chitosan scaffold for 5-Fluorouracil (Denkbaş et al. 2000), chitosan-pectin cross-linked scaffold with pentoxifylline incorporated films (Lin and Yeh 2010a), pentoxifylline loaded chitosan-alginate scaffold (Lin and Yeh 2010b), ketoprofen loaded Chitosan-carboxymethyl β–CD (Prabaharan and Jayakumar 2009), ampicillin-loaded alginate microspheres in chitosan-nano-hydroxyapatite scaffolds (Shi et al. 2007), dexamethasone impregnated Chitosan scaffold (Duarte et al. 2009), tetracycline loaded chitosan-hydroxyapatite scaffolds (Teng et al. 2009), amikacin and vancomycin loaded chitosan sponge (Noel et al. 2010), collagen-chitosan for transdermal drug delivery (Thacharodi and Panduranga Rao 1996), vitamin B2 as a model drug in glucose-cross-linked N-alkylated chitosan membranes (Li et al. 2002), nifedipine incorporated chitosan membrane (Thacharodi and Rao 1993), collagen-chitosan as a carrier of propranolol (Thacharodi and Panduranga Rao 1995) and the results demonstrated better release, stability with lowered enzymatic degradation. Carboxymethyl chitin has also been used for drug delivery (Jayakumar et al. 2010).

Besides, the delivery of growth factors may be attributed to chitin and chitosan. The incorporation of bFGF into chitosan was in periodontal regeneration (Tığlı et al. 2009). bFGF was also incorporated in chitosan-alginate scaffold and found to be a potential carrier system for tissue repair and regeneration (Ho et al. 2009). Chitosan microsphere scaffolds have been used for the evaluation of carriers of ALP and BMP-2 (Reves et al. 2009). CS composite has also been used for the delivery of VEGF for hard tissue regeneration (Riva et al. 2010). The administration of rhBMP-2 for bone regeneration has been successfully shown using the chitosan-collagen scaffold (Shi et al. 2009).

Chitosan-based DNA/siRNA complexes sometimes have the problem of early and immature release of nucleic acids (Buyens et al. 2012) and less penetration to cells. Hence, complexes with improved performances have been investigated with promising results. Enhanced cell penetration has been achieved by surface modifications of chitosan by ligands like transferrin (Mao et al. 2001; Chan et al. 2007), folate (Kim et al. 2006), mannose (Gao et al. 2003), and galactose (Gao et al. 2003; Park et al. 2001) have yielded good results. For increased stability, chitosan has been modified by Quaternization (Kean et al. 2005; Thanou et al. 2002), glycosylation

(Thanou et al. 2002; Strand et al. 2008), and hydrophobic modification (Lee et al. 2012). Fusogenic peptides and pH-sensitive neutral lipids have been added to several research projects to improve DNA/siRNA. Plans to mute RANK signaling utilizing chitosan hydrogel as a siRNA reservoir and vector have also been examined (Kang et al. 2017).

6 Application in Wound Healing

Chitin and different composite membranes have been tested from the perspective of wound healing and found to possess excellent biocompatibility, minimal or less tissue reaction, and regenerative and antibacterial properties (Singh et al. 2008; Azad et al. 2004). Similar results were also obtained with chitosan membrane (Santos et al. 2013), chitosan composites (Pang et al. 2008), silver sulfadiazine incorporated chitosan alone (Mi et al. 2003), and chitosan-alginate composite (Meng et al. 2010), argon plasma treated chitosan membranes in fibroplasia (Zhu et al. 2005). Different wound dressing materials based on chitosan hydrogel composite gave excellent results (Queiroz et al. 2003). Porous chitosan membrane proved to possess good hemostatic as well as excellent epithelialization properties (Mi et al. 2001). The different composite sheets of chitosan (Wang et al. 2012), diverse sponges of it (Lee et al. 2000), and composite bandages in non-cytotoxic in nature (Sudheesh Kumar et al. 2012). According to all of the aforementioned studies, chitin and chitosan, as well as their composites, have excellent bioacceptability, good regenerative properties, epithelialization, hemostasis, adhesion, stability, antibacterial and anti-inflammatory properties, and, most importantly, have improved wound healing.

7 Application in Cancer Diagnosis

Chitin and chitosan are frequently employed in the diagnosis and treatment of cancer among other purported biological purposes. In cancer imaging heavy metal-free luminescent zinc sulfide (ZnS) is considered bio-friendly to healthy and cancer cells over the use of heavy metal-containing nanocrystals such as cadmium sulfide, cadmium selenide, and zinc selenide (Derfus et al. 2004). However, in an in-vitro study chitosan encapsulated ZnS nanoparticle is used in cancer imaging (Higuchi et al. 2008) and the study shows the yield of mannosylated ZnS of size 120 nm after functionalizing with D-Mannose and its low cytotoxicity toward healthy and cancer cells. Moreover, its specificity of the binding property toward mannose-bearing KB cells under fluorescence microscopy encourages receptor-mediated imaging with nanoparticles as well as in cancer therapy (Higuchi et al. 2008).

The report of various researchers from across the world clearly demonstrates the importance of chitin and chitosan as therapeutic agent in cancer biology through the direct death of malignant cell lines and as a vehicle for anticancer medication delivery systems. Chitosan's anti-cancerous properties are mostly attributed to the maturation and infiltration of cytolytic T-lymphocytes via enhanced interleukin-1 and interleukin-2 production (Lin et al. 2007). Some researchers have shown anticancer efficacy through the direct destruction of tumor cells by triggering apoptosis (Gibot et al. 2015).

Chitin inhibits an elevated serum level of proinflammatory mediators chitinase-3 like protein-1 (CHI3L1) in breast cancer, colorectal cancer, ovarian cancer, leukemia, lymphoma, metastatic prostate cancer, lung cancer, and glioblastoma by blocking the synthesis of vascular endothelial growth factor C. (VEGF-C). Additionally, other forms of chitin, such as chitin-glucan-aldehyde-quercetin conjugation and silver-embedded chitin nanocomposites, exhibit biocompatible cytotoxicity in human breast cancer (MCF-7) cells (Solairaj et al. 2017) and in macrophage cancer cell line (J774) respectively (Singh et al. 2018). Further, in the drug delivery system doxorubicin coated with chitin-based poly L lactic acid composite nano gel and ellagic acid in chitin nanoparticle shows cytotoxic effect in the liver (Arunraj et al. 2014) and breast (Pirzadeh-Naeeni et al. 2020) carcinoma respectively. Similarly, chitin, chitosan, and chitosan oligosaccharides are also used in cancer therapy. A 2015 study by Gibot et al. found that the apoptotic impact increased in RPMI7951 cells, cell proliferation decreased in SKMEL38 cells, and cell adhesion quality decreased in A 375 cells (Gibot et al. 2015). Moreover, chitosan oligosaccharide in colorectal cancer abolished tumor progression (Mattaveewong et al. 2016), and modulated cell autophagy in the A549 lung cancer cell line in oral squamous cell carcinoma producing the cytotoxic effect by arresting cell cycle and apoptosis without any adverse effect on adjacent noncancerous cells (Wimardhani et al. 2014). This chitosan also shows anti-cancerouse effects in liver carcinoma in mice (Jiang et al. 2015), and lung cancer A549 cell line (Gao et al. 2020). In lung cancer, the A549 cell line chitosan selenate shows its potent activity in the apoptosis of cancer cells. In addition, many chitosan nanoparticles are employed in anti-cancerous drug delivery systems, including chitosan oligosaccharide conjugated to 5-fluorouracil and vanillin, chitosan oligosaccharide conjugated to indomethacin, and chitosan graphene oxide attached to thioguanine (Hasanzade and Raissi 2019; Lee et al. 2017; Li et al. 2016).

8 Application in Dentistry

Chitosan is been widely studied in dental engineering for its biocompatibility, non-toxicity, biodegradability, and antimicrobial activities. Apart from that, it has the property to form film and gel which is very helpful for application in the dental field (Fiorillo 2019; Husain et al. 2017). Chitosan has widely been used for caries prevention as well as in conservative dentistry (Ortiz and Boyce 2008). One remarkable study report denotes better periodontal health and condition with chitosan brushing (Zeza et al. 2017). Chitosan has also been proven to reduce the bleeding time after the extraction of teeth (Pippi et al. 2017). It has been investigated for anti-inflammatory and pain relief properties with significant outcomes in dental affection and engineering (Lope-Lopez et al. 2015). Chitosan also proved to combat demineralization in dental affection cases (Uysal et al. 2011). The form of toothpaste of chitosan has been used to reduce the oral bacterial count (Mohire and Yadav 2010).

9 Other Applications of Chitin and Chitosan Apart from Biomedical Spheres

Chitin and chitosan are employed in a variety of industries, including food and cosmetics, nutraceuticals, water treatment, and bioremediation, in addition to biomedicine. The usage of chitin and chitosan in bioremediation is significant due to the conversion of contaminants and interaction with heavy metals and removal in an aqueous solution (Hayes et al. 2008; Barriada et al. 2007). Papayafish scale collagen has been found to have iron-clearing properties from groundwater (Irawan et al. 2018). Chitin and chitosan are utilized in the food industry as food additives and have also been included in food packaging materials because of their antioxidant and antibacterial properties. Silver thiosulfate and actinides are two industrial pollutants that have been treated using chitin-based biomaterials (Kosyakov et al. 2002). Chitin has also been used in the paper industry for increasing strength.

10 Conclusion and Future Scope

Fish processing is now becoming a major industry in many countries and fish by-products are being widely used in many sectors including biomedical applications. The vast source of collagen, proteins, oils, chitin, and chitosan is derived from the fish industry. The encouraging biomedical applications of chitin and chitosan have attracted the attention of researchers to explore their use in a more intensive and easier application, as well as making the research into a new dimension for application in biomedical spheres. Still, the user has yet to be investigated in light of biocompatibility, degradability, acceptability, and bioactivity. Further, the research program should be oriented to develop composites of chitin and chitosan more fruitfully in various biomedical applications. The same has to reach the hand of clinicians at a reachable price. The research should be aimed to transfer from laboratories to clinics.

Acknowledgements The vice chancellor of West Bengal University of Animal and Fishery Sciences, in Kolkata, India, is gratefully acknowledged by the authors. The authors also acknowledge the kind support of the ICAR National Professor project of the Indian Council of Agricultural Research, New Delhi.

References

Coppola D, Lauritano C, Palma Esposito F, Riccio G, Rizzo C, de Pascale D (2021) Fish Waste: From Problem to Valuable Resource. Mar Drugs 19:116. https://doi.org/10.3390/md19020116

FAO (2018) The state of world fisheries and aquaculture 2018: Meeting the sustainable development goals. FAO, Rome, Italy

Ferraro V, Cruz IB, Jorge R, Malcata F, Pintado M, Castro P (2010). Valorisation of natural extracts from marine source focused on Marine by-Products: a review. https://doi.org/10.1016/J.FOODRES.2010.07.034

Mo WY, Man YB, Wong MH (2018) Use of food waste, fish waste and food processing waste for China's aquaculture industry: Needs and challenge. Sci Total Environ 613–614:635–643. https://doi.org/10.1016/j.scitotenv.2017.08.321

Shahidi F, Varatharajan V, Peng H, Senadheera R (2019) Utilization of marine by-products for the recovery of value-added products. J Food Bioact 6:. https://doi.org/10.31665/JFB.2019.6184

Shavandi A, Hou Y, Carne A, McConnell M, Bekhit A (2019) Marine Waste utilization as a source of functional and health compounds. Adv Food Nutr Res. https://doi.org/10.1016/bs.afnr.2018.08.001

Rinaudo M (2006) Chitin and chitosan: Properties and applications. Prog Polym Sci 31:603–632. https://doi.org/10.1016/j.progpolymsci.2006.06.001

Sankararamakrishnan N, Sanghi R (2006) Preparation and characterization of a novel xanthated chitosan. Carbohydr Polym 66:160–167. https://doi.org/10.1016/j.carbpol.2006.02.035

Kurita K (2006) Chitin and chitosan: functional biopolymers from marine crustaceans. Mar Biotechnol N Y N 8:203–226. https://doi.org/10.1007/s10126-005-0097-5

Dash M, Chiellini F, Ottenbrite RM, Chiellini E (2011) Chitosan—A versatile semi-synthetic polymer in biomedical applications. Prog Polym Sci 36:981–1014. https://doi.org/10.1016/j.progpolymsci.2011.02.001

Roberts GAF (1992) Chitin chemistry. Macmillan, London

Khor E, Lim LY (2003) Implantable applications of chitin and chitosan. Biomaterials 24:2339–2349. https://doi.org/10.1016/S0142-9612(03)00026-7

Khor E (2014) Chitin: Fulfilling a Biomaterials Promise. Elsevier

Ehrlich H, Maldonado M, Hanke T, Meissner H, Born R, Scharnweber D, Worch H Spongins: nanostructural investigations and development of biomimetic material model. 4

Ehrlich H, Maldonado M, Spindler K, Eckert C, Hanke T, Born R, Goebel C, Simon P, Heinemann S, Worch H (2007) First evidence of chitin as a component of the skeletal fibers of marine sponges. Part I. Verongidae (demospongia: Porifera). J Exp Zoolog B Mol Dev Evol 308B:347–356. https://doi.org/10.1002/jez.b.21156

Brunner E, Ehrlich H, Schupp P, Hedrich R, Hunoldt S, Kammer M, Machill S, Paasch S, Bazhenov VV, Kurek DV, Arnold T, Brockmann S, Ruhnow M, Born R (2009a) Chitin-based scaffolds are an integral part of the skeleton of the marine demosponge Ianthella basta. J Struct Biol 168:539–547. https://doi.org/10.1016/j.jsb.2009.06.018

Brunner E, Richthammer P, Ehrlich H, Paasch S, Simon P, Ueberlein S, van Pée K-H (2009b) Chitin-Based organic networks: an integral part of cell wall biosilica in the diatom Thalassiosira pseudonana. Angew Chem Int Ed 48:9724–9727. https://doi.org/10.1002/anie.200905028

Jeuniaux C, Voss-Foucart MF (1991) Chitin biomass and production in the marine environment. Biochem Syst Ecol 19:347–356. https://doi.org/10.1016/0305-1978(91)90051-Z

Stepnowski P, Olafsson G, Helgason H, Jastorff B (2004) Recovery of astaxanthin from seafood wastewater utilizing fish scales waste. Chemosphere 54:413–417. https://doi.org/10.1016/S0045-6535(03)00718-5

Kim I-Y, Seo S-J, Moon H-S, Yoo M-K, Park I-Y, Kim B-C, Cho C-S (2008) Chitosan and its derivatives for tissue engineering applications. Biotechnol Adv 26:1–21. https://doi.org/10.1016/j.biotechadv.2007.07.009

Rinaudo M (2008) Main properties and current applications of some polysaccharides as biomaterials. Polym Int 57:397–430. https://doi.org/10.1002/pi.2378

Abe M, Takahashi M, Tokura S, Tamura H, Nagano A (2004) Cartilage-scaffold composites produced by bioresorbable beta-chitin sponge with cultured rabbit chondrocytes. Tissue Eng 10:585–594. https://doi.org/10.1089/107632704323061942

Menon D, Vidyapeetham AV, Profile S, Furuike T, Profile S, Rangasamy J 144 Publications 5,749 Citations

Mikhailov G, Lebedeva M, Nud'ga L, Petrova VA, (2001) Composite fibers based on chitin and cellulose. Russ J Appl Chem 74:1573–1576. https://doi.org/10.1023/A:1013734008473

Shamshina JL, Zavgorodnya O, Choudhary H, Frye B, Newbury N, Rogers RD (2018) In search of stronger/cheaper chitin nanofibers through electrospinning of Chitin-Cellulose composites using an ionic liquid platform. ACS Sustain Chem Eng 6:14713–14722. https://doi.org/10.1021/acssuschemeng.8b03269

Li Z, Ma J, Li R, Yin X, Dong W, Pan C (2018) Fabrication of a blood compatible composite membrane from chitosan nanoparticles, ethyl cellulose and bacterial cellulose sulfate. RSC Adv 8:31322–31330. https://doi.org/10.1039/C8RA05536J

Bahmani E, Zonouzi H, Koushkbaghi S, Hafshejani F, Fassadi Chimeh A, Irani M (2020) Electrospun polyacrylonitrile/cellulose acetate/MIL-125/TiO2 composite nanofibers as an efficient photocatalyst and anticancer drug delivery system. Cellulose 27:1–17. https://doi.org/10.1007/s10570-020-03459-1

Zhu K, Wang Y, Lu A, Fu Q, Hu J, Zhang L (2019) Cellulose/Chitosan composite multifilament fibers with Two-Switch shape memory performance. ACS Sustain Chem Eng 7:6981–6990. https://doi.org/10.1021/acssuschemeng.8b06691

Hu D, Wang H, Wang L (2016) Physical properties and antibacterial activity of quaternized chitosan/carboxymethyl cellulose blend films. LWT - Food Sci Technol 65:398–405. https://doi.org/10.1016/j.lwt.2015.08.033

Youssef AM, El-Sayed SM, El-Sayed HS, Salama HH, Dufresne A (2016) Enhancement of Egyptian soft white cheese shelf life using a novel chitosan/carboxymethyl cellulose/zinc oxide bionanocomposite film. Carbohydr Polym 151:9–19. https://doi.org/10.1016/j.carbpol.2016.05.023

Shamshina JL, Gurau G, Block LE, Hansen LK, Dingee C, Walters A, Rogers RD (2014) Chitin–calcium alginate composite fibers for wound care dressings spun from ionic liquid solution. J Mater Chem B 2:3924–3936. https://doi.org/10.1039/C4TB00329B

Shen X, Shamshina JL, Berton P, Bandomir J, Wang H, Gurau G, Rogers RD (2016) Comparison of hydrogels prepared with Ionic-Liquid-Isolated versus Commercial Chitin and Cellulose. ACS Sustain Chem Eng 4:471–480. https://doi.org/10.1021/acssuschemeng.5b01400

Azadi MDA, Hassanjili S, Zarrabi K, Sarkari B (2018) Solidification of hydatid cyst fluid with an injectable chitosan/carboxymethylcellulose/β-glycerophosphate hydrogel for effective control of spillage during aspiration of hydatid cysts. Prog Biomater 7:35–54. https://doi.org/10.1007/s40204-018-0082-5

Wang T, Chen L, Shen T, Wu D (2016) Preparation and properties of a novel thermo-sensitive hydrogel based on chitosan/hydroxypropyl methylcellulose/glycerol. Int J Biol Macromol 93:775–782. https://doi.org/10.1016/j.ijbiomac.2016.09.038

Kabir SMF, Sikdar PP, Haque B, Bhuiyan MAR, Ali A, Islam MN (2018) Cellulose-based hydrogel materials: chemistry, properties and their prospective applications. Prog Biomater 7:153–174. https://doi.org/10.1007/s40204-018-0095-0

Mathur NK, Narang CK (1990) Chitin and chitosan, versatile polysaccharides from marine animals. J Chem Educ 67:938. https://doi.org/10.1021/ed067p938

Li Z, Ramay HR, Hauch KD, Xiao D, Zhang M (2005) Chitosan-alginate hybrid scaffolds for bone tissue engineering. Biomaterials 26:3919–3928. https://doi.org/10.1016/j.biomaterials.2004.09.062

Jiang T, Abdel-Fattah WI, Laurencin CT (2006) In vitro evaluation of chitosan/poly(lactic acid-glycolic acid) sintered microsphere scaffolds for bone tissue engineering. Biomaterials 27:4894–4903. https://doi.org/10.1016/j.biomaterials.2006.05.025

Jiang T, Nukavarapu SP, Deng M, Jabbarzadeh E, Kofron MD, Doty SB, Abdel-Fattah WI, Laurencin CT (2010) Chitosan-poly(lactide-co-glycolide) microsphere-based scaffolds for bone tissue engineering: in vitro degradation and in vivo bone regeneration studies. Acta Biomater 6:3457–3470. https://doi.org/10.1016/j.actbio.2010.03.023

Tsai W-B, Chen Y-R, Li W-T, Lai J-Y, Liu H-L (2012) RGD-conjugated UV-crosslinked chitosan scaffolds inoculated with mesenchymal stem cells for bone tissue engineering. Carbohydr Polym 89:379–387. https://doi.org/10.1016/j.carbpol.2012.03.017

Wu H, Wan Y, Dalai S, Zhang R (2010) Response of rat osteoblasts to polycaprolactone/chitosan blend porous scaffolds. J Biomed Mater Res A 92:238–245. https://doi.org/10.1002/jbm.a.32376

Thuaksuban N, Nuntanaranont T, Pattanachot W, Suttapreyasri S, Cheung L (2011) Biodegradable polycaprolactone-chitosan three-dimensional scaffolds fabricated by melt stretching and multi-layer deposition for bone tissue engineering: Assessment of the physical properties and cellular response. Biomed Mater Bristol Engl 6:015009. https://doi.org/10.1088/1748-6041/6/1/015009

Im SY, Cho SH, Hwang JH, Lee SJ (2003) Growth factor releasing porous poly (epsilon-caprolactone)-chitosan matrices for enhanced bone regenerative therapy. Arch Pharm Res 26:76–82. https://doi.org/10.1007/BF03179936

Yilgor P, Tuzlakoglu K, Reis RL, Hasirci N, Hasirci V (2009) Incorporation of a sequential BMP-2/BMP-7 delivery system into chitosan-based scaffolds for bone tissue engineering. Biomaterials 30:3551–3559. https://doi.org/10.1016/j.biomaterials.2009.03.024

Shi S, Cheng X, Wang J, Zhang W, Peng L, Zhang Y (2009) RhBMP-2 microspheres-loaded chitosan/collagen scaffold enhanced osseointegration: an experiment in dog. J Biomater Appl 23:331–346. https://doi.org/10.1177/0885328208090013

Gümüşderelioğlu M, Aday S (2011) Heparin-functionalized chitosan scaffolds for bone tissue engineering. Carbohydr Res 346:606–613. https://doi.org/10.1016/j.carres.2010.12.007

Prabaharan M, Rodriguez-Perez MA, de Saja JA, Mano JF (2007) Preparation and characterization of poly(L-lactic acid)-chitosan hybrid scaffolds with drug release capability. J Biomed Mater Res B Appl Biomater 81:427–434. https://doi.org/10.1002/jbm.b.30680

Chang C, Peng N, He M, Teramoto Y, Nishio Y, Zhang L (2013) Fabrication and properties of chitin/hydroxyapatite hybrid hydrogels as scaffold nano-materials. Carbohydr Polym 91:7–13. https://doi.org/10.1016/j.carbpol.2012.07.070

Kumar P, Srinivasan S, Lakshmanan V-K, Tamura H, Nair S, Jayakumar R (2011) Synthesis, characterization and cytocompatibility studies of α-chitin hydrogel/nano hydroxyapatite composite scaffolds. Int J Biol Macromol 49:20–31. https://doi.org/10.1016/j.ijbiomac.2011.03.006

Sudheesh Kumar PT, Srinivasan S, Lakshmanan V-K, Tamura H, Nair SV, Jayakumar R (2011) β-Chitin hydrogel/nano hydroxyapatite composite scaffolds for tissue engineering applications. Carbohydr Polym 85:584–591. https://doi.org/10.1016/j.carbpol.2011.03.018

Madhumathi K, Binulal NS, Nagahama H, Tamura H, Shalumon KT, Selvamurugan N, Nair SV, Jayakumar R (2009) Preparation and characterization of novel β-chitin–hydroxyapatite composite membranes for tissue engineering applications. Int J Biol Macromol 44:1–5. https://doi.org/10.1016/j.ijbiomac.2008.09.013

Li X, Feng Q, Cui F (2006) In vitro degradation of porous nano-hydroxyapatite/collagen/PLLA scaffold reinforced by chitin fibres. Mater Sci Eng C 26:716–720. https://doi.org/10.1016/j.msec.2005.06.062

Frohbergh ME, Katsman A, Botta GP, Lazarovici P, Schauer CL, Wegst UGK, Lelkes PI (2012) Electrospun hydroxyapatite-containing chitosan nanofibers crosslinked with genipin for bone tissue engineering. Biomaterials 33:9167–9178. https://doi.org/10.1016/j.biomaterials.2012.09.009

Ge Z, Baguenard S, Lim LY, Wee A, Khor E (2004) Hydroxyapatite–chitin materials as potential tissue engineered bone substitutes. Biomaterials 25:1049–1058. https://doi.org/10.1016/S0142-9612(03)00612-4

Liu X, Li X, Fan Y, Zhang G, Li D, Dong W, Sha Z, Yu X, Feng Q, Cui F, Watari F (2010) Repairing goat tibia segmental bone defect using scaffold cultured with mesenchymal stem cells. J Biomed Mater Res B Appl Biomater 94B:44–52. https://doi.org/10.1002/jbm.b.31622

EBSCOhost 65959103 Biocompatible β-chitin Hydrogel/Nanobioactive Glass Ceramic Nanocomposite Scaffolds for Periodontal Bone Regeneration. https://web.p.ebscohost.com/abstract?direct=true&profile=ehost&scope=site&authtype=crawler&jrnl=09711198&AN=65959103&h=ODSD8capDjdUQBltuvKLKIRbAbe2kFIu%2fbfN7tlNnCZfDfwJiqSyEBXOWgh%2bqDo%2fjHok8mu61W5D9Zn4ZKCYxQ%3d%3d&crl=c&resultNs=AdminWebAuth&resultLocal=ErrCrlNotAuth&crlhashurl=login.aspx%3fdirect%3dtrue%26profile%3dehost%26scope%3dsite%26authtype%3dcrawler%26jrnl%3d09711198%26AN%3d65959103. Accessed 10 Jul 2022

Hench LL, Splinter RJ, Allen WC, Greenlee TK (1971) Bonding mechanisms at the interface of ceramic prosthetic materials. J Biomed Mater Res 5:117–141. https://doi.org/10.1002/jbm.820050611

Jones JR (2013) Review of bioactive glass: From Hench to hybrids. Acta Biomater 9:4457–4486. https://doi.org/10.1016/j.actbio.2012.08.023

Toskas G, Cherif C, Hund R-D, Laourine E, Mahltig B, Fahmi A, Heinemann C, Hanke T (2013) Chitosan(PEO)/silica hybrid nanofibers as a potential biomaterial for bone regeneration. Carbohydr Polym 94:713–722. https://doi.org/10.1016/j.carbpol.2013.01.068

Jongwattanapisan P, Charoenphandhu N, Krishnamra N, Thongbunchoo J, Tang I-M, Hoonsawat R, Smith SM, Pon-On W (2011) In vitro study of the SBF and osteoblast-like cells on hydroxyapatite/chitosan–silica nanocomposite. Mater Sci Eng C 31:290–299. https://doi.org/10.1016/j.msec.2010.09.009

Potential Use of Chitosan as a Cell Scaffold Material for Cartilage Tissue Engineering Tissue Engineering. https://www.liebertpub.com/doi/abs/https://doi.org/10.1089/107632702320934100. Accessed 10 Jul 2022

Griffon DJ, Sedighi MR, Schaeffer DV, Eurell JA, Johnson AL (2006) Chitosan scaffolds: Interconnective pore size and cartilage engineering. Acta Biomater 2:313–320. https://doi.org/10.1016/j.actbio.2005.12.007

Kim SE, Park JH, Cho YW, Chung H, Jeong SY, Lee EB, Kwon IC (2003) Porous chitosan scaffold containing microspheres loaded with transforming growth factor-β1: Implications for cartilage tissue engineering. J Controlled Release 91:365–374. https://doi.org/10.1016/S0168-3659(03)00274-8

Lee JE, Kim KE, Kwon IC, Ahn HJ, Lee S-H, Cho H, Kim HJ, Seong SC, Lee MC (2004) Effects of the controlled-released TGF-β1 from chitosan microspheres on chondrocytes cultured in a collagen/chitosan/glycosaminoglycan scaffold. Biomaterials 25:4163–4173. https://doi.org/10.1016/j.biomaterials.2003.10.057

Yan L-P, Wang Y-J, Ren L, Wu G, Caridade SG, Fan J-B, Wang L-Y, Ji P-H, Oliveira JM, Oliveira JT, Mano JF, Reis RL (2010) Genipin-cross-linked collagen/chitosan biomimetic scaffolds for articular cartilage tissue engineering applications. J Biomed Mater Res A 95A:465–475. https://doi.org/10.1002/jbm.a.32869

Hoemann CD, Sun J, McKee MD, Chevrier A, Rossomacha E, Rivard G-E, Hurtig M, Buschmann MD (2007) Chitosan–glycerol phosphate/blood implants elicit hyaline cartilage repair integrated with porous subchondral bone in microdrilled rabbit defects. Osteoarthritis Cartilage 15:78–89. https://doi.org/10.1016/j.joca.2006.06.015

Hao T, Wen N, Cao J-K, Wang H-B, Lü S-H, Liu T, Lin Q-X, Duan C-M, Wang C-Y (2010) The support of matrix accumulation and the promotion of sheep articular cartilage defects repair in vivo by chitosan hydrogels. Osteoarthritis Cartilage 18:257–265. https://doi.org/10.1016/j.joca.2009.08.007

Xi LuJ, Prudhommeaux F, Meunier A, Sedel L, Guillemin G (1999) Effects of chitosan on rat knee cartilages. Biomaterials 20:1937–1944. https://doi.org/10.1016/S0142-9612(99)00097-6

Jin R, Moreira Teixeira LS, Dijkstra PJ, Karperien M, van Blitterswijk CA, Zhong ZY, Feijen J (2009) Injectable chitosan-based hydrogels for cartilage tissue engineering. Biomaterials 30:2544–2551. https://doi.org/10.1016/j.biomaterials.2009.01.020

Tan H, Chu CR, Payne KA, Marra KG (2009) Injectable in situ forming biodegradable chitosan–hyaluronic acid based hydrogels for cartilage tissue engineering. Biomaterials 30:2499–2506. https://doi.org/10.1016/j.biomaterials.2008.12.080

Park H, Choi B, Hu J, Lee M (2013) Injectable chitosan hyaluronic acid hydrogels for cartilage tissue engineering. Acta Biomater 9:4779–4786. https://doi.org/10.1016/j.actbio.2012.08.033

Subramanian A, Lin H-Y, Vu D, Larsen G (2004) Synthesis and evaluation of scaffolds prepared from chitosan fibers for potential use in cartilage tissue engineering. Biomed Sci Instrum 40:117–122

Neves SC, Moreira Teixeira LS, Moroni L, Reis RL, Van Blitterswijk CA, Alves NM, Karperien M, Mano JF (2011) Chitosan/Poly(ε-caprolactone) blend scaffolds for cartilage repair. Biomaterials 32:1068–1079. https://doi.org/10.1016/j.biomaterials.2010.09.073

Majima T, Funakosi T, Iwasaki N, Yamane S-T, Harada K, Nonaka S, Minami A, Nishimura S-I (2005) Alginate and chitosan polyion complex hybrid fibers for scaffolds in ligament and tendon tissue engineering. J Orthop Sci 10:302–307. https://doi.org/10.1007/s00776-005-0891-y

Funakoshi T, Majima T, Iwasaki N, Yamane S, Masuko T, Minami A, Harada K, Tamura H, Tokura S, Nishimura S-I (2005a) Novel chitosan-based hyaluronan hybrid polymer fibers as a scaffold in ligament tissue engineering. J Biomed Mater Res A 74A:338–346. https://doi.org/10.1002/jbm.a.30237

Funakoshi T, Majima T, Iwasaki N, Suenaga N, Sawaguchi N, Shimode K, Minami A, Harada K, Nishimura S (2005b) Application of tissue engineering techniques for rotator cuff regeneration using a Chitosan-Based hyaluronan hybrid fiber scaffold. Am J Sports Med 33:1193–1201. https://doi.org/10.1177/0363546504272689

Irie T, Majima T, Sawaguchi N, Funakoshi T, Nishimura S, Minami A (2011) Biomechanical and histologic evaluation of tissue engineered ligaments using chitosan and hyaluronan hybrid polymer fibers: A rabbit medial collateral ligament reconstruction model. J Biomed Mater Res A 97A:111–117. https://doi.org/10.1002/jbm.a.32938

Funakoshi T, Majima T, Suenaga N, Iwasaki N, Yamane S, Minami A (2006) Rotator cuff regeneration using chitin fabric as an acellular matrix. J Shoulder Elbow Surg 15:112–118. https://doi.org/10.1016/j.jse.2005.05.012

Sato M, Maeda M, Kurosawa H, Inoue Y, Yamauchi Y, Iwase H (2000) Reconstruction of rabbit Achilles tendon with three bioabsorbable materials: histological and biomechanical studies. J Orthop Sci 5:256–267. https://doi.org/10.1007/s007760050161

Tendon healing in vivo and in vitro: chitosan improves range of motion after flexor tendon repair—Minerva Ortopedica e Traumatologica 2013 August;64(4):445–52. https://www.minervamedica.it/en/journals/minerva-orthopedics/article.php?cod=R14Y2013N04A0445. Accessed 11 Jul 2022

Shao H-J, Chen CS, Lee Y-T, Wang J-H, Young T-H (2010a) The phenotypic responses of human anterior cruciate ligament cells cultured on poly(ϵ-caprolactone) and chitosan. J Biomed Mater Res A 93A:1297–1305. https://doi.org/10.1002/jbm.a.32629

Shao H-J, Lee Y-T, Chen C-S, Wang J-H, Young T-H (2010b) Modulation of gene expression and collagen production of anterior cruciate ligament cells through cell shape changes on polycaprolactone/chitosan blends. Biomaterials 31:4695–4705. https://doi.org/10.1016/j.biomaterials.2010.02.037

Taravel MN, Domard A (1995) Collagen and its interaction with chitosan: II. Influence of the physicochemical characteristics of collagen. Biomaterials 16:865–871. https://doi.org/10.1016/0142-9612(95)94149-F

Taravel MN, Domard A (1996) Collagen and its interactions with chitosan: III. some biological and mechanical properties. Biomaterials 17:451–455. https://doi.org/10.1016/0142-9612(96)89663-3

Ma J, Wang H, He B, Chen J (2001) A preliminary in vitro study on the fabrication and tissue engineering applications of a novel chitosan bilayer material as a scaffold of human neofetal dermal fibroblasts. Biomaterials 22:331–336. https://doi.org/10.1016/S0142-9612(00)00188-5

Ueno H, Yamada H, Tanaka I, Kaba N, Matsuura M, Okumura M, Kadosawa T, Fujinaga T (1999) Accelerating effects of chitosan for healing at early phase of experimental open wound in dogs. Biomaterials 20:1407–1414. https://doi.org/10.1016/S0142-9612(99)00046-0

Mizuno K, Yamamura K, Yano K, Osada T, Saeki S, Takimoto N, Sakurai T, Nimura Y (2003) Effect of chitosan film containing basic fibroblast growth factor on wound healing in genetically diabetic mice. J Biomed Mater Res A 64A:177–181. https://doi.org/10.1002/jbm.a.10396

Yan X, Khor E, Lim L-Y (2000) PEC Films Prepared from Chitosan-Alginate Coacervates. Chem Pharm Bull (tokyo) 48:941–946. https://doi.org/10.1248/cpb.48.941

Wang L, Khor E, Wee A, Lim LY (2002) Chitosan-alginate PEC membrane as a wound dressing: Assessment of incisional wound healing. J Biomed Mater Res 63:610–618. https://doi.org/10.1002/jbm.10382

Ma L, Gao C, Mao Z, Zhou J, Shen J, Hu X, Han C (2003) Collagen/chitosan porous scaffolds with improved biostability for skin tissue engineering. Biomaterials 24:4833–4841. https://doi.org/10.1016/S0142-9612(03)00374-0

Burkatovskaya M, Castano AP, Demidova-Rice TN, Tegos GP, Hamblin MR (2008) Effect of chitosan acetate bandage on wound healing in infected and noninfected wounds in mice. Wound Repair Regen 16:425–431. https://doi.org/10.1111/j.1524-475X.2008.00382.x

Guo R, Xu S, Ma L, Huang A, Gao C (2011) The healing of full-thickness burns treated by using plasmid DNA encoding VEGF-165 activated collagen–chitosan dermal equivalents. Biomaterials 32:1019–1031. https://doi.org/10.1016/j.biomaterials.2010.08.087

Kumar PTS, Raj NM, Praveen G, Chennazhi KP, Nair SV, Jayakumar R (2013) In vitro and in vivo evaluation of microporous chitosan Hydrogel/Nanofibrin composite bandage for skin tissue regeneration. Tissue Eng Part A 19:380–392. https://doi.org/10.1089/ten.tea.2012.0376

Li J, Pan J, Zhang L, Guo X, Yu Y (2003a) Culture of primary rat hepatocytes within porous chitosan scaffolds. J Biomed Mater Res A 67A:938–943. https://doi.org/10.1002/jbm.a.10076

Chupa JM, Foster AM, Sumner SR, Madihally SV, Matthew HWT (2000) Vascular cell responses to polysaccharide materials: in vitro and in vivo evaluations. Biomaterials 21:2315–2322. https://doi.org/10.1016/S0142-9612(00)00158-7

Wang XH, Li DP, Wang WJ, Feng QL, Cui FZ, Xu YX, Song XH, van der Werf M (2003) Crosslinked collagen/chitosan matrix for artificial livers. Biomaterials 24:3213–3220. https://doi.org/10.1016/S0142-9612(03)00170-4

Wang X, Yan Y, Lin F, Xiong Z, Wu R, Zhang R, Lu Q (2005) Preparation and characterization of a collagen/chitosan/heparin matrix for an implantable bioartificial liver. J Biomater Sci Polym Ed 16:1063–1080. https://doi.org/10.1163/1568562054798554

Park I-K, Yang J, Jeong H-J, Bom H-S, Harada I, Akaike T, Kim S-I, Cho C-S (2003) Galactosylated chitosan as a synthetic extracellular matrix for hepatocytes attachment. Biomaterials 24:2331–2337. https://doi.org/10.1016/S0142-9612(03)00108-X

Li J, Pan J, Zhang L, Yu Y (2003b) Culture of hepatocytes on fructose-modified chitosan scaffolds. Biomaterials 24:2317–2322. https://doi.org/10.1016/S0142-9612(03)00048-6

She Z, Liu W, Feng Q (2009) Self-assembly model, hepatocytes attachment and inflammatory response for silk fibroin/chitosan scaffolds. Biomed Mater 4:045014. https://doi.org/10.1088/1748-6041/4/4/045014

Fan J, Shang Y, Yuan Y, Yang J (2010) Preparation and characterization of chitosan/galactosylated hyaluronic acid scaffolds for primary hepatocytes culture. J Mater Sci Mater Med 21:319–327. https://doi.org/10.1007/s10856-009-3833-y

Piscioneri A, Campana C, Salerno S, Morelli S, Bader A, Giordano F, Drioli E, Bartolo LD (2011) Biodegradable and synthetic membranes for the expansion and functional differentiation of rat embryonic liver cells. Acta Biomater 7:171–179. https://doi.org/10.1016/j.actbio.2010.07.039

Yuan Y, Zhang P, Yang Y, Wang X, Gu X (2004) The interaction of Schwann cells with chitosan membranes and fibers in vitro. Biomaterials 25:4273–4278. https://doi.org/10.1016/j.biomaterials.2003.11.029

Cao W, Cheng M, Ao Q, Gong Y, Zhao N, Zhang X (2005) Physical, mechanical and degradation properties, and Schwann cell affinity of cross-linked chitosan films. J Biomater Sci Polym Ed 16:791–807. https://doi.org/10.1163/1568562053992496
Mingyu C, Kai G, Jiamou L, Yandao G, Nanming Z, Xiufang Z (2004) Surface modification and characterization of chitosan film blended with Poly-L-Lysine. J Biomater Appl 19:59–75. https://doi.org/10.1177/0885328204043450
Cheng M, Deng J, Yang F, Gong Y, Zhao N, Zhang X (2003) Study on physical properties and nerve cell affinity of composite films from chitosan and gelatin solutions. Biomaterials 24:2871–2880. https://doi.org/10.1016/S0142-9612(03)00117-0
Chiono V, Pulieri E, Vozzi G, Ciardelli G, Ahluwalia A, Giusti P (2008) Genipin-crosslinked chitosan/gelatin blends for biomedical applications. J Mater Sci Mater Med 19:889–898. https://doi.org/10.1007/s10856-007-3212-5
Yang Z, Mo L, Duan H, Li X (2010) Effects of chitosan/collagen substrates on the behavior of rat neural stem cells. Sci China Life Sci 53:215–222. https://doi.org/10.1007/s11427-010-0036-1
Guo X, Zahir T, Mothe A, Shoichet MS, Morshead CM, Katayama Y, Tator CH (2012) The effect of growth factors and soluble Nogo-66 receptor protein on transplanted neural Stem/Progenitor survival and axonal regeneration after complete transection of rat spinal cord. Cell Transplant 21:1177–1197. https://doi.org/10.3727/096368911X612503
Denkbaş EB, Seyyal M, Pişkin E (2000) Implantable 5-fluorouracil loaded chitosan scaffolds prepared by wet spinning. J Membr Sci 172:33–38. https://doi.org/10.1016/S0376-7388(00)00314-8
Lin H-Y, Yeh C-T (2010a) Controlled release of pentoxifylline from porous chitosan-pectin scaffolds. Drug Deliv 17:313–321. https://doi.org/10.3109/10717541003713733
Lin H-Y, Yeh C-T (2010b) Alginate-crosslinked chitosan scaffolds as pentoxifylline delivery carriers. J Mater Sci Mater Med 21:1611–1620. https://doi.org/10.1007/s10856-010-4028-2
Prabaharan M, Jayakumar R (2009) Chitosan-graft-β-cyclodextrin scaffolds with controlled drug release capability for tissue engineering applications. Int J Biol Macromol 44:320–325. https://doi.org/10.1016/j.ijbiomac.2009.01.005
Shi PJ, Li YB, Zhang L, Zuo Y, Wu M, Wang HN (2007) Fabrication of n-HA/CS Porous Scaffold Carrying Ampicillin-Loaded Microspheres. Mater Sci Forum 544–545:941–944. https://doi.org/10.4028/www.scientific.net/MSF.544-545.941
Duarte ARC, Mano JF, Reis RL (2009) Preparation of chitosan scaffolds loaded with dexamethasone for tissue engineering applications using supercritical fluid technology. Eur Polym J 45:141–148. https://doi.org/10.1016/j.eurpolymj.2008.10.004
Teng S-H, Lee E-J, Wang P, Jun S-H, Han C-M, Kim H-E (2009) Functionally gradient chitosan/hydroxyapatite composite scaffolds for controlled drug release. J Biomed Mater Res B Appl Biomater 90B:275–282. https://doi.org/10.1002/jbm.b.31283
Noel SP, Courtney HS, Bumgardner JD, Haggard WO (2010) Chitosan sponges to locally deliver amikacin and vancomycin: a pilot in vitro evaluation. Clin Orthop Relat Res 468:2074–2080. https://doi.org/10.1007/s11999-010-1324-6
Thacharodi D, Panduranga Rao K (1996) Collagen-chitosan composite membranes controlled transdermal delivery of nifedipine and propranolol hydrochloride. Int J Pharm 134:239–241. https://doi.org/10.1016/0378-5173(96)04453-5
Li F, Liu WG, Yao KD (2002) Preparation of oxidized glucose-crosslinked N-alkylated chitosan membrane and in vitro studies of pH-sensitive drug delivery behaviour. Biomaterials 23:343–347. https://doi.org/10.1016/S0142-9612(01)00111-9
Thacharodi D, Rao KP (1993) Release of nifedipine through crosslinked chitosan membranes. Int J Pharm 96:33–39. https://doi.org/10.1016/0378-5173(93)90209-X
Thacharodi D, Panduranga Rao K (1995) Collagen-chitosan composite membranes for controlled release of propranolol hydrochloride. Int J Pharm 120:115–118. https://doi.org/10.1016/0378-5173(94)00423-3

Jayakumar R, Prabaharan M, Nair SV, Tokura S, Tamura H, Selvamurugan N (2010) Novel carboxymethyl derivatives of chitin and chitosan materials and their biomedical applications. Prog Mater Sci 55:675–709. https://doi.org/10.1016/j.pmatsci.2010.03.001

Tığlı RS, Akman AC, Gümüşderelıoğlu M, Nohutçu RM (2009) In vitro release of dexamethasone or bFGF from Chitosan/Hydroxyapatite scaffolds. J Biomater Sci Polym Ed 20:1899–1914. https://doi.org/10.1163/156856208X399945

Ho Y-C, Mi F-L, Sung H-W, Kuo P-L (2009) Heparin-functionalized chitosan–alginate scaffolds for controlled release of growth factor. Int J Pharm 376:69–75. https://doi.org/10.1016/j.ijpharm.2009.04.048

Reves BT, Bumgardner JD, Cole JA, Yang Y, Haggard WO (2009) Lyophilization to improve drug delivery for chitosan-calcium phosphate bone scaffold construct: A preliminary investigation. J Biomed Mater Res B Appl Biomater 90B:1–10. https://doi.org/10.1002/jbm.b.31390

De la Riva B, Sánchez E, Hernández A, Reyes R, Tamimi F, López-Cabarcos E, Delgado A, Évora C (2010) Local controlled release of VEGF and PDGF from a combined brushite–chitosan system enhances bone regeneration. J Controlled Release 143:45–52. https://doi.org/10.1016/j.jconrel.2009.11.026

Buyens K, De Smedt SC, Braeckmans K, Demeester J, Peeters L, van Grunsven LA, de Mollerat du Jeu X, Sawant R, Torchilin V, Farkasova K, Ogris M, Sanders NN, (2012) Liposome based systems for systemic siRNA delivery: Stability in blood sets the requirements for optimal carrier design. J Controlled Release 158:362–370. https://doi.org/10.1016/j.jconrel.2011.10.009

Mao H-Q, Roy K, Troung-Le VL, Janes KA, Lin KY, Wang Y, August JT, Leong KW (2001) Chitosan-DNA nanoparticles as gene carriers: synthesis, characterization and transfection efficiency. J Controlled Release 70:399–421. https://doi.org/10.1016/S0168-3659(00)00361-8

Chan P, Kurisawa M, Chung JE, Yang Y-Y (2007) Synthesis and characterization of chitosan-g-poly(ethylene glycol)-folate as a non-viral carrier for tumor-targeted gene delivery. Biomaterials 28:540–549. https://doi.org/10.1016/j.biomaterials.2006.08.046

Kim TH, Nah JW, Cho M-H, Park TG, Cho CS (2006) Receptor-Mediated gene delivery into antigen presenting cells using Mannosylated Chitosan/DNA nanoparticles. J Nanosci Nanotechnol 6:2796–2803. https://doi.org/10.1166/jnn.2006.434

Gao S, Chen J, Xu X, Ding Z, Yang Y-H, Hua Z, Zhang J (2003) Galactosylated low molecular weight chitosan as DNA carrier for hepatocyte-targeting. Int J Pharm 255:57–68. https://doi.org/10.1016/S0378-5173(03)00082-6

Park IK, Kim TH, Park YH, Shin BA, Choi ES, Chowdhury EH, Akaike T, Cho CS (2001) Galactosylated chitosan-graft-poly(ethylene glycol) as hepatocyte-targeting DNA carrier. J Controlled Release 76:349–362. https://doi.org/10.1016/S0168-3659(01)00448-5

Kean T, Roth S, Thanou M (2005) Trimethylated chitosans as non-viral gene delivery vectors: Cytotoxicity and transfection efficiency. J Controlled Release 103:643–653. https://doi.org/10.1016/j.jconrel.2005.01.001

Thanou M, Florea BI, Geldof M, Junginger HE, Borchard G (2002) Quaternized chitosan oligomers as novel gene delivery vectors in epithelial cell lines. Biomaterials 23:153–159. https://doi.org/10.1016/S0142-9612(01)00090-4

Strand SP, Issa MM, Christensen BE, Vårum KM, Artursson P (2008) Tailoring of chitosans for gene delivery: Novel Self-Branched glycosylated chitosan oligomers with improved functional properties. Biomacromol 9:3268–3276. https://doi.org/10.1021/bm800832u

Lee SJ, Huh MS, Lee SY, Min S, Lee S, Koo H, Chu J-U, Lee KE, Jeon H, Choi Y, Choi K, Byun Y, Jeong SY, Park K, Kim K, Kwon IC (2012) Tumor-Homing Poly-siRNA/Glycol Chitosan Self-Cross-Linked nanoparticles for systemic siRNA delivery in cancer treatment. Angew Chem Int Ed 51:7203–7207. https://doi.org/10.1002/anie.201201390

Kang SH, Revuri V, Lee S-J, Cho S, Park I-K, Cho KJ, Bae WK, Lee Y (2017) Oral siRNA delivery to treat colorectal liver metastases. ACS Nano 11:10417–10429. https://doi.org/10.1021/acsnano.7b05547

Singh R, Chacharkar MP, Mathur AK (2008) Chitin membrane for wound dressing application—preparation, characterisation and toxicological evaluation. Int Wound J 5:665–673. https://doi.org/10.1111/j.1742-481X.2008.00482.x

Azad AK, Sermsintham N, Chandrkrachang S, Stevens WF (2004) Chitosan membrane as a wound-healing dressing: Characterization and clinical application. J Biomed Mater Res B Appl Biomater 69B:216–222. https://doi.org/10.1002/jbm.b.30000

Santos TC, Höring B, Reise K, Marques AP, Silva SS, Oliveira JM, Mano JF, Castro AG, Reis RL, van Griensven M (2013) In vivo performance of Chitosan/Soy-Based membranes as Wound-Dressing devices for acute skin wounds. Tissue Eng Part A 19:860–869. https://doi.org/10.1089/ten.tea.2011.0651

Pang HT, Chen XG, Ji QX, Zhong DY (2008) Preparation and function of composite asymmetric chitosan/CM-chitosan membrane. J Mater Sci Mater Med 19:1413–1417. https://doi.org/10.1007/s10856-007-3168-5

Mi F-L, Wu Y-B, Shyu S-S, Chao A-C, Lai J-Y, Su C-C (2003) Asymmetric chitosan membranes prepared by dry/wet phase separation: a new type of wound dressing for controlled antibacterial release. J Membr Sci 212:237–254. https://doi.org/10.1016/S0376-7388(02)00505-7

Meng X, Tian F, Yang J, He C-N, Xing N, Li F (2010) Chitosan and alginate polyelectrolyte complex membranes and their properties for wound dressing application. J Mater Sci Mater Med 21:1751–1759. https://doi.org/10.1007/s10856-010-3996-6

Zhu X, Chian KS, Chan-Park MBE, Lee ST (2005) Effect of argon-plasma treatment on proliferation of human-skin–derived fibroblast on chitosan membrane in vitro. J Biomed Mater Res A 73A:264–274. https://doi.org/10.1002/jbm.a.30211

de Queiroz AAA, Ferraz HG, Abraham GA, del Mar FM, Bravo AL, Román JS (2003) Development of new hydroactive dressings based on chitosan membranes: Characterization and in vivo behavior. J Biomed Mater Res A 64A:147–154. https://doi.org/10.1002/jbm.a.10265

Mi F-L, Shyu S-S, Wu Y-B, Lee S-T, Shyong J-Y, Huang R-N (2001) Fabrication and characterization of a sponge-like asymmetric chitosan membrane as a wound dressing. Biomaterials 22:165–173. https://doi.org/10.1016/S0142-9612(00)00167-8

Wang T, Zhu X-K, Xue X-T, Wu D-Y (2012) Hydrogel sheets of chitosan, honey and gelatin as burn wound dressings. Carbohydr Polym 88:75–83. https://doi.org/10.1016/j.carbpol.2011.11.069

Lee YM, Kim SS, Park MH, Song KW, Sung YK, Kang IK (2000) β-Chitin-based wound dressing containing silver sulfurdiazine. J Mater Sci Mater Med 11:817–823. https://doi.org/10.1023/A:1008961730929

Sudheesh Kumar PT, Lakshmanan V-K, Anilkumar TV, Ramya C, Reshmi P, Unnikrishnan AG, Nair SV, Jayakumar R (2012) Flexible and microporous chitosan Hydrogel/Nano ZnO composite bandages for wound dressing. In vitro and in vivo evaluation. ACS Appl Mater Interfaces 4:2618–2629. https://doi.org/10.1021/am300292v

Derfus AM, Chan WCW, Bhatia SN (2004) Probing the Cytotoxicity of Semiconductor Quantum Dots. Nano Lett 4:11–18. https://doi.org/10.1021/nl0347334

Higuchi Y, Oka M, Kawakami S, Hashida M (2008) Mannosylated semiconductor quantum dots for the labeling of macrophages. J Controlled Release 125:131–136. https://doi.org/10.1016/j.jconrel.2007.10.007

Lin S-Y, Chan H-Y, Shen F-H, Chen M-H, Wang Y-J, Yu C-K (2007) Chitosan prevents the development of AOM-induced aberrant crypt foci in mice and suppressed the proliferation of AGS cells by inhibiting DNA synthesis. J Cell Biochem 100:1573–1580. https://doi.org/10.1002/jcb.21152

Gibot L, Chabaud S, Bouhout S, Bolduc S, Auger FA, Moulin VJ (2015) Anticancer properties of chitosan on human melanoma are cell line dependent. Int J Biol Macromol 72:370–379. https://doi.org/10.1016/j.ijbiomac.2014.08.033

Solairaj D, Rameshthangam P, Arunachalam G (2017) Anticancer activity of silver and copper embedded chitin nanocomposites against human breast cancer (MCF-7) cells. Int J Biol Macromol 105:608–619. https://doi.org/10.1016/j.ijbiomac.2017.07.078

Singh A, Dutta PK, Kumar H, Kureel AK, Rai AK (2018) Synthesis of chitin-glucan-aldehyde-quercetin conjugate and evaluation of anticancer and antioxidant activities. Carbohydr Polym 193:99–107. https://doi.org/10.1016/j.carbpol.2018.03.092

Arunraj TR, Sanoj Rejinold N, Ashwin Kumar N, Jayakumar R (2014) Bio-responsive chitin-poly(l-lactic acid) composite nanogels for liver cancer. Colloids Surf B Biointerfaces 113:394–402. https://doi.org/10.1016/j.colsurfb.2013.09.023

Pirzadeh-Naeeni S, Mozdianfard MR, Shojaosadati SA, Khorasani AC, Saleh T (2020) A comparative study on schizophyllan and chitin nanoparticles for ellagic acid delivery in treating breast cancer. Int J Biol Macromol 144:380–388. https://doi.org/10.1016/j.ijbiomac.2019.12.079

Mattaveewong T, Wongkrasant P, Chanchai S, Pichyangkura R, Chatsudthipong V, Muanprasat C (2016) Chitosan oligosaccharide suppresses tumor progression in a mouse model of colitis-associated colorectal cancer through AMPK activation and suppression of NF-κB and mTOR signaling. Carbohydr Polym 145:30–36. https://doi.org/10.1016/j.carbpol.2016.02.077

Wimardhani YS, Suniarti DF, Freisleben HJ, Wanandi SI, Siregar NC, Ikeda M-A (2014) Chitosan exerts anticancer activity through induction of apoptosis and cell cycle arrest in oral cancer cells. J Oral Sci 56:119–126. https://doi.org/10.2334/josnusd.56.119

Jiang Z, Han B, Li H, Yang Y, Liu W (2015) Carboxymethyl chitosan represses tumor angiogenesis in vitro and in vivo. Carbohydr Polym 129:1–8. https://doi.org/10.1016/j.carbpol.2015.04.040

Gao J, Zhao Y, Wang C, Ji H, Yu J, Liu C, Liu A (2020) A novel synthetic chitosan selenate (CS) induces apoptosis in A549 lung cancer cells via the Fas/FasL pathway. Int J Biol Macromol 158:689–697. https://doi.org/10.1016/j.ijbiomac.2020.05.016

Hasanzade Z, Raissi H (2019) Assessment of the chitosan-functionalized graphene oxide as a carrier for loading thioguanine, an antitumor drug and effect of urea on adsorption process: Combination of DFT computational and molecular dynamics simulation studies. J Biomol Struct Dyn 37:2487–2497. https://doi.org/10.1080/07391102.2018.1496140

Lee J-Y, Termsarasab U, Lee MY, Kim D-H, Lee SY, Kim JS, Cho H-J, Kim D-D (2017) Chemosensitizing indomethacin-conjugated chitosan oligosaccharide nanoparticles for tumor-targeted drug delivery. Acta Biomater 57:262–273. https://doi.org/10.1016/j.actbio.2017.05.012

Li P-W, Wang G, Yang Z-M, Duan W, Peng Z, Kong L-X, Wang Q-H (2016) Development of drug-loaded chitosan–vanillin nanoparticles and its cytotoxicity against HT-29 cells. Drug Deliv 23:30–35. https://doi.org/10.3109/10717544.2014.900590

Fiorillo L (2019) Chlorhexidine Gel Use in the Oral District: A Systematic Review. Gels 5:31. https://doi.org/10.3390/gels5020031

Husain S, Al-Samadani KH, Najeeb S, Zafar MS, Khurshid Z, Zohaib S, Qasim SB (2017) Chitosan biomaterials for current and potential dental applications. Materials 10:602. https://doi.org/10.3390/ma10060602

Ortiz C, Boyce MC (2008) Bioinspired structural materials. Science 319:1053–1054. https://doi.org/10.1126/science.1154295

Zeza B, Wohlfahrt C, Pilloni A (2017) Chitosan brush for professional removal of plaque in mild peri-implantitis. Minerva Stomatol 66:163–168. https://doi.org/10.23736/s0026-4970.17.04040-7

Pippi R, Santoro M, Cafolla A (2017) The Use of a Chitosan-Derived hemostatic agent for postextraction bleeding control in patients on antiplatelet treatment. J Oral Maxillofac Surg 75:1118–1123. https://doi.org/10.1016/j.joms.2017.01.005

Lope-Lopez J, Jan-Pallí E, González-Navarro B, Jané-Salas E, Estrugo-Devesa A, Milani M (2015) Efficacy of chlorhexidine, dexpanthenol, allantoin and chitosan gel in comparison with bicarbonate oral rinse in controlling post-interventional inflammation, pain and cicatrization in subjects undergoing dental surgery. Curr Med Res Opin 31:2179–2183. https://doi.org/10.1185/03007995.2015.1108909

Uysal T, Akkurt MD, Amasyali M, Ozcan S, Yagci A, Basak F, Sagdic D (2011) Does a chitosan-containing dentifrice prevent demineralization around orthodontic brackets? Angle Orthod 81:319 325. https://doi.org/10.2319/062910-359.1

Mohire NC, Yadav AV (2010) Chitosan-based polyherbal toothpaste: As novel oral hygiene product. Indian J Dent Res 21:380. https://doi.org/10.4103/0970-9290.70808

Hayes M, Carney B, Slater J, Brück W (2008) Mining marine shellfish wastes for bioactive molecules: Chitin and chitosan ndash; Part A: extraction methods. Biotechnol J 3:871–877. https://doi.org/10.1002/biot.200700197

Barriada JL, Herrero R, Prada-Rodríguez D, de Vicente MES (2007) Waste spider crab shell and derived chitin as low-cost materials for cadmium and lead removal. J Chem Technol Biotechnol 82:39–46. https://doi.org/10.1002/jctb.1633

Irawan C, Nata IF, Putra MD, Marisa R, Asnia M, Arifin YF (2018) Biopolymer of Chitosan from Fish Scales as Natural Coagulant for Iron–Contaminated Groundwater Treatment. J Rekayasa Kim Lingkung 13:93–99. https://doi.org/10.23955/rkl.v13i2.10601

Kosyakov VN, Yakovlev NG, Veleshko IE (2002) Application of chitin-containing fiber material "Mycoton" for actinide absorption. J Nucl Sci Technol 39:508–511. https://doi.org/10.1080/00223131.2002.10875518

Utilization of Fish Waste and By-Products for Fish Meal Production as a Potential Feed Ingredient, Fish Waste to Valuable Products: Recent Applications and Research Update

Satheesh Muniasamy, Bharathipriya Rajasekaran, Bharathi subramaniam, Subashini Muniasamy, and Gour Hari Pailan

Abstract Fishmeal is an important component in animal feeds for a variety of reasons. They serve as a primary protein source for animal feeds. In addition to acting as a strong attractant in feed, it is nutrient-dense and provides a highly digestible source of amino acids and essential omega-3 fatty acids. Fishmeal remains one of the most nutritionally balanced components for aquafeed, piglet, and pet diets, despite fast ingredient innovation around the world. Protein is required in all types of animal feed, and the quality of protein distinguishes one ingredient from another. Essential amino acids must be provided in the meal to promote growth. Plant-based concentrates and meals may be high in protein, but they lack some critical amino acids like lysine, methionine, and cysteine, which are abundant in fishmeal. The excellent fatty acid profile of fishmeal is another reason for utilizing fishmeal as a feed ingredient. When raw fish is processed into fishmeal, the majority of the fish oil is removed, leaving between 6 and 10% fat by weight in the meal. This percentage supplies important polyunsaturated fatty acids, particularly vital omega-3 fatty acids (EPA and DHA). Fishmeal output, which is crucial to the animal feed industry, has remained relatively steady year over year, while demand continues to increase rapidly. The amount of ocean-caught fish meal has dropped, while the amount of fishmeal

S. Muniasamy (✉)
Fish Nutrition Biochemistry and Physiology Division, ICAR – Central Institute of Fisheries Education, Versova, Mumbai, Maharashtra 400061, India
e-mail: satheesh.fntmb006@cife.edu.in

B. Rajasekaran
Faculty of Agro-Industry, International Center of Excellence in Seafood Science and Innovation, Prince of Songkla University, Hat Yai, Songkhla 90110, Thailand

B. subramaniam · S. Muniasamy · G. H. Pailan
Dr. M.G.R Fisheries College and Research Insitute, Tamil Nadu Dr. J Jayalalithaa Fisheries University, Thiruvallur District, Ponneri, Tamil Nadu 601204, India
e-mail: ghpailan@cife.edu.in

S. Maqsood et al. (eds.), *Fish Waste to Valuable Products*, Sustainable Materials and Technology, https://doi.org/10.1007/978-981-99-8593-7_13

made from processing wastes, filet trims, and offal has increased. Around 30% of the fishmeal produced globally is made from trimmings that have been up-cycled into sustainable fishmeal, contributing to 51% of all fishmeal produced globally. To meet up the demand for fish meals for the feed industry, efficient utilization of fish wastes, bycatch, and processing wastes for fish meal production via energy-efficient technologies are needed.

Keywords Fish waste · Fish meal · Aquafeed · Poultry feed · Animal feed · Wet rendering · Dry rendering · Waste utilization · Fish by-products

1 Introduction

Nutritionists perceive fishmeal as a superior, highly digestible feed ingredient that is preferred for inclusion in the diet of most animal production, particularly cattle, poultry, swine, fish, and shrimp. The calorie content of fishmeal is high per unit weight, and it is a great source of protein, lipids, minerals, and vitamins with less amount of carbohydrates. Fishmeal has been used as an additive in feed formulation to enhance the growth and performance of farmed animals (Ashbrook 1917). It is mostly prepared from small, wild-caught marine fishes that are bony and oily and considered unfit for direct human consumption due to their low market values. Since the majority of these fish are caught for the sole purpose of producing fishmeal and fish oil and it is regarded as "industrial" fish. On the other hand, by-catch from other fisheries and discards like head, skin, trimmings, etc., from fish processing industries are also used as raw material for fish meal production. In general, 90% of fish meal is produced from oily fish species such as sardine, anchovy, capelin, and menhaden whereas, white fish offal from cod and haddock contributes less than 10%. Currently, the fish meal supply is constant between 6.0 and 6.5 million tons per year (FAO 2019). Approximately, 4–5 tons of raw material are required to produce a ton of dry fish meal. Countries such as Peru, Chile, China, Thailand, Iceland, Norway, Denmark, the United States, and Japan contribute one-third of the world's fishmeal production. Fishmeal supplements showed a synergistic effect with plant and animal proteins in the feed, which promotes growth and enhances disease resistance by boosting the immune system. The increasing demand for fish meal as a feed additive has created competition in the animal feed industry and is becoming an expensive protein source. Several reports have been documented to partially or fully replace the fish meal with a less expensive and widely available protein source in feed formulation. In this chapter, global trends, nutritional value, and effective production techniques of the fish meal were reviewed. Furthermore, the potential benefits of fish meal supplements in livestock, poultry, and aquafeed feed were revisited. Finally, present challenges associated with the application of fish meal and its alternative solutions were addressed.

2 Global Status of Fish Meal Production

The production of fishmeal represents a large but dwindling share of global fisheries. Whole fish or leftovers from the fish processing industries are used for the production of fishmeal. Primarily, small pelagic fishes including *Peruvian anchoveta* (which accounts for the largest part), menhaden, blue whiting, capelin, sardine, mackerel, and herring are utilized as whole fish. Fishmeal and fish oil production vary in accordance with variations in catches of certain species, particularly anchoveta, which is influenced by the El Nino-southern oscillation that influences stock abundance. FAO report documented that *Peruvian anchoveta* catches reduced from 18 million tons (2018) to 16 million tons (2020), respectively. This equates to around 20% of overall marine catch fisheries. The gradual decrease in fishmeal supply has been accompanied by the increasing demand due to the expansion of the aquaculture industry, as well as pig and poultry farming, pet food, and pharmaceutical industries. According to the International Fishmeal and Fish Oil Organization (2020), approximately 86% of fishmeal was utilized in aquafeed, 9% in pig feed, 1% in poultry feed, and 4% in pet food. Whereas, 73% of fish oil was intended for aquaculture, 16% for human consumption, and 11% for other applications (including pet food and biofuel).

In 2030, it is anticipated that the production of fishmeal will increase by 11–13% as compared to 2020. The proportion of fish catches conversion into fishmeal is expected to decrease from 18% in 2020 to 17% in 2030, respectively. However, the production of fishmeal is anticipated to rise due to the overall growth in capture fisheries and the utilization of fish waste from processing sectors. The total fishmeal produced from fish waste is anticipated to rise from 27% in 2020 to 29% in 2030. According to IFFO, by-products from processing industries contribute to 27% of the world's total production of fishmeal in 2020. Fishmeal is regarded as the most nutrient-dense and easily absorbed component in animal diets and aquafeed. However, there has been a noticeable decline in their inclusion rates in feed formulation as a result of supply and price volatility and steadily rising demand from the feed industries. In recent days, fishmeal and fish oil are being utilized more and more selectively at particular points in the production process, such as for diets for hatcheries, brood stock, and finishing, while their inclusion in grower diets is declining. Therefore, nutritionists have been working to replace fishmeal, particularly with insect meal without compromising nutritive benefits (Hua et al. 2019).

3 Production Process of Fish Meal

The most common and commercial methods to produce fish meal include the wet reduction process also known as wet rendering and the dry reduction process or dry rendering. The process methods for the production of fish meal will be selected based on the type of raw materials (whole fish, by-catch, or by-product) and lipid content (fatty or lean fish). The wet reduction process is the traditional method for the production of fishmeal. The process involves cooking the raw material, pressing of cooked material, and finally drying it to a lower moisture content suitable for transportation and storage without bacterial and mold growth. The separation of oil and water from the solid portions is the fundamental processing step for producing high-quality fishmeal. Firstly, raw material passes through a lengthy, steam-jacketed screw conveyor cylinder for cooking. The cooking temperature is around 95 to 100 °C for 15 to 20 min. The cooking process denatures and/or coagulates the protein and ruptures the cell wall of tissue thereby oil and water can be separated from the fish muscle. This crucial step sterilizes the raw material and helps to remove the mixture of water, oil, and soluble protein in the form of "liquor." Secondly, the pressing operation aids to separate the cooked material into two distinct phases. After being cooked, the liquid (press liquor) is extracted by pressing, leaving behind a solid residue known as "press cake." Pressing can be done using a screw press. The double screw press is the most preferable compared to the single screw press which removes the maximum amount of oil and water from the press cake. Approximately, the press cake contains about 45–55% moisture and 2–3% oil after passing through the screw press. Then, press liquor is centrifuged to separate oil and stick water. The extracted oil is refined more than twice before being delivered to storage tanks. It is crucial to stabilize the oil to preserve its quality before storing it in the storage tank. The extracted oil should not be exposed to air, heat, or light while being stored. Stick water is a term that refers to the liquid that remains after the separation of oil and suspended particles from the press liquor (about 65% of the raw material). Stick water contains vitamins, minerals, leftover oil, and soluble and undissolved proteins (20%). Stick water is evaporated until it reaches the consistency of a thick liquid that contains 30–50% solids. This substance can be added back to the press cake or it can be marketed as "condensed fish soluble." Consequently, one has the option to buy a "press cake" meal or a "full" dinner (where all of the solubles have been added back). Pressing is frequently skipped when a small quantity of oil is present in raw material, e.g., lean fish. The solid press cake turns into large lumps after pressing, which reduces the efficiency of drying. The fluffing process must be carried out before drying to break down the lumps into small particles (<1 cm). Finally, the drying can be done using a direct drier (flame drier) or indirect drier (steam-heated drier). The temperature of hot air is around 600 °C, however, particles of press cake reach only 80 °C due to evaporative cooling produced by rapid evaporation of water from their surface. The wet meal was dried in a dryer to < 10% moisture content within 15–20 min.

The dry reduction process is a batch operation and is normally used for the production of fishmeal from non-oily fish and fish offal. In this process, the raw material is coarsely grounded via a hacker or mixer. The grounded material was cooked in a steam-jacked cooker and it also acts as a drier which is referred to as cooker-drier. After drying, the fish meal is cooled to room temperature by the cooler. Then, fish meal was subjected to milling, sieving, and packed in polyethylene line jute bags or polyethylene line HDPE woven sacks. This process was not been commercially used in recent times due to high production costs. The produced fish meal is a dry product that has a brownish-grey color. The fishmeal may become burned and lose some of its nutritious content if the drying process is not adequately regulated or if it is over-dried. It is recommended to store the fish meal in a cool, dry location away from animals and birds.

4 Nutritive Value of Fishmeal

Fish meal is an essential component of the diet for the majority of aquaculture species as well as many land-based farm animals due to its high quality and balanced nutrients, particularly essential amino acids, key fatty acids, vitamins, and minerals. Fish meal is well-known for its nutritious richness, excellent digestibility, and palatability.

4.1 Proteins in Fish Meal

On average, the high-quality fishmeal has a crude protein content of about 60–72%. Protein is a necessary component of any diet; however, its nutritional value depends on the balance of its amino acids and absorption. Fishmeal is a well-known protein supplement in animal feeds due to its amino acid composition. The limiting amino acids such as lysine and methionine are typically lacking in the proteins found in cereal grains and other plant concentrates. However, soybean and other legume meals are excellent sources of lysine and tryptophan but are deficient in sulfur-containing amino acids (methionine and cystine). Even when adequately prepared, plant-based proteins typically do not have the same level of digestibility as fishmeal, and their use in diets is frequently constrained since it reduces feed intake and development rates. Fishmeal has an overall protein digestibility level exceeding 95%. In contrast, depending on the species of plant, the digestibility of plant-based proteins varies substantially from 77 to 96% due to the presence of non-digestible carbohydrates (oligosaccharides) or structural fiber components (cellulose). The application of fish meal as supplementation is preferable more than plant proteins due to the absence of nutritional inhibitors or anti-nutritional elements. Anti-nutritional factors are harmful and interfere with nutrient digestion, absorption, or metabolism. For instance, the trypsin-inhibitor, a naturally occurring anti-nutritional component in raw soybean retards the action of the digestive enzyme trypsin from breaking down meal proteins in animals' intestines. Chickpea lectrogens prevent collagen formation, which is the prevalent protein in connective tissue and offers structural support. Another antinutritional component in cottonseed meal/oil named gossypol negatively affects animals by decreasing fertility.

4.2 Lipids in Fish Meal

Even though the majority of the oil is removed during the processing of fish meal, the leftover lipid accounts for about 6 to 10% of the total weight, however, it can range from 4 to 20%. Fish lipids are easily absorbed and are a great source of n-3 fatty acids such as linolenic acid, eicosapentaenoic acid, and docosahexaenoic acid. Small-size algae and zooplankton that are eaten by fish transmit the n-3 fatty acids through the food chain. In general, essential fatty acids are required for growth and reproduction as well as appropriate larval development. Also, PUFA seems to support the immune system in fighting off and reducing the stress response and disease agents. Beneficial phospholipids, fat-soluble vitamins, and steroid hormones are also present in fishmeal. However, species, physiology, sex, age, and feeding habits impact the amount and quality of oil in the fish meal. All animals, including fish, shrimp, poultry, pigs, and ruminants like cows, sheep, and goats can easily digest the lipids in fishmeal and fish oil. The digestibility of lipids in these animals is more than 90%.

4.3 Mineral and Vitamins in Fish Meal

A good quality fish meal typically has an ash content of about 17–25%. Higher mineral content, particularly calcium, phosphorus, and magnesium are present in fishmeal. Most of the ash in fishmeal is made up of calcium and phosphorus. Unlike phosphorus found in plants, which is not readily available to most animals. Whereas, fish meal contains phosphorus largely found in the organic form known as phytate. Furthermore, phosphorus in plants is not easily absorbed by monogastric animals (those with a one-compartment stomach, such as pigs, dogs, and humans). Whereas, the microbial community in the rumen's stomach allows ruminants (cows, sheep, and goats) to use the phosphorus in phytate. Fish origin and composition, meal processing technique and product freshness are a few variables that might affect the vitamin content of fishmeal. Fishmeal has comparatively low fat-soluble vitamin content due to the removal of oil during the production process. Cobalamin, niacin, choline, pantothenic acid, and riboflavin are among the B-complex vitamins that are rich in fish meal with moderate concentration.

4.4 Fishmeal Quality Analysis and Its Management

One of the most important criteria for meeting the nutrient requirements of growing animals is the quality of fishmeal. Because the growth performance of the animals depends on the quality of the fishmeal. Generally, the quality of the fishmeal is determined by the freshness of the fish and processing conditions (De Koning 2002). Some analytical methods, such as total volatile nitrogen and free fatty acids, are used to express the quality. The total volatile nitrogen content depends on the temperature and duration. The enzymes generated by the free fatty acids initially increase to their maximum and then decrease, so the FFA is not used as a stable index. Because the long-chain FFA is turned into short-chain volatile and water-soluble FFA. Additionally, phospholipid levels are also used to detect the quality of fishmeal. Normally, the fresh fish (wet) phospholipid level ranges between 0.5 and 0.8%. During processing, most of the phospholipids end up with the meal (Sandfeld 1983; De Koning et al. 1990, 1985). Apart from that, some of the general points should also be remembered while processing fishmeal. (a) Raw material quality; (b) processing temperature; (c) fat stability; and (d) plant, machine, and utensil hygiene. The following physical properties of the fishmeal are very important for preparing good quality feed: The color of the fishmeal should be light brown to dark brown in color. The odor of the fishmeal should be free from rancidity. Fine granules and powder form are important texture properties of the fishmeal, and it should also be free from any contamination. Based on the above quality parameters and the physical properties, normally there are three grades of fishmeal available.

Nutrients	Grade 1	Grade 2	Grade 3
Moisture (% maximum)	10	10	10
Crude protein (% minimum)	60	50	40
Crude lipid (%, maximum)	8	10	11
Salinity (%, maximum)	2	3	5
Ash (%, maximum)	2	3	4
Hard and sharp solid materials	Not permitted		
Total volatile nitrogen (mg/100 g, maximum)	150	250	350

Source (FAO, Feed and Feed Ingredient Standards)

Proper management practices should be used to produce fishmeal of higher grade. One of the most crucial steps in maintaining freshness is handling the raw materials. The biological breakdown of raw materials is typically slowed down by cooling and icing them. Moisture must be kept out of the fishmeal. To prevent condensation and drip at night, which would lead to localized mold growth and lumping in the fish meal, the inner surface of the roof should, if required, be insulated. Ceilings should also be installed in the stores. Fishmeal should only be kept outdoors in arid areas. Whether treated with an antioxidant or kept after curing, the fish meal must be kept from excessive self-heating. Because of this, the width of bulk storage units shouldn't be greater than 5 m (FAO 1971).

4.5 Unknown Growth Factor in Fish Meal

It had been hypothesized that fishmeal has unknown growth factors that remain unexplained, frequently referred to as UGF. In numerous feeding trials, animals fed diets with a comparable amino acid composition to a fish meal have developed less well than those fed with fishmeal itself. Despite this fact, the UGF component has not been isolated yet. In some reports, it has been documented that carefully supplemented plant protein diets have shown similar outcomes as comparable to fishmeal. The UGF trait might be attributed to the dietary balance rather than the presence of some unidentified component. Fishmeal from commercial sources is unlikely to be able to keep up with the anticipated growth in global production of aqua and animal feeds. Aquaculture has recently consumed about 46% of the yearly fishmeal production; this percentage is anticipated to rise over the next ten years as demand for aquaculture products rises. The best possible use of fishmeal in animal diets is to reduce feeding costs, which contribute 40% or more of operating costs. In recent years, fishmeal has become one of the most expensive feedstuffs due to its high concentration of protein. Since 2000, the price of high-quality fishmeal ($\leq$65% protein) has varied between $385 and $554 per ton and it is 2.0–3.5 times more costly than soybean meal.

5 Fish Meal as Feed Ingredients

Fishmeal is a great source of digestible energy for farmed animals including pigs, sheep, cattle, poultry, and fish. It has a gross energy of 21.9 MJ kg of dry matter and is nutritionally dense. Due to high digestibility, fishmeal is frequently used to improve feed conversion ratio (FCR) and faster growth rate in young pigs and poultry. Further, reduction in unabsorbed nutrients leads to lower levels of nitrogen and phosphorus excretion. Many of the micronutrients in fishmeal are known to have positive effects on the health and welfare of farm animals, as well as the quality of the final product.

5.1 Fish Meal in Aquafeed

The finfish and crustacean aquaculture sector are still highly dependent upon marine capture fisheries for sourcing key dietary nutrient inputs, including fish meal and fish oil. This dependency is particularly strong within compound aquafeeds. The inclusion level of fishmeal in aquafeeds varies between the same species, e.g., 5–40% in shrimp, 20–50% in salmon, 15–55% in trout, 40–80% in eel, 7–70% in marine fishes, 5–20% in tilapia, 1–5% in milkfish, 5–25% in freshwater prawn, 0–20% in Chinese carps, and 3–40% in catfish. This variation depends greatly on stocking density, water and feed management, natural food availability, species differences, etc. Fishmeal is a proteinaceous flour added as a primary protein source in aquafeed. However, the inclusion of fishmeal as the primary protein source in compounded aquafeeds is now threatening feed manufacturers to depend on it. Accordingly, feed formulators look for alternative feedstuffs that can replace fishmeal with no adverse effects on fish performance (Daniel 2018). In the meantime, fish wastes meal is readily available, less expensive than fishmeal, considered to be suitable, and have a sustainable supply for replacing fishmeal in commercial aquafeeds (Baeza-Ariño et al. 2016).

The efficacy of using trout-offal as a protein ingredient in gilt-head bream (*Sparus aurata*) diets was researched by Kotzamanis et al. (2001). Trout-offal (intestine, frames, heads, skeletons, and fins) was finely minced and thoroughly mixed with the other dietary components to form a homogenous paste, then pelleted out. The study concluded that trout-offal is a suitable, non-polluting alternative for fishmeal in seabream diets. European sea bass (*Dicentrarchus labrax*) was evaluated with different animal protein sources as dietary fishmeal replacers including fermented prawn waste liquor at 30% in juvenile feeds (Nor et al. 2011), and a mixture of shrimp and tilapia protein hydrolysates in combination with poultry by-products meal at 15% (Robert 2014). For other marine fish species, tuna muscle by-product powder replaced 50% of dietary fishmeal. Meanwhile, whole tuna by-products replaced 30% without reducing the growth performance of olive flounder (*Paralichthys olivaceus*) (Uyan et al. 2006; Kim et al. 2014a, b). Recently, Muttharasi et al. (2019) suggested that *Rastrelliger kanagurta*, *Sphyraena barracuda*, and *Fenneropenaeus* indicus disposal meals can be used as complete fishmeal alternates to obtain low-cost feeds for farmed carp, *Cyprinus carpio*. In addition, Saleh et al. (2020) concluded that tilapia by-products replaced 30% of fishmeal in the European sea bass (*D. labrax*) diet without compromising fish growth or health.

5.2 *Fish Meal in Animal Feed*

Fishmeal is also fed to dogs, cats, pigs, cattle, and other terrestrial animals. Typically, fishmeal makes up around 10% of pigs' diet. In pig starter fed, soybean meal was substituted with menhaden fishmeal had better growth performance and average daily feed intake in weanling pigs (Stoner et al. 1990). Pigs fed with a diet incorporated with fishmeal (22.5%) resulted in higher weight gain as compared to soybean meal and rapeseed meal (Kyriazakis and Emmans 1993). Sanderson et al. (2001) documented that the inclusion of fishmeal in the daily diet is an effective way to enhance protein adsorption and deposition in young cattle. Heifers fed with a diet supplemented with fishmeal showed higher average daily gain as compared to those fed with a diet containing urea and soybean meal (Oldham and Smith 1982). Further, milk yield was noticed higher in early lactating cows when fed with the fishmeal-added diet (Miller et al. 1972). Intake of fishmeal at 1.19 kg/day from a diet supplemented with 5% of fishmeal resulted in increased milk and protein yields of Holstein cows within

10–20 weeks (Adachi et al. 2000). Esteban et al. (2007) reported that fresh fish can be utilized as a substitute ingredient in swine feed along with other conventional protein sources such as soybean meal to partially fulfill protein requirements. Generally, fishmeal is included at a range of 5–20% in dairy cows' feeds to increase the yield of milk due to the presence of balanced amino acids. Beef requires high-quality protein compared to calves to achieve satisfactory rates of growth in a particular period immediately after weaning. Further, fishmeal is generally given to the pregnant ewe to meet the difficulty in late pregnancy, particularly when carrying twins or triplets. Therefore, it is essential to provide a high-quality diet to meet the requirements for developing fetuses. Research has shown that the inclusion of fish meals can result in heavier lambs at birth, therefore its use has welfare benefits for ewes and lambs.

5.3 Fish Meal in Poultry Feed

In order to achieve economic rates of growth or egg production, it is necessary to feed poultry with high-quality diets with high concentrations of essential nutrients. Islam et al. (1997) reported that fishmeal substituted broiler feed (12%) resulted in higher body weight gain, feed intake, and feed efficiency. Feeding fishmeal or fish oil has been shown to increase the PUFA content of poultry tissue without adversely affecting the eating quality. Also, increasing the n-3 content of eggs has been of particular interest which significantly contributes to the additional n-3 required in the human diet. Ockerman (1992) postulated that the use of fishmeal as a supplement increases the n-3 content in the poultry flesh, thus making these tissues to be another source of n-3 fatty acids for the human diet. It is recognized that fishmeal added to poultry feed can reduce infection, improve immune status and reduce inflammation in poultry farming. A desirable ratio of n-6 to n-3 fatty acids is about 4:1. However, commonly used feed has an oversupply of n-6 and an undersupply of n-3 fatty acids e.g., 1:8, 1:11, and 1:10 which causes an imbalance in the diet. The inclusion of fishmeal into the diet increases the n-3 fatty acids thereby the overall desired ratio (4:1) can be obtained. Fishmeal is recognized as safe and natural and is an effective feed ingredient for poultry. The Lion Eggs standards permit the use of fishmeal in pullet and layer diets, although diets for commercial laying hens should be free from any feed material that is likely to produce taint (or toxins) in eggs. It has been recommended that an inclusion of 15 g/kg fish meal in the diet is the maximum used to achieve a favorable taste assessment and unchanged shelf life of the eggs.

6 Replacement of Fishmeal with Alternative Protein Sources

Due to the rising production of aquaculture, the demand for feed resources has increased, especially for fishmeal with a high protein content (Tacon and Metian 2008). Fishmeal has been used as a crucial protein source for both carnivorous and omnivorous species cultured in aquaculture as well as for terrestrial animals and poultry (Sammadar 2018). Considering its increasing demand and decreasing supply urges researchers to search for alternatives to fishmeal in diets (Hardy 2010). The alternative ingredient must have high protein content, good amino acid profile, high nutrient digestibility, and low fiber level (Gatlin et al. 2007).

6.1 Animal-Based Protein Sources

Meat and bone meal is an effective source of protein and minerals for the production of feed. According to Moutinho et al. (2017), the growth and feed efficiency of juvenile gilthead seabream remains unaffected when meat and bone meal replaced 50% of fish meal in the starter diet. Sato and Kikuchi (1997) documented that Japanese flounder substituted with 60% meat meal does not alter weight gain, feed efficiency, and protein efficiency ratio. Grouper (*Epinephelus coioides*) were fed with a mixture of meat and bone meal (4:1) at varied inclusion levels, in which up to 80% of fishmeal replacement had no significant differences in the growth results (Millamena 2002). Yang et al. (2004) studied the effect of meat and bone meal on carp (*Carrassius auratus gibelius*) and concluded that fishmeal can be replaced for up to 50% without affecting the growth rate. For *C. auratus*, feed containing a partial replacement of fish meal with meat and the bone meal alone or in combination with lysine and methionine supplements demonstrated the best growth results (Hu et al. 2008). According to Keramat et al. (2014), poultry by-products have the ability to replace fishmeal in rainbow trout diets by up to 33% without negatively affecting growth performance. Fowler (1991) reported that 50% of fishmeal was replaced with poultry by-products in chinook salmon's feed (*Oncorhynchus tshawytscha*). Kureshy et al. (2000) found that at different inclusion levels poultry by-products could efficiently substitute 67% of fishmeal in the diet of red drum juveniles. Seabass growth was unaffected when replacing fish meal (100%) with turkey meal in feeds (Muzinic et al. 2006). Another study by Bureau et al. (2000) studied the feather meal and bone meals from different origins as a protein source in rainbow trout (*Oncorhynchus mykiss*) diets and showed that it did not affect the protein and energy retention of the fish.

6.2 Plant-Based Protein Sources

The efficiency of various plant protein resources (soybean, linseed, canola, sunflower, cottonseed, etc.) at appropriate levels as a partial or complete replacement for fish meal has been evaluated in fish diets to reduce the cost of feed (Antolovic et al. 2012; Badwy et al. 2008; Soltan et al. 2008). According to Ajani et al. (2016), soybean meal can be used to either partially or entirely replace fishmeal in the diet of *O. niloticus* without affecting growth or nutrient utilization. Rab et al. (2008) studied the effect of soybean meal in the channel catfish diet (*Ictalurus punctatus*) with increasing levels (10 to 40%). The results showed that fish fed with a 30% substitution of soybean meal had desirable fish growth and development. Kasper et al. (2007) replaced up to 47.6% of the fish meal in yellow perch diets with soybean meal won't have an impact on the fish's feed intake, weight gain, feed efficiency ratio, or survival. Zhou et al. (2005) investigated the growth performance of young cobia (*Rachycentron canadum*) fed diets containing varying levels of soybean meal from 0 to 60%. The outcomes demonstrated that the 30% soybean meal can be used without affecting fish growth. Jindal et al. (2007) examined the growth performance of Channa punctatus, in which the fingerlings fed diets containing soybean meal at high (75–100%) inclusion levels exhibit improved live weight gain, percent weight gain, specific growth rate, apparent protein digestibility, protein efficiency ratio, and energy retention.

7 Future Prospects

The diminishing global supplies of wild caught and increasing market price of small pelagic fish due to increasing fishing costs as well as increasing demand for fish products in human diets destined for reduction into fish meal. This directly results in the cost of fish meal and it consequently pressures feed manufacturers for dietary substitution so as to remain profitable. In upcoming years, the application of fishmeal as a protein source in feed will no longer be economically sustainable. Therefore, in agreement with IFFO, fish meal used in the long term will be increasingly targeted as a specialty feed ingredient in the higher-value starter, finisher, and broodstock feeds (Jackson 2009). Utilization of fish waste meal, animal by-products, poultry by-products, and balanced plant-based protein sources can be used widely as a substitution for fish meal. This reduces the cost of feed as well as environmentally friendly techniques.

8 Conclusion

Fishmeal and fish oil, which were once thought of as commodities, are now correctly positioned as important elements in the market. They are increasingly being employed in tailored diets for specialty goods or to suit the more demanding nutritional needs of young or brood animals. Their great significance in animal feeds goes beyond just improving the animal's growth performance, with several micronutrients, in particular, known to support healthy physiological function. The bioavailability of the essential amino acids is excellent, and lysine and methionine are particularly abundant in fishmeal. For a feed ingredient, this is a comparably broad and rich nutritional spectrum that supports growth and ideal physiological function during the most delicate periods of the life cycle. Overall, the advantages of using fishmeal and fish oil in terrestrial animal feed go much beyond their obvious benefits as a source of dietary energy and crude protein. With regard to their potential contribution as functional components to the growth and health of the farmed animal, they are more valuable than the bulk of other ingredients.

Acknowledgements ICAR – National Talent Scholarship for Satheesh Muniasamy from the Indian Council of Agricultural Research, India and PSU President Scholarship for Bharathipriya Rajasekaran from Prince of Songkla University, Thailand were gratefully acknowledged.

Author Contribution **Satheesh Muniasamy**: Conceptualization, Data curation, Writing original draft. **Bharathipriya Rajasekaran**: Conceptualization, Data curation, Resources, Supervision Writing-review & editing. **Bharathi Subramaniam:** Conceptualization, Data curation, Resources, Writing-review & editing. **Subashini Muniasamy**: Conceptualization, Data curation, Writing-review & editing, **Gour Hari Pailan:** Conceptualization, Data curation, Writing-review & editing.

Conflict of Interest The authors have declared no conflicts of interest.

References

Adachi N, Suzuki K, Kasai K, Hiroki M, Kume S, Nonaka I, Abe A (2000) Effect of supplemental fish meal on milk yield and milk composition of Holstein cows during early lactation. Asian Australas J Anim Sci 13(3):329–333

Ajani EK, Orisasona O, Omitoyin BO, Osho EF (2016) Research article total replacement of fishmeal by soybean meal with or without methionine fortification in the diets of Nile Tilapia, Oreochromis niloticus. Oreochromis niloticus

Antolovic N, Kožul V, Antolovi63 M, a Bolotin J (2012) Effects of partial replacement of fish meal by soybean meal on growth of juvenile saddled bream (Sparidae). Turkish J Fisher Aquatic Sci 12(2)

Ashbrook FG (1917) Fish meal as a feed for swine

Badwy TM, Ibrahim EM, Zeinhom MM (2008) Partial replacement of fishmeal with dried microalga (Chlorella spp. and Scenedesmus spp.) in Nile tilapia (Oreochromis niloticus) diets. In *8th International Symposium on Tilapia in Aquaculture* (Vol. 2008, pp. 801–810)

Baeza-Ariño R, Martínez-Llorens S, Nogales-Mérida S, Jover-Cerda M, Tomás-Vidal A (2016) Study of liver and gut alterations in sea bream, S parus aurata L., fed a mixture of vegetable protein concentrates. Aquacul Res 47(2):460–471

Bureau DP, Harris AM, Bevan DJ, Simmons LA, Azevedo PA, Cho CY (2000) Feather meals and meat and bone meals from different origins as protein sources in rainbow trout (Oncorhynchus mykiss) diets. Aquaculture 181(3–4):281–291

Daniel N (2018) A review on replacing fish meal in aqua feeds using plant protein sources. Intern J Fish Aquatic Stud 6(2):164–179

De Koning AJ (2002) Quantitative quality tests for fish meal. II. An investigation of the quality of South African fish meals and the validity of a number of chemical quality indices. Intern J Food Prop 5(3):495–507

De Koning AJ, Milkovitch M, Wessels JPH (1985) The free fatty acid content of fish oil. FFA formation during pre-processing storage of anchovy at different temperatures. Fishing Industry Research Institute Progress Report 244, 175

De Koning AJ, Mol TH, Przybylak PF, Thornton SJ (1990) The free fatty acid content of fish oil. Part II. The effect of anchovy quality on the free fatty acid content of the resulting anchovy oil and meal. Fat Sci Technol 92:1937197

Esteban MB, Garcia AJ, Ramos P, Márquez MC (2007) Evaluation of fruit–vegetable and fish wastes as alternative feedstuffs in pig diets. Waste Manage 27(2):193–200

FAO (1971) The production of fish meal and oil. FAO Fisheries Technical p 142. Rome

Fowler LG (1991) Poultry by-product meal as a dietary protein source in fall chinook salmon diets. Aquaculture 99(3–4):309–321

Gatlin DM III, Barrows FT, Brown P, Dabrowski K, Gaylord TG, Hardy RW, …, Wurtele E (2007) Expanding the utilization of sustainable plant products in aquafeeds: a review. Aquac Res 38(6):551–579

Hardy RW (2010) Utilization of plant proteins in fish diets: effects of global demand and supplies of fishmeal. Aquac Res 41(5):770–776

Hu M, Wang Y, Wang Q, Zhao M, Xiong B, Qian X, … Luo Z (2008) Replacement of fish meal by rendered animal protein ingredients with lysine and methionine supplementation to practical diets for gibel carp, Carassius Auratus Gibelio. Aquaculture 275(1–4):260–265

Hua K, Cobcroft JM, Cole A, Condon K, Jerry DR, Mangott A,…, Strugnell JM (2019) The future of aquatic protein: implications for protein sources in aquaculture diets. One Earth 1(3):316–329

Islam KMS, Tareque AMM, Howlider MAR (1997) Complete replacement of dietary fish meal by duckweed and soybean meal on the performance of broilers. Asian Australas J Anim Sci 10(6):629–634

Jackson A (2009) The continuing demand for sustainable fishmeal and fish oil in aquaculture diets. Intern Aquafeed 12(5):32–36

Jindal M, Garg SK, Yadava NK, Gupta RK (2007) Effect of replacement of fishmeal with processed soybean on growth performance and nutrient retention in Channa punctatus (bloch.) fingerlings. Livestock Res Rural Develop 19(11)

Kasper CS, Watkins BA, Brown PB (2007) Evaluation of two soybean meals fed to yellow perch (Perca flavescens). Aquac Nutr 13(6):431–438

Keramat Amirkolaie A, Shahsavari M, Hedayatyfard M (2014) Full replacement of fishmeal by poultry by–product meal in rainbow trout, Oncorhynchus mykiss (Walbaum, 1972) diet

Kim HS, Jung WG, Myung SH, Cho SH, Kim DS (2014a) Substitution effects of fishmeal with tuna byproduct meal in the diet on growth, body composition, plasma chemistry and amino acid profiles of juvenile olive flounder (Paralichthys olivaceus). Aquaculture 431:92–98

Kim KW, Lee JH, Bae KM, Kim KD, Lee BJ, Han HS, Kim SS (2014b) Comparison of Extruded and Moist Pellets for Whole-body Proximate Composition and Growth Performance of Olive Flounder Paralichthys olivaceus. Korean J Fish Aquatic Sci 47(6):810–817

Kotzamanis YP, Alexis MN, Andriopoulou A, Castritsi-Cathariou I, Fotis G (2001) Utilization of waste material resulting from trout processing in gilthead bream (Sparus aurata L.) diets. Aquac Res 32:288–295

Kureshy N, Davis DA, Arnold CR (2000) Partial replacement of fish meal with meat-and-bone meal, flash-dried poultry by-product meal, and enzyme-digested poultry by-product meal in practical diets for juvenile red drum. N Am J Aquac 62(4):266–272

Kyriazakis I, Emmans GC (1993) The effect of protein source on the diets selected by pigs given a choice between a low and high protein food. Physiol Behav 53(4):683–688

Millamena OM (2002) Replacement of fish meal by animal by-product meals in a practical diet for grow-out culture of grouper Epinephelus coioides. Aquaculture 204(1–2):75–84

Miller D, Soares JH Jr, Sanders M (1972) Growth and plasma amino acid pattern of chicks fed fish meal as sole source protein: effect of dietary levels of chloride, sulfate, glutamic acid and methionine. Poult Sci 51(1):171–177

Moutinho S, Peres H, Serra C, Martínez-Llorens S, Tomás-Vidal A, Jover-Cerdá M, Oliva-Teles A (2017) Meat and bone meal as partial replacement of fishmeal in diets for gilthead sea bream (Sparus aurata) juveniles: diets digestibility, digestive function, and microbiota modulation. Aquaculture 479:721–731

Muttharasi C, Muralisankar T, Uthayakumar V, Gayathri V, Thangal SH, Anandhan K (2019) Utilization of marine fisheries wastes for the production of the freshwater fish Cyprinus carpio. Trop Anim Health Prod 51(8):2305–2313

Muzinic LA, Thompson KR, Metts LS, Dasgupta S, Webster CD (2006) Use of turkey meal as partial and total replacement of fish meal in practical diets for sunshine bass (Morone chrysops × Morone saxatilis) grown in tanks. Aquac Nutr 12(1):71–81

Nor NM, Zakaria Z, Manaf MSA, Salleh MM (2011) The effect of partial replacement of dietary fishmeal with fermented prawn waste liquor on juvenile sea bass growth. J Appl Aquac 23(1):51–57

Ockerman HW (1992) Pharmaceutical and biological products. In Inedible Meat by-Products (pp 283–328). Springer, Dordrecht

Oldham JD, Smith T (1982) Protein–energy interrelationships for growing and for lactating cattle. In *Protein contribution of feedstuffs for ruminants* (pp 103–130). Butterworth-Heinemann

Pei Z, Xie S, Lei W, Zhu X, Yang Y (2004) Comparative study on the effect of dietary lipid level on growth and feed utilization for gibel carp (Carassius auratus gibelio) and Chinese longsnout catfish (Leiocassis longirostris Günther). Aquac Nutr 10(4):209–216

Rab A, Khan SU, Afzal M, Ali MR, Qayyum M (2008) Replacement of fishmeal with soybean meal in diets for channel catfish, Ictalurus punctatus fry introduced in Pakistan. Pakistan J Zool 40(5)

Robert M, Zatylny-Gaudin C, Fournier V, Corre E, Le Corguillé G, Bernay B, Henry J (2014) Transcriptomic and peptidomic analysis of protein hydrolysates from the white shrimp (L. vannamei). J Biotechnol 186:30–37

Saleh N, Wassef E, Barakat K, Helmy H (2020) Bioconversion of fish-wastes biomass into a fish meal alternate for European seabass (Dicentrarchus labrax) diets. Intern Aquatic Res 12(1):40–52

Samaddar A (2018) A review of fish meal replacement with fermented biodegradable organic wastes in aquaculture. Int J Fish Aquatic Stud 6(6):203–208

Sanderson R, Dhanoa MS, Thomas C, McAllan AB (2001) Fish-meal supplementation of grass silage offered to young steers: effects on growth, body composition and nutrient efficiency. J Agric Sci 137(1):85–96

Sandfeld P (1983) In quality criteria for FishMeal; Association of fish meal and fish oil manufacturers in Denmark, Copenhagen: Denmark, p 221

Sato T, Kikuchi K (1997) Meat meal as a protein source in the diet of Juvenile Japanese Flounde. Fish Sci 63(6):877–880

Soltan M, El-L S (2008) Effect of probiotics and some spices as feed additives on the performance and behaviour of the Nile tilapia, Oreochromis niloticus. Egyp J Aquatic Biol Fish 12(2):63–80

Stoner GR, Allee GL, Nelssen JL, Johnston ME, Goodband RD (1990) Effect of select menhaden fish meal in starter diets for pigs. J Anim Sci 68(9):2729–2735

Tacon AG, Metian M (2008) Global overview on the use of fish meal and fish oil in industrially compounded aquafeeds: Trends and future prospects. Aquaculture 285(1–4):146–158

Uyan O, Koshio S, Teshima SI, Ishikawa M, Thu M, Alam MS, Michael FR (2006) Growth and phosphorus loading by partially replacing fishmeal with tuna muscle by-product powder in the diet of juvenile Japanese flounder, Paralichthys Olivaceus. Aquaculture 257(1–4):437–445

Uyan O, Koshio S, Teshima S, Ishikawa M, Michael FR, Ren T, Laining A (2007) Effects of tuna muscle powder in diet on the growth and phosphorus loading of juvenile red sea bream, Pagrus Major. Aquac Sci 55(1):29–40

Zhou QC, Mai KS, Tan BP, Liu YJ (2005) Partial replacement of fishmeal by soybean meal in diets for juvenile cobia (Rachycentron canadum). Aquac Nutr 11(3)

Fish Waste as a Potential Feed Ingredient for Fish Meal Production

S. M. M. S. Afreen, S. L. Rasmiya Begum, S. M. M. S. Himaya, and M. S. M. Imthiyas

Abstract The fisheries and aquaculture sector has increased production in present, however, the products aren't utilized to their full economic potential, leading to the generation of waste. Disposal of seafood wastes because of not only substantial environmental pollution but also a loss of the prospective value of such goods stressed the prominence of finding suitable modalities to manage fish waste. The experiments have explored, that being a main source of protein, minerals, and vitamins, fish waste could be converted to fish meal as it is the costliest source of protein. Fish meal is one of the most valuable byproducts of the marine industry that isn't used for human food and can be utilized as a feed ingredient not only for fish but for other animals too. On the other hand, production of fish meal is challenged due to lower catch from capture fishing, increased cost, and put forward to utilize the alternative sources for fish meal production. Fish byproducts are employed in the creation of fish meals to address these difficulties and could reduce fishing pressure on meal-targeted species. Accordingly, the book chapter discusses elucidating the utilization of fish waste and byproducts as a source for fish meal production, factors influencing the production, and merits and demerits addressed by the production and how those can be overcome.

Keywords Animal feed · Byproducts · Environmental pollution · Immunity · Fish meal · Fish waste management · Nutritive value · Processing · Protein source

S. M. M. S. Afreen (✉) · S. L. R. Begum · S. M. M. S. Himaya
Faculty of Technology, Department of Biosystems Technology, South Eastern University of Sri Lanka, Oluvil, Sri Lanka
e-mail: Afreen0899@gmail.com

M. S. M. Imthiyas
Sri Lanka School of Agriculture, Palamunai, Sri Lanka

S. M. M. S. Afreen
KM/KM/Al-Manar Central College, Kalmunai, Sri Lanka

S. Maqsood et al. (eds.), *Fish Waste to Valuable Products*, Sustainable Materials and Technology, https://doi.org/10.1007/978-981-99-8593-7_14

1 Introduction

The fishing industry not only provides necessary food but also has the potential to bring in money for the government. Additionally, it can produce a range of waste, such as solid, liquid, and gas disposal (Jayathilakan et al. 2012). More than half of fish muscles, comprising, skin fins, heads, viscera, and filleting frame are thrown as "wastes," according to valuations. Every year, global fisheries throw-outs surpass 20 million tons, accounting for 25% of total marine fishing catch production (Caruso 2016). Discards include fish processing wastes, "non-target" species, and byproducts. Most fishermen do not use the waste from fish capturing and processing as a side product of their fishing business. If these waste materials are not dealt with, they contribute to pollution, community problems, and the spreading of illnesses like cholera. (Krishnamoorthy 2018), so fish waste management by converting them into valuable products has received attention.

Fish meal (FM) is the most valued byproduct of the marine sector due to its protein source (Gutasi 2021). For farmed fish meals, it is the greatest nutritional and digestible element (FAO 2018). It has a good essential amino acid (EAA) composition and nutrient digestibility is pleasant and is widely available. The manufacturing of aqua feeds has a substantial impact on total production costs; on average, one kg of fish meal needs four to six kg of fish (Shepherd and Jackson 2013). Due to the usage of an expensive protein source, fish meal, feed expenses represent more than half of overall aquaculture production costs (Coyle et al. 2004). The fishing industry consumes around 60% of the world's total fish meal production each year, and this percentage is likely to rise in the future to support future aquaculture expansion (Tacon, Hasan and Metian 2011). Several fish nutritionists have been working to find new strategies to lessen the demand for fish meals. As a result, fish byproducts and wastes should be turned into fish meals to remedy the problems. Meanwhile, by forming feed from these undesired goods, the value of the fishery byproducts could be raised to a lesser level (Ferraro et al. 2010).

The sections that follow describe how fishing wastes and byproducts are utilized in the creation of fish meals, the importance of fish meals, the factors that affect fish meals, and recommendations in this context.

2 Fish Waste and Byproducts

The processing of fish before it is sold results in the production of byproducts. In presence stands not any universally accepted delineation of fish byproducts. Typically, we refer to heads, viscera, bone, cut-offs, skin, and fish that are injured or unhealthy for human intake, as well as by-catch. Byproducts are well-defined in Norway as goods that are not considered to be normal profitable goods (round, fillet, and beheaded fish), but could be reused after treatment (Bergman 2015). The term "waste" refers to products that must be composted, burned, or otherwise destroyed since they cannot be used for feeds or value-added products. (Bergman 2015). The number of byproducts accessible is estimated differently. Only 40% of fish goods used by humans are generated by the entire fish industry, with the remaining 60% consisting of unwanted byproducts (Dekkers et al. 2011). 9.1 million tons of fish waste are thrown away annually, according to United Nations Food and Agriculture Organization (FAO) (Pérez Roda et al. 2019).

Fish byproducts vary in composition and stability, an excellent source of lipid, protein, and other worthwhile components like nucleic acids, calcium, phosphorous, and bioactive compounds. In particular, fish skin is high in protein, especially gelatin and collagen, trimmings and bones are high in calcium, and the head, intestines, and bones are high in lipids (Kandyliari et al. 2020). The average protein content of fish byproducts ranges between 49.22 and 57.92% by dry weight, the ash content is 21.79–30.16%, and the fat content is 7.16–19.10%(Abbey et al. 2017).

Fish byproducts vary due to species, the extent of process, and environmental conditions because, each has a distinct and unique composition, size, and form, (Rustad et al. 2011). The processing of edible fish produces discards that primarily consist of heads (9–12%), skin and fins (1–3%), muscle trimmings (15–20%) viscera (12–18%), bones (9–15%), and scales (5%) (Martínez-Alvarez et al. 2015). Byproducts are classified into two broad categories based on their potential for spoilage and degradation. The first group of easily degradable byproducts includes fractions containing high concentrations of various endogenous enzymes while the second group includes more steady byproducts such in place of skin, heads, and bones (Rustad et al. 2011). The amount of fish byproducts available is enormous, and there are a lot of opportunities for different value-added goods to be made from this raw material.

3 Value-Added Fish Products

Waste and byproducts generated during fish processing aren't used to their full economic potential and possibly can be transformed into valued products such as fish meal, collagen, fish oil, silage, fish protein hydrolysate (FPH), etc. This creates a positive economic and environmental impact, allows for the viable usage of fish means, and increases the accessibility of beneficial marine proteins and lipids for the world's growing population. This section of the chapter briefly summarizes the value-added products that have been developed so far utilizing fish waste and byproducts. Each has its pros and cons (Coppola et al. 2021).

Fish meal: It is the more promising product available and an increasing trend is noticed for fish meal production from fish byproducts over several decades, where it is estimated to be about 1.5 million tonnes by 2030 (Guenard 2021). Several scholars have addressed the process of making fishmeal out of fish byproducts and waste (Ockerman and Basu 2014; Pagarkar et al. 2014; Masagounder et al. 2016).

Fish protein hydrolysates (FPH): These are the waste products produced when fish proteins are enzymatically broken down into small peptides with 2–20 amino acids. FPHs are also used as milk alternative for people as well as an animal feed and fish-based fertilizers (Pagarkar et al. 2014).

Fish oil: These are primarily composed of triglycerides of fatty acids, phospholipids, wax esters, and glycerol ethers (Afreen and Ucak 2020). The high interest in fish oil stems from the fact that it possesses two important Polyunsaturated fatty acids (PUFA) containing eicosapentaenoic acid and docosahexaenoic acid known as omega-3 fatty acids. These two most common PUFA are used as food complements and in the production of biofuels (Coppola et al. 2021).

Fish silage: In order to create this product, whole fish or leftover fish components are mixed with acids, enzymes, or microbes. Fish enzymes are used to carry out this process. Even on a large scale, fish silage is a low-cost and simple technique. It reduces industry odor and drainage issues. It also has the disadvantage of being a high-volume product that must be consumed at the same location where it is produced (Afreen and Ucak 2020).

Collagen: It is a prevalent structural protein that is complicated. Fish collagen absorbs into the body more efficiently and has a higher bioavailability (up to 1.5 times) than collagen from pigs and cows (Pagarkar et al. 2014). It's a fantastic functional molecule for the food, makeup products, biomaterial, pharmaceutical, and nutraceutical industries. The development of a strategic extraction technique for commercial exploitation is the primary challenge with the utilization of fish collagen (Martínez-Alvarez et al. 2015).

4 Fish Waste to Fish Meal

(i)(i) Fish meal

Fish meal has historically been employed as the main protein source in the aqua feed business, due to its high protein level and well-balanced essential amino acid (EAA) profile, (El-Sayed 2020). It is a dry, brownish grey, fine to a coarse powder. Fish meal is made by drying whole fish, fish byproducts or fish trimmings, to remove 90 to 95% of the water and grinding them into a powder (Krishnamoorthy 2018). One-third of the annual global catch of fish is utilized as raw resources for fish meal production; these are sustainable catches with no direct human consumption outlet (Barlow 2003). The nutritional composition of the fish meal is made up of protein (70%), fat (9%), water (8%), minerals (10%), and vitamins, reliant on the type of fish utilized (Afreen and Ucak 2020). In practice, fish meals have been divided into four categories: herring type, anchovy/pilchard, menhaden, and white fish (Ockerman and Basu 2014).

Fishmeal, with the biological value due to its richness in EAAs, mainly lysine and sulfur amino acids which the animal body cannot produce, makes itself an unrivaled feed constituent. In addition, the inclusion of FM in a well-balanced diet compensates for amino acid deficiencies in vegetable proteins (Pagarkar et al. 2014). FM is used in the manufacturing of feed for fish, ruminants, crustaceans, poultry, pigs, and

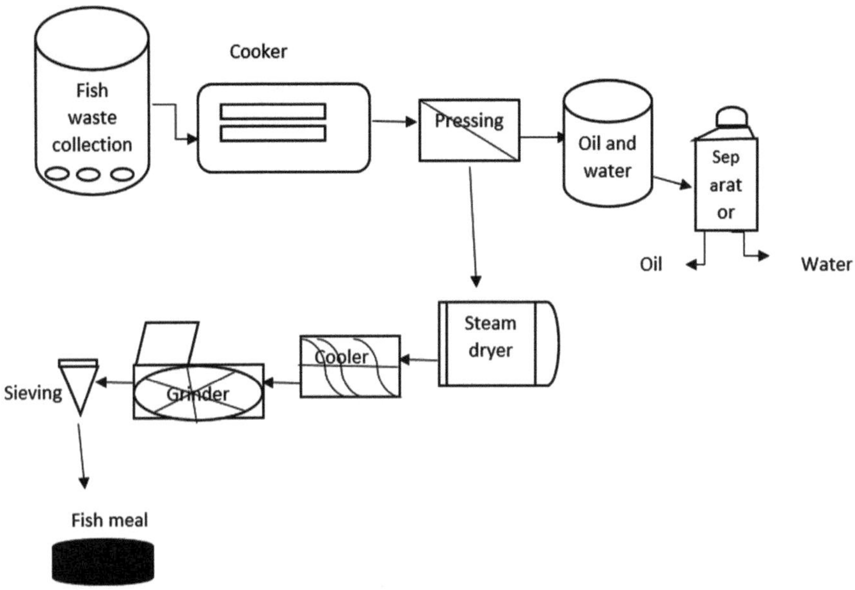

Fig. 1 Main unit procedures of fish meal

companion animals. This FM is combined with other components to make animal feed (Masagounder et al. 2016).

(ii)(ii) Manufacturing Process

The primary goal of FM production is to remove moisture from raw fish to less than 10%. The oil content of the FM must not exceed 10%. There are primarily two methods for producing FM known as wet and dry reduction, where the first one is widely utilized to produce the majority of the FM around the world. FM production comprises various manufacturing processes as indicated below (Fig. 1).

Cooking

Cooking for around 15 to 20 min at a temperature of 95 to 100 ° C, is used to coagulate fish proteins and rupture the cell walls of fish tissues, which aids in the parting of oil and water in fish. The most popular technique is to constantly convey the fish through a steam cooker, it continuously conveys it (Nissen 2003).

Pressing

Just after cooking, the mass is pressed to remove water level up to 45–55% and oil 2–3%. A single screw press or a double screw press can be used for pressing, whereas the latter one is preferred more because it removes the most oil and moisture from cooked fish. A mixture of water, oil, and solids is squished through the press perforations and later separated; the solids, water and oil are brought back to the press cake at the end of the press, and the separated oil is refined and stored as a separate product (Gunn 2003).

Drying

The press cake is then introduced into a drier either direct (cylindrical drum) or indirect (steam-jacketed cylinder or a cylinder with steam-heated disks), along with the retrieved solid particles and the stick water concentrate, to reduce the moisture level to 10% to make the product stable. The drying process can be accelerated by raising the temperature. There are, even so, certain critical limits to avoid quality loss, particularly of protein. When using the machinery and environmental conditions typical to the fishmeal business, the temperature of the drying material must not rise above 90 °C in order to preserve the nutritional value. Currently, large-scale fish meal producers hardly ever employ direct dryers. To eliminate steel impurities, the dried press cake is put through a magnetic separator (Krishnamoorthy, 2018).

Cooling

In the cooler, dried FM is chilled to room temperature.

Sieving

Prior to grinding, dry press cake is put through a vibrating screen to remove extraneous components including wood, fishing hooks, cloth, and nails.

Milling

Fish meal with small particles is produced by grinding.

Packing

Fish meal is typically packaged in jute bags lined with polyethylene (PE). The outside packaging is then labeled properly. Fishmeal storage must be moisture-proof. (Marvin et al., 2019).

(iii)(iii) Merits and demerits of Fishmeal Incorporated into Fish Diets

Merits

Nutritive value

Fish meal is well-known for its higher nutritive value with enriched energy, protein, minerals, and vitamins (choline, biotin, and vitamin B12, as well as vitamins (A, D, and E) contents. In addition, FM is a rich source of unsaturated fatty acids such as eicosapentaenoic acid and docosahexaenoic acid, in humans and laboratory animals, these fatty acids have a beneficial effect on autoimmune disease, heart disease, and inflammatory diseases (Simopoulos 2002). Particularly, high-quality fish meal provides between 60 and 72% protein contents by mass, making it one of the preferred animal protein additions in farm animal diets from a nutritional point of view (Cho and Kim 2011). FM with mackerel dried at 70 °C supplemented with synthetic amino acids (Lysine, Tryptophan, and Threonine) had the maximum dietary value when fed to weanling pigs until they were 29 days old. FM with mackerel matched the amino acid profile of the porcine plasma protein diet (Kim and Easter 2001).

Performance

The addition of fishmeal to animal diets improves feed efficiency and growth by improving food palatability, uptake of nutrients, digestion, and absorption (Olsen and Hasan 2012). Therefore, from 10 to 20 weeks postpartum, Holstein cows with a 5 percent FM consumption produced more milk and protein (Adachi et al. 2000). Weanling pig expansion was enhanced by replacing soybean meal in the beginning diet with menhaden FM, which also improved average daily gains, average daily feed intake, and gain/feed ratio (Stoner et al. 1990).

Immunity

The nutrients in fishmeal help to maintain a healthy functional immune system, in turn, boosting disease resistance and reducing the dependency on antibiotics and other drugs. For example, eicosapentaenoic acid in FM can help prevent cardiovascular disease and has a major effect on the retina, vascular and hemostatic systems, the brain and other body tissues (Visentainer et al. 2000).

Therefore, following an immune assault, rats' development is somewhat prevented by diets with a reasonably high n-3 PUFA to n-6 PUFA ratio (Jeffery et al. 1997). Due to the anti-inflammatory properties of n-3 fatty acids in FM, made the early weaned pigs consume more feed and fight better for disease (Cho and Kim2011). FM nutrients also aid in increasing phytohemagglutinin-induced proliferation and CD4 cell population in laying hen spleen cells (Babu et al. 2005).

Demerits

Toxicity

Being a protein-rich feed ingredient, fishmeal is subject to deterioration easily and quickly when overheated. During deterioration, amino acid decarboxylation, referring to protein retrogression generates toxic materials. In particular, a toxic substance known as gyserosinis formed during overheating of fish meal causes significant damage (ulceration) to the mucous membrane of poultry stomachs and tends to result in bleeding and vomiting (Cho and Kim 2011). Furthermore, histamine in FM causes poisoning and gizzard erosion in broilers (Pan and Yu 2014).

Expensive

The high cost of fishmeal limits its utilization as a feed ingredient. Feed expenses represent for more than half of total aquaculture outputs due to the use of such an expensive source of protein, fish meal (Gutasi 2021). The total global production of fishmeal in 2016 was 4,445,000 tonnes (Green 2016), and the price of FM has more than doubled in the recent past (Pavan Kumar et al. 2014) and it is expected to increase by 90% between 2010 and 2030 (World Bank 2013).

Environmental Pollution

Though conversion of fish waste into valuable products like fishmeal reduces the solid waste generation in the environment considerably, there is the possibility of polluting surface water. Wastewater derived from fishmeal processing plants may be disposed into water bodies which alters the water quality that becomes unsuitable for living beings and other purposes also key nutrients can leach into the body of water, causing water deterioration. Additionally, feed sediment accumulated on the seabed or pond bottom causes pollution, raising the risk of anoxia and mortality.

5 Factors Affecting the Production and Utilization of Fish Meal

Composition and type of fish

The raw materials composition, in terms of dry matter, fat, and protein determines the production rate as well as quality. For instance, the demand for fish meals produced from small, bony non-food fish species like anchovy would be high as protein supplement due to its relatively higher protein content than the fish meal produced from food fish species like mackerel. Additionally, the fat content plays a crucial role in plan as it not only forecasts oil yield but also determines if it is worthwhile to construct oil recovery systems.

Length of season

The profitability of the business depends on how many days a year the plant might have been in operation. Additionally, the longer the season, the more importance must be given to strategies for lowering variable costs and more money should be invested in initiatives that will save labor and power while ensuring higher returns and quality of products.

Economic

The production of the fish meal should be expanded to meet the growing demand and is challenged with a consistent supply of raw materials at a reasonable price, where continuous availability of fish waste and byproducts can be influenced by the seasonality of catches. Meanwhile, larger-scale production requires higher investment in terms of inputs, equipment, labor, etc.

Technical

The conventional method of fishmeal production employs many laborers which brought a challenge to production. The utilization of advanced and automated technologies though help counteract labor issue; the requirement for skilled and trained personnel remains still. There are a lot of laborers in developing countries who don't know how to maintain fish meal plants. Production plants are becoming larger, and production is being concentrated in fewer factories that combine raw materials from various sources or facilities.

Social

Aquatic meals like fish and other seafood are often regarded as healthful. The idea behind this approach is that fish includes healthy components such as highly digested proteins, marine oil, antioxidants, minerals, and vitamins. However, consumers' poor knowledge and understanding regarding the importance of marine byproducts and or value-added products and their nutritive values impede their usage at a larger scale of byproducts in commodities.

Environmental and ethical issues

The environmental and ethical issues being raised by non-governmental organizations and the media are significant, and they may have an important effect on the actual use of water resources in aqua feeds such as a fish meal. For smooth functioning, the placement of the plant in relation to homes and closed harbor is essential. Public policies safeguarding the environment from unfavorable air and water pollution are important to thoroughly review that influence the choice of equipment, factory premises, and finance required for odor abatement and wastewater emissions cleaning [6, 15, and 25].

6 Suggestions and Conclusion

Fish waste and byproducts have been found the potential to be utilized for the process of fish meals as a valorization approach to fish waste. The production of fish waste fish meals become popular over years due to their nutritional value and rather good performance compared to other feed ingredients worldwide. Though, its production and consumption fluctuate because of a lack of awareness, drawbacks, and challenges associated with, the unavailability of resources, facilities, technologies, etc. To cope with this, various measures have to be undertaken at multi-levels. Foremost, people should be made aware of the possibility of fish waste and byproducts as a constituent for FM production and its subsequent benefits. As private sectors play a key part in food and feed production, their participation has to be encouraged further at country levels to invest their capital in FM production, which requires appreciative regulations and legislation. Hence, globally, the developed countries should join hands together with developing countries to facilitate production via the provision of skilled, trained personnel, and advanced technologies. Eventually, the studies should be explored more and in-depth to evaluate the potential of fish waste-based fish meal as a feed ingredient in various aspects among different sets of animals.

References

Abbey L et al (2017) Nutrient content of fish powder from low value fish and fish byproducts. Food Sci Nutr 5(3):374–379. https://doi.org/10.1002/fsn3.402

Adachi N et al (2000) Effect of supplemental fish meal on milk yield and milk composition of holstein cows during early lactation. Asian Australas J Anim Sci 13(3):329–333. https://doi.org/10.5713/ajas.2000.329

Afreen M, Ucak I (2020) Fish processing wastes used as feed ingredient for animal feed and aquaculture feed. J Surv Fish Sci 6(2):55–64. https://doi.org/10.18331/SFS2020.6.2.7

Babu US et al (2005) Babu 2005 immune.pdf. Int J Poult Sci 4(9):652–656

Barlow SM (2003) Fish meal. International Association of Fish Meal and Oil Manufacturers. https://doi.org/10.2307/1443094

Bergman K (2015) Co-products in the Swedish Seafood Processing Industry Quantification and present uses. Available at: https://bioenv.gu.se/digitalAssets/1591/1591301_kristinabergman.pdf

Caruso G (2016) Fishery waste and by-productos: a resource to be valorised. J Fish Sci 10(1):12–15

Cho JH, Kim IH (2011) Fish meal - nutritive value. J Anim Physiol Anim Nutr 95(6):685–692. https://doi.org/10.1111/j.1439-0396.2010.01109.x

Coppola D et al (2021) Fish waste: from problem to valuable resource. Mar Drugs 19(2):1–39. https://doi.org/10.3390/md19020116

Coyle SD et al (2004) Evaluation of growth, feed utilization, and economics of hybrid tilapia, Oreochromis niloticus x Oreochromis aureus, fed diets containing different protein sources in combination with distillers dried grains with solubles. Aquac Res 35(4):365–370. https://doi.org/10.1111/j.1365-2109.2004.01023.x

Dekkers E et al (2011) Oxidative stability of mahi mahi red muscle dipped in tilapia protein hydrolysates. Food Chem 124(2):640–645. https://doi.org/10.1016/j.foodchem.2010.06.088

El-Sayed A-FM (2020) Nutrition and feeding, in *Tilapia Culture*, pp 135–172. https://doi.org/10.1016/b978-0-12-816509-6.00007-0

FAO (1988) The production of fish meal and oil. FAO Fish Tech Pap 1(142):27

FAO (2018) World fishery and aquaculture, biologiya morya-marine biology. Rome, Italy: FAO

Ferraro V et al (2010) Valorisation of natural extracts from marine source focused on marine by-products: a review. Food Res Int 43(9):2221–2233. https://doi.org/10.1016/j.foodres.2010.07.034

Green K (2016) Seafish fish meal and fish oil facts and figures, in *Seafish*, pp. 1–33

Guenard R (2021) Poisson from a petri dish. Inform. https://doi.org/10.4060/ca9229en

Gunn M (2003) The use of fish by-products in aquaculture, Scientific Committee on Animal Health and Animal Welfare

Gutasi A (2021) Benefit and drawbacks of fish meal substitution in aquaculture diets.

Jayathilakan K et al (2012) Utilization of byproducts and waste materials from meat, poultry and fish processing industries: a review. J Food Sci Technol 49(3):278–293. https://doi.org/10.1007/s13197-011-0290-7

Jeffery NM, Newsholme EA, Calder PC (1997) Level of polyunsaturated fatty acids and the n-6 to n-3 polyunsaturated fatty acid ratio in the rat diet alter serum lipid levels and lymphocyte functions. Prostaglandins Leukot Essent Fatty Acids 57(2):149–160. https://doi.org/10.1016/S0952-3278(97)90005-3

Kandyliari A et al (2020) Nutrient composition and fatty acid and protein profiles of selected fish by-products. Foods 9(2):1–14. https://doi.org/10.3390/foods9020190

Kim SW, Easter RA (2001) Nutritional value of fish meals in the diet for young pigs. J Anim Sci 79(7):1829–1839. https://doi.org/10.2527/2001.7971829x

Krishnamoorthy E (2018) Fish Meal and Oil from Fish Waste: An Industrial Perspective Introduction. Indian Council of Agricultural Research, pp 53–61

Martínez-Alvarez O, Chamorro S, Brenes A (2015) Protein hydrolysates from animal processing by-products as a source of bioactive molecules with interest in animal feeding: a review. Food Res Int 73(1069):204–212. https://doi.org/10.1016/j.foodres.2015.04.005

Marvin I et al (2019). Nordic Centre of Excellence Network in Fishmeal and Fish Oil. https://doi.org/10.5281/zenodo.3243334

Masagounder K et al (2016) Optimizing nutritional quality of aquafeeds. Elsevier Inc., Aquafeed Formulation. https://doi.org/10.1016/B978-0-12-800873-7.00006-3

Nissen H (2003) Trends in the utilization and production of seafood byproducts. Adv Seafood Byprod, 2002 Conference Proceedings, 03(01):351–358

Ockerman HW, Basu L (2014) Inedible. Encycl Meat Sci. https://doi.org/10.1016/B978-0-12-384731-7.00032-5

Olsen RL, Hasan MR (2012) A limited supply of fishmeal: impact on future increases in global aquaculture production. Trends Food Sci Technol 27(2):120–128. https://doi.org/10.1016/j.tifs.2012.06.003

Pagarkar AU, Shaikh SM, Saiyad A (2014) 'Fish waste: its ecofriendly utilization and uses. *Fishcoops*, pp 3–9

Pan D, Yu Z (2014) Intestinal microbiome of poultry and its interaction with host and diet. Gut Microbes 5(1):108–119

Pavan Kumar B, Ramudu KR, Devi BC (2014) Mini review on incorporation of cotton seed meal, an alternative to fish meal in aquaculture feeds. Int J Biol Res 2(2):99–105. https://doi.org/10.14419/ijbr.v2i2.3274

Pérez Roda MA et al. (2019) A third assessment of global marine fisheries discards, FAO Fisheries and Aquaculture Technical Paper. Available at: https://www.fao.org/documents/card/en/c/ca2905en/

Rustad T, Storrø I, Slizyte R (2011) Possibilities for the utilisation of marine by-products. Int J Food Sci Technol 46(10):2001–2014. https://doi.org/10.1111/j.1365-2621.2011.02736.x

Shepherd CJ, Jackson AJ (2013) Global fishmeal and fish-oil supply: Inputs, outputs and marketsa. J Fish Biol 83(4):1046–1066. https://doi.org/10.1111/jfb.12224

Simopoulos AP (2002) Omega-3 fatty acids and cardiovascular disease: the epidemiological evidence. Environ Health Prev Med 6(4):203–209. https://doi.org/10.1007/BF02897971

Stoner GR et al (1990) Effect of select menhaden fishmeal in starter diets for pigs. J Anim Sci 68:2729–2735

Tacon AGJ, Hasan MR, Metian M (2011) Demand and supply of feed ingredients for farmed fish and crustaceans : Trends and prospects, FAO Fisheries and Aquaculture Technical Paper. Available at: http://www.fao.org/docrep/015/ba0002e/ba0002e.pdf

Visentainer J et al (2000) Concentração de ácido eicosapentaenóico (EPA) e ácido docosahexaenóico (DHA) em peixes da costa brasileira. Ciencia e Tecnologia De Alimentos 20:90–97

World Bank (2013) Fish to 2030 prospects for fisheries and aquaculture, Agriculture and environmental services discussion paper. Available at: http://documents.worldbank.org/curated/en/2013/12/18882045/fish-2030-prospects-fisheries-aquaculture

Fish Wastes as Source of Fertilizers and Manures

Inain Jaies, Imtiyaz Qayoom, Farheen Saba, and Sameena Khan

Abstract Solid wastes such as fish carcasses, viscera, skin, and heads, as well as liquid wastes, are produced in huge amounts during fish processing. Leaving these wastes unattended is a concern for the environment as well as the human populace residing in its proximity. Apprehension for environmental degradation, especially of aquatic ecosystems is associated with high organic content in fish wastes which, if discharged, will increase the productivity of waterbodies leading to pollution. On the other hand, if these wastes are leftover or dumped improperly, they result in a nauseating smell, thus making it arduous for humans to live nearby. There is however an effective alternate method by which fish wastes can be utilized, resulting in its scientific disposal and putting less stress on the environment. Fish waste can be a good source of manure and fertilizers. Crops chiefly require nitrogen, phosphorous, and potassium for growth as essential nutrient elements which can be mainly drawn from fish wastes. Moreover, they can be a good source of calcium for soils that are deficient in calcium levels. Fish-based fertilizers typically include substantial levels of N, P, and Ca and essential minerals. Phosphate rock, which is a restricted resource, provides more than 85 percent of the phosphorus used in agriculture. As a result, using fish waste as a phosphorus-rich organic waste is a better option. Fish fertilizer is an organic liquid fertilizer that works quickly and is manufactured from by-products of the fishing business. It contains a lot of nitrogen, phosphorus, and potassium, as well as trace minerals like calcium, magnesium, sulfur, chlorine, and sodium. Fish emulsion, when used correctly, can increase crop yields, especially in cool climes. This is because, in the cooler months, manure breaks down slowly, whereas fish emulsion fertilizes at a constant rate. The commercial fertilizers, particularly

I. Jaies
Division of AAHM, Faculty of Fisheries, SKUAST-K Rangil Ganderbal, Jammu and Kashmir 190006, India

I. Qayoom (✉) · S. Khan
Division of AEM, Faculty of Fisheries, SKUAST-K Rangil Ganderbal, Jammu and Kashmir 190006, India
e-mail: drsheikhimtiyaz@gmail.com

F. Saba
Department of Zoology, School of Sciences, MANUU Gachibowli, Hyderabad, Telangana 500032, India

S. Maqsood et al. (eds.), *Fish Waste to Valuable Products*, Sustainable Materials and Technology, https://doi.org/10.1007/978-981-99-8593-7_15

fish based, are being utilized in crops of agriculture and horticulture and a number of fertilizers can be produced by fish waste. Fish waste composting is a fantastic approach to reduce waste volume, and compost supplemented with fish waste could be a useful organic fertilizer. This book chapter will provide insight into the utilization of fish wastes that are a boon to agriculture and soothe the environment as well.

Keywords Fish waste · Fertilizers · Manures · Environmental pollution

1 Introduction

The term "fish waste" refers to a variety of fish species or by-catch items that have no or very little commercial value, are damaged and underdeveloped species, have economic worth but are not collected in sufficient numbers to justify sale. According to Environmental Protection Agency (EPA), "fish waste can include bits of flesh, shells, skin, guts, bones, or liquid stick water, among other things." In other words, fish waste includes several fish species or by-catch items having no or very low commercial value, commercial species being undersized or damaged, and commercial species not collected in sufficient quantities to merit sale (Caruso 2016).

Between 1961 and 2017, global food fish consumption climbed at an average yearly rate of 3.1 percent, nearly double the rate of annual world population growth (1.6 percent) (FAO 2020). It is believed that roughly two-thirds of fish are discarded as garbage, posing significant environmental and economic implications (Coppola et al. 2021). As a result, a significant increase in the quantity of production of fish waste production around the world has been observed. The release of fish wastes at such an enormous level has impressed upon the concerned stakeholders, policymakers, processing unit heads, and environmentalists to dispose and recycle the wastes on scientific guidelines or otherwise, which may lead to serious environmental concerns. Every year, a large amount of biomass is thrown, most commonly burnt, which increases the energy consumption for its management and has its own environmental impact. Fish waste has mostly been used as fish meal as it contains the same quantity of proteins as present in fish meat (Mo et al. 2018). Moreover, fish waste contains a sizable amount of biodegradable organic matter, that might be repurposed to make waste-activated sludge more appealing as a co-substrate to boost the generation of methane during anaerobic co-digestion (Wu and Song 2021). Because of its nutritional makeup, fish wastes can also enhance compost or provide plant nutrients. Fish wastes can modify this sentence (Løes et al. 2018). Fish processing industries generate solid wastes such as skin, viscera, fish heads, fish bones, and flesh leftovers. Among liquid wastes, blood, water, and brine are released from drained storage tanks, washing and cleaning of flesh, and blood and soluble compounds from the gut. In addition to these wastes, detergents and other cleaning agents are also released as liquid wastes from processing units. Some gaseous wastes such as carbon dioxide and hydrogen sulfide are also liberated on account of various chemical reactions carried out during processing, storing, or freezing processes.

Aquaculture wastes can be divided into two categories:

- **Solid wastes:** Unused feed and feces from farmed fish are the primary sources of solid waste (Akinwole et al. 2016). Sometimes fishes that fail to succeed in the culturing process and retard or die underway also form huge wastes in large culturing units. Suspended and settled solids are two more types of solid wastes monitored in culturing units. Suspended solids are tiny particles that remain floating in water unless coagulated or settled down by sedimentation. The removal of these solids from culture systems is the most challenging (Cripps and Bergheim 2000). On the other hand, settled solids also form a source of waste, although these settle down quickly but can easily be removed from the culture column on account of their bigger particle size (Ebeling and Timmons 2012).
- **Dissolved wastes:** These wastes are by-products of fish metabolism or degraded leftover fish feed. Dissolved wastes contain nitrogen (N) and phosphorous (P) as components of concern (Boyd and Massaut 1999). The retention of nitrogen ranges from 25 to 30% (Boyd 2003) to 10–49% (Piedrahita 2003) while phosphorous retention ranges between 17 and 40% in different fish species (Boyd 2003).

2 Proximate Composition of Fish Waste

The term "proximate composition" is commonly used in the food industry to refer to the components of crude protein, moisture, ether extract, crude ash, crude fiber, and nitrogen-free extracts, which are given as percentages in the feed. The proximate composition of fish waste is given in Table 1 (Palkar et al. 2018).

Fertilizers made from fish usually have high levels of N, P, and Ca. The macronutrients N, P, K, and Ca concentration varies between species and also relies on the type of waste used. Fish scales are abundant in nitrogen and as rich source of calcium (Ca) and phosphorous (P) as well. Fish bones consist of 60–70% minerals, primarily calcium and phosphorus in the form of hydroxyapatite (Ghaly et al. 2013). In general, the typical N–P–K values for inland fish are 120:11:13, and for marine fish, they are 130:16:11 (Bogard et al. 2015).

Table 1 Proximate composition of fish waste

S.No	Component	Percentage
1	Moisture	77.09 ± 0.14%
2	Crude protein	15.20 ± 0.15%
3	Fat	4.03 ± 0.07%
4	Ash	3.30 ± 0.11%
5	Nitrogen free extracts	0.38 ± 0.06%

3 Synthetic Fertilizers and Their Hazardous Effects

"Synthetic fertilizers are chemically manufactured products containing one or more of the primary elements that are essential for plant growth, viz., nitrogen, phosphorus, and potassium," (Anonymous, 2015, University of California Agriculture and Natural Resources). They usually have various ratios of nitrogen, phosphorus, potassium, calcium, magnesium, and other components. Unlike organic fertilizers, synthetic fertilizers provide the necessary nutrients to the soil right away. There are various benefits of using synthetic fertilizers compared to natural manures given in Table 2.

Synthetic fertilizers are essential for revitalizing the soil by delivering the nutrients that plants require for successful growth. Nitrogen, potassium, and phosphate are the most prevalent nutrient sources in mineral fertilizers. They work by giving the appropriate mix of nutrients to the soil, which increases output and ensures nutritious food. The soil would be depleted if fertilizers were not used, making it harder for plants to thrive. Modern synthetic fertilizers are primarily made up of nitrogen, phosphorous, and potassium compounds, with some additional nutrients mixed in combination which varies between plant varieties in accordance with their nutritional requirements. Nitrogen is an essential plant nutrient that aids in the plant's development and metabolic functions which impresses the incorporation of synthetic chemicals such as ammonium nitrate, ammonium phosphate, and potassium sulfate for making synthetic fertilizers. Synthetic fertilizers however have negative long-term consequences. Continuous synthetic fertilizer use can degrade soil quality. It disrupts the natural soil ecosystem, altering its pH and nutrient balance, causing soil acidification, imbalanced nutrient levels, and reduced microbial activity, ultimately diminishing soil fertility. In the long run, synthetic fertilizers harm the natural makeup of soil. Iron, carotene, zinc, copper, vitamin C, and protein are all inadequate in plants that grow in highly fertilized soil. Studies have also shown that beneficial soil bacteria that convert organic remains from dead plants and animals into nutrient-rich organic materials are killed by synthetic fertilizers. Nitrogen and phosphorous-based fertilizers drain into groundwater leading to an increase in water contamination and toxicity. Some phosphates contain modest levels of radionuclides (Tirado and Allsopp 2012) and some studies suggest increased radioactivity around phosphate mining regions. The salt concentration in chemical fertilizers is one of the

Table 2 Benefits of synthetic fertilizers

1	Synthetic fertilizers provide the soil with precise amounts of nutrients in a constant manner
2	Synthetic fertilizers are simple to use and have practically immediate results
3	Unlike organic fertilizers, which must break down before being absorbed, they operate immediately on the soil
4	They are less expensive than organic fertilizers in general
5	They can be handled easily

Sabry (2015)

profound triggers that damage plants as well as soil, thus likely to impair agricultural productivity. They raise nitrate content in soils. Plants grown in such soils turn to poisonous nitrites in the intestines when eaten. According to Usry (2013), nitrate is transformed into a very poisonous chemical (nitrite) in the digestive system of living organisms. These toxic nitrites combine with hemoglobin in the bloodstream to induce methemoglobin, which affects the vascular and respiratory systems, resulting in asphyxia and, in extreme circumstances, death. Synthetic fertilizers are prone to leaching into groundwater and surface water. Rainfall or excessive irrigation can wash these chemicals into rivers and lakes, causing water pollution and contributing to eutrophication, a phenomenon where excess nutrients disrupt aquatic ecosystems by promoting algae overgrowth. In addition to this, the production and application of synthetic fertilizers release ammonia and nitrous oxide into the atmosphere, leading to air pollution and contributing to greenhouse gas emissions, further aggravating climate change. Thus, counting all these problems related to the practice of synthetic fertilizers, it becomes imperative to supplement or replace these chemicals with environmentally safe nutrient-supplying products.

4 Application of Fish Waste as Fertilizer and Manure

The leftover fish can be utilized as a raw material in manufacturing organic fertilizers. Compost made from fish waste will give the added benefit of containing nutrients such as calcium, magnesium, and potassium. It helps to keep compost in good shape, sanitary, and without contaminants like phytotoxic chemicals and heavy metals (Kinnunen et al. 2005). The benefit of using fish waste as compost is that it can solve issues with fish waste discarding by transforming it into completed products with a good market value and the least stress on the environment. In horticultural crops, the compost created from seaweed and fish waste works well as soil amendment as shown by the yields of lettuce and tomatoes (Illera-vives et al. 2015). When compared to crops getting mineral fertilizers or no fertilizers, the use of compost at a rate of 66 tha^{-1} considerably enhanced the production of tomatoes and was related to larger diameter of fruit and higher fruit weight. When compared to the control and mineral fertilizer treatments, the compost residual effect was noteworthy and led to greater commercial yields of lettuce. According to Enviro-Fish Africa., 2006 fish waste from tilapia beneficiation after composting with crushed grass, was tested as a source of humic acids to stimulate the growth of the lettuce plant (Busato et al. 2018). In lettuce, the humic acid extract (20 mg L^{-1}) greatly enhanced the amount of dry and wet root matter.

5 Procedure for Making Fish Waste as Fertilizers and Manures

Fish waste compositing is a relatively new and ecologically beneficial method of disposing of fish waste to get rid of unwanted fish mass. This also allows us to get rid of their awful odor and convert waste into a valuable marketable product. Fish waste is collected and layered with carbon, nitrogen, air, and moisture for composting. Fish wastes such as viscera comprising of the digestive system, kidney, air bladder, liver, ovary, and testis, and leftover organs like scales, fins, and gills are put into a trench for decomposition. The digestive system, fins, scales, gills, liver, kidney, air bladder, ovary, testis, and other waste organs are placed as the first layer. For the carbon supply, the second layer is sawdust or any other wood product. After dumping the fish waste into the pit, care is to be taken for the first seven days since the pit emits a foul odor, increasing the risk of disturbance from dogs, cats, and other predators. As the microorganisms require moisture for the decomposition process, watering is one of the significant factors during the process. Temperature is an important aspect of the decomposition process, and shade is essential to prevent the pit from drying out. Aeration of the well is also required for the breakdown process. *Bacillus*, *Rhizobium*, and *Azotobacter* (Phosphate solubilizing bacteria, PSB), are some of the bacteria that can be employed to speed up the breakdown process (Fig. 1). All layers are blended after one month for well aeration, and this process is to be repeated every 15 days until the fertilizer is ready. Every month, the pit is monitored. Excellent quality fertilizer is made in 180 days (Balkhande 2020). After the fertilizer has been prepared, it is filtered through the mesh, and a sample is sent to any KVK for NPK and other micronutrient analyses (Balkhande and Chavhan 2020).

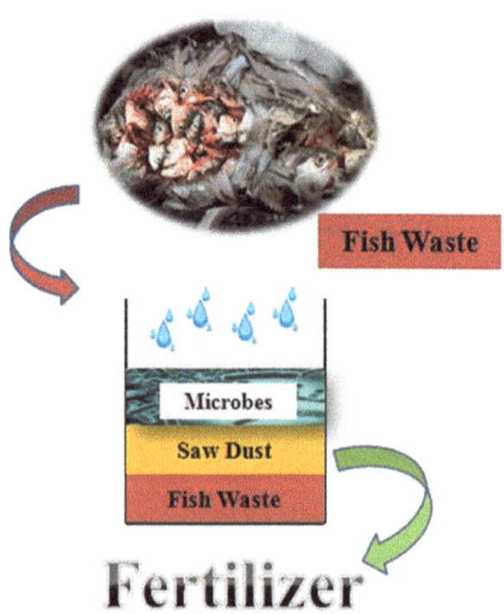

Fig. 1 Simple procedure for production of fertilizer from fish waste

6 Advantages of Using Fish Waste: Economic Benefits and Prevention of Environmental Pollution

Every year 20 million tons of fish are discarded annually from global fisheries, accounting for 25% of total marine fishery catch production (Kim and Mendis 2006). Among the items that are discarded constitute fish processing wastes, by-products, and non-target species. More than half of all fish tissues, including skin, fins, heads, and viscera are predicted to be thrown away as "waste." Furthermore, fish processing sector discards are estimated to account for up to 75% of total product volume (Olsen et al. 2014). Because fish waste causes so many problems, it should be recycled as much as possible. Fishery wastes have become a global concern in recent years, influenced by a variety of operational, biological, and technical elements and socioeconomic factors as well (Arvanitoyannis and Kassaveti 2008). Disposing of this detritus has long been a challenge for fish merchants who clean and process fish, from enormous commercial food processors to small sport fishing operations. Fish waste cannot be kept for longer than 24 h due to its high protein content. Furthermore, fish processing sector discards are estimated to account for up to 75% of total product volume (Olsen et al. 2014). Gonçalves et al. 2007 observed significant trammel net discards in the Mediterranean Sea, with a total of 137 species (79.7% of the total) thrown away and an overall discard rate varied from 15 to 49% for Greece to Portugal, respectively.

The environmental impact of fish wastes on aquatic ecosystems is an essential feature, as the organic wastes released into the water bodies could drastically alter the benthic assemblage biodiversity and community structure. (Vezzulli et al. 2008). Furthermore, fish wastes cause a huge economic loss, despite the fact that they may be used as a source of nutrition for farm fishes. As a result, the management of fish discards encompasses a variety of issues, the most important of which is the need to reduce this source of pollution and to find the best solutions to the problem. As a result, better fish waste management is required to address environmental concerns while also making complete use of biomass for high-value commercial reasons. Fish waste management is an appealing issue because it indicates a way to address the environmental implications of fishery discards while also providing a tool to utilize them as a source of feed for farmed fish, so supporting aquaculture expansion in a sustainable manner in the future. The use of fisheries wastes and by-products has the potential to reduce waste that would otherwise be thrown, resulting in nutrient enrichment and eutrophication of aquatic bodies. As a source of high-value chemicals, fish wastes, and by-products can be used effectively as fish manure, which would not only protect the environment but also give economic security.

7 Fish Wastes: A Successful Source of Manures—A Case Study

Fish can be utilized to generate both liquid and solid fertilizers (Davis et al. 2004.). Fertilizer made from fish source material can boost the output of some fruit plants by up to 60% (Glogoza 2007). Organic fertilizers are an excellent choice. Vanny et al. (2019) employed. Solid Organic Fertilizer (SOF) is made from Tilapia (*Oreochromis mossambicus*) fish waste by utilizing Bakasang traditional fermentation technology. In TambakLorok Market, Kusuma and Yulianto (2019) investigated the use of fish waste processing as a compost raw material. They claim that the TambakLorok Market is a market where locals sell their own fish catches. However, due to the enormous number of fish that are caught, not all of the fish captured can be used. In the TambakLorok market and at the fish auction site as well, the fish waste is normally only stacked in a few spots. It is vital to process fish waste in order to lessen the environmental impact, one of which is the use of fish waste as a compost raw material. Compost made from fish waste has the added benefit of containing nutrients such as potassium, calcium, and magnesium. Furthermore, compost from fish waste helps to ensure that the compost is hygienic and free of pollutants such as heavy metals and phytotoxic chemicals (Kinnunen et al. 2005).

8 Constraints in Using Fish Waste as Fertilizers and Manures

The major disadvantages in using fish waste as fertilizer or manure include:

- Because the amount of nutrients in organic fertilizer is typically low, using it is more of a process than an event, requiring a large volume of fertilizer to achieve the necessary nutrient dose.
- In addition, the composition varies.
- The process of producing fish excrement is unpleasant and time-consuming.
- Mineralization and the release of nutrients, particularly nitrogen, are both slow processes.

9 Conclusion

A large amount of fish waste is generated by fish processing plants due to the rapid expansion of aquaculture. If left untreated, this waste has the potential to radically alter the community structure and biodiversity of an ecosystem. Because the fertilizer is organic and high in nitrogen, phosphorus, and potassium, it may be transformed into fish dung and fertilizer with ease. This organic fish-based fertilizer can be used to boost compost or offer plant nutrients. It will also safeguard the ecosystem from the negative effects of synthetic fertilizers.

References

Akinwole AO, Dauda AB, Ololade AO (2016) Haematological response of *Clarias gariepinus* juveniles reared in treated wastewater after waste solids removal using alum or Moringaoleifera seed powder. Int J Aquaculture 6(11):1–8

Arvanitoyannis IS, Kassaveti A (2008) Fish industry waste: treatments, environmental impacts, current and potential uses. Int J Food Sci Technol 43(4):726–745

Balkhande JV (2020) Devising of organic fertilizer from fish and crab wastes: waste to best technology. Int J Fish Aquatic Stud 8(2):01–05

Balkhande JV, Chavhan B (2020) Preparation of organic fertilizer from fish waste: a new avenue in employment generation. Int Res J Sci Eng A10:21–24

Bogard JR, Thilsted SH, Marks GC, Wahab MA, Hossain MA, Jakobsen J, Stangoulis J (2015) Nutrient composition of important fish species in Bangladesh and potential contribution to recommended nutrient intakes. J Food Compos Anal 42:120–133

Boyd CE (2003) Guidelines for aquaculture effluent management at the farm-level. Aquaculture 226(1–4):101–112

Boyd CE, Massaut L (1999) Risks associated with the use of chemicals in pond aquaculture. Aquacult Eng 20(2):113–132

Busato JG, de Carvalho CM, Zandonadi DB, Sodré FF, Mol AR, de Oliveira AL, Navarro RD (2018) Recycling of wastes from fish beneficiation by composting: chemical characteristics of the compost and efficiency of their humic acids in stimulating the growth of lettuce. Environ Sci Pollut Res 25(36):35811–35820

Caruso G (2016) Fishery wastes and by-products: A resource to be valorised. J Fish Sci 10(1):0–0

Coppola D, Lauritano C, Palma Esposito F, Riccio G, Rizzo C, de Pascale D (2021) Fish waste: from problem to valuable resource. Mar Drugs 19(2):116

Cripps SJ, Bergheim A (2000) Solids management and removal for intensive land-based aquaculture production systems. Aquacult Eng 22(1–2):33–56

Davis JG, Brown MAP, Evans C, Mansfield J (2004) The integration of foliar applied seaweed and fish products into the fertility management of organically grown sweet pepper. Organic Farming Research Foundation Project Report. North Carolina State University

Ebeling JM, Timmons MB (2012) Recirculating aquaculture systems Tidwell JH (Ed.) Aquaculture production systems. A publication of World Aquaculture Society. John Willey & Sons Inc. 245–278

Enviro-Fish Africa (2006) Recommendations on beneficiation and commercialization of fishing activities in the BCLME region. BCLME Project LMR/SE/03/02, Benguela Current Large Marine Ecosystem Programme, Windhoek, Namibia, p. 31

Ghaly AE, Ramakrishnan VV, Brooks MS, Budge SM, Dave D (2013) Fish processing wastes as a potential source of proteins. Amino acids and oils: A critical review. J Microb Biochem Technol 5(4):107–129

Glogoza P (2007) Effect of foliar applied compost tea and fish emulsion on organically grown soybean. U of MN extension service

Gonçalves JMS, Stergiou KI, Hernando JA, Puente E, Moutopoulos DK, Arregi L, Erzini K (2007) Discards from experimental trammel nets in southern European small-scale fisheries. Fish Res 88(1–3):5–14

Hapsari N, Welasih T (2010) Utilization of fish waste into organic fertilizer. UPN Veteran, JawaTimur

Illera-Vives M, Labandeira SS, Brito LM, López-Fabal A, López-Mosquera ME (2015) Evaluation of compost from seaweed and fish waste as a fertilizer for horticultural use. Sci Horticultur 186:101–107

Kim SK, Mendis E (2006) Bioactive compounds from marine processing byproducts–a review. Food Res Int 39(4):383–393

Kinnunen RE, Gould MC, Cambier P (2005) Composting commercial fish processing waste from fish caught in the Michigan waters of the Great Lakes. Michigan State University Extension, USA

Kusuma IM, Yulianto B (2019) Utilization of fish waste processing as compost raw material in TambakLorok market. In E3S Web of Conferences 125:07004, EDP Sciences

Løes AK, Katsoulas N, Caceres R, de Cara M, Cirvilleri G, Kir A, Knebel L, Malinska K, Oudshoorn F, Raskin B (2018) Current use of peat, plastic and fertiliser inputs in organic horticultural and arable crops across Europe. European Commission, Brussels, Belgium, 2018

Mo WY, Man YB, Wong MH (2018) Use of food waste, fish waste and food processing waste for China's aquaculture industry: Needs and challenge. Sci Total Environ 613:635–643

Olsen RL, Toppe J, Karunasagar I (2014) Challenges and realistic opportunities in the use of by-products from processing of fish and shellfish. Trends Food Sci Technol 36(2):144–151

Palkar ND, Koli JM, Gund DP, Patange SB, Shrangdher ST, Sadawarte RK, Akhade AR (2018) Preparation of co-dried fish silage by using fish market waste and its comparative study. Int J Pure App Biosci 6(2):1567–1577

Piedrahita RH (2003) Reducing the potential environmental impact of tank aquaculture effluents through intensification and recirculation. Aquaculture 226(1–4):35–44

Production of Liquid and Solid Organic Fertilizer from Tilapia Fish (Oreochromismossambicus) Wasteusing "Bakasang" Traditional Fermentation Technology

Sabry AK (2015) Synthetic fertilizers; role and hazards. Fertil Technol 1:110–133

Selvya H, Nainggolan G, Jamahir B, Wirjosentono, (2013) Study of the utilization of fish waste from fish auction sites (tpi) and sibolga traditional markets as compost raw material. Unimal Chem Technol J 2(2):90–99

Tirado R, Allsopp M (2012) Phosphorus in agriculture: problems and solutions. Greenpeace Research Laboratories Technical Report (Review) 2

Usry M (2013) The health effects of synthetic fertilizer. www.southlandorganics.com

Vanny MA, Tiwow A, Abram PH, NurHopiyanti. International Journal of Engineering and Advanced Technology (IJEAT) ISSN: 2249–8958, Volume-8, Issue-3S, February 2019 885–888

Vezzulli L, Moreno M, Marin V, Pezzati E, Bartoli M, Fabiano M (2008) Organic waste impact of capture-based Atlantic bluefin tuna aquaculture at an exposed site in the Mediterranean Sea. Estuar Coast Shelf Sci 78(2):369–384

Wu Y, Song K (2021) Anaerobic co-digestion of waste activated sludge and fish waste: Methane production performance and mechanism analysis. J Clean Prod 279:123678

Using Fish Waste and By-Products for Manufacturing Organic Fertilizers and Manures

Menakshi Dhar, Raman Jasrotia, and Seema Langer

Abstract Fish waste is a rising concern because with the exponential rise in the population; the rising demand for fish has increased the production of fish waste. The management of fish waste is one of the main problems that have a great impact both on the economy and the environment. Approximately 30–70% of the fish waste is generated from the consumed fish. Recycling fisheries waste is important as this can be helpful in the prevention of nutrient loss from the food chain. Environmental pollution can be controlled by converting this waste into organic manure and this can be done by utilizing simple techniques with a low budget. Fish waste is rich in nutrients such as nitrogen, potassium, and calcium which can be useful in organic farming systems. Composting fish waste is a cheap method for its disposal which obliterates the vector-borne diseases and larvae of flies. Fertilizers made from fish waste are useful to increase the content of soil nutrients, control plant diseases, eliminate unwanted plants, and decline the growth of parasites. Fish waste can be used in the form of manure and fertilizer in farms, kitchen gardens, and field crops to enhance the organic matter and nutrients in the soil, increase the moisture-holding capacity and promote the production and quality of the soil. Production of fertilizers from fish by-products promotes nutrient recycling from the aquatic to the terrestrial ecosystem. Various fertilizers made from the fish meal are available in the marketplace, some of which are used in organic agriculture. Fish waste composting will be beneficial for the reduction of fisheries by-products and waste.

Keywords Economy · Fish waste · Manure · Fertilizers · Population · Recycling

---

M. Dhar · R. Jasrotia (✉) · S. Langer
Department of Zoology, University of Jammu, Jammu 180006, India
e-mail: ramanjasrotia18@gmail.com

R. Jasrotia
Department of Zoology, Cluster University of Jammu, Jammu, India

S. Maqsood et al. (eds.), *Fish Waste to Valuable Products*, Sustainable Materials and Technology, https://doi.org/10.1007/978-981-99-8593-7_16

1 Introduction

Farming of aquatic animals such as molluscs, finfish, shellfish, aquatic plants in seawater, brackish water, saline water, or fresh water is known as aquaculture. There has been a constant increase in the consumption of fish products and the rate of fish consumption has surged in the last few years as it has been recognized as an essential component of a healthful lifestyle and balanced diet. Per capita consumption of fish across the globe rose to 20.2 kg in 2015 which was just 9 kg in 1961 (FAO 2018). With the increase in the human population and the change in behavior of food consumption, there has been a growing demand for fishery products which in turn has led to the generation of huge amounts of waste products. Such organic matter is regarded as post-harvest losses and is a cause of concern for the fish management practices (Blanco et al. 2017). Fish wastes should be managed properly to curb the environment-related problems and at the same time, this waste should be utilized for the preparation of value-added products. As the business of fishing is growing at a rapid rate which often leads to the production of waste, chiefly the fish waste that is released from the fillet activities (Tugiyono et al. 2020). It has been estimated that approximately 25% of the fish catch is discarded annually (Rustad 2009). According to FAO 2014; fisheries production is consistently increasing at the rate of 3.2% every year. Only 40% of the catch is utilized for consumption by humans and about 60% is thrown away, which causes severe environmental issues (Chalamaiah et al. 2012; Caruso 2015). Fish wastes produced from the aquaculture sector have significant amounts of various plant nutrients, that can be harvested and restored in the productive chains. Various valuable components such as proteins, bioactive peptides, chitin, collagen, gelatin, and fertilizers can be prepared from fish wastes by utilizing different methods (Arvanitoyannis and Kassaveti 2008). The transformation of nutrient-rich fish wastes into products that can be sold in markets can even provide economic benefits to society (van der Wiel 2020).

2 Fisheries Production

Asia is the largest producer of aquaculture and in this continent, aquaculture is much more diverse than in any other region (FAO 2019; Bush et al. 2019). With the rapid development and advancement in fishing technologies, a considerable increase has been seen in aquaculture production from 1954 to 2014. In 2014, fisheries production was approximately 93.4 million tonnes all over the world. Aquaculture productivity depends on net primary productivity and the pathway by which this yield moves through the food web in fresh and marine water ecosystems and enters the human food chain (Iverson 1990). Aquaculture production increased by three times from 1997 to 2017, as it was about 34 million tonnes in 1997 while in 2017 the live weight volume increased to 112 million tonnes. The main fish species that accorded for 75% of production in the aquaculture sector in the year 2017 included catfish, carps, tilapia, bivalves, and seaweeds (Naylor et al. 2020). In 2018, the production in the aquaculture sector increased to 114.5 million tonnes in live weight (FAO 2020). According to the report published it has been found that production, consumption, and trade in aquaculture reached their zenith in the year 2018. According to the reports published by Food and Agriculture Organisation, 2020; it has been found that in the year 2018, total fish production was 179 million tonnes, out of which 88% was used directly for consumption by humans. With the rapid rise in aquaculture production, it has been estimated that by 2030, aquaculture production will be similar to capture production. In 2002, 76% of fisheries produced across the world were used for consumption by humans directly, and leftover 24% were utilized in the preparation of fish by-products such as fish meal and oil (Brander 2007). In 2016, production from the aquaculture sector accounted for 53% of the fish products that were used directly for consumption by humans. Tsani and Koundouri 2018 evaluated that the aquaculture sector has depicted a rapid growth in the food industry throughout the world. Besides providing food to millions of people, this sector also provides financial benefits to the community, thus forming an important pillar of Blue Growth Targets.

3 Fish Waste and By-Products

With the rapid rise in population across the globe and the subsequent augmentation in industrialization and urbanization, there has been a tremendous increase in aquaculture production due to the blooming of fishing technologies. Consequently, the amount of fish waste produced in fisheries production has shown a remarkable increase. It has been evaluated that approximately two-thirds of the fish is thrown away as waste, which is responsible for the creation of economic losses and environmental degradation (Coppola et al. 2021). Highly valuable nutritive elements are present in the waste generated from aquaculture production. In recent years, problems related to the generation of fishing waste have surged and it is becoming a global issue that is influenced by various biological, socio-economic, and technical factors (Arvanitoyannis and Kassaveti 2008). The processing of fish waste is essential to lessen its impact on the environment (Kusuma et al. 2019). The concentration and volume of waste generated from fish processing operations are determined by various factors such as the composition of raw material, processing of water source, additive utilized, and the processing of unit. According to Dubey et al. 2021; waste can be categorized as solid waste and liquid waste.

(a) Solid waste: Waste produced from the shunned parts of fish such as head, skin, scales, liver, viscera, trimmings, and bones is classified as solid waste. In the fishery industry, the fish that gets decomposed during the fish processing is also included in the category of solid waste. Stale fish or the fish rejected during the fish processing operations also combines for the solid waste.

(b) Liquid waste: Huge amounts of phosphorus and nitrogen present in the water released from the fish processing wastes cause algal proliferation and have a great influence on the aquatic plants and animals inhabiting the water body. Besides these elements, suspended solids sway the water-living plant and animal life by decreasing the intensity of light passing through the water body. The results of their study were indicative of the fact that an effective strategy for fish waste management includes three steps viz., minimization of fish waste, characterization of fish waste produced, and utilization of the waste generated from the fish processing units.

Dauda et al. 2019 categorized the wastes originating from aquaculture practices into four different classes: Solid waste derived from the discarded part of the fish, organic matter including elements such as nitrogen, phosphorus, and potassium, chemical compounds, and infectious microbes. Rustad et al. 2011 estimated that 75% of the fish raw material contributes to all the by-products obtained in the fish processing units, which are either discarded or utilized directly as a feed. Although some parts of the fish wastes are utilized for the production of fish oil and fish meal, with the advancements in technology, more value-added products can be reclaimed from fish wastes (Blanco et al. 2015).

30% of the by-products generated during fish processing include bones, scales, and skin that are rich in collagen (Gomez-Guillen et al. 2002; Karayannakidis et al. 2014). Viscera of the fish serve as an excellent source of protein and polyunsaturated fatty acids (Raa et al. 1983) and a value-added product can be obtained from the fish waste by converting it into fish protein hydrolysate. These fish protein hydrolysates have a wide range of applications such as they can be used as a fertilizer, animal feed, and peptone for the microbial culture (Alder-Nissen 1986; Gildberg et al. 1989; Vecht-Lifshitz et al. 1990; Kurbanogtu and Algur 2002). Viscera and carcasses of fish are a good source of bioactive molecules i.e., proteins and proteases (Silva et al. et al. 2014) that are used to prepare the fish protein hydrolysates. These protein hydrolysates can also be used as an alternative to commercially available nitrogen sources for the growth media in bacterial culture (Aspmo et al. 2005).

50% of the captured fish is discarded and this leads to the production of 32 million tonnes of waste every year (Kristinsson and Rasco 2000). Collagen has been produced from the skin and bones of fish that are released as waste products from fish processing units (Nalinanon et al. 2007; Wang et al. 2007). Conventional sources of collagen are poultry, pig, and cow, but these sources may act as carriers of prion and viral diseases (Silva et al. 2014). So nowadays, collagen obtained from fish by-products can be used as an alternative to mammalian collagen (Gomez-Guillen et al. 2002). Valorization of fish waste is pivotal for the implementation of Sustainable Development 2030 goals that aim at 'zero discards' (Thirukumaran et al. 2020).

4 Processing of Fish Waste

Huge amounts of fish processing waste are discarded away from fish markets and processing plants. It has been estimated that during 2006–07, approximately 3, 02, 750 tonnes of waste were released from pre-processing and processing industries in India (Mohanty et al. 2020). During these years, the highest amount of waste was released from shrimp processing followed by finfish processing industries. Thus, the generation of waste from fish processing industries and their improper disposal is a matter of great concern as they are one of the leading causes of environmental pollution. The use of fish waste aims to get rid of the environmental problems generated by the disposal of fish waste. With the advancement in technology, the use of these fish wastes has found its way to be used for the benefit of mankind.

The discarded solid waste, liquid waste, and under-utilized fish are rich in essential nutrients viz., fatty acids such as poly-unsaturated fatty acids, proteins, minerals, and carotenoids. Therefore, waste generated from fish processing units is accountable for nutrient loss and severe environmental problems (Venugopal 2021). Studies conducted have shown that fish by-products are apt for consumption by humans and have numerous other applications that are highly valuable in the market (Halim et al. 2016a, b). Such by-products are a good source of collagen, polyunsaturated fatty acids, chitin, and peptides. Among various fish by-products, the main structural protein found in the skin and scales of fish is collagen which represents approximately 70% of the dry weight of fish (Blanco et al. 2017). During the processing of fish, viscera is also generated as a major by-product and it is a potential source of proteins, lipids, and protein hydrolysates (Villamil et al. 2017).

The products obtained by processing fish waste are known to have bioactive and functional properties which are essential for the food, cosmetics, nutraceuticals, feed, agricultural and pharmaceutical industries. Different methods can be utilized to recover the bioactive materials from by-products released from fish processing units (a) microbial fermentation which involves the use of enzymes obtained from the microbes to procure the bioactive materials (b) physical or chemical methods that include the treatment of fish by-products with physical or chemical agents (c) enzymatic methods that use various enzymes to hydrolyze fish by-products. The best method among all these to valorize the fish by-products is an enzymatic method (Blanco et al. 2017; Villamil et al. 2017).

There is a long history behind the composting of fish waste and the preparation of compost is one of the most appropriate methods for recycling biodegradable fish waste that has potential application in modern agricultural practices. Turnover of organic fish waste into compost has a great economic value and proves beneficial to the environment (Castaldi et al. 2004; Deguchi et al. 2009). Fertilizer rich in organic matter is obtained by the processing of fish waste, and it has proven advantageous for the production of crops in agricultural fields (Ramalli et al. 2001). The method utilized for the preparation of fertilizers from fish wastes is similar to the decomposition process that is carried out by species of worms (Balkhande and Chavhan 2020). A huge amount of fish waste is generated from the fish markets that are responsible for the creation of numerous economic and environmental issues. So, one of the best ways for the management of this waste is to prepare fertilizer from it, which will be profitable both for the fish sellers and farmers.

A method for the preparation of fertilizer from fish waste has been described by various workers. One of the methods has been illustrated by Balkhande and Chavhan, 2020. In this method, they prepared a pit for the decomposition of fish waste. Various layers were made in this pit, of which the first layer included the scales, gills, liver, testis, fins, ovaries, viscera, bones, and kidneys of the fish. Temperature played a key role in the decomposition of this fish waste in the compost pit and shady areas were generally preferred by them. The well-aerated system was also essential for the process of decomposition. Water sprinkling in the compost pit was necessary because microbes involved in decomposing the fish waste require humidity. Various bacteria such as *Bacillus, Rhizobium*, and phosphorus solubilizing bacteria (PSB) were utilized for accelerating the decomposition process. According to them, much care was required to be taken for the first few days as the fish waste is susceptible to predation. After a period of 30 days, all layers were mixed well for proper aeration. According to their study, the best quality fertilizer was obtained after six months. In another study Balkhande 2020 collected fish wastes such as gills, fins, scales, liver, kidney, testis, ovary, and viscera from the Bhokhar fish market in Maharashtra. For processing the waste, a pit was prepared, and fish waste was placed in the first layer. After this, a source of carbon such as saw dust was added to the second layer. Waste collected from the garden and soil was placed in the third layer. Partially decomposed dung was placed above the garden waste and soil. To avoid predation, the pit was kept covered. In the fish composting unit, watering was done for 3 days a week as water is an essential component for the growth of microbes. For accelerating the process of decomposition, four bacteria viz., *Bacillus, Rhizobium, Azatobacter,* and phosphate solubilizing bacteria (PSB) were put in fish comporting unit Mixing of all the layers was done after thirty days for the proper aeration of the compost. In his experiment, a green sun shade mat was used to keep away the sunlight and animals from the composting unit. From the above study, it was ascertained that it took six months to prepare organic fertilizer from fish waste. The fish compost that is ready for use has a pH between 7 and 8.5 (Day and Shaw 2001). According to Gray et al. 1971, for the rapid degradation of any biodegradable waste, the incorporation of bacteria in the compost pit is essential as they are involved in the generation of heat and initial decomposition in compost. Balkhande 2020 evaluated the percentage of free carbonate in the fertilizer obtained from fish waste and it was found to be 4.8% which is classified under the good category of fertilizers. The nitrogen that is an essential component of fertilizer was found to be 1.18% which was very high when compared with the nitrogen content present in vermicompost (Manna et al. 2015).

Radziemska et al. 2019 also prepared a compost by mixing 20% fine bark and 80% fish waste. Pine bark served as a source of carbon and various micro and macro-elements. Another method for the preparation of fish compost has been proposed by López-Mosquera et al. 2011. They used the windrow method for the creation of compost in a piece of land by collecting waste from different fish species viz., tuna, sardine, and mackerel. They intermixed this fish waste with the seaweed (*Laminaria sps.* and *Cystoseira sps.*). Pine bark was also added to elevate the ratio of carbon and nitrogen. Fish waste, seaweed, and pine bark were mixed in a ratio of 1:1:3. The compost was prepared in a trapezium-shaped pile which was 6 m in length, 1 m in height, and 2 m in width. They concluded that the total time for the completion of the process of composting was four months. Similarly, Kusuma et al. 2019 collected fish waste from the Tambak Lorak Fish Market in Indonesia and prepared fish compost from the collected waste. They prepared compost by placing 2 kg of fish waste in a permeable plastic container having 15.08% moisture content. It was left as such for two weeks and then mixed with fruits and vegetables and placed in a bucket. Luthfiati and Junianto 2021 proposed a bakasang method for processing fish waste into liquid organic fertilizer. According to them, the process of preparing an organic fertilizer from the fish waste can be split up into four stages. In the first step, fish waste is cleaned and any blood, or dirt attached to this is removed. After this, the fish gut is cut to make a smooth paste. Then the 200 g of this crushed fish waste is put in a 1.5 l bottle, which is tightly closed and kept for sun drying for about ten days. After drying up, levels of various elements that are essential for plants such as phosphorus, nitrogen, and potassium were analyzed. This way they prepared an organic liquid fertilizer by using the fish waste.

Ilham 2012 described a method for the preparation of organic liquid fertilizer. For this, 4 kg of fish waste was finely chopped and mixed with 1000 ml of coconut water, 50 ml of EM-4, and 7000 ml of clean water in a plastic container, and the mixture was stirred very well, till the solution became homogenous. To this, finely chopped banana peels were added and the mixture was allowed to undergo anaerobic fermentation for thirteen days. After this, the solution was filtered and placed in another bottle. Another method for the preparation of biofertilizer from fish waste was given by Waryanti and Sudarno 2013. They put 200 g of fish waste in four containers containing 1000 ml of distilled water and 100 ml of molasses. Local microorganism solution (MOL) of banana hump was added in four different concentrations i.e., 0 ml, 50 ml, 100 ml, and 150 ml in all the containers, and was kept for anaerobic fermentation for fourteen days. Tiwow et al. 2019, obtained liquid organic fertilizer by fermenting 10 kg of *Tilapia* fish waste for fifteen days, using the fermentation method. To assess the effect of fertilizer prepared from fish waste on the yield and growth of *Amaranthus* and *Trigonella foenum-graecum*, an experiment was conducted by Thankachand and Chitra (2021) in which they collected fish wastes from the local fish markets. 5000 ml of water was filled in 10 l of the clean clay pot. To this, they added 2 kg of jaggery, 2 kg of fish waste, and 2 kg properly mashed bananas, and the mixture was stirred very well. For the prevention of any kind of contamination, the clay pot was covered with a cloth. The contents of the pot were mixed daily and after a fortnight, the filtrate was obtained by the filtering of contents, which was utilized as fertilizer after a proper dilution.

5 Application of Fish Waste as Fertilizers

Fish waste composting is a natural and environment-friendly process and a manageable substitute for the management of fish discards. The problem of fish waste and pollution can be eradicated by composting the fish waste. It is the cheapest method of disposing off fish waste. Fertilizers formulated from the waste products (Fig. 1), such as fins, skin, gills, kidney, viscera, bones, ovaries, testis, scales, are beneficial for increasing the content of nutrients in the soil, decreasing the prevalence of parasitic infections, eliminating the seeds of unwanted plants and suppressing the various diseases associated with plants. Fish waste manure can be used as a supplement as

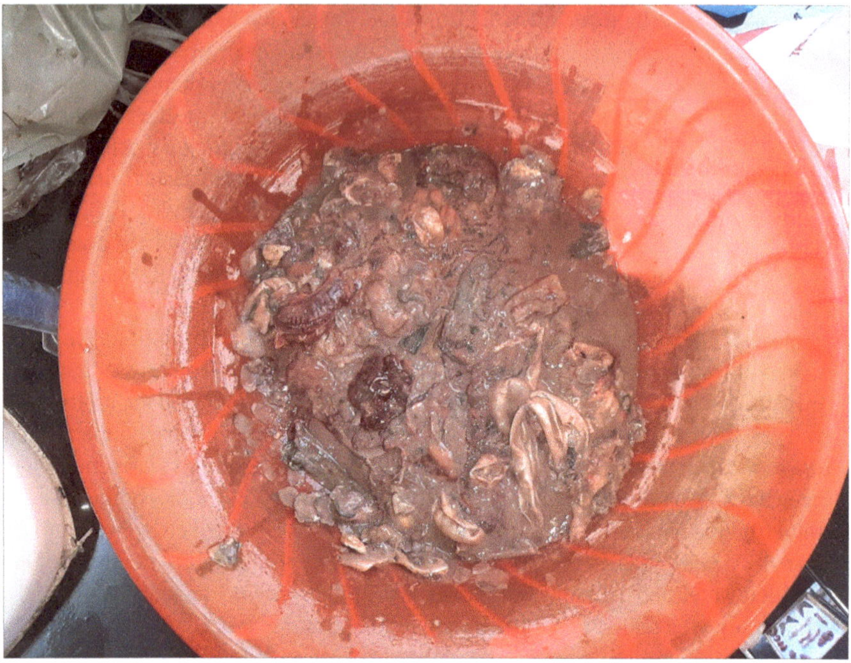

Fig. 1 Fish waste consisting of blood, viscera, gills, fins, liver, air bladder and flesh

it elevates the concentration of organic matter and nutrients in the soil and fosters the moisture-holding capability of soil this way it helps to escalate the quality and production of crops (Balkhande and Chavhan 2020). Fish waste is an excellent source of phosphorus that is necessary for the growth of plants (Radziemska et al. 2019). Although the fish waste compost is not used extensively but is rich in phosphorus, it is appropriate for crops that need phosphorus in the available form directly (Lanno et al. 2021).

Bioorganic fertilizers are cheap, and eco-friendly and are utilized to ameliorate the fertility and quality of the soil. These bio-fertilizers serve as an excellent alternative to chemical fertilizers (Vessey 2003). Rishitha and Rao (2019) assessed the growth of plants *Trigonella foenum-graecum* and *Pennisetum glaucum* that were grown by using the biofertilizers derived from the wastes of *Catla catla, Labeo rohita,* and mixed fish waste. In an attempt to evaluate the effect of marine waste on the growth of pea and tomato seedlings a study was conducted by Rebecca et al. (2015). They collected the fish waste from the local fish market and used this as manure for the plants. They concluded that the application of fish waste as a fertilizer has a significant effect on the time of seedlings. Balkhande (2000) reported that organic manure prepared from fish waste can prove advantageous even for the fish sellers and farmers who discard away the fish waste. Selling the manure prepared from the fish waste can also be helpful in the creation of employment which will provide economic benefits to the society. Fish waste is an excellent source of nitrogen, phosphorus, and potassium and is beneficial to an agronomist for organic farming. Fish waste is suitable for use in the agricultural fields as it is rich in calcium, phosphorus, and nitrogen (Illera-Vives et al. 2015) Various fertilizers derived from fish waste are available in markets these days, a few of which are certified for the use in organic farming (EC Regulation 1991). According to Illera-Vives et al. (2015), Castro et al. (2006) fish waste can be utilized for the irrigation of *Solanum lycopersicum.* Initiative for using the fish waste has been taken across the globe in pursuit of the techniques available for the transformation of fish discards into appropriate products that can be used in agricultural activities (López-Mosquera et al. 2011). Compost prepared from fish waste can be used for the amendment of soil as it enhances the texture and fertility of the soil, and this lessens the application of chemical fertilizers in the soil. The same can be done by the employment of new technologies (Vandecasteele et al. 2016).

The efficiency of applying the fish waste as an organic fertilizer was tested by Radziemska et al. (2019) by experimenting on the ice lettuce crop. They found that the fertilization of soil with the compost prepared from the fish waste increases the yield of leaves in the ice lettuce crop and it also had profound effects on the concentration of phosphorus, sodium, magnesium, calcium, nitrogen, and potassium in the leaves of ice lettuce. Their study proposed that the compost prepared from fish waste can be a valuable fertilizer for the proper growth of crops. López-Mosquera et al. (2011) proposed that the product obtained after composting fish waste can be applied as a fertilizer in agricultural fields as it is prepared from natural components and therefore can be used without any limitations. Compost prepared from fish waste is advantageous as it is rich in nutrients and contains calcium, potassium, and magnesium. Besides this, compost from fish waste is devoid of any contaminants such as phytotoxic compounds and heavy metals (Kinnunen et al. 2005). Fish waste can be used as a raw material for the preparation of organic fertilizer as it contains potassium, phosphorus, and nitrogen thus plays a vital role in providing nutrition to the plants (Hapsari and Welasih 2010). Fish compost contains 8.25% nitrogen and 4.73% phosphorus. The percentage of phosphorus in the compost from fish waste is best for any fertilizer as plants require phosphorus in PO_4^{3-} form. The presence of potassium in plants influences the intake of other elements. The percentage of potassium in fish compost is 1.56, which is the best concentration because the higher the level of potassium, the better will be the growth and development of plants (Kusuma et al. 2019).

Studies done by Ahuja et al. 2020 predicted that in Norway the fertilizers produced from the amount of available fish waste can help in the creation of a commercial product that can act as a substitute for commonly used dried manure derived from poultry farms that are used in agricultural farms. Luthfiati and Junianto (2021) proposed that fish waste can be used to make liquid organic fertilizer, as it contains all the essential nutrients required by the plants for their normal growth. Suartini et al. 2018 found that liquid organic fertilizer obtained from the waste of *Katsuwonus pelamis* by using bakasang method contains appropriate levels of phosphorus and nitrogen. Studies conducted by Pati (2017) depicted that liquid organic fertilizer obtained from the waste of Tilapia when used in agricultural fields has a remarkable result on the number of flowers, leaves, and the height of the stem of a long bean plant. Similarly, a study done by Zahroh et al. (2018) found that the fertilizer concentration of 4.5% has a significant effect on the stem height and leaves of red chilli plants. Liquid organic fertilizer ameliorates the fertility of the soil and provides essential nutrients such as potassium, phosphorus, and sodium that can enhance the productivity of crops (Luthfiati and Junianto, 2021). Wyatt and Mc Gourty (1990) reported that fish fertilizer prepared from fish waste increases both the vigor and growth of plants. The fertilizers prepared from the head, gills, and intestines of fish waste supply essential nutrients that can satiate the requirements of *Amaranthus dubius* and *Trigonella foenum graecum* (Thankachan and Chitra 2021). Fish waste fertilizer increased the height of the plant, length of root, shoot, the surface area of the leaf, number of branches, and diameter of stem and decreased the germination time of seeds in the studied plants. Shahsavani et al. (2017) found that the use of fish waste had a notable effect on the yield of *Vigna sinensis*. An increase in the utilization of fish waste as fertilizer creates an additional opportunity for the establishment of the fish waste-derived fertilizer industry.

Utilization of fish waste has an added advantage, besides increasing the growth performance of the plants when used as fertilizer, it also helps to reduce the environmental pollution which may occur due to the improper disposal of the waste and their by-products. Results of the study conducted by Thanh et al. (2023) has also revealed that the use of fish waste collected from the ponds inhabited by snake head fish, along with other agricultural wastes reduces the cost involved in using the chemical fertilizers and also improves the productivity of crops. Nutrient management is essential for increasing the growth of crops and with the increase in production of crops after Green Revolution, rise in the use of chemical fertilizers has been reported and this has led to the depletion of soil fertility over the time and has caused numerous other problems in the ecosystem. In order to overcome this problem of environment degradation, alternative solutions are required and one of them is to shift towards the use of organic fertilizers. Fertilizers derived from the fish waste that includes the viscera, gills, blood, flesh, fins are rich in nitrogen, phosphorus and potassium and in order to avoid the growth of fungus in the fertilizer, acids are added so as to maintain the pH. The use of fish waste and by-products in the form of fertilizers supply essential nutrients to the soil and also reduces its negative impact on the environmental surroundings (Suge et al. 2011). A study was conducted by Jubin and Radzi 2022 to assess the effect of fish fertilizer along with the inorganic fertilizer on the growth of *Zea mays*. The supplement of the fish waste and by-products in the inorganic fertilizer enhances the physical, biological and chemical properties of the soil and also improves the microbial growth in soil that increases the rate of decomposition of organic matter. They found that the use of inorganic fertilizer along with the fertilizers prepared from the fish waste and by-products depicts significant effect on the area of the leaf and girth of the stem.

Fish waste is generally thrown in open places by the fish sellers and this creates numerous problems. According to the studies conducted by Balkhande and Chavhan

(2020), preparation of fish fertilizer from the fish waste do not require a lot of expenditure and involves the simple procedure. Compositing of fish waste is an excellent method and is considered as environment friendly as it promotes the use of waste to the best. Basic steps involved in the preparation of organic fertilizer from the fish waste are: preparation of compost pit, fish waste collection of fish waste, addition of saw dust, black soil, kitchen waste, few bacteria such as *Bacillus, Rhizobium,* requires weekly watering, mixing of layers, monitoring monthly and after approximately 180 days the fertilizers are prepared in the pit and after carrying out the analysis of the nutrient composition of the fertilizer prepared out of the waste, it can be sold.

6 Conclusion

Wastes obtained from the fish processing units have a huge prospective. At present, the popularisation of the technique among common people, fish farmers, agronomists, and researchers for the commercial production of organic fertilizers and manures from the discarded fish waste is the need of an hour. By utilizing this technique, today's waste can be converted into tomorrow's best as fish waste disposal is one of the major problems that the society is facing. Fish waste composting serves as an effectual source of fertilizers, rich in various nutrients required by the plants for their proper growth. Transforming fish waste into useful by-products such as fertilizers is a logical solution for the sustainable management of this waste. Besides maintaining the fertility of the soil, these bio-fertilizers are economical, eco-friendly, and easily manufactured as compared to chemical fertilizers. The use of fish waste and by-products permits the decline of these wastes that otherwise have the potential to give rise to various social, and economic environmental implications. Valorization of fish waste into advantageous products necessitates tight cooperation between research communities, industrialists, and governance bodies.

References

Adler-Nissen J (1986) Enzymic hydrolysis of food proteins. Elsevier Applied Science Publishers, London, UK, pp 1–427

Ahuja I, Dauksas E, Remme JF, Richardsen R, Løes AK (2020) Fish and fish waste-based fertilizers in organic farming—with status in Norway: a review. Waste Manage 115:95–112. https://doi.org/10.1016/j.wasman.2020.07.025

Arvanitoyannis IS, Kassaveti A (2008) Fish industry waste: treatments, environmental impacts, current, and potential uses. Int J Food Sci Technol 43(4):726–745. https://doi.org/10.1111/j.1365-2621.2006.01513.x

Aspmo SI, Horn SJ, Eijsink VG (2005) Use of hydrolysates from Atlantic cod (*Gadus morhua L.*) viscera as a complex nitrogen source for lactic acid bacteria. FEMS Microbiol Lett 248:65–68

Balkhande JV (2020) Devising of organic fertilizer from fish and crab wastes: waste to best technology. Int J Fish Aquatic Stud 8(2):01–05

Balkhande JV, Chavhan AB (2020) Preparation of organic fertilizer from fish waste: a new avenue in employment generation. Int Res J Sci Eng A10:21–24

Blanco M, Fraguas J, Sotelo CG, Pérez-Martín RI, Vázquez JA (2015) Production of Chondroitin sulphate from head, skeleton and fins of *Scyliorhinus canicula* by-products by combination of enzymatic, chemical precipitation and ultrafiltration methodologies. Mar Drugs 13:3287–3308

Brander KM (2007) Global fish production and climate change. Proc Natl Acad Sci 104(50):19709–19714

Bush SR, Belton B, Little DC, Islam MS (2019) Emerging trends in aquaculture value chain research. Aquaculture 498:428–434

Caruso G (2015) Fishery wastes and by-products: a resource to be valorised. J Fish Sci 9:080–083

Castaldi P, Garau G, Melis P (2004) Influence of compost from sea weeds on heavy metal dynamics in the soil-plant system. Fresenius Environ Bull 13(11b):1322–1328

Castro RS, Azevedo CMSB, Bezerra-Neto F (2006) Increasing cherry tomato yield using fish effluent as irrigation water in Northeast Brazil. Sci Hortic 110:44–50

Chalamaiah M, Kumar BD, Hemalatha R, Jyothirmayi T (2012) Fish protein hydrolysates: proximate composition, amino acid composition, antioxidant activities and applications: a review. Food Chem 135:3020–3038. https://doi.org/10.1016/j.foodchem.2012.06.100

Coppola D, Lauritano C, Esposito PF, Riccio G, Rizzo C, de Pascale D (2021) Fish waste: from problem to valuable resource. Mar Drugs 19(2):116. https://doi.org/10.3390/md19020116

Dauda AB, Ajadi A, Tola-Fabunmi AS, Ayoola Olusegun Akinwole AO (2019) Waste production in aquaculture: Sources, components and managements in different culture systems. Aquacult Fish 4(3):81–88. https://doi.org/10.1016/j.aaf.2018.10.002

Day M, Shaw K (2001) Biological, chemical and physical processes of composting. In: Compost utilization in horticultural cropping systems. Boca Raton, FL, Lewis Publishers, USA, pp 17–50

Deguchi S, Kawamoto H, Tanaka O, Fushimi S, Uozumi S (2009) Compost application increases the soil temperature on bare Andosol in a cool climate region. Soil Sci Plant Nutrit 55(6):778–782

Dubey S, Meher P, Shetty A, Umtol A, Kirloskar S (2021) Waste Management in Fishery Industry: A Review. Int J Eng Res Technol 9(3):206–209

FAO (2014) The state of world fisheries and aquaculture—opportunities and challenges. FAO, Rome

FAO (2016) The State of World Fisheries and Acquaculture. Available online: http://www.fao.org/publications/sofia/2016/en/#:~{}:text=The%20State%20of%20World\T1\textquoterights%20Fisheries%20and%20Aquaculture%202016&text=As%20always%2C%20the%20scope%20is,%2C%20utilization%2C%20trade%20and%20consumption

FAO (2018) The state of food security and nutrition in the world. building climate resilience for food security and nutrition. FAO, Rome

FAO (2019) Fisheries and aquaculture software. FishStatJ: software for fishery and aquaculture statistical time series. http://www.fao.org/fishery/statistics/software/fishstatj/en

FAO (2020) The state of world fisheries and aquaculture 2020. Sustainability in Action. Rome: FAO

Gildberg A, Batista I, Strom E (1989) Preparation and characterization of peptones obtained by a two-step enzymatic hydrolysis of whole fish biotechnol. Appl Biochem 11:413–423

Gomez-Guillen MC, Turnay F, Fernández-Diaz MD, Ulmo N, Lizarbe MA, Montero P (2002) Structural and physical properties of gelatin extracted from different marine species: a comparative study. Food Hydrocoll 16:25–34. https://doi.org/10.1016/S0268-005X(01)00035-2

Gray KR, Sherman K, Biddlestone AJ (1971) A review of composting-Part I. Process Biochem 6:32–36

Halim NRA, Yusof HM, Sarbon NM (2016a) Functional and bioactive properties of fish protein hydrolysates and peptides: a comprehensive review. Trends Food Sci Technol 51:24–33

Halim NRA, Yusof HM, Sarbon NM (2016b) Functional and bioactive properties of fish protein hydrolysates and peptides: a comprehensive review. Trends Food Sci Technol 51:24–33

Hapsari N, Welasih T (2010). Utilization of fish waste into organic fertilizer, UPN Veteran, Jawa Timur

Ilham D (2012) Preparation of Liquid Organic Fertilizer from the Stomach of Jambal Siam Fish by Using EM-4 as a Starter. Faculty of Fisheries and Marine Science Riau University

Illera-Vives M, Labandeira SS, Brito LM, López-Fabal A, López-Mosquera ME (2015) Evaluation of compost from seaweed and fish waste as a fertilizer for horticultural use. Sci Hortic 186:101–107

Iverson RL (1990) Control of marine fish production. Limnol Oceanogr 35:1593–1604

Jubin JJ, Radzi NM (2022) Application of Fish Waste Fertilizer on the Growth of Maize (Zea mays) IOP Conf. Ser.: Earth Environ. Sci. 1059 012070

Karayannakidis PD, Chatziantoniou SE, Zotos A (2014) Effects of selected process parameters on physical and sensorial properties of yellowfin tuna (*Thunnus albacares*) skin gelatin. J Food Process Eng 37:461–473

Kinnunen RE, Gould MC, Cambier P (2005) Composting commercial fish processing waste from fish caught in the Michigan waters of the great lakes. Michigan State University Extension, USA

Kristinsson HG, Rasco BA (2000) Fish protein hydrolysates: production, biochemical, and functional properties. Crit Rev Food Sci Nutr 40(1):43–81. https://doi.org/10.1080/10408690091189266

Kurbanoglu EB, Algur OF (2002) The influence of ram horn hydrolysate on the crop yield of the mushroom *Agaricus bisporus*. Sci Hortic 94:351–357

Kusuma IM, Syafrudin, Yulianto B (2019) Utilization of fish waste processing as compost raw material in Tambak Lorok Market. E3S Web of Conferences 125:07004. https://doi.org/10.1051/e3sconf/20191250 ICENIS 2019 7004

Lanno M, Kriipsalu M, Shanskiy M, Silm M, Kisand A (2021) Distribution of phosphorus forms depends on compost source material. Resources 10:102. https://doi.org/10.3390/resources10100102

López-Mosquera ME, Fernández-Lema E, Villares R, Corral R, Alonso B, Blanco C (2011) Composting fish waste and seaweed to produce a fertilizer for use in organic agriculture. Proc Environ Sci 9:113–117

Luthfiati N, Junianto (2021) Review articles; utilization of fish offal for organic liquid fertilizer. Global Sci J 9(10)

Manna MC, Sahu Asha, Patra AK, Khanna SS, Chaudhari SK, Sikka AK (2015). Rapid composting technique, way to enhance soil organic carbon, productivity and soil health. ICAR-IISS technology folder Bhopal. http://www.iiss.nic.in/news%20and%20event/Rapid%20Composting%20Technique.pdf

Mohanty B, Hauzoukim SS, Swain S (2020) A review on fish processing wastes generation in India and its further utilization prospects into different value added compounds. Indian J Nat Sci 10(60):24177–24182

Nalinanon S, Benjakul S, Visessanguan W, Kishimura H (2007) Use of pepsin for collagen extraction from the skin of bigeye snapper (*Priacanthus tayenus*). Food Chem 104:593–601 doi: https://doi.org/10.1016/j.foodchem.2006.12.035

Naylor RL, Hardy RW, Buschmann AH, Bush SR, Cao L, Klinger DH, Little DC, Lubchenco J, Shumway SE, Troell M (2021) A 20-year retrospective review of global aquaculture. Nature 591(7851):551–563

Pati BJK (2017) The effect of liquid organic fertilizer for Tilapia Fish waste on the growth and production of long bean plants. Faculty of Teacher Training and Education, Sanata Dhama University. Yogyakarta, Essay

Raa J, Gildberg A, StrømT (1983) Silage production theory and practice. In: Ledward D, Taylor A, Lawrie R, editors. Upgrading waste for feeds and food. London: Butterworths, pp 1–11732

Radziemska M, Vaverková MD, Adamcová D, Brtnický M, Mazur Z (2019) Valorization of fish waste compost as a fertilizer for agricultural use. Waste Biomass Valoriz 10(9):2537–2545

Ranalli G, Botturea G, Taddei P, Garavni M, Marchetti R, Sorlini G (2001) Composting of solid and sludge residues from agricultural and food industries. Bioindicators of monitoring and compost maturity. J Environ Sci Health 36:415–436

Rebecca LJ, Anbuselvi S, Sharmila S, Medok P, Sarkar D (2015) Effect of marine waste on plant growth. Pharm Lett 7(10):299–301

Regulation EC (1991) Regulation 2092/91/EEC of 22 July 1991 on organic production of agricultural products and indications referring thereto on agricultural products and foodstuffs. Off J Eur Commun L198(15):198

Rishitha MD, Rao M (2019) Bioconversion of fish waste into a liquid fertilizer and its impact on semi- arid tropical crops. Res J Life Sci, Bioinform, Pharmaceut Chem Sci 5(1):903–912

Rustad T, Storrø I, Slizyte R (2011) Possibilities for the utilisation of marine by-products. Int J Food Sci Technol 46:2001–2014. https://doi.org/10.1111/j.1365-2621.2011.02736.x

Rustad T (2007) Physical and chemical properties of protein seafood by-products. In F. Shahidi (Ed.), Maximising the value of marine by-products, Woodhead Publishing, pp 3–21

Shahsavani SH, Abaspour A, Parsaeian M, Unesi Z (2017) Effect of fish waste, chemical fertilizer and biofertilizer on yield and yield components of bean (*Vigna sinensis*) and some soil properties. Iran J Pulses Res 8(1):45–59

Silva JFX, Ribeiro K, Silva JF, Cahú TB, Bezerra RS (2014) Utilization of tilapia processing waste for the production of fish protein hydrolysate. Anim Feed Sci Technol 196:96–106. https://doi.org/10.1016/j.anifeedsci.2014.06.010

Suartini K, Paulus HA, Minarni RJ (2018) Production of liquid organic fertilizer from the offal waste of Skipjack fish (*Katsuwonus pelamis*). J Acad Chem 7(2):70–74; Environ Eng 2(1):1–6

Suge JK, Omunyin ME, Omami EN (2011) Effect of organic and inorganic sources of fertilizer on growth, yield and fruit quality of eggplant (*Solanum melongena* L). Arch Appl Sci Res 3(6):470–479

Thanh DT, Ty NM, Hien NV, Berg H, Nguyen TKO, Vu PT, Minh VO, Da CT (2023) Effects of organic fertilizers produced from fish pond sediment on growth performances and yield of Malabar and *Amaranthus* vegetables. Front Sustain Food Syst 7:1045592

Thanh DT, Ty NM, Hien NV, Berg H, Vu PT Minh VO Da CT (2023)

Thankachan M, Chitra G (2021) The potential effect of fish waste fertilizer on the growth and yield of *Amaranthus dubius* and *Trigonella foenum-graecum*. Int J Adv Res 9(04):406–411

Thirukumaran R, Priya VKA, Krishnamoorthy S, Ramakrishnan P, Moses JA, Anandharamakrishnan C (2022) Resource recovery from fish waste: Prospects and the usage of intensified extraction technologies. Chemosphere 299:134361

Tiwow VMA, Adrianton APH, Hopiyanti N (2019) Production of Liquid and Solid Organic Fertilizer from Tilapia Fish (*Oreochromis mossambicus*) Wasteusing “Bakasang” Traditional Fermentation Technology. Int J Eng Adv Technol 8(3S):2249–8958

Tsani S, Koundouri P (2018) A methodological note for the development of integrated aquaculture production models. Front Mar Sci 4:406. https://doi.org/10.3389/fmars.2017.00406

Tugiyono T, Febryano IG, Yuwana P, Suharso S (2020) Utilization of fish waste as fish feed material as an alternative effort to reduce and use waste. Pak J Biol Sci 23(5):701–707

van der Wiel BZ, Weijma J, van Middelaar CE, Kleinke M, Buisman CJ, Wichern F (2021) Restoring nutrient circularity in a nutrient-saturated area in Germany requires systemic change. Nutr Cycl Agroecosyst 121(2):209–226

Vandecasteele B, Sinicco T, D'Hose T, Nest V, Mondini TC (2016) Biochar amendment before or after composting affects compost quality and N losses, but not P plant uptake. J Environ Manage 168:200–209

Vecht-Lifshitz SE, Almas KA, Zomer E (1990) Microbial growth on peptones from fish industrial wastes. Lett Appl Microbiol 10:183–186

Venugopal V (2021) Valorization of seafood processing discards: Bioconversion and bio-refinery approaches. Front Sustain Food 5

Vessey JK (2003) Plant growth promoting rhizobacteria as biofertilizers. Plant Soil 255:571–586. https://doi.org/10.1023/A:1026037216893

Villamil O, Váquiro H, Solanilla JF (2017) Fish viscera protein hydrolysates: Production, potential applications and functional and bioactive properties. Food Chem 224:160–171. https://doi.org/10.1016/j.foodchem.2016.12.057

Wang L, An X, Xin Z, Zhao L, Hu Q (2007) Isolation and characterization of collagen from the skin of deep-sea redfish (*Sebastes mentella*). J Food Sci 72(8):E450–E455

Waryanti A, Sudarno SE (2013) Study of the effect of adding coconut husk on the manufacture of liquid fertilizer from fish washing waste on the quality of macro nutrients (CNPK). J Environ Eng 2(2):1–7
Wyatt B, Mc Gourty G (1990) Use of marine by products on agricultural crops. International byproducts conferences, Anehorage, Alaska.
Zahroh F, Kusrinah, Siti, MS 2018.Comparison of Variations in Concentration of Liquid Organic Fertilizer from Fish Waste on the Growth of Red Chili (*Capsicum annum* L.) Plants. J Biol Appl Biol 1(1):50–57

Quintessential Utilization of Non-edible Aquatic Biowaste: In Pursuit of a Paradigm Shift Toward Wealth (from Waste) in Aquaculture"

Ramjanul Haque, Paramita Banerjee Sawant, Jitendra Kumar Sundaray, Rajesh Kumar, Narinder Kumar Chadha, Soibam Ngasotter, and K. A. Martin Xavier

Abstract Aquaculture plays an increasing role in future food security. Thirty-four calories per person per day are provided by fish and fish products globally. However, the rearing, harvesting, and processing of fish produces enormous amounts of trash, which is an issue for the entire world. For every ton of fish consumed, approximately the same quantity of fish waste (FW) is disposed of either through ocean dumping or land disposal. Unutilized waste has an impact on a larger coastal zone at many ecosystem levels, reducing benthos, plankton, and nekton biomass, variety, and density, and altering the structure of natural food webs. Alternatives to pricey feed additives should be investigated in order to meet the sustainable development goals (SDGs) of preventing the depletion of valuable aquatic resources. Wastes from the fisheries sector could be treated with various methods and can be utilized for pigments, chitosan, and collagen, which can be used in fish feed, biomedical, and

R. Haque (✉) · P. B. Sawant · N. K. Chadha · S. Ngasotter · K. A. Martin Xavier
ICAR-Central Institute of Fisheries Education, Mumbai 400061, India
e-mail: ramjanul.aqcpa903@cife.edu.in

P. B. Sawant
e-mail: paromita@cife.edu.in

N. K. Chadha
e-mail: nkchadha@cife.edu.in

S. Ngasotter
e-mail: soibam.phtpb004@cife.edu.in

K. A. Martin Xavier
e-mail: martinxavier@cife.edu.in

J. K. Sundaray · R. Kumar
ICAR-Central Institute of Freshwater Aquaculture, Bhubaneswar 751002, India
e-mail: jsundaray@gmail.com

R. Kumar
e-mail: rajeshfishco@yahoo.co.uk

S. Maqsood et al. (eds.), *Fish Waste to Valuable Products*, Sustainable Materials and Technology, https://doi.org/10.1007/978-981-99-8593-7_17

pharmaceutical industries. Currently, the production of biogas, biodiesel, biofertilizer, and bioplastic from non-recyclable fish waste is widely practiced. The various waste processing activities need additional inputs and outputs in order to recover energy and separate the necessary components from aquatic waste. The primary goal of our study towards sustainable aquaculture is the conversion of these wastes while also recovering important materials before disposal, which would also help in boosting the circular economy.

Keywords Fish waste · Circular economy · Environment pollution · Green technology · Utilization · Bioactive compounds · Sustainable aquaculture

1 Introduction

In the twenty-first century, the fisheries and aquaculture sectors have received greater recognition for their vital role in ensuring global food security and nutrition. Nearly 20% of the average per capita animal protein consumption of the world's 3.2 billion people came from aquaculture (FAO 2021). Thirty-four calories per person per day are provided by fish and fish products globally. Choe et al. (2020) stated that large amounts of trash are produced during fish farming, fishing, and processing, which now became a global concern. Fish trimmings and some particular parts, such as fish heads, fish guts, fish tails and fish fins, fish skins, fish scales, and fish bones, are all included under the concept of "fish waste." The terms "fish waste," "fish processing waste," "by-products," "raw materials," and "rest raw materials" have all been used in various research studies (Choe et al. 2020). According to Illera-Vives et al. (2015) and Karim et al. (2015), for every ton of fish consumed and disposed of via ocean dumping or land disposal, about the same quantity of fish waste (FW) was produced. Fish farm waste has the potential to alter natural food webs by negatively affecting the biomass, density, and variety of benthos, plankton, and nekton. Waste from fish farms can also have an impact on the neighborhood and be directly impacted by the effluent (Gowen 1991; Pillay 1991). To achieve sustainable and equitable global fisheries and aquaculture, revolutionary changes in policy, management, and innovative technology must be accelerated in order to utilize the growing amount of waste produced by the aquaculture industry.

2 Fish Waste as a Secondary Source of Resource

Depending on the region and species, the waste produced by aquaculture has a widely varied range of characteristics. Nearly 32 million tons of waste are produced from the residuals from the total amount of fish caught (more than 50%), which are not consumed as food (Arvanitoyannis et al. 2008). Large volumes of soluble-inorganic excretory waste and particulate organic waste are produced by aquaculture farms

(Ackefors 1994). The average yield in the fish processing industry is calculated using a gutted fish with the head on, which is approximately 40%½ (Marsh and Bechtel 2012). During the processing of fish, only 35–40% of the flesh is edible; the remaining consists of bones, skin/scales, swim bladders, intestines, roes, liver, and blood (Sachindra and Mahendrakar 2015). Fish offals like heads, frames, tails, skin, bones, fins, and viscera are included in the disposal portion. Fishmeal is an essential component of commercial and formulated diets but it is also a major pollution source. Fishmeal contamination looks to be a global issue. Fishmeal can either be, intentionally or accidently, contaminated with heavy metals, persistent organic pollutants (POPs), and pesticides. Alternatives to raw fishmeal protein sources should be researched by utilizing the waste in order to supply the large demand for fish while minimizing the current dependency on marine water and freshwater fishing resources for sustainable aquaculture.

3 Waste Generated from Fisheries Sector

Fish waste is some portion of fish tissue, such as bones, guts, heads, and tails, which is not suitable for human food but can be utilized to make fishmeal. A survey claims that more than half of the fish captured are not consumed (Kristinsson and Rasco 2000). Common by-products of finfish include trims, fish skins, fish heads, fish frames (bones with attached flesh), fish viscera (guts), and blood. According to Stevens et al. (2018), the following by-product fractions were present in the total wet weight of Atlantic salmon: viscera (12.5%), heads (10%), frames (10%), skins (3.5%), blood (2%), and belly flap (2%). Both raw and cooked shrimp, a significant part of the seafood business, are edible. In either event, only around 40% of the shrimp are fit for human consumption, and the remaining 60% are processed trash (shrimp shells) in the commercial shrimp processing sector (Barratt and Montano 1986; Dayakar et al. 2021).

4 Challenges and Negative Impact of Fish Waste

The bulk of by-products and wastes are produced during the processing of large quantities of fish, shrimp, and other aquatic species. Fish and shrimp processing effluents have very high levels of organic matter, nutrients, total suspended particles, fat, oil, pathogenic and other microorganisms. Therefore, the receiving coastal and marine habitats are quite likely to have negative consequences from fish and shrimp processing effluents. There are significant environmental issues as a result of the coastal region receiving about 40% of the oyster shell debris (Zhu et al. 2020). The garbage from shrimp is subsequently put into landfills, dumped in the ground, and dumped into the ocean, which causes significant surface pollution with an unpleasant odor in coastal areas, and raises serious environmental pollution concerns. In any event, it is commonly acknowledged that disposing of shrimp waste has a huge ecological impact (Kelleher 2005). The loss of valuable living resources makes shrimp waste a severe environmental issue. Environmental contamination hampers the healthy ecosystem and curses for endangered species (Morgan and Chuenpagdue 2003). The material's high susceptibility to spoilage is a significant issue with shrimp biomaterial valorization. Within an hour of processing, breakdown starts to occur in tropical temperatures, producing biogenic amines, which have a highly unpleasant odor. The biomaterial decomposes into actual waste if this decay cannot be prevented or stopped and becomes an expensive financial burden if it is not properly disposed of due to its high protein content.

5 Need for Waste Management

It is obvious that appropriate technology should be used to stop degradation and turn the biomaterial into useful goods. This is good for both environmental and economic reasons. Technology ought to offer methods for fractionation as well as techniques for delaying or stopping deterioration. Consequently, there is a lot of interest in recycling fish waste. It's an innovative idea to turn waste from the fish processing industry into marketable organic feed and fertilizer products. After the proper treatment, those biomaterials or biowaste which contain a variety of useful substances, can significantly increase overall profitability. Fishmeal made from fish waste contains crude protein (58%), which is lower than the 60–70% found in high-quality fishmeal, but it is still a wholesome product that might be used as a source of fishmeal for fish at lower trophic levels. According to a study, fish processing waste is currently used to make up to 25% of fishmeal (Chiu et al. 2013). Fish waste does include a lot of monounsaturated, palmitic, and oleic acids and is a good source of fat (19% dry matter) and nutrients (Esteban et al. 2007). When compared to other fish oils, shrimp shell waste contains n-3 fatty acids in lipid that also contains additional beneficial components, like carotenoids (Amiguet et al. 2012; Sowmya and Sachindra 2012). The bulk of the total fatty acids in shrimp oil are polyunsaturated fatty acids (PUFA), particularly eicosapentaenoic acid (EPA) and docosahexaenoic acid (DHA) (Gulzar and Benjakul 2019; Takeungwongtrakul et al. 2012). The two main n-3 fatty acids, i.e., EPA and DHA in PUFA are well known for their therapeutic and nutraceutical uses. Shrimp oil contains phospholipids, cholesterol, and carotenoids in addition to fatty acids (Raju et al. 2022). Different species of shrimp have different amounts of lipid components and carotenoids. In comparison to saturated and monounsaturated fatty acids, shrimp oil contains higher polyunsaturated fatty acids (PUFA) (Gulzar and Benjakul 2019). According to some reports, *P. monodon* meat and *L. vannamei* waste both have higher PUFA concentrations such as 44.3 and 43.57%, respectively (Gómez-Estaca et al. 2017). Astaxanthin is present in large amounts in a number of sources of shrimp oil or shrimp processing by-products (SPBP). According to Yang et al. (2022), astaxanthin monoester made up 59% of the carotenoid content in *L. vannamei*, with free astaxanthin making up 33% and astaxanthin diester (8%). Crustacea are an excellent source of the dietary fat-soluble vitamins that adults require (Stancheva and Dobreva 2013). The preservation of human health depends on these fat-soluble vitamins, including vitamin A (retinol), vitamin D, and vitamin E (gamma-tocopherol). López et al. (2006) reported that vitamin A and vitamin E concentrations in oil extracted from the *L. vannamei* cephalothorax ranged from 0.9 to 1.6 mg/100 g. According to Gómez-Estaca et al. (2017), the oil from *Litopenaeus vannamei* waste contains up to 65 mg/g of cholesterol. In this context, shrimp industry by-products must be given to aqua feed as a source of protein and a rich supply of carotenoids (particularly astaxanthin) to promote and augment overall growth, build muscle, improve skin pigmentation, and improve fish health thanks to its antioxidant properties (Haque et al. 2021, 2023). However, the remaining trout intestines from smoking fish were mentioned by Kotzamanis et al. (2001) as a potential source of fatty acids for gilthead bream. Utilizing trout offal in sea bream diets is an alternate, non-polluting method of employing fish industry by-products. The squid protein hydrolysate (SPH) contained 61–64 (%) hydrophilic amino acids, crude lipids, 84–88 (%) crude protein, 6–7 (%) ash, 3 (%) sugar, and trace levels of NaCl, according to Kotzamanis et al. (2001).

6 Bioactive Compounds from Fish and Shellfish Waste

6.1 *Chitin and Chitosan*

The shells of crustaceans like shrimp, crabs, and others, as well as fungi, insects, algae, and mushrooms, are plentiful with chitin, the second-most abundant polysaccharide in the world (Arcidiacono and Kaplan 1992). One of the most prevalent renewable biopolymers, chitin resembles cellulose and is mostly made up of unbranched chains of 1,4-N-acetyl-D-glucosamine (Ngasotter et al. 2023a). Chitin is not only an essential component of invertebrates; vertebrates also contain chitin. Contrary to cellulose, chitin has a carbon to nitrogen ratio of 8 to 1 (Struszczyk 2006). Chitin comes in three varieties: chitin A, chitin B, and chitin C. The form, which is usually derived from crab and shrimp shells, is frequently used. Chitin is commercially marketed, too. Chitin's chains are arranged anti-parallel to one another. Strong hydrogen bonds are present in α-chitin due to its anti-parallel structure, which boosts its stability (Sikorski et al. 2009).

Chitin's intermolecular hydrogen bonding prevents it from dissolving in water (Minke and Blackwell 1978). However, derivatives of chitin can be created that are soluble in water, such as chitosan or carboxymethyl chitin. Chitosan, a naturally occurring carbohydrate polymer that has been altered, is produced when chitin is deacetylated (Yeul and Rayalu 2013). Nitrogen makes about 6–7% of chitin and 7–9.5% of chitosan in its deacetylated state. Numerous extremely beneficial features of chitin and chitosan, such as immunological function, hemostasis and wound healing, antioxidant activity, antibacterial activity, and the removal of heavy metals and other impurities, are present in these two substances.

Recently, there has been a lot of interest in the separation and application of chitin in both its micro and nano forms, especially nano chitin in the form of nanocrystals or nanowhiskers (100–800 nm in length and 6–60 nm in width) and nanofibers (several m in length and 10–100 nm in width) (Ngasotter et al. 2022, 2023b; Sampath et al. 2022). There are two methods for converting native chitin into nano chitin: (I) Top-down technique, which uses physical or chemical processes such as acid hydrolysis, high-pressure homogenization, ultra-sonication, grinding, and TEMPO-mediated oxidation. (ii) Bottom-up method, which converts chitin solutions or gels into nano chitin via electrospinning, self-assembly, and dissolution-regeneration (Yang et al. 2020). For some valuable characteristics, it led to an increase in the use of nano chitin in the fields of packaging, food, biomedical, biological, and cosmetics. Chitin nanocrystals, for instance, have been utilized to successfully stabilize Pickering oil-in-water emulsions (Cheikh et al. 2021). Nano chitin, which functions as dietary fiber, can reportedly block the breakdown of fat, according to several in vitro research on digestive systems (Zhou et al. 2020, 2021). Other potential uses for nano chitin in food include enhancing saltiness and serving as a reinforcing nano filler in a variety of packaging films (Somsak et al. 2021).

6.2 *Pigment Composition*

Carotenoid is obtained after processing shrimp, crab, trout, lobster, crayfish, salmon, snapper, and tuna industry waste. The most common pigments found in both plants and animals, ranging from red to yellow, are carotenoids, which are found in the lipids of fish waste. In a racemic combination, astaxanthin contains three stereoisomers that combine to create a complex with a protein that builds up in the exoskeleton of crustaceans (Haque et al. 2021). Due to its unique binding properties, astaxanthin is primarily found in crustacean waste in combination with other substances. With proteins (carotenoproteins) or lipoproteins (carotenolipoproteins), the pigment forms a chemical compound (Higuera-Ciapara et al. 2006). Carotenoids are extracted from the head, body carapace, and leftover shrimp waste using a variety of organic solvents (Sachindra 2006). The extracted residue can be used to create chitin and/or chitosan, and the recovered carotenoids can successfully substitute synthetic carotenoids in formulations for aquaculture feed (Haque et al. 2021). The level of redness in seafood directly affects its price or quality. The antioxidant action is reportedly ten times more powerful than carotene (Naguib 2000). It is utilized in the culinary, cosmetic, and salmonid and crustacean feed industries (De Holanda and Netto 2006).

6.3 *Polyunsaturated Fatty Acids*

Shrimp waste has a substantial amount of mono- and poly-unsaturated fatty acids, which combined account up 34% of the product's total fatty acids, according to the fatty acid composition of the waste. Furthermore, it seems to contain a lot of saturated fatty acids. According to reports from India, the acetone extract made from shrimp waste contains a lot of saturated fatty acids (Sachindra et al. 2006). According to Bragagnolo and Rodriguez-Amaya, penaeid shrimp from the Brazilian region had a significant concentration of unsaturated fatty acids, demonstrating that the composition of fatty acids varies depending on the kind of shrimp (2001). According to Senphan and Benjakul (2012) and Takeungwongtrakul et al., the cephalothorax and hepatopancrease of shrimp are important sources of highly unsaturated omega-3 fatty acids, such as eicosapentaenoic acid (EPA, 20:5n3) and docosahexaenoic acid (DHA, 22:6n3) (2012). Following these acids, saturated fatty acids and monounsaturated fatty acids (which together constituted up 37.5% of the lipid extracted from the cephalothorax of the Pacific white shrimp, *L. vannamei*) are found (Gulzar and Benjakul 2019).

6.4 Essential Amino Acids

All necessary amino acids, with the exception of tryptophan, are present in both the original shrimp waste and the powder created after fermented shrimp waste was lyophilized (Bhaskar et al. 2010). Glutamic acid and aspartic acid were found to be dominated as amino acids in caratenoproteins, extracted from shrimp waste using enzymatic extraction (Simpson and Haard 1985). In aquaculture, by-product from fish waste can be utilized as an immunostimulant and growth promoter (Amar et al. 2000). Fish waste protein hydrolysates are known to be superior in terms of nutrition as feed ingredients because they contain a high concentration of important amino acids (Gildberg and Stenberg 2001). These traits obviously indicate the material's potential relevance as a dietary element that will support wellness in the diets of young fish and penaeid shrimps.

6.5 Alpha-Tocopherol

Tocopherol content in fish waste varied depending on the species, age, sex, fish waste component, and type of extraction process (Afonso et al. 2016; Gómez-Estaca et al. 2017; Gulzar and Benjakul 2018). Brown shrimp meat and fermented shrimp waste (head and cephalothoraxes) had tocopherol values of 7.73 mg per 100 g and 50.5 mg per 100 g, respectively (Merdzhanova et al. 2018). Gomez-Estaca et al. (2017) found a greater tocopherol concentration (1.26 g/100 g) in the waste extract from the cephalothorax, cuticles, tails, and pleopods of *L. vannamei*. Fish's muscular and reproductive systems need tocopherol, a fat-soluble vitamin with antioxidant properties (vitamin E) (Afonso et al. 2016). Additionally, tocopherol is necessary to prevent lipid peroxidation in the food system as well as the oxidation of low-density lipoprotein in living things (Mathur et al. 2015).

6.6 Fish Calcium

Calcium deficiency in the diet can be treated with calcium powder made from the tuna's backbone. The main calcium-containing ingredients are dolomite, bone meal, and oyster shell. In the building business, calcium is utilized to generate early strengthening agents. In the culinary and agricultural industries, calcium is used as a food antiseptic to keep fruits and vegetables from going bad and to make cheese-making simpler.

6.7 Carotenoproteins

The processing waste from shrimp and other crustaceans can be utilized to make carotenoproteins, which are high-density lipoproteins connected to stable carotenoid complexes (Dayakar et al. 2022). In their ovaries and eggs, carotenoprotein is found as carotenolipoproteins, while in the exoskeletons of crustaceans, it is found as chitinocarotenoids and crustacyanins (Pattanaik et al. 2020). According to research by Sowmya et al. (2011), carotenoproteins are bioactive natural colorants that have the ability to boost farmed species' growth, coloration, and immunity (Pattanaik et al. 2021). In order to fully use these species, carotenoprotein can be produced from shrimp shells and head debris and used as a functional addition in foods, drinks, and animal feeds to promote growth (Dayakar et al. 2023).

7 Role of Bioactive Substances from Fish Waste

7.1 Antioxidant Activity

Oxidation is the term for the typical physiological process that occurs in living organisms. Chitosan, protein hydrolysate, carotenoprotein, astaxanthin, and tocopherol are examples of bioactive fish waste products that exhibit potent antioxidant effects via a variety of methods (Ambigaipalan and Shahidi 2017; Chintong et al. 2019). Chitosan demonstrated lowering potential, DPPH radical scavenging activity, and prevention of carotene bleaching (Younes et al. 2014). By using shrimp shell hydrolysates (SSH) and shrimp shell protein hydrolysates (SPH), researchers were able to lessen the oxidative deterioration of cholesterol, the bleaching of beta-carotene caused by cupric ions, and the DNA damage brought on by peroxyl and hydroxyl radicals (Ambigaipalan and Shahidi 2017). Likewise, Pangasius viscera spray-dried protein hydrolysate showed strong antioxidant activity (Hassan et al. 2019). According to Sila et al. (2013), deep-water pink shrimp shell waste-derived astaxanthin showed superior antioxidant activity to commercial antioxidant BHA. Astaxanthin demonstrated strong anti-oxidant activity against the DPPH and ABTS radicals, as well as the ability to prevent the bleaching of β-carotene and quenched singlet oxygen. The amount of conjugated double bonds, the hydroxyl (OH) group, the keto (C=O) group, and the chemical makeup of astaxanthin all have an impact on the antioxidant activity (Chintong et al. 2019).

7.2 Activities Against Microbes

Chitosan from the shell of *L. vannamei* has demonstrated antibacterial efficacy against Gram-positive and Gram-negative bacteria, according to Vilar et al. (2016). *S. maltophilia*, *B. subtilis*, and *E. cloacae* can all be inhibited by chitosan at concentrations as low as 78, 625, and 156 g/mL, respectively. Astaxanthin showed a strong inhibitory impact on *E. coli*, *S. mutans*, *P. auriginosa*, *S. typhi*, and *S. aureus*. Astaxanthin's ability to interact and break bacterial cell membranes due to its lipophilic nature may contribute to its ability to have an antibacterial impact (Sukmawati et al. 2020).

7.3 Anti-inflammatory Activity

The body's physiologically necessary defense mechanism against pathogens, free radicals, and dead cells is inflammation. Pro-inflammatory cytokines including NF-, IL-1, and IL-6 are blocked by anti-inflammatory substances (Santos et al. 2015). The inflammatory response of astaxanthin isolated from L. vannamei waste in rat alveolar macrophages stimulated by phorbol myristate and lipopolysaccharide (LPS) was studied (Santos et al. 2015). Astaxanthin was isolated from the shell of an Asian tiger shrimp using a solvent extraction approach, and it demonstrated anti-inflammatory effects that increased the stability of the erythrocyte membrane (Sukmawati et al. 2020).

8 Green Technologies for Efficient Utilization of Fish Waste

8.1 As Feed Ingredients

According to Yang et al. (2006), lactic acid fermentation of biowaste could cause the fiber to be broken down and increase the amount of water-soluble carbohydrates in the fermented products. Biowaste frequently contains biological components such as chitin, protein, lipids, pigments, flavorings, and calcium carbonate (Bueno-Solano et al. 2009). "Fish waste to Wealth" is an approachable way to deal with fish processing waste and turn it into organic fertilizer and feed additives that are self-stabilizing. The use of fish waste in animal feed is currently a highly sought-after alternative because it not only lowers the cost of animal feed and production but also benefits the environment and ecosystem (Westendorf 2000). Fish waste can be added to poultry feed as a probiotic supplement and nitrogen source (Hammoumi et al. 1998).

8.2 Fish Protein Concentrate (FPC)

A stable protein concentration made from entire fish is called fish protein concentrate (FPC). Removing water, oil, bones, and other materials increased the protein concentration. A variety of whole fish may now be processed into protein concentrate, which has little similarity to the original raw material.

Types of FPC: There are mainly three major types of FPC, which were defined by the Food and Agriculture Organization of the United.

Type A: A powder with a maximum total fat concentration of 0–75% that is almost tasteless and odorless. **Type B**: A powder with a maximum fat level of 3% and an odor or flavor that is unrestricted but unquestionably fishy. **Type C**: Common fish meal produced in an environment that meets acceptable hygiene standards.

8.3 Bioremediation Agent

Diverse aquatic pollutants can be removed from water and wastewater using derivatives of chitin and chitosan, which has proved to have good potential. Metal cations, metal anions, radionuclides, dyes, phenol-substituted protein anions, and other contaminants are removed by using derivatives of chitin and chitosan (Bhatnagar and Silanpaa 2009).

8.4 Nutraceuticals and Flavoring Agent

Leucine, an important amino acid, has been found in shrimp head hydrolysates, which are suitable in the animal feed industry. Glutamic acid, aspartic acid, alanine, and glycine in shrimp head hydrolysates also act as flavor enhancers (Randriamahatody et al. 2011).

8.5 Removal of Metal and Dye from Wastewater

Wastewater from a range of industries, including mining, textile, leather, paper, and plastic, contains metal, acid, and dye (He et al. 2020). Through the food supply chain, these heavy metals can be consumed by living things and passed on to people, posing health risks (Nunez-Gomez et al. 2017). Since these poisons are poisonous and challenging to remove from contaminated water, they represent a serious risk to living things and the ecosystem as a whole (Druzian et al. 2019). To remove Fe, Al, Mn, Co, and Ni from mine-affected water, the idea of using shrimp shell powder as a biopolymer was taken into consideration (Nunez-Gomez et al. 2017). Due to the high chitin and calcium carbonate content of shrimp shell powder, it successfully removed heavy metals from mine-affected water (Nunez-Gomez et al. 2017). A cheap and efficient adsorbent for extracting heavy metals from wastewater may be debris from shrimp shells. Shrimp waste was used to extract chitosan, and ionic gelation techniques were used to create chitosan nanoparticles (Ali et al. 2018). The efficiency of the freshly created nano-chitosan particles in removing Fe (II) and Mn (II) ions from water was tested. The outcomes demonstrated that nano-chitosan had a 99.8 and 95.3% efficiency in removing Fe (II) and Mn (II) ions, respectively, with adsorption capacities of 116.2 and 74.1 mg/g (Ali et al. 2018). Shrimp shell waste was processed into hydro char (SHC) utilizing the deproteinization and deacetylation method, then hydrothermal carbonization (Nirmal et al. 2020). This carbon-rich hydrochar created from shrimp shell waste has the potential to be a starting point for energy- and carbon-sequestering technologies (Kannan et al. 2017). Additionally, shrimp shell waste and its active components have a variety of uses. For example, shrimp waste's chitosan can be used to remove radioactive materials, create artificial fish bites, destroy benzene, and act as an oil spill dispersion (Rostamian et al. 2019).

8.6 Plankton Production in Aquaculture Ponds

Using a natural fermentation technique, the underutilized fish processing waste was cost-effectively converted into fish hydrolysate. Utilizing them as bio-organic manure, liquid organic fertilizer, feed additive, feed supplement, and feed binder during feed technology has improved their value. According to Sahu et al. (2014), the nutrients include calcium (2.24%), magnesium (1.75%), phosphorus (1.98%), potassium (0.65%), sulfur (1.52%), boron (10.4 ppm), and nitrogen (2.95%). For growth and optimal production, plankton and fish food organisms need both macronutrients and micronutrients. A liquid organic fertilizer called fish hydrolysate (Planktofert) has all the necessary nutrients in exactly the right amounts. Micronutrients and macronutrients used in pond ecosystems at the recommended fertilization rate are economical, environmentally benign, and have no negative effects on the water's quality or fish growth (Sahu et al. 2014). Both macro and micronutrients are present in fish hydrolysate in a balanced way. It has macronutrients like N:P:K::1.5:0.5:0.4

and micronutrients like copper, magnesium, iron, and zinc (Sahu et al. 2014). At low inclusion levels, the application of fish hydrolysate generally has a positive impact on growth performance and feed consumption.

8.7 Energy Conversion Strategy

Because it offers a significant surface area for electrochemical processes, the porous carbon material is a useful electrode material for energy applications (Kannan et al. 2017). Chitin is a nitrogen-based polymer found in large quantities in shrimp shell waste that may be a rich source of porous carbon. The inclusion of heteroatoms (such as S and P) may also boost the electrochemical activity, catalytic effectiveness, and adsorption capacity of this supply of carbon with nitrogen (Zheng et al. 2021). Waste from shrimp shells was used to produce catalysts with a high specific surface area that were co-doped with phosphorus (Zheng et al. 2021). Moreover, the catalyst made from shrimp shells demonstrated superior long-term stability compared to its commercial counterpart. Therefore, N, P-doped catalysts based on shrimp shells may be effective in air–cathode microbial fuel cells for generating electricity (Zheng et al. 2021).

8.8 Biodegradable Plastic Production

Polyethylene and polypropylene, two plastic materials generated from petroleum, take a very long time to break down and are particularly bad for the environment (Elhussieny et al. 2020). It is therefore exciting to create biodegradable plastic from a natural biopolymer that may be broken down by bacteria, such as chitosan (Wang et al. 2018). In this case, glycerol was employed as a plasticizer while shrimp shell waste that had been extracted for chitosan and cassava peel starch were combined to create bioplastic (Saridewi and Malik 2019). The freshly created bioplastic was mechanically and physically robust and included 7% chitosan (Saridewi and Malik 2019). In a different study, Thammahiwes et al. (2017) used either calcified or uncalcified shrimp shell powder (2.5%) as filler for the production of bioplastics based on wheat gluten (WG). Consequently, trash from rice straws and shrimp shells could be used to produce biodegradable bioplastic from natural sources (Elhussieny et al. 2020).

8.9 Biogas Production

Biogas is made up of a variety of substances that are broken down during anaerobic digestion, namely methane CH_4, carbon dioxide CO_2, hydrogen sulfide H_2S, and hydrogen H_2. Due to its high level of organic carbon, fish waste has the potential to be an acceptable source for the creation of methane. Biogas generation is however constrained by the high ammonia nitrogen content in fishery biomass. Co-digestion can be used to handle fish waste anaerobically. The right co-substrate mixture composition is the main problem with the co-digestion process. The C:N ratio, macro- and micronutrients, pH, biodegradable organic matter, hazardous chemicals, and dry matter are a few of the critical factors that must be in balance (Tomczak-Wandzel et al. 2013).

8.10 Biofertiliser Production

Anaerobic digestion waste is also recycled as a vital supply of a nutrient-rich substance as biogas facilities which become more popular. Digestate can be used as a biofertilizer after a few unit operations. The digestate still contains elements like nitrogen, phosphorus, and potassium. Fish waste contains high-quality digestate, which can be used to make fertilizer for farms. To acquire the proper level of NPK, the final by-product of digestion is mixed with organic waste using the same minerals as the original transformation (nitrogen, phosphorous, and potassium). Biogas plant waste enables us to conserve energy, lessen our carbon impact, and use fossil fuels less frequently. As a result, co-digested waste and biofertilizer quality are connected (Koszel et al. 2015). Fish hydrolysate, also known as "Planktofert" and "Shelfifert," is a natural liquid fertilizer for fish that contains more than 40 trace minerals and elements. The underutilized fish processing waste is a great resource for making organic fertilizer with additional value and bio-additives.

8.11 Bio-oil/biodiesel Production

Fish oil is produced in significant amounts by the fish processing sector. This waste product might be converted into renewable energy. Due to the high hydrogen and low carbon content of fish oil, a lot of research has been done on its potential as a fuel. Fish bio-oil is a viable fuel for diesel engines due to its properties. It is higher quality and has a higher heating value than methyl-esterified vegetable oil waste, which is conventional diesel fuel. Diesel engines could be able to run on biodiesel made from fish waste, especially at low temperatures.

9 Recent Trends and Future Prospects in Aquaculture Research

The development of biotechnologies for the complete utilization of shrimp wastes still faces challenges and is constrained. The technical limitations of biocatalysts, biotransformation, and fermentation technologies include the restricted stability and reuse of enzymes, the difficulty of maintaining continuous reactions, and the constrained overall sample capacities. Although most shrimp waste is currently converted into animal feed, some of it is also turned into bioactive substances with additional value including chitin, chitosan, carotenoid, and protein. The fish meal, which was prepared from fish heads, skin, intestines, fins, gills, livers, kidneys, and scales, was determined to be mercury-free by Murthy et al. (2013). These compounds have also shown promise as base materials for catalysts, energy conversion, and wastewater remediation. Despite the fact that the recovered bioactive chemicals are known to provide a variety of biological benefits, including applications in food and medicine, their extraction calls for dangerous materials such as powerful acids and bases. Once more, the extraction procedure results in some dangerous wastes and effluents that endanger the environment. Therefore, the complete exploitation of waste without producing any new garbage is the direction that shrimp processing waste will go in the future.

10 Conclusion

Aquaculture waste can be treated by following a lot of techniques. Making biogas, biodiesel, and biofertilizer from non-recyclable fish waste is the best approach to handle sick or dead fish as well as mixed rubbish. The harmful biomass is also recycled and transformed into useful heat, power, or fuel. Fish viscera-based waste has the most potential for producing protein hydrolysate. Pigments, chitosan, and collagen separation for the cosmetics, culinary, biomedical, and pharmaceutical industries are some of the most well-liked contemporary uses of aquaculture waste. The various waste processing activities need additional inputs and outputs in order to recover energy and separate the necessary components from aquatic waste. A wide range of bioactive substances, such as chitin, chitosan, protein, carotenoids, polyunsaturated fatty acids, -tocopherol, and minerals, can be found in fish waste. These bioactive compounds, according to the literature, exhibit a variety of bioactivities, such as antioxidant, antimicrobial, anti-hypertensive, anti-inflammatory, and anti-proliferative ones. However, the feed industries frequently use these active substances to improve the nutritive content and practical qualities of foods. Functional foods can be made from bioactive ingredients that have healthful nutritional and nutraceutical properties, such as protein hydrolysate and astaxanthin. Fish waste has recently been transformed into hydro char, porous carbon, and nanopowder, all of which have uses in biochemical engineering fields such bioremediation, energy conversion, and the creation of bioplastics. Therefore, it is more likely that future fish waste use will concentrate on producing environmentally friendly energy and wastewater remediation.

Acknowledgements Authors are highly thankful to the Director & Vice-Chancellor of ICAR-Central Institute of Fisheries Education, Mumbai and the Director of ICAR-Central Institute of Freshwater Aquaculture, Bhubaneswar for the support and encouragement. We are thankful to the Indian Council of Agricultural Research (ICAR), New Delhi for a Doctoral Fellowship to Ramjanul Haque. Financial support to Dr. J. K. Sundaray from DBT-AEDF (BT/PR28608/AAQ/3/922/2018) and Dr. Rajesh Kumar from Centre for Agricultural Bioinformatics (CABin), IASRI, New Delhi (Grant ID: 1004469/2020) are well acknowledged.

References

Ackefors H, Enell M (1994) The release of nutrients and organic matter from aquaculture systems in Nordic countries. J Appl Ichthyol 10(4):225–241

Afonso CL, Amarasinghe GK, Bányai K, Bào Y, Basler CF, Bavari S, Bejerman N, Blasdell KR, Briand FX, Briese T, Bukreyev A (2016) Taxonomy of the order mononegavirales: update 2016. Adv Virol 161(8):2351–2360

Ali MEA, Aboelfadl MMS, Selim AM, Khalil HF, Elkady GM (2018) Chitosan nanoparticles extracted from shrimp shells, application for removal of Fe (II) and Mn (II) from aqueous phases. Sep Sci Technol 53(18):2870–2881

Amar A (2000) The Supreme Court, 1999 term–foreword: the document and the doctrine

Ambigaipalan P, Shahidi F (2017) Bioactive peptides from shrimp shell processing discards: antioxidant and biological activities. J Funct Foods 34:7–17

Ambigaipalan P, de Camargo AC, Shahidi F (2017) Identification of phenolic antioxidants and bioactives of pomegranate seeds following juice extraction using HPLC-DAD-ESI-MSn. Food Chem 221:1883–1894

Amiguet VT, Kramp KL, Mao J, McRae C, Goulah A, Kimpe LE, Blais JM, Arnason JT (2012) Supercritical carbon dioxide extraction of polyunsaturated fatty acids from Northern shrimp (Pandalus borealis Kreyer) processing by-products. Food Chem 130(4):853–858

Arcidiacono S, Kaplan DL (1992) Molecular weight distribution of chitosan isolated from Mucor rouxii under different culture and processing conditions. Biotechnol Bioeng 39(3):281–286

Arvanitoyannis IS, Kassaveti A (2008) Fish industry waste: treatments, environmental impacts, current and potential uses. Int J Food Sci Technol 43(4):726–745

Barratt A, Montano R (1986) Shrimp heads—a new source of protein. Infofish Market Dig 4(86):21–22

Bhaskar S, Tian F, Stoeger T, Kreyling W, de la Fuente JM, Grazú V, Borm P, Estrada G, Ntziachristos V, Razansky D (2010) Multifunctional Nanocarriers for diagnostics, drug delivery and targeted treatment across blood-brain barrier: perspectives on tracking and neuroimaging. Part Fibre Toxicol 7(1):1–25

Bhatnagar A, Sillanpää M (2009) Applications of chitin-and chitosan-derivatives for the detoxification of water and wastewater—a short review. Adv Coll Interface Sci 152(1–2):26–38

Bragagnolo N, Rodriguez-Amaya DB (2001) Total lipid, cholesterol, and fatty acids of farmed freshwater prawn (Macrobrachium rosenbergii) and wild marine shrimp (Penaeus brasiliensis, Penaeus schimitti, Xiphopenaeus kroyeri). J Food Compos Anal 14(4):359–369

Bueno-Solano C, López-Cervantes J, Campas-Baypoli ON, Lauterio-García R, Adan-Bante NP, Sánchez-Machado DI (2009) Chemical and biological characteristics of protein hydrolysates from fermented shrimp by-products. Food Chem 112(3):671–675

Cheikh FB, Mabrouk AB, Magnin A, Putaux JL, Boufi S (2021) Chitin nanocrystals as Pickering stabilizer for O/W emulsions: effect of the oil chemical structure on the emulsion properties. Colloids Surf, B 200:111604. https://doi.org/10.1016/j.colsurfb.2021.111604

Chintong S, Phatvej W, Rerk-Am U, Waiprib Y, Klaypradit W (2019) In vitro antioxidant, antityrosinase, and cytotoxic activities of astaxanthin from shrimp waste. Antioxidants 8(5):128

Chiu A, Li L, Guo S, Bai J, Fedor C, Naylor RL (2013) Feed and fishmeal use in the production of carp and tilapia in China. Aquaculture 414:127–134

Choe U, Mustafa AM, Lin H, Choe U, Sheng K (2020) Anaerobic co-digestion of fish processing waste with a liquid fraction of hydrothermal carbonization of bamboo residue. Biores Technol 297:122542

Daraghmeh N, Chowdhry BZ, Leharne SA, Al Omari MM, Badwan AA (2015) Co-processed chitin-mannitol as a new excipient for oro-dispersible tablets. Mar Drugs 13(4):1739–1764. https://doi.org/10.3390/md13041739

Dayakar B, Xavier KM, Das O, Layana P, Balange A, Nayak BB (2021) Application of extreme halophilic archaea as biocatalyst for chitin isolation from shrimp shell waste. Carbohydr Polym Technol Appl 100093. https://doi.org/10.1016/j.carpta.2021.100093

Dayakar B, Xavier KM, Ngasotter S, Layana P, Balange AK, Priyadarshini B, Nayak BB (2022) Characterization of spray-dried carotenoprotein powder from Pacific white shrimp (Litopenaeus vannamei) shells and head waste extracted using papain: antioxidant, spectroscopic, and microstructural properties. LWT 159:113188

Dayakar B, Xavier M, Ngasotter S, Dhanabalan V, Porayil L, Balange AK, Nayak BB (2023) Extraction optimization and functional quality evaluation of carotenoproteins from shrimp processing side streams through enzymatic process. Environ Sci Pollut Res. https://doi.org/10.1007/s11356-023-30232-1

De Holanda HD, Netto FM (2006) Recovery of components from shrimp (Xiphopenaeus kroyeri) processing waste by enzymatic hydrolysis. J Food Sci 71(5):C298–C303

Deepthi S, Venkatesan J, Kim SK, Bumgardner JD, Jayakumar R (2016) An overview of chitin or chitosan/nano ceramic composite scaffolds for bone tissue engineering. Int J Biol Macromol 93:1338–1353. https://doi.org/10.1016/j.ijbiomac.2016.03.041

Dobreva D (2013) Black sea and freshwater fish as valuable sources of vitamin D3. Scripta Scientifica Pharmaceutica 1(1):44–47

Druzian SP, Zanatta NP, Côrtes LN, Streit AFM, Dotto GL (2019) Preparation of chitin nanowhiskers and its application for crystal violet dye removal from wastewaters. Environ Sci Pollut Res 26(28):28548–28557

Elhussieny A, Faisal M, D'Angelo G, Aboulkhair NT, Everitt NM, Fahim IS (2020) Valorisation of shrimp and rice straw waste into food packaging applications. Ain Shams Eng J 11(4):1219–1226

Esteban MB, Garcia AJ, Ramos P, Márquez MC (2007) Evaluation of fruit–vegetable and fish wastes as alternative feedstuffs in pig diets. Waste Manage 27(2):193–200

FAO (2021) The state of world fisheries and aquaculture 2020. Sustainability in action. Rome

Gildberg A, Stenberg E (2001) A new process for advanced utilisation of shrimp waste. Process Biochem 36(8–9):809–812

Gildberg A, Arnesen JA, Sæther BS, Rauø J, Stenberg E (2011) Angiotensin I-converting enzyme inhibitory activity in a hydrolysate of proteins from Northern shrimp (Pandalus borealis) and identification of two novel inhibitory tri-peptides. Process Biochem 46(11):2205–2209

Gómez-Estaca J, Calvo MM, Sánchez-Faure A, Montero P, Gómez-Guillén MC (2017) Development, properties, and stability of antioxidant shrimp muscle protein films incorporating carotenoid-containing extracts from food by-products. LWT-Food Sci Technol 64(1):189–196

Gowen RJ (1991) Aquaculture and the environment, in aquaculture and the environment. In: DePauw N, Joyce J (eds) European aquaculture society special publications no 16, Ghent, Belgium, pp 30–38

Gulzar S, Benjakul S (2018) Effect of pre-treatments on yield and properties of lipid extracted from cephalothorax of Pacific white shrimp (Litopenaeus vannamei) by ultrasonic assisted process. LWT, 100, pp.106- 113.

Gulzar S, Benjakul S (2019) Effect of pre-treatments on yield and properties of lipid extracted from cephalothorax of Pacific white shrimp (Litopenaeus vannamei) by ultrasonic assisted process. LWT 100:106–113

Hammoumi A, Faid M, Amarouch H (1998) Characterization of fermented fish waste used in feeding trials with broilers. Process Biochem 33(4):423–427

Haque R, Sawant PB, Sardar P, Xavier KM, Varghese T, Chadha NK, Pattanaik SS, Jana P, Naik VA (2021) Synergistic utilization of shrimp shell waste-derived natural astaxanthin with its commercial variant boosts physio metabolic responses and enhances colouration in discus (*Symphysodon aequifasciatus*). Environ Nanotechnol Monitor Manage 15:100405

Haque, R., Sawant, P. B., Sardar, P., Varghese, T., Xavier, K. M., Chadha, N. K., ... & Pattanaik, S. S. (2023). Shrimp shell waste-derived astaxanthin in synergistic combination with its commercial variant augments gonadal maturation and upregulates vitellogenin gene expression of discus (Symphysodon aequifasciatus). Aquaculture, 562, 738828.

Hassan MA, Xavier M, Gupta S, Nayak BB, Balange AK (2019) Antioxidant properties and instrumental quality characteristics of spray dried Pangasius visceral protein hydrolysate prepared by chemical and enzymatic methods. Environ Sci Pollut Res 26(9):8875–8884

He Y, Yuan Q, Mathieu J, Stadler L, Senehi N, Sun R, Alvarez PJ (2020) Antibiotic resistance genes from livestock waste: occurrence, dissemination, and treatment. NPJ Clean Water 3(1):4

Higuera-Ciapara I, Felix-Valenzuela L, Goycoolea FM (2006) Astaxanthin: a review of its chemistry and applications. Crit Rev Food Sci Nutr 46(2):185–196

Illera-Vives M, Labandeira SS, Brito LM, López-Fabal A, López-Mosquera ME (2015) Evaluation of compost from seaweed and fish waste as a fertilizer for horticultural use. Sci Hortic 186:101–107

Kannan S, Gariepy Y, Raghavan GV (2017) Optimization and characterization of hydrochar derived from shrimp waste. Energy Fuels 31(4):4068–4077

Karim NU, Lee MFMA, Arshad AM (2015) The effectiveness of fish silage as organic fertilizer on post-harvest quality of pak choy (Brassica rapa L. subsp. chinensis). Eur Intl J Sci Technol 4(5):163–174

Kelleher K (2005) Discards in the world's marine fisheries: an update, vol 470. Food & Agriculture Organization

Koszel M, Lorencowicz E (2015) Agricultural use of biogas digestate as a replacement fertilizers. Agric Agric Sci Procedia 7:119–124

Kotzamanis YP, Alexis MN, Andriopoulou A, Castritsi-Cathariou I, Fotis G (2001) Utilization of waste material resulting from trout processing in gilthead bream (Sparus aurata L.) diets. Aquac Res 32:288–295

Kristinsson HG, Rasco BA (2000) Fish protein hydrolysates: production, biochemical, and functional properties. Crit Rev Food Sci Nutr 40(1):43–81

Lopez-Cervantes J, Sanchez-Machado DI, Rios-Vazquez NJ (2006) High-performance liquid chromatography method for the simultaneous quantification of retinol, α-tocopherol, and cholesterol in shrimp waste hydrolysate. J Chromatogr A 1105(1–2):135–139

Marsh L, Bechtel PJ (2012) Waste (by-product) utilization. Seafood Ind: Species, Prod Process Saf 128–135

Mathew P, KG RN (2006) Ensilation of shrimp waste by Lactobacillus fermentum. Fish Technol 43(1)

Mathur P, Ding Z, Saldeen T, Mehta JL (2015) Tocopherols in the prevention and treatment of atherosclerosis and related cardiovascular disease. Clin Cardiol 38(9):570–576

Merdzhanova A, Dobreva DA, Panayotova V (2018) The comparison of proximate composition, fatty acids and fat-soluble vitamins content of the Black Sea sprat (Sprattus sprattus L.) during catching seasons. Annals Food Sci Technol 19:191–198

Minke RAM, Blackwell J (1978) The structure of α-chitin. J Mol Biol 120(2):167–181

Morgan L, Chuenpagdue R (2003) Shifting gears addressing the collateral impacts of fishing methods in U.S. waters. Island Press, Washington

Murthy LN, Mohan CO, Ravishankar CN, Badonia R (2013) Biochemical quality and heavy metal content of fish meal and squid meal produced in Veraval, Gujarat. ICAR

Naguib YM (2000) Antioxidant activities of astaxanthin and related carotenoids. J Agric Food Chem 48(4):1150–1154

Ngasotter S, Sampath L, Xavier KM (2022) Nanochitin: an update review on advances in preparation methods and food applications. Carbohydr Polym 119627

Ngasotter S, Xavier KAM, Porayil L, Balange A, Nayak BB, Eapen S, Adarsh KJ, Sreekala MS, Sharma R, Ninan G (2023a) Optimized high-yield synthesis of chitin nanocrystals from shrimp shell chitin by steam explosion. Carbohydr Polym 316:121040. https://doi.org/10.1016/j.carbpol.2023.121040

Ngasotter S, Xavier KAM, Meitei MM, Waikhom D, Madhulika, Pathak J, Singh SK (2023b) Crustacean shell waste derived chitin and chitin nanomaterials for application in agriculture food and health – A review. Carbohydr Polym Technol Appl 6:100349. https://doi.org/10.1016/j.carpta.2023.100349

Nirmal NP, Santivarangkna C, Rajput MS, Benjakul S (2020) Trends in shrimp processing waste utilization: an industrial prospective. Trends Food Sci Technol 103:20–35

Núñez-Gómez D, de Almeida Alves AA, Lapolli FR, Lobo-Recio MA (2017) Aplication of the statistical experimental design to optimize mine-impacted water (MIW) remediation using shrimp-shell. Chemosphere 167:322–329

Ou X, Zheng J, Zhao X, Liu M (2018) Chemically cross-linked chitin nanocrystal scaffolds for drug delivery. ACS Appl Nano Mater 1(12):6790–6799. https://doi.org/10.1021/acsanm.8b01585

Pattanaik SS, Sawant PB, KA MX., Srivastava PP, Dube K, Sawant BT, Chadha NK (2021) Dietary carotenoprotien extracted from shrimp shell waste augments growth, feed utilization, physio-metabolic responses and colouration in Oscar, Astronotus ocellatus (Agassiz, 1831). Aquaculture 534:736303

Pattanaik SS, Sawant PB, Xavier KM, Dube K, Srivastava PP, Dhanabalan V, Chadha NK (2020) Characterization of carotenoprotein from different shrimp shell waste for possible use as supplementary nutritive feed ingredient in animal diets. Aquaculture 515:734594

Pillay TVR (1991) Aquaculture and the environment. Blackwell Scientific, London, pp 26–45

Raju N, Sae-Leaw T, Osako K, Benjakul S (2022) Improved cholesterol depletion with enhanced astaxanthin and polyunsaturated fatty acids of lipid from Pacific white shrimp cephalothorax using prior ethanolic separation of polar lipid and β-Cyclodextrin. J Food Sci Technol 59(6):2255–2262

Randriamahatody Z, Sylla KSB, Nguyen HTM, Donnay-Moreno C, Razanamparany L, Bourgougnon N, Bergé JP (2011) Proteolysis of shrimp by-products (Peaneus monodon) from Madagascar Proteólisis de derivados de langostino (Peaneus monodon) de Madagascar. CyTA-J Food 9(3):220–228

Rostamian R, Firouzzare M, Irandoust M (2019) Preparation and neutralization of 1100 forcespun chitosan nanofibers from shrimp shell waste and study on its uranium 1101 adsorption in aqueous media. React Funct Polym 143

Sachindra MM, Mahendrakar MS (2015) Fish processing by-products: quality assessment and applications

Sachindra NM, Bhaskar N, Mahendrakar NS (2006) Recovery of carotenoids from shrimp waste in organic solvents. Waste Manage 26(10):1092–1098

Sahu BB, Barik NK, Mohapatra BC, Sahu BN, Sahu H, Sahoo P, Jayasankar P (2014) Valorization of fish processing waste through natural fermentation with molasses for preparation of biofertilizer and biosupplement. J Environ Sci Comput Sci Eng Technol 3:1849–1856

Sampath L, Ngasotter S, Porayil L, Balange AK, Nayak BB, Eappen S, Xavier KM (2022) Impact of extended acid hydrolysis on polymeric structural and thermal properties of microcrystalline chitin. Carbohydr Polym Technol Appl 4:100252. https://doi.org/10.1016/j.carpta.2022.100252

Santos Savio A, Machado Diaz AC, Chico Capote A, Miranda Navarro J, Rodríguez Alvarez Y, Bringas Pérez R, Estévez del Toro M, Guillen Nieto GE (2015) Differential expression of pro-inflammatory cytokines IL-15Ralpha, IL-15, IL-6 and TNFalpha in synovial fluid from rheumatoid arthritis patients. BMC Musculoskelet Disord 16(1):1–8

Saridewi N, Malik M (2019) Food packaging development of bioplastic from basic waste of cassava peel (manihot uttilisima) and shrimp shell. In: IOP conference series: materials science and engineering, vol 602, no 1, p 012053. IOP Publishing

Senphan T, Benjakul S (2012) Compositions and yield of lipids extracted from hepatopancreas of Pacific white shrimp (Litopenaeus vannamei) as affected by prior autolysis. Food Chem 134(2):829–835

Sikorski P, Hori R, Wada M (2009) Revisit of α-chitin crystal structure using high resolution X-ray diffraction data. Biomacromol 10(5):1100–1105

Sila A, Ayed-Ajmi Y, Sayari N, Nasri M, Martínez-Alvarez O, Bougatef A (2013) Antioxidant and anti-proliferative activities of astaxanthin extracted from the shell waste of deep-water pink shrimp (Parapenaeus longirostris). Nat Prod J 3:82–89

Simpson BK, Haard NF (1985) The use of proteolytic enzymes to extract carotenoproteins from shrimp wastes. J Appl Biochem 7(3):212–222

Somsak P, Sriwattana S, Prinyawiwatkul W (2021) Ultrasonic-assisted chitin nanoparticle and its application as saltiness enhancer. Int J Food Sci Technol 56(2):608–617. https://doi.org/10.1111/ijfs.14715

Sowmya R, Sachindra NM (2012) Evaluation of antioxidant activity of carotenoid extract from shrimp processing byproducts by in vitro assays and in membrane model system. Food Chem 134(1):308–314

Sowmya R, Rathinaraj K, Sachindra NM (2011) An autolytic process for recovery of antioxidant activity rich carotenoprotein from shrimp heads. Mar Biotechnol 13(5):918–927. https://doi.org/10.1007/s10126-010-9353-4

Stancheva M, Galunska B, Dobreva AD, Merdzhanova A (2012) Retinol, alpha-tocopherol and fatty acid content in Bulgarian Black Sea fish species. Grasas Aceites 63(2):152–157

Struszczyk MH (2006) Global requirements for medical applications of chitin and its derivatives. Polish Chitin Society, Monograph XI; Polish Chitin Society: Łódz, Poland, pp 95–102

Sukmawati D, Andrianto MH, Arman Z, Ratnaningtyas NI, Al Husna SN, El-Enshasy HA, Dailin D, Kenawy AA (2020) Antagonistic activity of phylloplane yeasts from Moringa oleifera Lam. leaves against Aspergillus flavus UNJCC F-30 from chicken feed. Ind Phytopathol 73(1):79–88

Takeungwongtrakul S, Benjakul S, Aran H (2012) Lipids from cephalothorax and hepatopancreas of Pacific white shrimp (Litopenaeus vannamei): compositions and deterioration as affected by iced storage. Food Chem 134(4):2066–2074

Thammahiwes S, Riyajan SA, Kaewtatip K (2018) Effect of shrimp shell waste on the properties of wheat gluten based-bioplastics. J Polym Environ 26(5):1775–1781

Tomczak-Wandzel R, Levlin E, Ekengren Ö (2013) Biogas production from fish wastes in co-digestion with sewage sludge. In: IWA specialist conference holistic sludge management, pp 6–8

Vilar Junior JC, Ribeaux DR, Alves da Silva CA, Campos-Takaki D, Maria G (2016) Physicochemical and antibacterial properties of chitosan extracted from waste shrimp shells. Int J Microbiol (2016)

Wang H, Qian J, Ding F (2018) Emerging chitosan-based films for food packaging applications. J Agric Food Chem 66(2):395–413

Westendorf ML (2000) Food waste as animal feed: an introduction. Food Waste Anim Feed 3–16

Xia G, Liu Y, Tian M, Gao P, Bao Z, Bai X, Yu X, Lang X, Hu S, Chen X (2017) Nanoparticles/thermosensitive hydrogel reinforced with chitin whiskers as a wound dressing for treating chronic wounds. J Mater Chem B 5(17):3172–3185

Yang H, Wang H, Huang M, Cao G, Tao F, Shen Q, Zhou G, Yang H (2022) Repurposing fish waste into gelatin as a potential alternative for mammalian sources: a review. Compr Rev Food Sci Food Saf 21(2):942–963

Yang J, Feng X, Zhou Q, Cheng W, Shang C, Han P, Lin CH, Chen HSV, Quertermous T, Chang CP (2016) Pathological Ace2-to-Ace enzyme switch in the stressed heart is transcriptionally controlled by the endothelial Brg1–FoxM1 complex. Proc Natl Acad Sci 113(38):E5628–E5635

Yang SY, Ji KS, Baik YH, Kwak WS, McCaskey TA (2006) Lactic acid fermentation of food waste for swine feed. Biores Technol 97(15):1858–1864

Yang X, Liu J, Pei Y, Zheng X, Tang K (2020) Recent progress in preparation and application of nano-chitin materials. Energy & Environ Mater 3(4):492–515

Yeul VS, Rayalu SS (2013) Unprecedented chitin and chitosan: a chemical overview. J Polym Environ 21(2):606–614

Younes I, Sellimi S, Rinaudo M, Jellouli K, Nasri M (2014) Influence of acetylation degree and molecular weight of homogeneous chitosans on antibacterial and antifungal activities. Int J Food Microbiol 185:57–63

Zhang Q, Wang L, Liu S, Li Y (2021) Establishment of successive co-fermentation by Bacillus subtilis and Acetobacter pasteurianus for extracting chitin from shrimp shells. Carbohyd Polym 258:117720

Zhou H, Dai T, Liu J, Tan Y, Bai L, Rojas OJ, McClements DJ (2021) Chitin nanocrystals reduce lipid digestion and β-carotene bioaccessibility: an in-vitro INFOGEST gastrointestinal study. Food Hydrocolloids 113:106494. https://doi.org/10.1016/j.foodhyd.2020.106494

Zhou H, Tan Y, Lv S, Liu J, Mundo JLM, Bai L, McClements DJ (2020) Nanochitin-stabilized pickering emulsions: Influence of nanochitin on lipid digestibility and vitamin bioaccessibility. Food Hydrocolloids 106:105878. https://doi.org/10.1016/j.foodhyd.2020.105878

Zhu Z, Zhang B, Cai Q, Ling J, Lee K, Chen B (2020) Fish waste based lipopeptide production and the potential application as a bio-dispersant for oil spill control. Front Bioeng Biotechnol 8:734

Thermal Behavior and Physicochemical Properties of Fish Scales for the Generation of Value-Added Products

Arthur Vinicius Sousa Silva, Gabriela Morais da Costa, and Glauber Cruz

Abstract The thermal behavior of fish scales was evaluated by thermogravimetry (TG/DTG and DTA curves), which identified the main thermal degradation steps. The fish scales are composed of moisture content (14%), volatile materials (48%), fixed carbon (12%), ash (41%), carbon (20%), hydrogen (3%), nitrogen (6%), oxygen (69%), and sulfur (1%), and the inorganic phase contains CaO, MgO, Na_2O, P_2O_5, and $Ca_{10}(PO_4)_6(OH)_2$. Calorimetry showed 9 MJ kg^{-1} HHV, 0.49 g cm^{-3} apparent density, 8 MJ kg^{-1} UHV, and 4 GJ m^{-3} energy density. The combustion of the scales in a DTF revealed a performance compatible with those of other conventional biomasses, with approximately 90% burnout and lower CO_2, SO_2, and CO emissions, when compared with coffee and rice husks. Regarding co-combustion processes with coal, the scales showed better performance and lower CO emissions. Their composition and residues generated by their burning can potentially be applied for fertilization, soil conditioning, biomedicine, adsorbents, and asphalt production.

Keywords Fish wastes · Fish scales · Characterization · Thermal analysis · Combustion · Energy · Burnout · Emissions · Residues · Collagen · Hydroxyapatite

A. V. S. Silva
Department of Thermal and Fluid Engineering, School of Mechanical Engineering, University of Campinas, Cidade Universitária Zeferino Vaz, Campinas, São Paulo 13.083-970, Brazil

A. V. S. Silva · G. Cruz (✉)
Postgraduate Program in Mechanical Engineering, Department of Mechanics and Materials, Federal Institute of Education, Science and Technology of Maranhão, Avenida Getúlio Vargas 04, São Luís, Maranhão 65.030-005, Brazil
e-mail: cruz.glauber@ufma.br

G. M. da Costa · G. Cruz
Department of Mechanical Engineering, Process and Thermal Systems Laboratory, Federal University of Maranhão, Avenida dos Portugueses 1966, São Luís, Maranhão 65080-505, Brazil

S. Maqsood et al. (eds.), *Fish Waste to Valuable Products*, Sustainable Materials and Technology, https://doi.org/10.1007/978-981-99-8593-7_18

1 Introduction

According to Lustosa-Neto et al. (2018) and Ching-Velasquez et al. (2020), the world fish production in 2016 was approximately 170 million tons, of which the majority (88%) were intended for human consumption and the remainder (12%) were used in products as oils, hides, and silage. In such a scenario, the global fishing industry generates large amounts of waste, whose disposal is performed improperly, representing serious risks to both human health and environment, since the generation rate of residues and by-products is higher than the degradation rate (Castañeda et al. 2016; Arumugam and Ponnusami 2017; Kara et al. 2018).

The national fishing industry grew 4.9% in comparison with the previous year, reaching 758,000 tons of fish produced in 2019, of which tilapia was the main species, with 432,000 tons, thus consolidating Brazil as the 4th (fourth) largest worldwide producer, behind only China, Indonesia, and Egypt (ABP 2020). Fish as a whole exported approximately US$275 million in that year and fish farming was the second most important segment, representing US$12 million (4% of the total production).

Fish production in the State of Maranhão is one of the most economically representative activities on the Brazilian coast, both in terms of species diversity and quantity (in tons) (Almeida et al. 2011). For example, the productivity of yellow hake (*Cynoscion acoupa*), the species most appreciated by the local population, was estimated at 3,600 t year^{-1} (Almeida et al. 2011); that of tilapia (*Oreochromis niloticus*) was 4,000 t year^{-1} and native species such as tambaqui (*Colossoma macropomum*) and curimatã (*Prochilodus lineatus*) corresponded to 38,500 t year^{-1}. According to ABP (2020), the productivity of species such as carp (*Cyprinus carpio*) and trout (*Salmo trutta*) was estimated at 2,500 t year^{-1} (the 6th (sixth) highest production in the country with 45,000 tons). Fish farming, especially regarding native fish, in Maranhão State grew 15.2% in comparison to 2018 and the state is the 3rd (third) largest producer in Brazil.

In 2016, approximately 76 million tons of fish waste were generated (FAO 2018), and according to the literature (Sockalingam and Abdullah 2015; Bhagwat and Dandge 2016; Leite et al. 2016; Bermúdez-Penabad et al. 2017; Pickler and Filho 2017), depending on the type of fish and the level of treatment, the amount of organic matter discarded during fish processing (viscera, tail, skins, head, and scales) can reach 50–80% per weight of the total production. The tailings contain 40–65% oil and can be converted into biodiesel by transesterification, for example, Yahyaee et al. (2013) and Martins et al. (2015).

Approximately 1.5 million tons of fish scales were produced worldwide in 2016 (FAO 2018; Silva et al. 2022), which represents 4% of the dry residue (Ghaly et al. 2013). According to Silva et al. (2019) and Ribeiro et al. (2019), in general, such waste materials are considered worthless and impractical and are discarded in an unplanned manner in inappropriate places or even thrown overboard, thus consisting of an abundant and low-cost material. According to Wu et al. (2021), fish scales are also destined for landfills or incinerators. In the current context and with the transition from linear to circular economy, sustainable practices such as the use of alternative resources and the development of renewable processes must be promoted for increasing efforts toward waste reuse and recycling (Coppola et al. 2021).

The great importance of fish scales for several industrial sectors, including textile markets, production of detergents and cosmetics, and scientific and analytical research is well known due to their numerous valuable components, according to Fig. 1 (Pati et al. 2010; Lin et al. 2010; Holá et al. 2011; Daboor et al. 2012). Such residues are considered biocomposites, which can be comprised mainly of an organic and an inorganic phase. The former (from 40 to 55%) contains a large quantity of proteins such as collagen, fat, lecithin, and scleroprotein, several vitamins (A, E, and D), mucin, and keratin, and their contents are different for each individual fish species (Sockalingam and Abdullah 2015; Huang et al. 2016; Lustosa-Neto et al. 2018). The second phase, inorganic in nature (from 7 to 25%), is composed of calcium carbonate contents and hydroxyapatite (Sockalingam and Abdullah 2015; Martins et al. 2015; Chinh et al. 2019).

Most studies on fish scales have focused on the characterization of collagen, extensively used in the cosmetics industry (Chen et al. 2010; Chai et al. 2010; Pati et al. 2010; Chinh et al. 2019), and hydroxyapatite, due to its use in bone grafts and important osteoconductive and bioactive properties (Kongsri et al. 2013; Paul

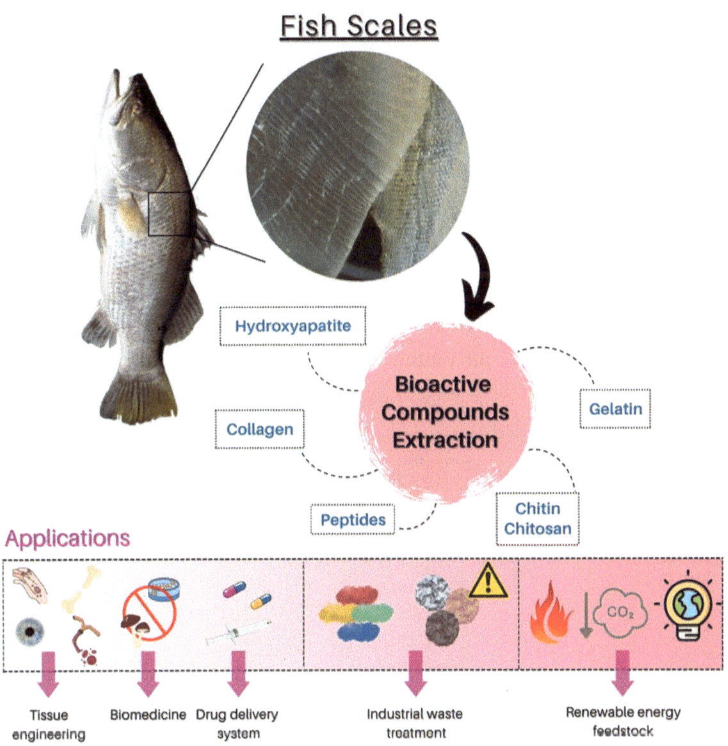

Fig. 1 Composition and applications of fish scales in several industrial sectors

et al. 2017). Furthermore, according to Khrunyk et al. (2020) and Qin et al. (2022), the transformation of bioactive compounds (e.g., proteins, peptides, and calcium phosphates) extracted from fish scales into products of high economic value by biotechnological routes has shown high potential for the production of high-quality biomaterials and bioadsorbents of heavy metals in waste industrial effluents and sediments, toxic dyes, and fertilizers.

This chapter addresses the use of fish scales (residues) in a sustainable way through their transformation into value-added products toward reducing socioenvironmental vulnerabilities and maximizing the economic activities of such a productive sector.

2 Thermal Behavior of Fish Scales

Silva et al. (2019) evaluated the steps of thermal degradation of fish scales in oxidizing (Fig. 2a) and inert (Fig. 2b) atmospheres using TG/DTG curves (Thermogravimetry and Derived Thermogravimetry) and observed an initial step of mass loss from 20 to 130 °C related to the removal of water content from the sample in an oxidizing atmosphere (80% N_2/20% O_2). The endothermic peak occurred due to the absorption of heat necessary for the evaporation of the moisture from the material at approximately 92 °C.

An accentuated mass loss (around 48%) was observed between 190 and 476 °C, which is associated with the loss of organic matter (mainly proteins). At this decomposition stage, the maximum temperature peak also occurred, thus representing the maximum combustion rate at approximately 350 °C, i.e., it was responsible for the greatest event of mass loss of the fish scale samples (Silva et al. 2019). Such an exothermic process results from the combustion of volatile material added to the solid carbonaceous content.

According to Silva et al. (2019), the last temperature range under oxidizing atmosphere for the scales (from 750 °C) showed a high percentage of remaining mass (≈42%), which was entirely attributed to the non-degraded inorganic matter in that

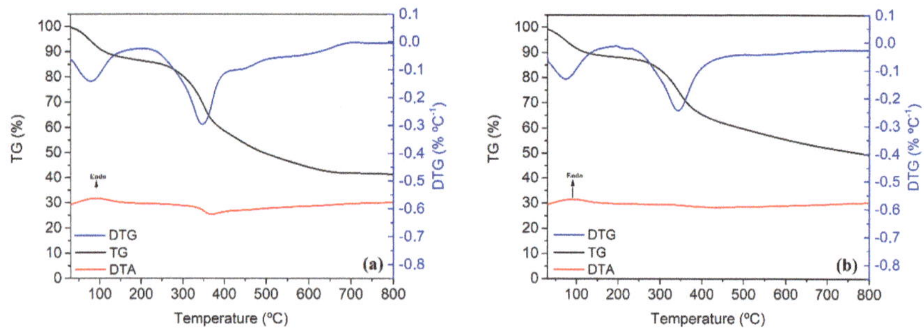

Fig. 2 TG/DTG and DTA curves for the fish scales: **a** air synthetic atmosphere (N_2/O_2: 80/20%) and **b** pyrolysis atmosphere (100% N_2)

range. When compared to the immediate analysis, this value was close to 45.91%, denoting the ash present in the material.

Furthermore, Silva et al. (2019) observed the water loss (12.1% per mass) of the scales occurred in the 35–220 °C temperature range under an inert atmosphere (100% N_2), and a 29.47% mass loss between 220 and 540 °C was attributed to the thermal degradation of organic matter. The remaining mass percentage (approximately 58.43%) was ascribed to inorganic and the remaining carbon residue (Santos et al. 2009).

Torres et al. (2012) evaluated *Araipaima gigas* scales under an inert atmosphere and identified three endothermic peaks, namely: the first (70 °C) is related to the collagen denaturation process, the second (225 °C) refers to the melting and decomposition of collagen, and the last (320 °C) is attributed to the thermal degradation of the organic component. The thermal degradation was confirmed through a TG curve, where a high degradation rate was observed at around 300 °C. At 900 °C, the remaining weight was approximately 58% of the initial value, at which the degradation thermal treatment was initiated, thus corroborating previous studies that used pure collagen and collagen–hydroxyapatite composites (Uskokovic et al. 2003).

Pati et al. (2010) evaluated the thermal behavior of *Labeorohite* and *Catla* scales through TG–DTA curves (Thermogravimetry and Differential Thermal Analysis simultaneous) and reported no significant weight loss before 100 °C and a large mass loss with peaks at 174 °C (where a loss of water molecules was observed), 411 °C, and 600 °C. The DTA curve revealed degradations at 142 and 286 °C (endothermic nature), and 535 °C (exothermic). At 286 °C, the macromolecule fragmented and a greater decomposition occurred at 535 °C due to the formation of gaseous elements. The ash content in the material was 11.6% per weight and no further degradation of the material occurred after 600 °C—only a large amount of waste was left.

Côrtes et al. (2019) analyzed freshwater fish scales by TG/DTG and observed an initial mass loss stage at approximately 100 °C related to the removal of moisture content. The greatest mass loss was observed from 250 to 500 °C, with a peak at 320 °C, and attributed to the removal of volatile compounds, mainly proteins. Around 60–65% of the initial mass remained in the samples and was related to the content of ash and fixed carbon. The authors also evaluated fish scale samples by immediate analysis and obtained the following compositions: 14% moisture, 48% volatile material, related to collagen proteins, 12% fixed carbon, and 41% ash content, related to the inorganic phase (hydroxyapatite).

According to Phillips et al. (1990) and Ferreira et al. (2017), thermochemical processes convert biomasses into biofuels with HHV (high heat value) ranging from 5 to 24 MJ kg^{-1}. An analysis in a bomb calorimeter conducted by Silva et al. (2019) revealed an experimental HHV of 9 MJ kg^{-1} and theoretical LHV (lower heat value) of 8.17 MJ kg^{-1} for the fish scales, useful heating value (UHV) of 8 MJ kg^{-1}, and 4 GJ m^{-3} of energy density (Fig. 3). Such values can be attributed to the contents of compositional carbon (20.3%) and hydrogen (3.5%), and the content of volatile materials (39.7%), as observed in ultimate and proximate analyzes. Another important parameter to be considered is the elemental oxygen content (69.1%), because in large quantities, it can reduce the heat value of fuel due to the high enthalpy values

Fig. 3 Proximate and ultimate analyzes, higher, lower, and useful heating values and energy density of fish scales (dry base and ash free). Adapted from Silva et al. (2019)

of the bonds, requiring a higher activation energy so that the bonds formed by the oxygen atoms can be broken, according to Atkins and Jones (2012) and Braz and Crnkovic (2014).

3 Physicochemical Properties of Fish Scales

Figure 3 shows the physical–chemical properties of fish scales obtained by ultimate and proximate analyzes. The low moisture value ($\approx$11%) found was considered adequate for application in combustion processes, since biofuels with high values of this component achieve low energy performance (García et al. 2012). Other authors have claimed the ash composition ($\approx$46%) of fish scales is composed mainly of calcium carbonate ($CaCO_3$), hydroxyapatite ($Ca_{10}(PO_4)_6(OH)_2$), and other metals (Sockalingam and Abdullah 2015; Martins et al. 2015; Chinh et al. 2019). The samples showed 0.49 g cm^{-3} apparent density and $C_{1.69}H_{3.52}O_{4.32}N_{0.45}S_{0.02}$ molecular formula (Silva et al. 2019) and their composition differs from those of other biomass and solid waste available in the literature and especially from coal (Zhou et al 2015; Mortari et al. 2018; Cruz et al. 2018; Sarkar and Wang 2020; Senneca et al. 2020). Solid biofuels display no uniform composition; therefore, the structural, proximate, and ultimate analyzes of the different residues and biomasses are distinct from each other (Coelho and Costa 2007; Wielgosinski et al. 2017), hampering the establishment of reference values for the samples.

The literature showed the compositions of carbon (20%), nitrogen (6%), and hydrogen (3%) provided by the ultimate analysis of fish scales are predominant

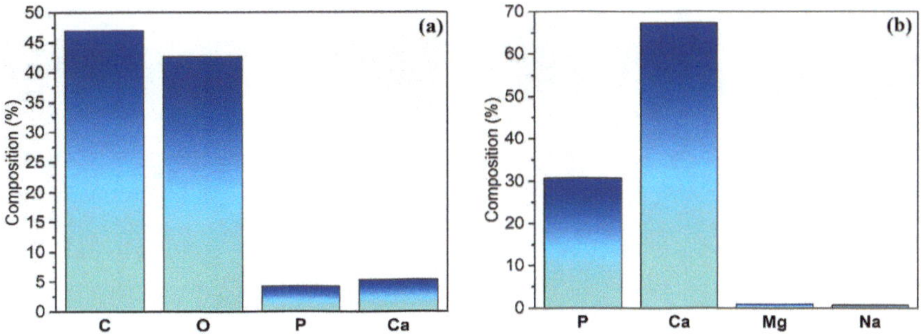

Fig. 4 Percentage of the main inorganic and metalic compounds of the fish scales obtained by **a** EDS and **b** ICP-OES. Adapted from Silva et al. (2019)

elements in the collagen structure (organic phase) (Santos et al. 2009; Sockalingam and Abdullah 2015; Huang et al. 2016; Lustosa-Neto et al. 2018). A lower than 1.0% sulfur content suggests low emissions of sulfur oxides (SO_x), since such pollutants are directly related to the amount of sulfur present in the sample (Silva et al. 2019). Reductions in its concentration lead to reductions in the emission of gases, especially greenhouse ones (Cruz and Crnkovic 2015).

Using Energy Dispersive Spectroscopy (EDS), Silva et al. (2019) observed the major components of fish scales are carbon (47.0%), oxygen (42.0%), calcium (5.0%), and phosphorus (4.0%), as showed in Fig. 4a. They also identified some alkali and alkaline earth metals (Na and Mg), as well as other inorganic elements (Al and S); however, their compositions were below 1.0%. The main inorganic components and metals identified and quantified by ICP-OES analysis (Fig. 4b) were Ca (68.0%), P (31.0%), Mg (1.0%), and Na (0.7%). Thus, the oxides present in the samples were CaO (40.6%), P_2O_5 (52.6%), MgO (1.8%), and Na_2O (1.7%), similar to those obtained by Ikoma et al. (2003), who qualitatively characterized fish scales of *Pagrus major*.

Santos et al. (2009) qualitatively evaluated the fibrous and mineral structures of *Piau* fish (*Leporinus elongatus*) scales by Energy Dispersive Spectroscopy (EDS) separately. The fibrous structure indicated the presence of elements such as carbon, nitrogen, oxygen, phosphorus, calcium, and magnesium, in which atoms of C, N, and O were related to the organic phase and the other components and, C and O were associated with the inorganic phase. The presence of this last step indicates small mineral crystals are distributed among the collagen fibers. Regarding the mineral structure, nitrogen was not identified and the inorganic composition was attributed to the minerals present in the material (Ikoma et al. 2003). Finally, Santos et al. (2009) concluded the mineral step is predominantly composed of apatite, specifically in the hydroxyapatite phase, since the calcium-phosphorus (Ca/P) ratio obtained by EDS analysis was approximately 1.67, which is in line with the values provided by the literature (Ikoma et al. 2003).

The FTIR spectra (Fourier transform infrared) and SEM images (scanning electron microscopy) confirmed the organic and inorganic phases through the presence

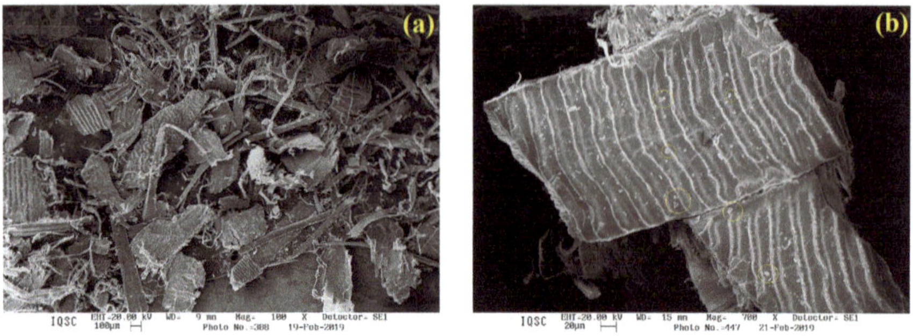

Fig. 5 SEM images for the fish scales with magnitudes **a** 100 and **b** 700 times

of structures belonging to type I collagen (type I, II, and III amides), carbonates, phosphates, and hydroxyapatite crystals, as showed in Fig. 5a, b and Table 1 (Silva et al 2019). The authors also observed an amorphous behavior, confirmed by X-Ray diffractograms (XRD), whose calculated crystallinity index value was approximately 11%. According to Xu et al. (2013) and Martinez et al. (2019), the amorphous product has a higher reactivity in thermal processes than the crystalline product, as the absence of crystallinity makes the material more susceptible to thermal degradation during a combustion process.

Santos et al. (2009) analyzed the aspects of *Piau* fish scales by SEM images before and after crushing and/or grinding. The unprocessed samples showed growth rings with a uniform appearance, characterizing a structure formed by lamellae and collagen fibers that grow in a co-aligned and compacted way, and a distribution of holes over the entire surface characteristic of the species under study. After milling, a set of disordered fibers resulting from the crushing process and a distribution of minerals among them were observed. The author also detected crystals with a hexagonal structure, which characterizes the presence of hydroxyapatite (Deer et al. 2000), and analyzed FTIR and XRD spectra. The large diffraction peaks obtained in the diffractograms indicated the crystals were of the order of micro or nanometers, with low crystallinity, or structurally disordered, or even showing the three characteristics (Ikoma et al. 2003).

Table 1 Main functional groups obtained by FTIR spectra of fish scales

Functional groups	Phosphate (PO_4^{3-})	Carbonate (CO_3^{2-})	OH group	Amide A	Amide I	Amide II	Amide III
Wave numbers (cm^1)	1060, 476	1439, 864	3597	3411	1688	1550	1245

4 Combustion Process of Fish Scales in DTF (Drop Tube Furnace)

Silva et al. (2022) evaluated the combustion profile (burning efficiency and gaseous emissions) of fish scales under the influence of three variables (residence time, temperature, and particle size) in a laboratory-scale DTF, showed in Fig. 6a, b, at six different temperatures (ranging 600–1100 °C) and six residence times (ranging 250–500 ms). The burning efficiency was measured through burnout (Cloke et al. 2002; Osório et al. 2006; Biswas et al. 2006) and the ash composition was defined by EDS and ICP-OES analysis. Pearson's correlation test (Manwatkar et al. 2021) determined the pairwise correlations of the parameters (e.g., residence time, temperature, granulometry, burnout, gaseous emissions, and ash composition).

According to Table 2, at 1100 °C and 250–500 ms residence times, an increasing trend in the firing efficiency due to an increase in the residence time of the samples in the reactor was observed. The lowest efficiency was reported for 250 ms (58.8%) and the highest one was achieved for the times of 450 (95.5%) and 500 ms (96.2%). The other measured values were 86.9% for 300 ms, 92.1% for 350 ms, and 93.3% for 400 ms. The increase in residence time contributes to both the efficiency of the combustion process and reductions in the concentration of emissions of gases such as carbon monoxide (Roy et al. 2014).

As showed in Table 3, the highest efficiency for 500 ms residence time and 600–1100 °C was detected at 1000 and 1100 °C (91.1% and 96.2%, respectively), whereas

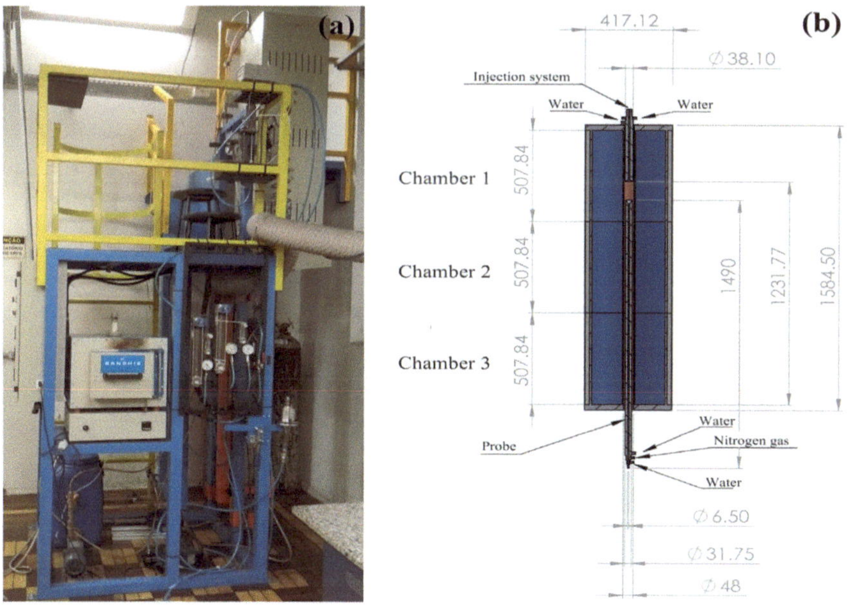

Fig. 6 **a** DTF system and **b** reactor diagram. Adapted from Silva et al. (2022)

Table 2 Burnout and gas species concentrations for different residence times

Residence times (ms)	250	300	350	400	450	500
Burnout (%)	58.8 ± 1.7	86.9 ± 1.5	92.1 ± 1.1	93.3 ± 2.9	95.5 ± 0.9	96.2 ± 1.0
CO_2 (%)	2.0 ± 0.2	2.6 ± 0.2	3.9 ± 0.1	3.7 ± 0.4	4.9 ± 0.4	4.8 ± 0.5
SO_2 (ppm)	47.3 ± 2.4	64.4 ± 3.6	60.0 ± 0.2	73.2 ± 3.6	107.4 ± 4.0	97.6 ± 1.8
CO (ppm)	3684.3 ± 96.9	2231.9 ± 60.6	1077.8 ± 23.5	701.5 ± 19.5	454.8 ± 15.9	356.5 ± 13.9
NO (ppm)	832.9 ± 54.9	623.6 ± 27.1	527.8 ± 21.6	553.9 ± 25.7	347.6 ± 7.1	330.8 ± 4.7

600 °C showed the worst performance (31.2%). The results for the other temperatures were 51.9% at 700 °C, 61.2% at 800 °C, and 63.1% at 900 °C. Therefore, adjustments in temperature contribute to avoiding incomplete combustion (Coelho and Costa 2007).

The fish scales burned in DTF with particle sizes ≈200 and <100 μm show similar burnout values for all adopted parameters, according to Table 4. The average burnout for both sizes was 85.8% for 300 ms, 91.2% for 350 ms, and 93.1% for 400 ms, respectively; while for the >300 μm samples the results were 24.4, 85.4 and 87.4% for 300, 350, and 400 ms, respectively.

According to Spliethoff and Hein (1998), a balance must be established between the residence time in the reactor and the granulometry available for combustion for the obtaining of better oxidation, since inadequate granulometries can delay the burning process due to the need for more heating and devolatilization of the particles and shorter residence times close to the burner (Ballester et al. 2015).

The results presented showed that the burnout of fish scale is comparable to conventional biomass under similar operating conditions of temperature, residence time, and airflow, among others. For example, Branco and Costa (2017) and Mortari et al. (2021) reported burnout in the range of 90% during the combustion of grape pomace and wheat straw at 1100 °C and 300 ms residence time. Branco and Costa (2017) also evaluated the combustion of wheat straw and rice husks with different residence times and sizes and reported burnout ranging from 76 to 98%, while fish scales showed burnout in the range of 60–96% under similar conditions. The fish scales also showed higher burnout than mineral coal observed by Biswas et al. (2006) and Zheng et al. (2020), whose values were 85% and 67%, respectively, when burned in DTF.

Regarding gaseous emissions, the concentrations of carbon dioxide, sulfur dioxide, carbon monoxide, and nitrogen monoxide (both corrected to 10% oxygen—O_2) were evaluated. Initially, an increase in the concentration of carbon dioxide and sulfur dioxide emissions was observed; it was proportional to the increase in the residence time of the samples in the DTF, i.e., from 2.0% to 4.9% for CO_2 and 47.3 ppm to 107.4 ppm for SO_2, with 250 ms and 500 ms, showing the lowest and highest values, respectively, as showed in Table 2. The behavior of the concentrations of carbon monoxide and sulfur monoxide was opposite to that of CO_2 and SO_2, decreasing as the residence time increased. The values ranged from 356.5 ppm to 3684.3 ppm for CO, and 330.8 ppm to 832.9 ppm for NO, with 500 ms and 250 ms, having the smallest and largest measurements, respectively.

No stable range was obtained in the gauging of gases produced at 600 and 700 °C due to the formation of an oily layer on the filter that caused clogging and compromised the reading by the gas analyzer. This occurred because incomplete combustion can result in volatile organic compounds, tar, and soot (Zellagui et al. 2017). The concentrations of gases generated at 800–1100 °C were 2.3–4.8% for carbon dioxide, 51.3–97.6 ppm for sulfur dioxide, 356.5–3875.4 ppm for carbon monoxide, and 330.8–800.4 ppm for nitrogen monoxide (Table 3).

According to Table 4, the CO_2 concentration levels for particle sizes <100 μm, 200 μm, and >300 μm were, respectively, 2.2%, 2.6%, and 1.7%, for 300 ms, 2.7%,

Table 3 Burnout and gas species concentrations for different combustion temperatures

Temperatures (°C)	600	700	800	900	1000	1100
Burnout (%)	31.2 ± 15.7	51.9 ± 8.9	61.2 ± 27.2	63.1 ± 4.8	91.1 ± 2.6	96.2 ± 1.0
CO_2 (%)	n.d.	n.d.	2.3 ± 0.2	2.3 ± 0.3	3.8 ± 0.3	4.8 ± 0.5
SO_2 (ppm)	n.d.	n.d.	51.3 ± 5.0	54.0 ± 2.5	60.9 ± 9.5	97.6 ± 1.8
CO (ppm)	n.d.	n.d.	3875.4 ± 66.9	3230.3 ± 65.9	1052.0 ± 60.8	356.5 ± 13.9
NO (ppm)	n.d.	n.d.	800.4 ± 26.0	667.0 ± 17.2	480.4 ± 9.1	330.8 ± 4.7

Table 4 Burnout and gas species concentrations for different particle sizes and residence times

Residence times (ms)	300			350			400		
Particle sizes	<100	≈200	>300	<100	≈200	>300	<100	≈200	>300
Burnout (%)	84.6 ± 2.1	86.9 ± 1.5	24.4 ± 12.9	90.2 ± 0.5	92.1 ± 1.1	85.4 ± 7.0	92.8 ± 1.1	93.3 ± 2.9	87.4 ± 3.4
CO_2 (%)	2.2 ± 0.1	2.6 ± 0.2	1.7 ± 0.1	2.7 ± 0.3	3.9 ± 0.1	2.9 ± 0.1	3.8 ± 0.5	3.7 ± 0.4	3.3 ± 0.1
SO_2 (ppm)	56.0 ± 0.2	64.4 ± 3.6	39.5 ± 1.2	69.4 ± 2.6	60.7 ± 0.2	56.9 ± 2.1	67.5 ± 3.9	73.2 ± 3.6	57.6 ± 0.3
CO (ppm)	2118.8 ± 56.1	2231.9 ± 60.6	4008.3 ± 124.9	1392.9 ± 38.4	1077.8 ± 23.5	2654.6 ± 67.6	1227.8 ± 11.5	701.5 ± 19.5	1907.4 ± 9.1
NO (ppm)	566.9 ± 73.3	623.6 ± 27.1	1067.0 ± 19.9	482.6 ± 11.1	527.8 ± 21.6	691.2 ± 14.8	454.4 ± 10.9	553.9 ± 25.7	692.1 ± 12.5

3.9%, and 2.9% for 350 ms, and 3.8%, 3.7%, and 3.3% for 400 ms. The values measured for the same conditions for SO_2, were 56.0, 64.4, and 39.5 ppm for 300 ms, 69.4, 60.7, and 56.9 ppm for 350 ms, and 67.5, 73.2, and 57.6 ppm for 400 ms. An alternation in the maximum values was observed between granulometries <100 μm and 200 μm for the different parameters, reflecting a similar burnout and confirming the quality of their combustion in relation to the burning of particles >300 ms.

For CO emissions, the concentration levels measured for the three granulometry were, respectively, 2118.8, 2231.9, and 4008.3 ppm for 300 ms, 1392.9, 1077.8, and 2654.6 ppm for 350 ms, and 1227.8, 701.5, and 1907.4 ppm for 400 ms. Regarding NO, the values for the parameters studied were 566.9, 623.6, and 1067.0 ppm for 300 ms, 482.6, 527.8, and 691.2 ppm for 350 ms, and 454.4, 553.9, and 692.1 ppm for 400 ms. Differently from CO_2 and SO_2, in all conditions, the maximum values for carbon monoxide and nitrogen monoxide were obtained for granulometry > 300 μm. Therefore, samples of fish scales of larger particle sizes (<300 μm) showed a worse performance in the reactor, hence higher levels of pollutant emissions in comparison with samples with smaller granulometry (<100 and 200 μm).

The Pearson's correlation between gaseous emissions and burnout for the experiments (Tables 5, 6 and 7) indicated a positive correlation among burnout, CO_2, and SO_2, and a negative one for NO and CO; the values of carbon dioxide and sulfur dioxide are inversely proportional to those of carbon monoxide and nitrogen monoxide, except for sulfur dioxide at 200 μm, which showed a weak correlation (around 0.4) due to oscillations in the residence time range (300–400 ms).

Table 5 Pearson's correlation coefficient among gaseous emissions and burnout for different residence times

Variables	Burnout	CO_2	SO_2	CO	NO
Burnout	1.0				
CO_2	0.8	1.0			
SO_2	0.7	0.9	1.0		
CO	−0.9	−0.9	−0.8	1.0	
NO	−0.9	−0.9	−0.9	0.9	1.0

Table 6 Pearson's correlation coefficient among gaseous emissions and burnout for different combustion temperatures

Variables	Burnout	CO_2	SO_2	CO	NO
Burnout	1.0				
CO_2	0.9	1.0			
SO_2	0.8	0.9	1.00		
CO	−0.9	−0.9	−0.8	1.0	
NO	−0.9	−0.9	−0.9	0.9	1.0

Table 7 Pearson's correlation coefficient among gaseous emissions and burnout for different particle sizes and residence times (300, 350, and 400 ms)

Variables	Burnout	CO_2	SO_2	CO	NO
Particle size < 100 μm					
Burnout	1.0				
CO_2	0.9	1.0			
SO_2	0.9	0.6	1.0		
CO	−0.9	−0.8	−0.9	1.0	
NO	−1.00	−0.9	−0.9	1.0	1.0
Particle size ≈ 200 μm					
Variables	Burnout	CO_2	SO_2	CO	NO
Burnout	1.0				
CO_2	0.9	1.0			
SO_2	0.4	0.1	1.0		
CO	−1.0	−0.9	−0.4	1.0	
NO	−0.9	−1.0	0.1	0.9	1.0
Particle size > 300 μm					
Variables	Burnout	CO_2	SO_2	CO	NO
Burnout	1.0				
CO_2	1.0	1.0			
SO_2	1.0	1.0	1.0		
CO	−0.9	−0.9	−0.9	1.0	
NO	−1.0	−0.9	−1.0	0.9	1.00

Adapted from Silva et al. (2022)

The formation of carbon dioxide as well as sulfur dioxide in thermal systems are indications of complete or partially complete combustion, being released entirely as volatile gases and/or in the combustion of biochar (Williams et al. 2012; Sartor et al. 2014; Cruz and Crnkovic 2015; Krumal et al. 2019). It is important to note that higher sulfur dioxide emissions depend on the sample's sulfur composition and oxygen concentration (Kazanc et al. 2011; Zhao et al. 2016; Yang et al. 2018).

Carbon monoxide gas results from poor or incomplete combustion; it effectively contributes to atmospheric pollution and represents energy loss. More CO is formed under poor fuel conditions, due to an improper air–fuel mixture and lack of oxygen to oxidize carbon monoxide to carbon dioxide (Carvalho and Lacava 2003; Coelho and Costa 2007). As an example, the high concentration of carbon monoxide for 250 ms confirms the low efficiency of the burning of fish scales in shorter residence times.

According to Zhao et al. (2016) and Yang et al. (2018), emissions of nitric oxides (NO_x) are mainly due to the nitrogen content of the sample, i.e., a higher content of elemental nitrogen results in high amounts of NO. However, there are some cases in which NOx emissions do not directly reflect the nitrogen consumed in the burning process, depending on parameters such as fixed carbon content, residence time, and reactor temperature (Martins and Ferreira 2010; Kazanc et al. 2011; Duan et al. 2015; Zhao et al. 2016; Yang et al. 2018).

Comparing the burning of fish scales and coal assessed by Mortari et al. (2020) under similar conditions (combustion at 1000 °C and synthetic air atmosphere), fish scales showed lower concentrations of emissions (15 and 4% CO_2; 1600 and 1000 ppm CO; and 700 and 60 ppm SO_2 for coal and fish scales, respectively). However, nitrogen monoxide emissions were higher for fish scales due to their higher nitrogen content (1.3% for coal and 6.3% for fish scales) (Mortari et al. 2020; Silva et al. 2022). The same occurred for biomass such as sugarcane bagasse (Mortari et al. 2020), coffee husks, and rice husks (Cruz and Crnkovic 2019), where fish scales also had lower CO_2, SO_2 and CO emissions, and higher NO concentrations (50–200 ppm for woody biomass; and concentrations above 300 ppm for fish scales).

According to Coelho and Costa (2007) and Tan (2014), the formation of residues in burning processes depends on the composition of the fuel, the operating conditions of the equipment, and the nature of the flame. Table 8 shows the main constituent groups for burned fish scales at different temperatures (600, 700, 800, 900, 1000, and 1110 °C) obtained by FTIR (Silva et al. 2022). Initially, the increase in temperature caused the elimination of some wavelengths related to the organic phase of the material *in natura*, such as amide I and amide A (degraded at temperatures above 700 °C); in addition to an increase in the intensity of the peaks of phosphate and carbonate ions at temperatures above 600 °C.

Three main constituent groups were observed in the residues, namely: –OH, CO_3^{2-}, and PO_4^{3-}. According to Ikoma et al. (2003), their presence corroborated the hydroxyapatite formation at high temperatures (>600 °C). The wavelengths in Table 8 show absorption ranges similar to those obtained by Paul et al. (2017), Deb et al. (2019), Santana et al. (2019), and Buraiki et al. (2020), who evaluated the fish scales residues after heat treatment in different operating conditions and reactors.

Table 8 FTIR spectra for the residues of fish scales generated at different combustion temperatures

Functional groups	Wave numbers (cm^{-1})					
	600 °C	700 °C	800 °C	900 °C	1000 °C	1100 °C
Phosphate (PO_4^{3-})	472, 554, 600, 958, 1023	476, 560, 611, 958, 1036	476, 565, 605, 959, 1029	477, 565, 605, 958, 1022	472, 565, 594, 958, 1029	472, 548, 600, 957, 1022
Carbonate (CO_3^{2-})	874	873	873	872	874, 1496	874
OH group	667, 3483	3478	668	667	668	668
Amide I	1615	1631	n.d.	n.d.	n.d.	n.d.
Amide A	3397	3410	n.d.	n.d.	n.d.	n.d.

Adapted from Silva et al. (2022)

According to the ICP-OES analysis and Energy Dispersion Spectroscopy (EDS) (Fig. 7a, b), the residues showed an increase in the concentration of compounds such as Ca, P, Na, and Mg with the increase in the combustion temperature of the reactor. The highest amounts of calcium (37.4%), phosphorus (17.1%), magnesium (0.7%), and sodium (0.6%) were observed for ash from scales burned at 1100 °C. The concentration of the other metals (Fe, Si, Al, Zn, and B) also increased in relation to the *in natura* sample; however, potassium was not detected in the residues; according to Jenkins et al. (1998), the potassium vaporizes at temperatures around 500 °C. Metals such as nickel, molybdenum, cadmium, and lead were not quantified by ICP-OES analysis. Thus, the main oxides present in the ashes were P_2O_5 (ranging from 62.2 to 71.1%), CaO (ranging from 21.7 to 31.0%), MgO (around 4%), and Na_2O (around 3.5%). The Pearson correlations among elements obtained by EDS and ICP-OES and burnout (Table 9) showed positive correlations for Ca, P, Na, and Mg, confirming the growth trend in the concentration of compounds for higher burnouts. The other elements presented weak correlations for burnout.

According to San Miguel et al. (2012) and Cruz et al. (2018), biomass residues containing calcium, phosphorus, and magnesium can be used in agriculture as fertilizers, or to condition soils that have lost the main nutrients. Furthermore, recent studies have evaluated the use of fish scale residues in biomedical applications, and in civil industry for asphalt production (Lv et al. 2021; Qin et al. 2022).

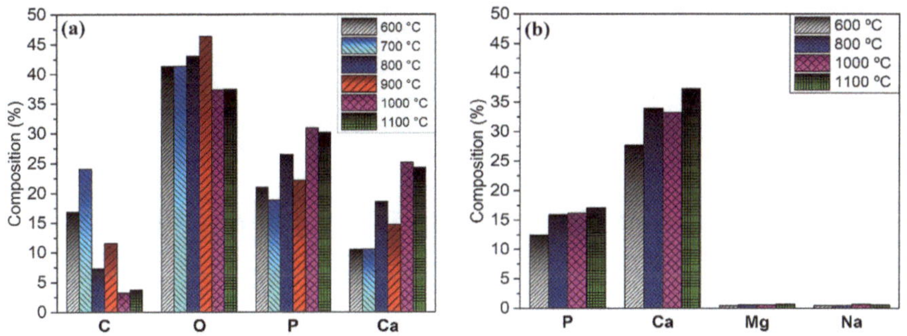

Fig. 7 Percentage of the main elements (organic and metallics) obtained by **a** EDS and **b** ICP-OES of fish scales combustion residues

Table 9 Pearson's correlation between burnout and inorganic elements

Variables	Burnout	P	Ca	Mg	Na	Others
Burnout	1.0					
P	0.9	1.0				
Ca	0.9	1.0	1.0			
Mg	0.9	1.0	1.0	1.0		
Na	0.9	0.6	0.56	0.6	1.0	
Others	−0.2	−0.5	−0.6	−0.5	0.3	1.0

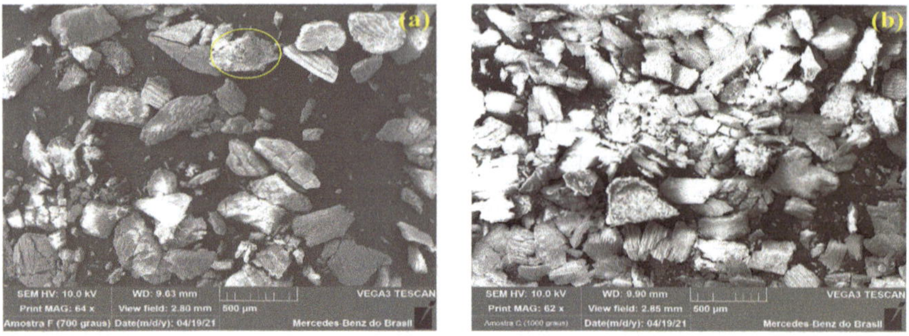

Fig. 8 SEM images of fish scale residues with amplitudes 60 times

The SEM images (Fig. 8a, b) of fish scale residues after combustion at different temperatures (700, 900, and 1100 °C) showed the set of collagen fibers and lamellae in the *in natura* samples was transformed into an agglomerate of compacted particles and probably melted due to the burning process, suggesting a total elimination of the organic phase, confirmed by FTIR analysis (Silva et al. 2022). Some grain boundaries displayed several meso and micropores on the surfaces, possibly due to the evaporation of moisture and removal of volatile matter (Cruz and Crnkovic 2015). According to Ikoma et al. (2003), after heat treatment of *Pragus major* fish scales, calcium phosphate crystals were sintered and aggregated on their surfaces. ICP-OES analyses, FTIR, and EDS confirmed the residues are comprised of only one mineral phase, composed predominantly of hydroxyapatite (Ikoma et al. 2003; Paul et al. 2017; Deb et al. 2019; Santana et al. 2019).

5 Co-combustion of Fish Scales and Coal in Drop Tube Furnace (DTF)

The coal samples and blends with fish scales in 10% (B10), 30% (B30), and 50% (B50) proportions were burned in a Drop Tube Furnace (DTF) at 1100 °C and 500 ms residence time (Silva et al. 2021). According to Fig. 9a–d, initially, the addition of fish scales increased the physicochemical properties of the samples (e.g., moisture content (4.1–7.7%), volatile materials (27.7–33.7%), and ash content (32.1–42.0%)) and reduced the amounts of fixed carbon (40.3–24.5%). Despite the increase, the moisture resulting from the blends remained below 10%, which is ideal for applications in firing processes (García et al. 2012; Yao et al. 2005). Moreover, the increase in volatile materials increased the scattering of combustible gases during combustion, considerably enriching the process (García et al. 2012). The addition of fish scales also increased the amount of nitrogen (0.9–3.6%) and reduced carbon content (53.8–37.1%). The amounts of hydrogen (3.5%) and sulfur (≈0.8%) do not change significantly due to the similar compositions between the charcoal used and the fish scales.

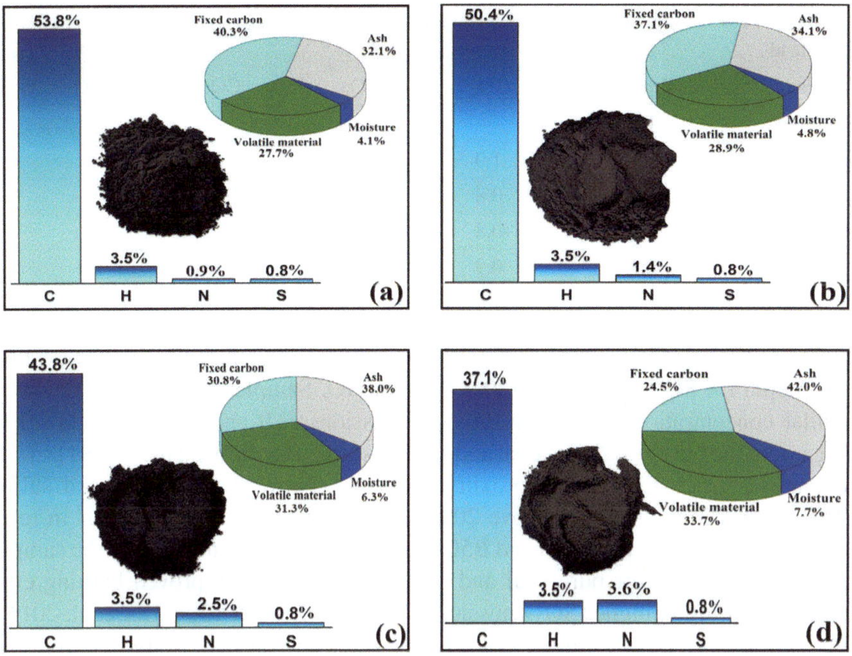

Fig. 9 Composition of **a** coal samples and blends with fish scales: **b** B10, **c** B30, and **d** B50 (dry base and ash free). Adapted from Silva et al. (2022)

A strong trend was observed in the burning efficiency with the increase in the amount of fish scales, according to Table 10. The lowest efficiency was observed for burning pure coal, with 84.4%, and the highest was achieved for B50, with 92.9% (approximately 10% higher than coal). The other measured values were 89.9% for B10 and 91.2% for B30. The best repeatability occurred for B10 and B30; however, all errors remained below 5%. The higher burnouts obtained are related to the higher content of volatile material (VM) and lower content of fixed carbon (FC) of the mixture in relation to pure coal (Mortari et al. 2021).

Regarding gaseous emissions, the concentrations of carbon dioxide, sulfur, and nitrogen monoxide increased as the amounts of fish scales were added to the mixture

Table 10 Burnout and gas species concentrations for coal samples and blends

Samples	Coal	B10	B30	B50
Burnout (%)	84.4 ± 4.4	89.9 ± 0.5	91.2 ± 1.0	92.9 ± 2.4
CO_2 (%)	6.8 ± 0.1	7.9 ± 0.3	7.4 ± 0.2	9.3 ± 0.2
SO_2 (ppm)	54.6 ± 3.3	72.2 ± 5.5	133.8 ± 7.3	182.5 ± 9.2
CO (ppm)	324.8 ± 16.2	287.5 ± 19.0	221.4 ± 18.7	160.6 ± 21.7
NO (ppm)	222.0 ± 2.9	353.9 ± 5.6	459.9 ± 4.2	780.9 ± 3.3

Table 11 Pearson's correlation coefficient among gaseous emissions and burnout for coal samples and blends

Variables	Burnout	CO_2	SO_2	CO	NO
Burnout	1.0				
CO_2	0.8	1.0			
SO_2	0.9	0.8	1.0		
CO	−0.9	-0.8	−1.0	1.0	
NO	0.9	0.9	1.0	−1.0	1.0

(Table 10). The variations for the gases evaluated were 6.8–9.3% for CO_2, 54.6–182.5 ppm for SO_2, and 222.0–780.9 ppm for NO. Samples B10 and B30 showed similar concentrations of carbon dioxide emissions (7.9 and 7.4%, respectively). Regarding the burning of pure coal and B50 mixture, CO_2 showed the lowest growth among the gases analyzed (approximately 37%). Nitrogen monoxide and sulfur dioxide emissions increased above 200% in the respective concentrations in relation to the burning of pure coal and B50, due to the amount of nitrogen in the sample composition (0.9% for pure coal and 3.6% for B50), and improved burning efficiency, resulting in higher SO_2 emissions (Williams et al. 2012; Sartor et al. 2014). The Pearson correlation among gaseous emissions and burnout (Table 11) for coal and blends indicated a positive correlation between burnout, CO_2, SO_2, and NO, and a negative one for CO, confirming the trend observed in Table 10.

The values obtained for CO ranged from 160.6 to 324.8 ppm; B50 and pure coal showed the lowest and the highest values, respectively (an approximately 50% reduction with the addition of fish scales). The decrease in the amount of CO emitted in the thermal process confirmed the improvement in the burning efficiency. Despite the higher burning efficiency and lower carbon monoxide emissions, B50 resulted in high concentrations of carbon dioxide and nitrogen monoxide in relation to pure coal and other mixtures. B10 and B30 were sufficient to guarantee burnouts close to 90% and lower concentrations of the main atmospheric pollutants cited in this chapter.

6 Other Applications for the Fish Scales Residues

This section presents recent data on the Circular Economy approach that emphasize the valorization of the most interesting compounds derived from fish scales, namely collagen, peptides, gelatin, and hydroxyapatite in various fields of application with substantial economic and ecological advantages.

6.1 Fish Scales in Tissue Engineering

6.1.1 Collagen

Markets and Markets (2022) reports that the Marine Collagen Market is estimated to reach $1,137 million by 2026, registering a 7.9% Compound Annual Growth Rate (CAGR) associated with the use of collagen in cosmetic, pharmaceutical, biomaterial, food, and nutraceutical industries. Hence, fish waste is a readily available and inexpensive source of collagen for these industries (Coppola et al. 2021).

Collagen is the most abundant structural protein found in vertebrates, representing approximately 30% of the total protein content in the body (Sherman et al. 2015; Coppola et al. 2020; Salvatore et al. 2020). It is indispensable for cell signaling, resilient for multicellular organisms, and resistant to mechanical stress (Naomi et al. 2021) and displays antigenic, anti-inflammatory, biocompatible, and biodegradable properties (Srinivasan and Durairaj 2021). Type I (COL-I), its most common form, is fibril-forming and the main component of the extracellular matrix. Fibers and fibrils act as a mechanical structural support for various connective tissues, including bones, skin, tendons, cartilage, dentin, cornea, blood vessels, and nerves (Sherman et al. 2015; Tang and Saito 2015).

COL-I has been extensively utilized as a biomaterial in the field of biomedicine, particularly for the purpose of tissue regeneration (Kuttappan et al. 2016; Pal et al. 2016; Su et al. 2020; Szychlinska et al. 2020). The current industrial production of collagen depends mainly on the extraction of skin, muscle, and bone from mammals such as cows and pigs (Salvatore et al. 2020). However, risks of zoonotic infectious diseases (e.g., bovine spongiform encephalopathy and foot and mouth disease), triggering of adverse immunological and inflammatory responses, and religious objections to the use of porcine and bovine collagen have led to constant searches for alternative natural sources that combine biocompatibility, low cost, availability, and ease of handling (Govindharaj et al. 2019; Qin et al. 2022).

Among the wastes discarded by the fish processing industry, fish scales (FS) have gained increasing interest because they are considered an excellent source of type I collagen (Matsumoto et al. 2015). FS are made of an extracellular matrix and composed of $(\alpha_1)_2(\alpha_2)$ triple helix and hydroxyapatite, which form highly ordered collagen fibers with tightly cross-linked regions (Oslan et al. 2022). COL-I extracted from fish scales has shown biocompatible, healing, immune-modulatory, osteogenic, and antimicrobial properties (Matsumoto et al. 2015; Tang and Saito 2015; Oh et al. 2016; Feng et al. 2020; Shalaby et al. 2020). Furthermore, marine collagen is free of cultural or religious restrictions and shows a low risk of zoonotic infection; therefore, it may become a suitable substitute for mammalian collagen (Salvatore et al. 2020).

Fish scales collagen-based biomaterials have been used in tissue engineering usually as hydrogels, scaffolds, membranes, and matrices for making wound dressings, bone graft composite materials, dental and vascular tissues, etc. (Tang and Saito 2015; Pal et al. 2016; Wang et al. 2017; Feng et al. 2020; Li et al. 2020). For example, type I collagen (COL-I) derived from the scales of Nile tilapia, *Oreochromis niloticus* (Linnaeus 1758), promotes the proliferation of a mouse odontoblast-like cell line (MDPC-23) and accelerates matrix mineralization for tissue regeneration in the oral maxillofacial area (Tang and Saito 2015). In vitro experiments using human bone marrow-derived mesenchymal stem cells have showed higher alkaline phosphatase (ALP) activity and osteocalcin (OCN) expression levels when cultured in a 3D porous scaffold of tilapia scale collagen (TC) cross-linked by mTGase.

The combination of TC and mTGase has exerted a strong effect on the process of osteoblast differentiation (Oh et al. 2016). Feng et al. (2020) developed a hybrid hydrogel dressing composed of collagen (*Larimichthys crocea* fish scales), sodium alginate, polymyxin B sulfate, and bacitracin (AC/OSA-PB), which produced an excellent antibacterial effect on *Staphylococcus aureus* (*S. aureus*, ATCC-14458) and *Escherichia coli* (*E. coli*, ATCC-8739) and a positive effect on wound healing capacity and biocompatibility in vitro (Feng et al. 2020). Arun Karthick et al. (2022) successfully produced a nanofibrous matrix made of fish scale-extracted Collagen/PVA (FCO-PVA) using the electrospinning method for wound dressing applications. The matrix displayed not only good antibacterial activity against *E. coli* and *S. aureus*, but also significantly improved water retention properties, providing a moist environment at the wound site that facilitated faster healing (Arun Karthick et al. 2022).

COL-I from fish scales can also be used in corneal and vascular tissue engineering applications. In recent years, collagen from such a biowaste has shown great potential for applications for corneal stromal regeneration due to its biocompatibility, excellent light transmission, low immunogenicity, and good biosafety after decellularization (Hsueh et al. 2019; Cheng et al. 2021). Li et al. (2020) successfully developed collagen membranes from decellularized and decalcified fish scales (*Oreochromis niloticus*) with curvature, transparency, and similar structure to those of natural human cornea. Wang et al. (2017) used COL-I extracted from fish scales as blood and lymphatic vessels, after modifying it into methylated collagen, and cross-linked it using methylation and 1,4-butanediol diglycidyl ether (BDE). In vivo studies revealed the interaction between collagen and the surrounding tissue was favorable. The authors suggested the use of the scaffold for several biomedical applications, such as drug delivery, wound healing, and treatment of conditions related to inflammation.

6.1.2 Gelatin

Gelatin is a natural polymer that can be obtained by the partial hydrolysis and thermal denaturation of collagen (Coppola et al. 2021) and, since it is a collagen-derived macromolecule, displays almost the same characteristics as collagen (Naomi et al. 2021). It has proven biocompatible with human tissue, flexible, favorable for film formation, non-toxic, and biodegradable; therefore, it can be applied in biomedicine (Naomi et al. 2021; Qin et al. 2022). Due to such unique properties, the global gelatine market is expected to reach US$3.6 billion by 2023, at a 6.6% CAGR, driven by the food and beverage, medical, pharmaceutical, nutraceutical, and sports nutrition industries (Markets and Markets 2022).

Gelatin is commonly extracted from pork-skin; however, due to its similar characteristics to porcine gelatin, fish gelatin can also be considered as an alternative source for the food and pharmaceutical industry as a component in packaging or delivery systems for drugs, medicine, and cosmetics (Caruso et al. 2020; Khrunyk et al. 2020). Beishenaliev et al. (2019) produced electrospun gelatin nanofibers from scales of three fish species, namely, *Eleutheronema tetradactylum*, *Epinephelus moara*, and *Lutjanus johnii,* varying the UV light cross-linking time by 5 min (UGN5), 10 min (UGN10), and 20 min (UGN20) for improving their water resistance. After a 14-day incubation in MEM, the surface areas of UGN5 and UGN10 scaffolds were dissolved by only 32% and 30%, respectively. The preliminary in vitro HaCaT cell study showed UGN5 scaffolds induced rapid keratinocyte migration, stimulating 107% cell proliferation after 24 h, thus indicating they might become a promising material for skin wound treatments (Beishenaliev et al. 2019).

Sghayyar et al. (2019) manufactured scaffolds by combining phosphate-based-glass fibers (PGFs) with gelatin extracted from the fish scales of java tilapia, cross-linked using glutaraldehyde (GTA) at various concentrations to produce a novel, biocompatible, and bioabsorbable material for potential wound healing applications. The results showed GTA-cross-linked scaffolds promoted cell adhesion and proliferation and accelerated in vitro artificial wound closure. Gelatin from *Mullet* scales (FSG) was combined with modified PLA (MPLA) and a natural antibacterial agent derived from freshwater clam (*Corbicula fluminea Estefanía*) shell powder (FCSP) for the production of biocompatible and antibacterial nanofibers through the electrospinning method (Wu et al. 2020). The MPLA/FCSP/FSG electrospun nanofibers can have a wide range of applicability in wound healing, food packaging, and bioprotective materials (Wu et al. 2020).

6.1.3 Hydroxyapatite

Hydroxyapatite (HAp) is a hexagonal mineral belonging to a group of calcium phosphates called apatites (Prado et al. 2021). It can be naturally found in mineralized tissues such as bone, dental enamel, and dentin, contributing to their high mechanical resistance (Prado et al. 2021). According to Pang et al. (2015), synthetic HAp can stimulate the mineralization of neoformed bone tissue in different bone defect models, constituting an excellent choice for use as a bone substitute and coating for dental implants (Pang et al. 2015). However, the high manufacturing costs of this ceramic material represent a limitation for its use in clinical applications (Prado et al. 2021).

Numerous papers have focused on the extraction of HAp from different sources for the purpose of hard tissue regeneration, according to Fig. 10 (Amna 2018; Granito et al. 2018; Odusote et al. 2019; Prado et al. 2021). HAp from fish waste has been widely used for the obtaining of bone tissue engineering biomaterials due to the similarity of its structure and chemical composition to those of the mineral elements of

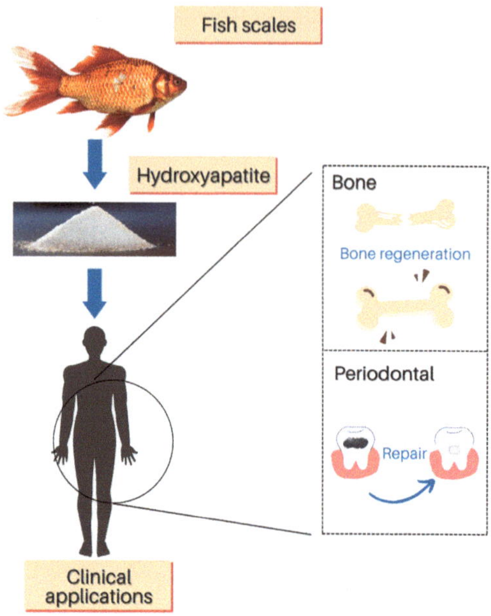

Fig. 10 Fish scales-derived hydroxyapatite and its clinical applications. Adapted from Granito et al. (2018) and Qin et al. (2022)

human bones and teeth (Shi et al. 2018; Surya et al. 2021). Moreover, the conversion of fish waste into HAp is an environmentally friendly process and can lead to cost reductions in bone repair treatments (Granito et al. 2018). According to Racioppo et al. (2021), hydroxyapatite extracted from natural resources shows excellent bioactivity, osteoconductivity, and osteoinductivity and generally better characteristics as a biomaterial when compared with synthetic materials. Knowing fish scales contain an extracellular hydroxyapatite-rich matrix, Muhammad et al. (2016) synthesized HAp from fish scales waste in 1-butyl-3-methylimidazolium acetate ionic liquid and obtained a $32 \pm 2\%$ yield, and Zainol et al. (2019) obtained a 36 wt% HAp yield from tilapia scales after an alkaline treatment with 5 M NaOH and 1200 °C calcination.

The biological responses of HAp from fish scales have also been thoroughly evaluated. In particular, crystalline nanostructured hydroxyapatite (nHAp) powders from *Cirrhinus mrigala* scales showed improved cellular activity and alkaline phosphatase (ALP) activity in MG-63 cells when compared with commercial HAp (Sathiskumar et al. 2019). Moreover, this marine biomaterial can also be combined with other materials such as synthetic polymers (PLA, PCL, PMMA, etc.) and chitosan. Composites of fish scale-derived hydroxyapatite (FHAp), polylactic acid (PLA), and a natural antibacterial agent from eggshell (EGS) have been employed for the preparation of a 3D-printing filament composite for use in biomedical material-related products such as bone joints, bone screws, tooth roots, and braces (Wu et al. 2021).

SEM results showed PLA/FHAP and PLA/EFHAP composites provide the highest cell adhesion of mouse embryonic fibroblasts (NIH 3T3 FB), whereas PLA and control show the lowest one, suggesting their implantation into the body would attract in vivo stem cells for repairing damaged tissues (Wu et al. 2021). Bioactive fish scales incorporated into the chitosan (CH) matrix were prepared by a combination of *Sparus aurata* decellularized FS with ultrasonic homogenized CH. CH/FS porous scaffolds were fabricated by lyophilization technique and in vitro cell experimental results showed collagen and HAp content of FS-induced proliferation of SaOS-2 osteoblastic cells after 14-day incubation and HAp from FS microparticles significantly induced biomineralization in CH (Kara et al. 2019).

6.1.4 Chitosan

Chitosan (CS) is a hydrophilic biopolymer derived from chitin by the N-deacetylation method (Santos et al. 2020). Chitosan is a low-cost and bio-friendly material that can be used in various industrial and biomedical applications. Its non-toxicity, biocompatibility, antimicrobial, antitumor, anticancer, anti-inflammatory, and biodegradability are some of its properties that are attracting widespread attention (Aboudamia et al. 2020; Santos et al. 2020). In the field of biomedicine, chitosan in its micro and nanoscale form has been investigated for its use in controlled drug and protein release therapy (Chaudhry et al. 2021; Ssekatawa et al. 2021). Besides that, this material can be produced by the chemical or biological method through several raw materials (Aboudamia et al. 2020). The main source of chitosan and chitin for industrial purposes is the biomass from shellfish, for example, crabs, shrimp, lobsters, and fish scales (Haj et al. 2020).

The exploitation of fish scales as a commercial source for the production of chitin and chitosan is a field that has not yet received adequate attention (Iber et al. 2022). Recent studies have focused efforts on the isolation and characterization of chitosan from several fish species such as *Labeo rohita*, *Labeo catla* (Kumar et al. 2021), *Sardina pilchardus* (Aboudamia et al. 2020), *Prochilodus magdalenae* (Molina-Ramírez et al. 2021), *Catla*, and *Cirrhinus mrigala* (Satpathy et al. 2021). Figure 11 shows the chemical structures of chitin and chitosan extracted from fish scales. Based on these findings, the extraction of chitosan from fish scales comes under three main stages: demineralization, deproteinization, and deacetylation (Molina-Ramírez et al. 2021). Deproteinization entails the removal of protein using a strong base, while the demineralization process involves the removal of calcium phosphates and calcium carbonates using a strong acid. The deacetylation is carried out by using a strong alkali such as NaOH (Iber et al. 2022).

The application of chitosan extracted from fish scales in various fields of biomedicine has been reported and has achieved promising results. For example, Putri et al. (2022) investigated the effectiveness of chitosan from Haruan fish scales (*Channa striata*) as an antibiofilm agent against *Porphyromonas gingivalis* biofilms. It was found that chitosan from scales of this species at concentrations of 2.5, 10, 20, and 40% proved to be capable of disrupting the structural integrity of the *P. gingivalis* bacterial cell wall. According to this study,

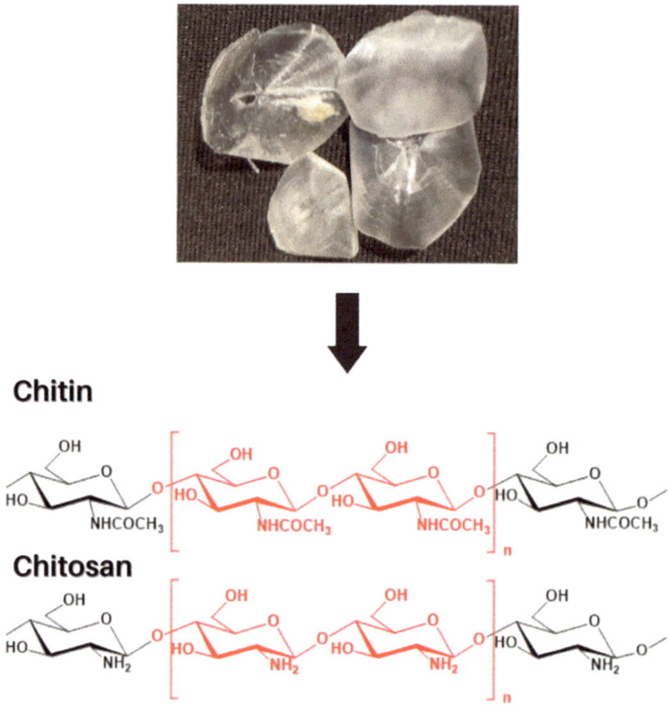

Fig. 11 Chitin and chitosan extracted from fish scales. Adapted from Mohammed et al. (2017)

the positively charged amine groups (–NH_2) of this polysaccharide are responsible for interacting with anionic groups on the cell wall, forming an impermeable layer around the bacterial cell that hinders the transportation of essential molecules for bacterial metabolism (Putri et al. 2022). Narendhran and Rajiv (2021) extracted chitosan from fish scales and fabricated Amphotericin B conjugated chitosan nanomaterial to evaluate its antifungal potential against *Candida parapsilopsis*. The chitosan nanomaterial (CS NM) and modified AmB-conjugated CS NM exhibited the best antifungal activity against *Candida parapsilopsis* and also showed good antibiofilm activity (Narendhran and Rajiv 2021).

Chitosan obtained from fish scales has also been used as a specific transport agent for anticancer drugs. Mohammed et al. (2017) developed and evaluated a polymeric prodrug based on 5-fluorouracil-1-acetic acid (FUAC) and loaded it onto chitosan (CS) derived from fish scales as a potential therapeutic product for the treatment of colon cancer. The in vitro cytotoxicity of CS-FUAC and pyrimidine analog (5-FU) conjugates were tested against HT-29. Cell viability results showed that CS-FUAC conjugates exhibited better cytotoxicity against colon cancer cells compared to 5-FU. In addition, CS-FUAC conjugates also showed greater selectivity for colon cancer cells, being almost 2 times more cytotoxic in colon cancer cells than in normal cells (CCD-18Co) (Mohammed et al. 2017).

6.2 *Bioadsorbent for Mineral/Element Extraction*

The discharge of industrial chemicals and heavy metals into aquatic ecosystems has been regarded as one of the most significant environmental threats in recent years (Ighalo and Eletta 2021). Concerns over the presence of pollutants such as dyes, heavy metals, pesticides, insecticides, and drugs in water bodies at levels higher than those established by the World Health Organization (WHO) and environmental agencies have arisen from their non-biodegradability, inhibition of photosynthesis, eutrophication, impact on the reproduction of animal species, and high toxicity (Elgarahy et al. 2021; Rezaei et al. 2022). Therefore, such effluents must be properly treated prior to being released into the environment (Neves et al. 2018).

Several techniques can be used to remove toxic substances from aqueous media (e.g., chemical precipitation, coagulative precipitation, ion exchange, membrane filtration, reverse osmosis, surface adsorption, among others (Liu et al. 2017). However, such traditional methods have disadvantages, such as ineffective pollution removal, excessive energy usage, production of toxic sludge to be safely disposed, management of generated wastes, and high cost (Liu et al. 2017; Rezaei et al. 2022). Over the past three decades, the generation of cost-effective and eco-friendly materials has been a major factor in the increasing number of techniques employed for removing pollutants from the environment (Rezaei et al. 2022).

Biosorption is a hopeful approach for the elimination of harmful substances from both industrial wastewater and natural bodies of water (Rezaei et al. 2022). The chemistry involved in the process is related to physicochemical interactions among various functional groups (e.g., phosphates, carboxyl, amines, sulfhydryl, phosphonates, sulphonates, and amides) available on the surface of bioadsorbent and metal ions (Pal and Maiti 2020). Over the past few years, the use of bioadsorbent materials has drawn more attention due to their eco-friendly, abundant, inexpensive, and effective nature for removing contaminants from wastewater (Niero et al. 2019). For instance, fish scales have been demonstrated to be a promising adsorbent material as they require no complex treatment methods or further modifications (Neves et al. 2018). As illustrated in Fig. 12, they are mainly constituted of a collagen fiber-hydroxyapatite complex (Qin et al. 2022), which exhibits adsorption properties that allow for the removal of both organic and inorganic pollutant species from various wastewater sources (Ribeiro et al. 2019).

Several studies have reported fish scales bioadsorbent can be used as an industrial waste treatment agent (Ribeiro et al. 2015; Das et al. 2016; Liu et al. 2017; Ighalo and Eletta 2020). Ighalo and Eletta (2020) successfully adsorbed Pb (II) and Zn (II) from aqueous solutions using *Micropogonias undulatus* fish scales and confirmed the adsorption process is spontaneous, i.e., adsorption occurs as soon as the bioadsorbent is added to the pollutant-containing solution and stirred. Furthermore, the fish

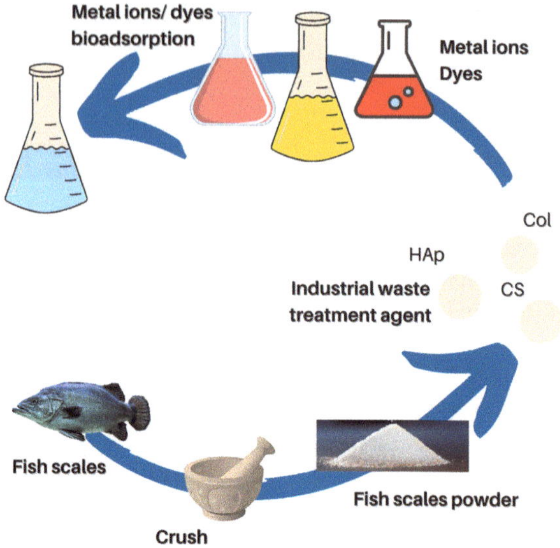

Fig. 12 Application of fish scales as a bioadsorbent for industrial wastewater treatment. Adapted from Liu et al. (2021)

scales biosorbent's monolayer adsorption capacities for Zn(II) and Pb(II) were 555.6 and 909.9 mg g^{-1}, respectively, indicating that the biosorbent from *Micropogonias undulatus* scales is an exceptional material for removing heavy metals from aqueous solutions (Ighalo and Eletta 2020).

Liu et al. (2017) showed that hydroxyapatite powder derived from exhibited strong abilities in removing lead ions from the solution, achieving an above 90% removal in 10 min. Such powders also exhibited a maximum adsorption capacity of 208.3 and 344.8 mg g^{-1} at pH = 5 and pH = 2.2, respectively, pointing to enhanced lead ion removal in more acidic conditions.

Teshale et al. (2020) focused on the adsorption of chromium (III) metal ions from tannery wastewater by waste fish scales, achieving a 99.75% chromium metal ion efficiency with the use of activated fish scales at 0.8 g adsorbent dosage, pH 5, 90 min contact time, and 150 mg L^{-1} initial concentration. In addition to heavy metal ions, research has also been conducted on the adsorption of various dyes using scales from several fish species. The modified scales of *Sardinella brasiliensis* (SSb) was used to extract Reactive Turquoise Blue 15(RTB15) (CAS 12225-39-7) and Reactive Red 120 (RR120) (CAS 61951-82-4) dyes (Niero et al. 2019). The adsorption kinetics experiments were performed with an initial dye concentration of 150 mg L^{-1} at 25 °C and yielded impressive results with 90% removal of both dyes within the first 15 min (Niero et al. 2019). Adsorption isotherms studies conducted at three different temperatures (25, 35, and 55 °C) indicated that temperature does affect the adsorption capacity of the adsorbent, as the thermodynamic parameters for interactions between the two dyes (RTB15 and RR120) and the SSb bioadsorbent were found to be an endothermic, yet spontaneous process in the case of RR120. In this scenario, the endothermic process is attributed to a heightened number of active sites and increased movement of the dye molecules, leading to the formation of an environment conducive to adsorption (Niero et al. 2019). Marrakchi et al. (2017a, b) analyzed the impact of *L. rohita* scales on the adsorption of reactive orange dye and reported the maximum adsorption capacities of the bioadsorbent at 30, 40, and 50 °C were 105.8, 107.2, and 114.2 mg g^{-1}, respectively.

7 Final Considerations

From an environmental viewpoint, improper disposal of fishing waste causes bad odors, decomposition of organic matter, and consequent proliferation of diseases and venomous animals. The effective use of fish scales (burning processes, for example) can reduce socioenvironmental vulnerabilities, minimize waste and inappropriate disposal, maximize economic activities, and promote bioenergy production.

According to the results presented in this chapter, fish scales can be a biomass alternative, mainly in coastal cities, through direct application in thermochemical processes and no further damage to thermal systems. Its use in combustion processes is a viable way of treating this waste, as it generates energy and avoids socioenvironmental problems arising from improper disposal. The feedstock composition and ashes showed potential applications for fertilization and soil conditioning, biomedicine, adsorbents, and civil industry.

Finally, the versatility of the use of the feedstock of fish waste (*in natura*, pretreated and/or residues) in several industrial sectors and fields of applied science can effectively contribute to minimizing recurrent socioenvironmental problems faced by modern society, which strongly seeks to solve them.

Acknowledgements The authors gratefully acknowledge the National Council for Scientific and Technological Development—CNPq (426162/2018-8) and Maranhão State Foundation for Scientific and Technological Development—FAPEMA (05447/19, 00957/19, 01661/21, and 06776/22) for the financial support; Federal University of Maranhão (UFMA), Combustion Laboratory (LC) of Federal University of Rio Grande do Sul (UFRGS), and Mercedes-Benz Brazil for the technical and professional support.

References

Aboudamia FZ et al (2020) Potential of discarded sardine scales (*Sardina pilchardus*) as chitosan sources. J Air Waste Manag Assoc (Taylor & Francis) 70(11):1186–1197. https://doi.org/10.1080/10962247.2020.1813840

Almeida ZS, Isaac VJ, Paz AC et al (2011) Assessment of potential fishery production system hake-yellow (*Cynoscion acoupa*) captured by fleet of commercial Araçagi, Raposa, Maranhão. Boletim do Laboratório de Hidrobiologia 24:35–42. https://doi.org/10.18764

Amna T (2018) Valorization of bone waste of Saudi Arabia by synthesizing hydroxyapatite. Appl Biochem Biotechnol 186(3):779–788. https://doi.org/10.1007/s12010-018-2768-5

Arumugam A, Ponnusami V (2017) Production of biodiesel by enzymatic transesterification of waste sardine oil and evaluation of its engine performance. Heliyon 3(December). https://doi.org/10.1016/j.heliyon.2017.e00486

Arun Karthick S et al (2022) A study on collagen-PVA and chitosan-PVA nanofibrous matrix for wound dressing application. Mater Today: Proc (Elsevier Ltd.) 56:1347–1350. https://doi.org/10.1016/j.matpr.2021.11.421

Atkins P, Jones L (2012) Princípios de Química. Bookman.

Ballester J, Barroso J, Cerecedo LM et al (2015) Comparative study of semi-industrial-scale flames of pulverized coals and biomass. Combust Flame 141:204–215. https://doi.org/10.1016/j.combustflame.2005.01.005

Beishenaliev A et al (2019) Fabrication and preliminary *in vitro* evaluation of ultraviolet-crosslinked electrospun fish scale gelatin nanofibrous scaffolds. J Mater Sci: Mater Med (Springer US) 30(6). https://doi.org/10.1007/s10856-019-6264-4

Bermúdez-Penabad N, Kennes C, Veiga MC (2017) Anaerobic digestion of tuna waste for the production of volatile fatty acids. Waste Manag 68:96–102. https://doi.org/10.1016/j.wasman.2017.06.010

Bhagwat PK, Dandge PB (2016) Isolations, characterization and valorizable applications of fish scale collagen in food and agriculture industries. Biocatal Agric Biotechnol 7:234–240. https://doi.org/10.1016/j.bcab.2016.06.010

Biswas S et al (2006) Studies on the combustion behavior of blends of Indian coals by TGA and drop tube furnace. Fuel Process Technol 87:191–199. https://doi.org/10.1016/j.fuproc.2005.05.002

Branco V, Costa M (2017) Effect of particle size on the burnout and emissions of particulate matter from the combustion of pulverized agricultural residues in a drop tube furnace. Energy Convers Manag 149:774–780. https://doi.org/10.1016/j.enconman.2017.03.012

Braz CEM, Crnkovic PM (2014) Physical-chemical characterization of biomass samples for application in pyrolysis process. Chem Eng Trans 37:523–528. https://doi.org/10.3303/CET1437088

Brazilian Association of Fish Farming—ABP (2020) Anuário Peixe BR da Piscicultura. https://www.peixebr.com.br/anuario-2020. Acessed 28 June 2021

Buraiki NS, Albadri B, Alsheriq S et al (2020) Characterization of *Catla catla* and *Oreochromis niloticus* fish scales derived hydroxyapatite scaffolds for regenerative medicine. Mater Today: Proc 27:2609–2616. https://doi.org/10.1016/j.matpr.2019.11.074

Caruso G, Floris R, Serangeli C, Di Paola L (2020) Fishery wastes as a yet undiscovered treasure from the sea: biomolecules sources, extraction methods and valorization. Mar Drugs 18(12):622(1–30). https://doi.org/10.3390/md18120622

Carvalho Jr JA, Lacava PT (2003) Emissões em processos de combustão. Editora UNESP

Castañeda D, Lozano JM, Suárez H (2016) Microstructural changes and the effect on myofibril proteins in yamu (*Brycon amazonicus*) fish meat during cold storage. Agronomia Colombiana 34:403–414. https://doi.org/10.15446/agron.colomb.v34n3.61316

Chai HJ, Li JH, Huang HN, Li TL, Chan YL, Shiau CY, Wu CJ (2010) Effects of sizes and conformations of fish-scale collagen peptides on facial skin qualities and transdermal penetration efficiency. J Biomed Biotechnol. https://doi.org/10.1155/2010/757301

Chaudhry G-S, Thirukanthan CS, Murni NurIslamiah K, Sung YY, Sifzizul TSM (2021) Characterization and cytotoxicity of low-molecular-weight chitosan and chito-oligosaccharides derived from tilapia fish scales. J Adv Pharm Technol Res 12(4):373–377. https://doi.org/10.4103/japtr.japtr

Chen S, Hirota N, Okuda M et al (2010) Microstructures and rheological properties of tilapia fish-scale collagen hydrogels with aligned fibrils fabricated under magnetic fields. Acta Biomater 7(2):644–652. https://doi.org/10.1016/j.actbio.2010.09.014

Cheng G et al (2021) Preliminary study on fish scale collagen lamellar matrix as artificial Cornea. Membranes 11(10). https://doi.org/10.3390/membranes11100737

Ching-Velasquez J, Fernández-Lafuente R, Rodrigues RC et al (2020) Production and characterization of biodiesel from oil fish waste by enzymatic catalysis. Renew Energy 153:1346–1354. https://doi.org/10.1016/j.renene.2020.02.100

Chinh NT, Manh VQ, Trung VQ et al (2019) Characterization of collagen derived from tropical freshwater carp fish scale wastes and its amino acid sequence. Nat Product Commun 14 (July). https://doi.org/10.1177/1934578X19866288.

Cloke M, Lester E, Thompson AW (2002) Combustion characteristics of coal using a drop-tube furnace. Fuel 81:727–735. https://doi.org/10.1016/S0016-2361(01)00199-5

Coelho P, Costa M (2007) Combustão. Edições Orion, Alfragide

Coppola D et al (2020) Marine collagen from alternative and sustainable sources: extraction, processing and applications. Mar Drugs 18(4):214. https://doi.org/10.3390/md18040214

Coppola D et al (2021) Fish waste: from problem to valuable resource. Mar Drugs 19(2):1–39. https://doi.org/10.3390/md19020116

Côrtes LN, Druzian SP, Streit A et al (2019) Biochars from animal wastes as alternative materials to treat colored effluents containing Basic Red 9. J Environ Chem Eng 7(December). https://doi.org/10.1016/j.jece.2019.103446

Cruz G, Crnkovic PM (2015) Evaluation of the combustion process of coffee husk samples in a drop tube furnace (DTF). Therm Eng 14(2):53–62. https://doi.org/10.5380/reterm.v14i2.62135

Cruz G, Crnkovic PM (2019) Assessment of the physical-chemical properties of residues and emissions generated by biomass combustion under N_2/O_2 an CO_2/O_2 atmospheres in a drop tube furnace (DTF). J Therm Anal Calorim 138:401–415. https://doi.org/10.1007/s10973-019-08238-0

Cruz G, Monteiro PAS, Braz CEM et al (2018) Investigation into the physical-chemical properties of chemically pretreated sugarcane bagasse. J Therm Anal Calorim 132:1039–1053. https://doi.org/10.1007/s10973-018-7041-1

Daboor SM et al (2012) Isolation and activation of collagenase from fish processing waste. Biosci Biotechnol 3(3):191–203. https://doi.org/10.4236/abb.2012.33028

Das A, Bhowal A, Datta S (2016) Biomass characterisation and adsorption mechanism of Cu (II) biosorption onto fish (*Catla catla*) scales. Int J Environ Eng 8(1):81. https://doi.org/10.1504/ijee.2016.078242

Deb P, Barua E, Lala S et al (2019) Synthesis of hydroxyapatite from Labeorohita fish scale for biomedical application. Mater Today: Proc 15:277–283. https://doi.org/10.1016/j.matpr.2019.05.006

Deer WA, Howie RA, Zussman J (2000) Minerais Constituintes das rochas: Uma introdução. Fundação Calouste Gulbenkian.

Duan L, Duan Y, Zhao C et al (2015) NO emission during co-firing coal and biomass in oxy-fuel circulating fluidizes bed combustor. Fuel 150:8–13. https://doi.org/10.1016/j.fuel.2015.01.110

Elgarahy AM et al (2021) A critical review of biosorption of dyes, heavy metals and metalloids from wastewater as an efficient and green process. Clean Eng Technol (Elsevier Ltd.) 4(November 2020):100209. https://doi.org/10.1016/j.clet.2021.100209

Feng X et al (2020) Preparation of aminated fish scale collagen and oxidized sodium alginate hybrid hydrogel for enhanced full-thickness wound healing. Int J Biol Macromol (Elsevier B.V.) 164:626–637. https://doi.org/10.1016/j.ijbiomac.2020.07.058

Ferreira S et al (2017) Biomass resources in Portugal: current status and prospects. Renew Sust Energy Rev 78:1221–1235. https://doi.org/10.1016/j.rser.2017.03.140

Food and Agriculture Organization of the United Nations—FAO (2018) The state of world fisheries and aquaculture: meeting the sustainable development goals. https://www.fao.org/publications/card/en/c/CA0191EN. Accessed 10 April 2020

García R et al (2012) Characterization of Spanish biomass wastes for energy use. Bioresour Technol 103(1):249–258. https://doi.org/10.1016/j.biortech.2011.10.004

Ghaly AE, Ramakrishnan VV, Brooks MS et al (2013) Fish processing wastes as a potential source of proteins, aminoacids and oils: a critical review. J Microb Biochem Technol 5:107–129. https://doi.org/10.4172/1948-5948.1000110

Govindharaj M, Roopavath UK, Rath SN (2019) Valorization of discarded Marine Eel fish skin for collagen extraction as a 3D printable blue biomaterial for tissue engineering. J Clean Product (Elsevier Ltd.) 230:412–419. https://doi.org/10.1016/j.jclepro.2019.05.082

Granito RN et al (2018) Hydroxyapatite from fish for bone tissue engineering: a promising approach. Int J Mol Cell Med Spring 7(2):80–90. https://doi.org/10.22088/IJMCM.BUMS.7.2.80

Haj NQ, Mohammed MO, Mohammood LE (2020) Synthesis and biological evaluation of three new chitosan schiff base derivatives (Dd). https://doi.org/10.1021/acsomega.0c01342

Holá M, Kalvoda J, Nová H et al (2011) Possibilities of LAICP-MS technique for the spatial elemental analysis of the recent fish scales: line scan vs. depth profiling. Appl Surf Sci 257(6):1932–1940. https://doi.org/10.1016/j.apsusc.2010.09.029

Hsueh YJ et al (2019) Extracellular matrix protein coating of processed fish scales improves human corneal endothelial cell adhesion and proliferation. Transl Vis Sci Technol 8(3). https://doi.org/10.1167/tvst.8.3.27

Huang CY, Kuo JM, Wu SJ et al (2016) Isolation and characterization of fish scale collagen from tilapia (*Oreochromis sp.*) by a novel extrusion-hydro extraction process. Food Chem 190:997–1006. https://doi.org/10.1016/j.foodchem.2015.06.066

Iber BT et al (2022) A review of various sources of chitin and chitosan in nature. J Renew Mater 10(4):1097–1123. https://doi.org/10.32604/JRM.2022.018142

Ighalo JO, Eletta OAA (2020) Response surface modelling of the biosorption of Zn(II) and Pb(II) onto Micropogonias undulatus scales: Box-Behnken experimental approach. Appl Water Sci (Springer International Publishing) 10(8):1–12. https://doi.org/10.1007/s13201-020-01283-3

Ighalo JO, Eletta OAA (2021) Recent advances in the biosorption of pollutants by fish scales: a mini-review. Chem Eng Commun (Taylor & Francis) 208(9):1301–1312. https://doi.org/10.1080/00986445.2020.1771322

Ikoma T, Kobayashi H, Tanaka J et al (2003) Microstructure, mechanical, and biomimetic properties of fish scales from Pagrus major. J Struct Biol 142:327–333. https://doi.org/10.1016/S1047-8477(03)00053-4

Jenkins BM, Baxter LL, Miles TR et al (1998) Combustion properties of biomass. Fuel Process Technol 54:17–46. https://doi.org/10.1016/S0378-3820(97)00059-3

Kara K, Ouanji F, Lotfi EM et al (2018) Biodiesel production from waste fish oil with high free fatty acid content from Moroccan fish-processing industries. Egypt J Pet 27:249–255. https://doi.org/10.1016/j.ejpe.2017.07.010

Kara A et al (2019) Bioactive fish scale incorporated chitosan biocomposite scaffolds for bone tissue engineering. Int J Biol Macromol (Elsevier B.V.) 130:266–279. https://doi.org/10.1016/j.ijbiomac.2019.02.067

Kazanc F, Khatami R, Crnkovic PM et al (2011) Emissions of NO_x and SO_2 from coals of various ranks, bagasse, and coal bagasse burning in O_2/N_2 and O_2/CO_2 environments. Energy Fuels 25:2850–2861. https://doi.org/10.1021/ef200413u

Khrunyk Y et al (2020) Progress in modern marine biomaterials research. Mar Drugs 18(12):589. https://doi.org/10.3390/md18120589

Kongsri S et al (2013) Nanocrystalline hydroxyapatite from fish scale waste: preparation, characterization and application for selenium adsorption in aqueous solution. Chem Eng J 215–216:522–532. https://doi.org/10.1016/j.cej.2012.11.054

Krumal K, Miluska P, Horak J et al (2019) Comparison of emissions of gaseous and particulate pollutants from the combustion of biomass and coal in modern and old-type boilers used for residential heating in the Czech Republic, Central Europe. Chemosphere 229:51–59. https://doi.org/10.1016/j.chemosphere.2019.04.137

Kumar M, Agrawal R, Pareek N (2021) Extraction and physicochemical properties assessment of chitin and chitosan from fish scales. Macromol Symp 399(1):1–6. https://doi.org/10.1002/masy.202100006

Kuttappan S, Mathew D, Nair MB (2016) Biomimetic composite scaffolds containing bioceramics and collagen/gelatin for bone tissue engineering—a mini review. Int J Biol Macromol (Elsevier B.V.) 93:1390–1401. https://doi.org/10.1016/j.ijbiomac.2016.06.043

Leite S et al (2016) Resíduos da comercialização de pescado marinho: volume de descarte e aspectos microbiológicos. Revista Brasileira de Tecnologia Agroindustrial 10(1):2112–2125. https://doi.org/10.3895/rbta.v10n1.2692

Li KY et al (2020) Fish-scale collagen membrane seeded with corneal endothelial cells as alternative graft for endothelial keratoplasty transplantation. ACS Biomater Sci Eng 6(5):2570–2577. https://doi.org/10.1021/acsbiomaterials.9b00562

Lin CC, Ritch R, Lin SM et al (2010) A new fish scale-derived scaffold for corneal regeneration. Eur Cell Mater 19:50–57

Liu WK et al (2017) From waste to health: synthesis of hydroxyapatite scaffolds from fish scales for lead ion removal. Jom 69(4):713–718. https://doi.org/10.1007/s11837-017-2270-5

Liu Y et al (2021) Preparation and application of hydroxyapatite extracted from fish scale waste using deep eutectic solvents. Ceram Int (Elsevier Ltd.) 47(7):9366–9372. https://doi.org/10.1016/j.ceramint.2020.12.067

Lustosa-Neto AD, Nunes ML, Maia LP et al (2018) The fishery and aquaculture products industry. Acta Fish Aquat Resour 6:28–48. https://doi.org/10.2312/Actafish.2018.6.2.28-48

Lv S, Hu L, Xia C et al (2021) Surface-treated fish scale powder with silane coupling agent in asphalt for performance improvement: conventional properties, rheology, and morphology. J Clean Prod 311(August). https://doi.org/10.1016/j.jclepro.2021.127772

Manwatkar P, Dhote L, Pandey RA et al (2021) Combustion of distillery sludge mixed with coal in a drop tube furnace and emission characteristics. Energy 221(April). https://doi.org/10.1016/j.energy.2021.119871

Markets and Markets (2022) Marine collagen market growth opportunities, development and industry overview. https://www.marketsandmarkets.com/Market-Reports/marine-collagen-market-155534506.html. Accessed 28 June 2022

Marrakchi F, Auta M et al (2017a) High-surface-area and nitrogen-rich mesoporous carbon material from fishery waste for effective adsorption of methylene blue. Powder Technol (Elsevier B.V.) 321:428–434. https://doi.org/10.1016/j.powtec.2017.08.023

Marrakchi F, Ahmed MJ et al (2017b) Mesoporous carbonaceous material from fish scales as low-cost adsorbent for reactive orange 16 adsorption. J Taiwan Instit Chem Eng (Elsevier B.V.) 71:47–54. https://doi.org/10.1016/j.jtice.2016.12.026

Martinez CLM, Rocha EPA, Carneiro ACO et al (2019) Characterization of residual biomasses from the coffee production chain and assessment the potential for energy purposes. Biomass Bioenergy 120:68–76. https://doi.org/10.1016/j.biombioe.2018.11.003

Martins CA, Ferreira MA (2010) Considerations on the formation of NO_x in combustion. Congresso Nacional de Engenharia Mecânica 6:1–11

Martins GI, Secco D, Rosa HA et al (2015) Physical and chemical properties of fish oil biodiesel produced in Brazil. Renew Sustain Energy Rev 42:154–157. https://doi.org/10.1016/j.rser.2014.10.024

Matsumoto R et al (2015) Rapid oriented fibril formation of fish scale collagen facilitates early osteoblastic differentiation of human mesenchymal stem cells. J Biomed Mater Res Part A 103(8):2531–2539. https://doi.org/10.1002/jbm.a.35387

Mohammed MO, Hussain KS, Haj NQ (2017) Preparation and bioactivity assessment of chitosan-1-acetic acid-5-flurouracil conjugates as cancer prodrugs. Molecules 22(11). https://doi.org/10.3390/molecules22111629

Molina-Ramírez C et al (2021) Characterization of chitosan extracted from fish scales of the colombian endemic species Prochilodus magdalenae as a novel source for antibacterial starch-based films. Polymers 13(13). https://doi.org/10.3390/polym13132079

Mortari DA, Torquato LDM, Crespi MS et al (2018) Co-firing of blends of sugarcane bagasse and coal. J Therm Anal Calorim 132:1333–1345. https://doi.org/10.1007/s10973-018-6996-2

Mortari DA, Pereira FM, Crnkovic PM (2020) Experimental investigation of the carbon dioxide effect on the devolatilization and combustion of a coal and sugarcane bagasse. Energy 204(August). https://doi.org/10.1016/j.energy.2020.117824

Mortari DA, Perondi D, Rossi GB et al (2021) The influence of water-soluble inorganic matter on combustion of grape pomace and its chars produced by slow and fast pyrolysis. Fuel 284(January). https://doi.org/10.1016/j.fuel.2020.118880

Muhammad N et al (2016) Extraction of biocompatible hydroxyapatite from fish scales using novel approach of ionic liquid pretreatment. Sep Purif Technol (Elsevier B.V.) 161:129–135. https://doi.org/10.1016/j.seppur.2016.01.047

Naomi R et al (2021) Natural-based biomaterial for skin wound healing (Gelatin vs. collagen): expert review. Polymers 13(14):1–20. https://doi.org/10.3390/polym13142319

Narendhran S, Rajiv P (2021) Synthesis of amphotericin B conjugated chitosan nanomaterial from fish scales and evaluation of its antifungal activity. J Cluster Sci (Springer, US) 3.https://doi.org/10.1007/s10876-021-02177-3

Neves CV et al (2018) Phenomenological modeling of reactive dye adsorption onto fish scales surface in the presence of electrolyte and surfactant mixtures. Environ Technol (United Kingdom) 39(19):2467–2483. https://doi.org/10.1080/09593330.2017.1356876

Niero G et al (2019) Using modified fish scale waste from *Sardinella brasiliensis* as a low-cost adsorbent to remove dyes from textile effluents. J Environ Sci Health Part A Toxic/Hazard Subst Environ Eng (Taylor & Francis) 54(11):1083–1090. https://doi.org/10.1080/10934529.2019.1631091

Odusote JK et al (2019) Synthesis and characterization of hydroxyapatite from bovine bone for production of dental implants. J Appl Biomater Funct Mater 17(2). https://doi.org/10.1177/2280800019836829

Oh HH et al (2016) Effect of enzymatically cross-linked tilapia scale collagen for osteoblastic differentiation of human mesenchymal stem cells. J Bioact Compat Polym 31(1):31–41. https://doi.org/10.1177/0883911515595240

Oslan SNH et al (2022) Extraction and characterization of bioactive fish by-product collagen as promising for potential wound healing agent in pharmaceutical applications: current trend and future perspective. Int J Food

Osório E, Gomes ML, Vilela CF et al (2006) Evaluation of petrology and reactivity of coal blends for use in pulverized coal injection (PCI). Int J Coal Geol 68:14–29. https://doi.org/10.1016/j.coal.2005.11.007

Pal D, Maiti SK (2020) An approach to counter sediment toxicity by immobilization of heavy metals using waste fish scale derived biosorbent. Ecotoxicol Environ Saf (Elsevier Inc.) 187(October 2019):109833. https://doi.org/10.1016/j.ecoenv.2019.109833

Pal P et al (2016) Accelerating full thickness wound healing using collagen sponge of mrigal fish (*Cirrhinus cirrhosus*) scale origin. Int J Biol Macromol (Elsevier B.V.) 93:1507–1518. https://doi.org/10.1016/j.ijbiomac.2016.04.032

Pang KM et al (2015) Biologic properties of nano-hydroxyapatite: an *in vivo* study of calvarial defects, ectopic bone formation and bone implantation. Bio-Med Mater Eng 25(1):25–38. https://doi.org/10.3233/BME-141244

Pati F, Adhikari B, Dhara S (2010) Isolation and characterization of fish scale collagen of higher thermal stability. Biores Technol 101:3737–3742. https://doi.org/10.1016/j.biortech.2009.12.133

Paul S, Pal A, Choudhury AR et al (2017) Effect of trace elements on the sintering effect of fish scale derived hydroxyapatite and its bioactivity. Ceram Int 43:15678–15684. https://doi.org/10.1016/j.ceramint.2017.08.127

Phillips VD, Kinoshita CM, Neill DR et al (1990) Thermochemical production of methanol from biomass in Hawaii. Appl Energy 35:167–175. https://doi.org/10.1016/0306-2619(90)90066-M

Pickler E, Filho V (2017) Evolução da piscicultura no Brasil: Diagnóstico e desenvolvimento da cadeia produtiva de tilápia. Livraria Ipea, Brasilia

Prado JPS et al (2021) *In vitro* and *in vivo* biological performance of hydroxyapatite from fish waste. J Mater Sci: Mater Med (Springer US) 32(9). https://doi.org/10.1007/s10856-021-06591-x

Putri DKT et al (2022) Activity test of chitosan haruan (*Channa striata*) fish scales as antibiofilm agent againts biofilm of Porphyromonas gingivalis. IOP Conf Ser: Earth Environ Sci 976(1). https://doi.org/10.1088/1755-1315/976/1/012014.

Qin D et al (2022) Development and application of fish scale wastes as versatile natural biomaterials. Chem Eng J 428(January). https://doi.org/10.1016/j.cej.2021.131102

Racioppo A et al (2021) Fish loss/waste and low-value fish challenges: state of art, advances, and perspectives. Foods 10(11):2725. https://doi.org/10.3390/foods10112725

Rezaei M, Pourang N, Moradi AM (2022) Removal of lead from aqueous solutions using three biosorbents of aquatic origin with the emphasis on the affective factors. Sci Rep (Nature Publishing Group UK) 12(1):1–20. https://doi.org/10.1038/s41598-021-04744-0

Ribeiro C et al (2015) Characterization of *Oreochromis niloticus* fish scales and assessment of their potential on the adsorption of reactive blue 5G dye. Colloids Surf A 482:693–701. https://doi.org/10.1016/j.colsurfa.2015.05.057

Ribeiro C, Scheufele FB, Alves HJ et al (2019) Evaluation of hybrid neutralization/biosorption process for zinc ions removal from automotive battery effluent by dolomite and fish scales. Environ Technol 40:2373–2388. https://doi.org/10.1080/09593330.2018.1441332

Roy Y et al (2014) Biomass combustion for greenhouse carbon dioxide enrichment. Biomass Bioenergy 66(1):186–196. https://doi.org/10.1016/j.biombioe.2014.03.001

Salvatore L et al (2020) Marine collagen and its derivatives: versatile and sustainable bio-resources for healthcare. Mater Sci Eng C (Elsevier) 113(December 2019):110963. https://doi.org/10.1016/j.msec.2020.110963

San Miguel G, Domínguez MP, Hernández M et al (2012) Characterization and potential applications of solid particles produced at a biomass gasification plant. Biomass Bioenergy 47:134–144. https://doi.org/10.1016/j.biombioe.2012.09.049

Santana CA, Piccirillo C, Pereira SIA et al (2019) Employment of phosphate solubilizing bacteria on fish scales—turning food waste into an available phosphorus source. J Environ Chem Eng 7 (October). https://doi.org/10.1016/j.jece.2019.103403

Santos EB, Vieira EFS, Cestari AR et al (2009) Characterization of the piau fish (*Leporinus elongatus*) scales and their applicatin to remove Cu (II) from aqueous solutions. Quim Nova 32:132–138. https://doi.org/10.1590/S0100-40422009000100026

Santos JRA et al (2020) Green adhesives for biomedical applications. Green Adhes 85–120. https://doi.org/10.1002/9781119655053.ch4

Sarkar JK, Wang Q (2020) Different pyrolysis process conditions of South Asian waste coconut shell and characterization of gas, bio-char, and bio-oil. Energies 13(8). https://doi.org/10.3390/en13081970

Sartor K, Restivo Y, Ngendakumana P et al (2014) Prediction of SO_x and NO_x emissions from a medium size biomass boiler. Biomass Bioenergy 65:91–100

Sathiskumar S et al (2019) Green synthesis of biocompatible nanostructured hydroxyapatite from *Cirrhinus mrigala* fish scale—a biowaste to biomaterial. Ceram Int (Elsevier Ltd and Techna Group S.r.l.) 45(6):7804–7810. https://doi.org/10.1016/j.ceramint.2019.01.086

Satpathy AA et al (2021) Functional and bioactive properties of chitosan from Indian major carp scale. Aquac Int 29(2):417–430. https://doi.org/10.1007/s10499-020-00622-0

Senneca O, Chirone R, Cortese L et al (2020) Pyrolysis and combustion of a solid refinery waste. Fuel 267(May). https://doi.org/10.1016/j.fuel.2020.117258

Sghayyar HNM et al (2019) Fish biowaste gelatin coated phosphate-glass fibres for wound-healing application. Eur Polym J (Elsevier) (October):109386. https://doi.org/10.1016/j.eurpolymj.2019.109386

Shalaby M et al (2020) Fish scale collagen preparation, characterization and its application in wound healing. J Polym Environ (Springer, US) 28(1):166–178. https://doi.org/10.1007/s10924-019-01594-w

Sherman VR, Yang W, Meyers MA (2015) The materials science of collagen. J Mech Behav Biomed Mater (Elsevier) 52:22–50. https://doi.org/10.1016/j.jmbbm.2015.05.023

Shi P et al (2018) Characterization of natural hydroxyapatite originated from fish bone and its biocompatibility with osteoblasts. Mater Sci Eng C (Elsevier B.V.). https://doi.org/10.1016/j.msec.2018.04.026

Silva AVS, Torquato LDM, Cruz G (2019) Potential application of fish scales as feedstock in thermochemical processes for the clean energy generation. Waste Manag 100:91–100. https://doi.org/10.1016/j.wasman.2019.09.007

Silva AVS, Mortari DA, Pereira F et al (2021) Assessment of the fish scales and coal blends combustion in a Drop Tube Furnace (DTF). ABCM https://doi.org/10.26678/ABCM.COBEM2021.COB2021-1040

Silva AVS, Mortari D, Conconi C et al (2022) Investigation of the combustion process of fish scales from Northeast Brazil in a Drop Tube Furnace (DTF). Environ Sci Pollut Res. https://doi.org/10.1007/s11356-022-20643-x

Sockalingam K, Abdullah HZ (2015) Extraction and characterization of gelatin biopolymer from black tilapia (*Oreochromis mossambicus*) scales. AIP Conf Proc 10(1063/1):4919191

Spliethoff H, Hein KRG (1998) Effect of co-combustion of biomass on emissions in pulverized fuel furnaces. Fuel Process Technol 54:189–205. https://doi.org/10.1016/S0378-3820(97)00069-6

Srinivasan S, Durairaj B (2021) Collagen isolation and characterization from *Sardinella longiceps*. J Adv Vet Animal Res 8(4):679–686. https://doi.org/10.5455/javar.2021.h560

Ssekatawa K et al (2021) Isolation and characterization of chitosan from Ugandan edible mushrooms, Nile perch scales and banana weevils for biomedical applications. Sci Rep (Nature Publishing Group UK) 11(1):1–14. https://doi.org/10.1038/s41598-021-81880-7

Su W et al (2020) Alginate-assisted mineralization of collagen by collagen reconstitution and calcium phosphate formation. ACS Biomater Sci Eng 6(6):3275–3286. https://doi.org/10.1021/acsbiomaterials.9b01841

Surya P et al (2021) Synthesis and characterization of nano-hydroxyapatite from *Sardinella longiceps* fish bone and its effects on human osteoblast bone cells. J Mech Behav Biomed Mater (Elsevier Ltd.) 119(March):104501. https://doi.org/10.1016/j.jmbbm.2021.104501

Szychlinska MA et al (2020) Evaluation of a cell-free collagen type I-based scaffold for articular cartilage regeneration in an orthotopic rat model. Materials 13:2369. https://doi.org/10.3390/ma13102369

Tan Z (2014) Air pollution and greenhouse gases. Springer, Canada

Tang J, Saito T (2015) Biocompatibility of novel type I collagen purified from tilapia fish scale: an *in vitro* comparative study. Biomed Res Int 2015:1–8. https://doi.org/10.1155/2015/139476

Teshale F, Karthikeyan R, Sahu O (2020) Synthesized bioadsorbent from fish scale for chromium (III) removal. Micron (Elsevier Ltd.) 130(December 2019):102817. https://doi.org/10.1016/j.micron.2019.102817

Torres FG, Troncoso OP, Amaya A (2012) The effect of water on the thermal transitions of fish scales from Arapaima Gigas. Mater Sci Eng 32:2212–2214. https://doi.org/10.1016/j.msec.2012.06.003

Uskokovic V, Ignjatovic N, Petranovic N (2003) Synthesis and characterization of hydroxyapatite-collagen biocomposite materials. Mater Sci Forum 413:269–274. https://doi.org/10.4028/www.scientific.net/MSF.413.269

Wang JK et al (2017) Fish scale-derived collagen patch promotes growth of blood and lymphatic vessels in vivo. Acta Biomaterialia (Acta Materialia Inc.) 63:246–260. https://doi.org/10.1016/j.actbio.2017.09.001

Wielgosinski G, Lechtanska P, Namiecinska O (2017) Emissions of some pollutants from biomass combustion in comparison to hard coal combustion. J Energy Inst 90(5):787–796. https://doi.org/10.1016/j.joei.2016.06.005

Williams A, Jones JM, Ma l et al (2012) Pollutants from the combustion of solid biomass fuels. Prog Energy Combust Sci 38:113–137. https://doi.org/10.1016/j.pecs.2011.10.001

Wu D, Wang S, Wu C (2020) Antibacterial properties and cytocompatibility of biobased nanofibers of fish scale gelatine, modified polylactide, and freshwater clam shell. Int J Biol Macromol (Elsevier B.V.) 165(Part A):1219–1228. https://doi.org/10.1016/j.ijbiomac.2020.10.002

Wu CS et al (2021) Novel composite 3D-printed filament made from fish scale-derived hydroxyapatite, eggshell and polylactic acid via a fused fabrication approach. Addit Manuf 46(March). https://doi.org/10.1016/j.addma.2021.102169

Xu F, Shi Y, Wang D (2013) X-Ray scattering studies of lignocellulosic biomass: a review. Carbohyd Polym 94:904–917. https://doi.org/10.1016/j.carbpol.2013.02.008

Yahyaee R, Ghobadian B, Najafi G (2013) Waste fish oil biodiesel as a source of renewable fuel in Iran. Renew Sustain Energy Rev 17:312–319. https://doi.org/10.1016/j.rser.2012.09.025

Yang B, Peng L, Wang Y et al (2018) The characteristics of air pollutants from the combustion of biomass pellets. Energy Sources Part A Recovery Util Environ Effects 40:351–357. https://doi.org/10.1080/15567036.2017.1419515

Yao BY, Changkook R, Adela K et al (2005) Effect of fuel properties on biomass combustion. Part II. Modelling approach identification of the controlling factors. Fuel 84(16):2116–2130. https://doi.org/10.1016/j.fuel.2005.04.023

Zainol I et al (2019) Extraction of natural hydroxyapatite from tilapia fish scales using alkaline treatment. Mater Today: Proc (Elsevier Ltd.) 16:1942–1948. https://doi.org/10.1016/j.matpr.2019.06.072

Zellagui S, Trouvé G, Schonnenbeck C et al (2017) Parametric study on the particulate matter emissions during solid fuel combustion in a drop tube furnace. Fuel 189:358–368. https://doi.org/10.1016/j.fuel.2016.10.104

Zhao BT, Su YX, Liu DY et al (2016) SO_2/NO_x emissions and ash formation from algae biomass combustion: process characteristics and mechanisms. Energy 113:821–830. https://doi.org/10.1016/j.energy.2016.07.107

Zheng S, Hu Y, Wang Z et al (2020) Experimental investigation on ignition and burnout characteristics of semi-coke and bituminous coal blends. J Energy Inst 93:1371–1381. https://doi.org/10.1016/j.joei.2019.12.007

Zhou H, Long YQ, Meng AH et al (2015) Classification of municipal solid waste components for thermal conversion in waste-to-energy research. Fuel 145:151–157. https://doi.org/10.1016/j.fuel.2014.12.015

Potential of Fish Waste for the Production of Sustainable Biogas and Biodiesel

Mehmood Ali, Faraz Sufyan, and Abdul Majeed Khan

Abstract The usage of fossil fuels in the production of energy and transportation sector is a contributing factor to global warming. In recent years, it has become clear that the use of biofuels like biodiesel, bioethanol, biogas, and other syn-fuels can help to mitigate climate change and global warming by reducing the usage of fossil fuels made from petroleum crude oil. The fish processing bio-waste fundamentally contaminates marine or aquatic ecosystems by reducing pH, temperature, and dissolved oxygen levels in water bodies. Fish waste can therefore be used in a variety of ways, such as the production of biofuels that can improve socio-economic stability, contribute to energy security, and increase resilience to environmental threats brought on by climate change by reducing GHG emissions. Gaseous fuel such as biogas is mainly composed of methane gas used for heating, cooking purposes, and the production of electricity through internal combustion engines, while biodiesel is a liquid biofuel used in diesel engines. An overview of fish waste produced by the aquaculture and fishing processing sectors is covered in this chapter, along with information on environmental issues related to the introduction of fishing waste into aquatic and marine environments. Additionally, it discusses the need of manufacturing biodiesel and biogas from fish waste and byproducts for long-term growth in the renewable energy sector and sustainable development. The characterization, engine performance, and emission characteristics of the biodiesel and biogas produced from fish waste are incorporated in the chapter's concluding section.

Keywords Fish waste · Biofuels · Biodiesel · Biogas · Transesterification · Anaerobic digestion · Biomethane

M. Ali (✉) · F. Sufyan
Department of Environmental Engineering, NED University of Engineering and Technology, Karachi 75270, Pakistan
e-mail: mehmood@neduet.edu.pk

F. Sufyan
Department of Petroleum Technology, University of Karachi, Karachi - 75270, Pakistan

A. M. Khan
Research Laboratory of Bioenergy (RLB), Department of Chemistry, Faculty of Science, Federal Urdu University of Arts, Science and Technology, Karachi 75300, Pakistan

S. Maqsood et al. (eds.), *Fish Waste to Valuable Products*, Sustainable Materials and Technology, https://doi.org/10.1007/978-981-99-8593-7_19

1 Introduction to Aquaculture and Fish Industry Waste Environmental Impacts

Aquaculture, also known as fish farming has grabbed huge attention in the recent years. The increase in population has led to an increase in demand of fish catch, which urged the need to improving the aquaculture production for food production (Dauda et al. 2019). Due to the increased aquaculture activities caused by the increase in fish demand, aquafarming may be necessary to fill the gap between supply and demand for food, particularly fish food (Choudhury et al. 2022). Therefore, there has been huge competition in terms of developing aquaculture infrastructure which includes land, natural resources, and water. It contributes to the global economy by creating more number of employment opportunities (Ahmad et al. 2022). The aquaculture does not include any food supplement that would cause harm to the environment. The waste can be reduced through appropriate feed and feeding methods during the production cycle. The surge in marine pollution loading and its biological-related changes with marine fish farming has caused greater concern over the protection coastal environment. One way of reducing fish waste is to create a market for byproducts. The utilization of byproducts would help generate additional revenue stream, cut down operating cost, and protect marine environment. The main methods that are used for the treatment of fish wastes include bioremediation, hydrolysis, anaerobic digestion, and transesterification. The fish production through aquaculture in 2018 was 0.178 billion tons, and as per UN Sustainable Development Goals (SDG) 2030, the usage of fish in food and nutrition holds huge significance (Thirukumaran et al. 2022). Fish waste used to be anaerobically digested, composted, and incinerated in the past, once this process was finished, the leftover waste was subsequently dumped into the sea, endangering the ecology (Wan Mahari et al. 2022).

The waste that is generated from fish farming if not being treated properly before its final disposal, then it would harm the surrounding land as well as coastal area. This has led the importance of developing the aquaculture and its waste disposal methods to protect the natural habitat as well as marine environment (Jayasinghe and Hawboldt 2012). Fish trap is the discarded part that comprises skin, trimmings, fins, bones, viscera, tails, and internal organs, with disposal of leftover pieces seriously harm the marine environment. Typically, for aquaculture and animal feed, fishmeal is processed through fish waste (Azad et al. 2019a). Organic waste obtained from fish farms can have a significant negative impact on sediment biogeochemistry because it stimulates sediment metabolism and results in reduced redox potential and increased oxygen consumption rates. In such circumstances, the pathways of anaerobic metabolism are strengthened, increasing the quantities of their harmful byproducts, like sulphides and methane, in the sediment. The macrofauna that has a significance in benthic metabolism via bio irrigation and bioturbation, is adversely affected by these conditions, as are benthic populations as a whole. Unconsolidated marine sediments are essential for recycling carbon and nutrients, while the crucial ecosystem function can be jeopardized through organic matter pollution. This function replaces ecosystem services like sustaining primary production, which benefits the environment by preventing climate change and supplying food (Sanz-Lazaro et al. 2021).

2 Utilization of Fish Waste and Its Byproducts to Make Biofuels

Fish processing plants generate billions of tonnes of fish waste each year. Fish wastes when dumped in the land, produce GHG's emissions that damage the environment. Therefore, it is much better to make use of waste and convert it into valuable byproducts, such as bio-fertilizer, biofuel, fodder, and other pharmaceutical components rather than disposing them off in the land or sea. The promising product from fish waste is biofuel, i.e., alternative to fossil fuels for its use in energy production and transportation. Biofuels can be produced from organic waste, seaweed, and microalgae via different thermochemical, biochemical, chemical, and physical extraction processes (Jayasinghe and Hawboldt 2012). The fish wastes for biodiesel production are motivated primarily by their non-toxic and biodegradable nature. Biodiesel emits fewer air toxins, CO_2, hydrocarbons, and other particulates as compared to mineral diesel. The fish oil extracted from fish wastes obtained from fish industry, is cleaned and purified, then through a chemical process using a catalyst in the transesterification and anaerobic digestion processes, the final products, i.e., biodiesel and biogas are produced respectively. The results from different tests such as gas chromatography-mass spectrometry (GCMS), Fourier transform nuclear magnetic resonance (FT-NMR), best management practices (BMP) assay, and proton nuclear magnetic resonance (H-NMR) analysis indicate that biofuels are efficiently converted and can be used in transportation and energy generation sectors (Yuvaraj et al. 2019). In particular, the development of alternative fuel from renewable sources has attracted a lot of attention due to Government institutions progressively encouraging the use of biofuels made from vegetable and animal oils. These initiatives will assist in introducing biodiesel to the market through required blends with regular diesel accessible at filling stations. The residues from tilapia processing plants for biodiesel production; thus, producing biodiesel from fish waste if properly disposed may benefit the environment by reducing residues that are harmful to the ecosystem (Martins et al. 2015). The utilization of industrial fish processing waste is the sole substrate that favors the biogas effectively, this was revealed in the past research study based on two different types of wastes examined in order to determine the link between the effectiveness of microorganism's activity for its conversion into biogas. Anaerobic digestion was used to produce biogas at a temperature of 35 °C, which showed biogas yield was higher with fish waste (FW) as compared to fish crude oil waste (FCOW). The produced methane yield was 426.3 CH_4 mL/g of VS from FW and 540.5 CH_4 mL/g of VS, from FCOW respectively. This suggested that fish wastes are more suitable sources in the mono-digestion process for biogas production (Bücker et al. 2020). According to a previous study, treated fish waste has numerous applications, such as dietic products (chitosan), biogas/biodiesel, food-packaging applications (chitosan), natural pigments, animal feed, enzyme isolation, cosmetics (collagen), soil fertilizer, moisture maintenance in foods (hydrolysates), and Cr immobilization. The burning of biogas can provide thermal or mechanical energy that can be derived if used as a fuel in cogeneration plant (Arvanitoyannis 2008). The biofuel potential of fish waste is determined by the characteristics of fish waste, type of fuel, and the size and location of the processing plant. The on-site processing of fish wastes can provide both the economic and environmental benefits in those areas where fish wastes are generated in abundance. This will also reduce the cost of transportation if processing plants are not far from the site. The effluent of fish waste was found to have heterogeneous and varied lipid content depending upon the fish type, processing techniques, harvesting season, etc. Furthermore, the waste has low oxidation and thermal stability, because due to the presence of active enzymes, and the high polyunsaturated fatty acid content, necessitating handling during transportation and storage. The recovery of the oil from the waste is the main barrier to the viability of using fish oil as a feedstock for producing biofuel. In existing combustors (boilers, furnace) and diesel engines, crude fish oils have been used in blending with No. 2 diesel and No.2/No.6 fuel oil as a partial and/or full replacement in existing stationary diesel engines and combustors (furnaces, boilers). It resulted in major benefits including cost savings and reduced emission, while other properties such as cold flow properties, density, viscosity, and specific fuel consumption have been enhanced with 50% blended with petroleum fuel. If fish oil is processed further removing proteins, sediments, and impurities using filtration technique, it will have long-term applicability for engines and boilers. Furthermore, through transesterification or other processes, the quality of refined fish biofuels can be improved significantly (Jayasinghe and Hawboldt 2012). Figure 1 depicts the two conversion routes of fish wastes into sustainable biofuels via triglyceride transesterification into biodiesel for use in transportation and power generation. While another route demonstrates the conversion of organic waste containing carbohydrates, fats, and proteins into biogas for heating, cooking, and power generation via anaerobic digestion.

According to a report by the Food and Agricultural Organization (FAO), 177.8 million tonnes of fisheries and aquaculture products were produced globally in 2019. However, between 50 and 70% of the fisheries and aquaculture products produced globally are considered waste and are dumped either at sea or in landfills (Khiari 2021). Table 1, presents the prospects of converting global fisheries and aquaculture to biomethane and biodiesel per year was estimated, with assuming 50% fisheries waste generated of total 177.8 MT /year comes to 88.9 MT/year. The lipid's extracted yield was 15.388 MT/year with17.31% by weight as mentioned in literature (Azad et al. 2019b), now assuming 90% conversion of extracted lipids into biodiesel through transesterification reaction yields 13.849 MT/year. Now, biomethane production has been calculated based on the amount of fisheries waste 88.9 MT per year contains 79.121×10^6 MT of volatile solids (VS) and that the yield of biomethane from fish waste is 426.3 mL/g of VS (Bücker et al. 2020), the total amount of biomethane production is found to be 33.729×10^9 (m^3 CH_4/year).

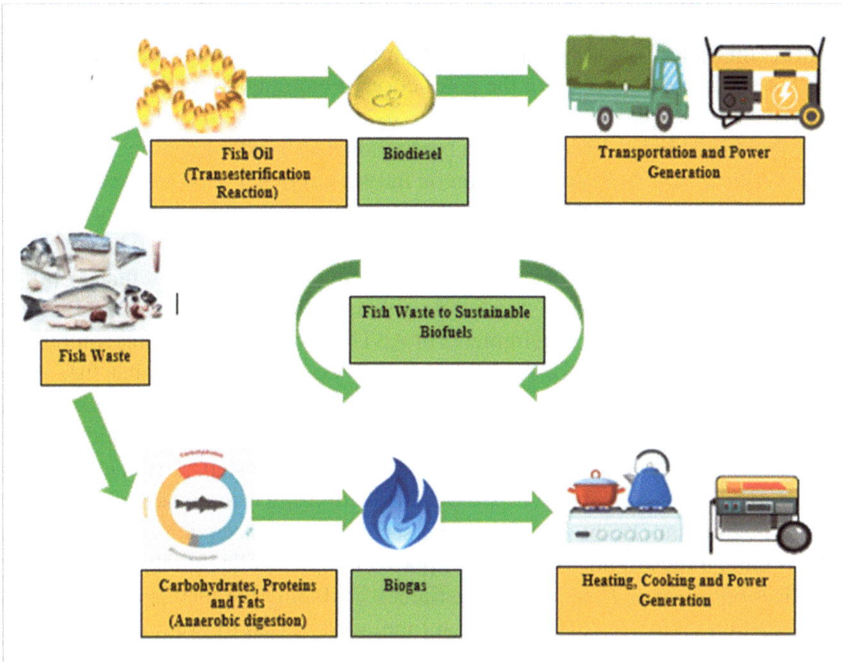

Fig. 1 Conversion routes of sustainable biogas and biodiesel production from fish waste

Table 1 Potential of global fisheries and aquaculture to produce biomethane and biodiesel per year (authors unpublished data)

Biodiesel		Biomethane	
Fisheries and aquaculture production (MT/year)	177.8	Fisheries and aquaculture production (MT/year)	177.8
Fisheries and aquaculture waste generated (MT/year)	88.9	Fisheries and aquaculture waste generated (MT/year)	88.9
Lipid yield from fisheries and aquaculture waste (MT/year)	15.388	Volatile content (VS) in fisheries and aquaculture waste (MT/year)	79.121×10^6
Biodiesel production (MT/year)	13.849	Biomethane production (m^3 CH_4/year)	$\times 10^9$

3 Production of Biodiesel and Characterization of Fish Waste

The alternative fuels are getting more importance due to increasing fossil fuel prices due to dwindling petroleum crude oil reserves and its adverse impact on the environment contributing greenhouse gases (GHGs). Biodiesel is made from microalgae, seaweed, and organic waste, and has attracted a lot of interest in recent years as an alternate fuel to mineral diesel. As a result, it is regarded as a fuel that produces less harmful emissions (Yuvaraj et al. 2019). There has been limited research conducted on fish waste as raw material for the production of biofuels, which has a potential to meet the energy requirements for the future using various valuable byproducts such as biodiesel and biogas. Fish waste has been studied for biodiesel production with oil extraction by water as solvent and then its fatty acid content has been reduced with its saponification with sodium hydroxide. The optimized experimental parameters, such as reaction temperature and catalyst concentration, were found to be 60 °C and 0.975 g respectively. The transesterification reaction was carried out under optimal conditions, yielding up to 92% biodiesel by weight. The produced biodiesel exhibited properties that met the requirements of ASTM D 6751 standard and combustion engines (Keshri et al. 2018). Another research work investigated that fish waste carrying high fatty acids potential as feedstock for biodiesel production based on cost-effectiveness. This requires two-step transesterification process which is used to convert high fatty acids fish oil into biodiesel. In the first step, fatty acid of fish oil is reduced through acid catalyzed-based esterification and it continues until fatty acid content is less than 2% by weight. In the second step, conversion of esterification process into biodiesel and glycerol takes place through alkaline catalyst-based transesterification process. Three process variables, including the methanol to oil molar ratio, reaction temperature, and reaction time, were varied to examine their effects on product yield. The maximum fatty acid methyl ester (FAME) conversion yield was found to occur at methanol to oil molar ratios of 15:1, reaction temperatures of 60 °C, and reaction times of 180 min, with FAME yields of 80.16%, 80.03%, and 80.39%, respectively (Samat, et al. 2018).

The production of ethylic biodiesel (EBD) and methylic/ethylic biodiesel (MEBD) through KOH-catalyzed transesterification reactions with ethanol and mixed methanol/ethanol, respectively, was also studied using a mixture of non-conventional feedstocks (40% castor bean oil+ 30% bitter almond oil, and 30% waste fish oil w/w). The effects of alcoholysis factors were investigated, including alcohol/oil molar ratio, reaction time, reaction temperature, and KOH quantity used. The results of the alcoholysis demonstrated that the type of alcohol influences the best conversion of the blended oils into biodiesel. The optimum FAME yield of 97.50% was obtained at 1% by weight of KOH, ethanol/oils (7:1) blend molar ratio, 65 °C, and a reaction duration of 60 min. The maximum yield of FAME 96.33% was attained at 0.75 by weight of KOH, alcohol/oils (6:1) blend molar ratio, and 45 min reaction time (Sedeeq et al. 2019).

A study was carried out to characterize the final product and optimize the enzymatic biodiesel production from fish waste oil using a composite central design and response surface approach. As a biocatalyst, *Thermomyces lanuginosus* lipase was immobilized on octadecyl metacrylate beads. 35 °C, 10% w/w biocatalyst concentration, and 216 rpm of agitation rate were the ideal conditions. After 24 h of reaction time, an experimental biodiesel yield of 75.3% was achieved under ideal conditions. While other metrics, such as density (0.89 ± 0.01 g/mL), viscosity (5.3 ± 0.004 mm^2/s), calorific value (38.1 ± 0.21 MJ/kg), and cloud point (10.5 ± 0.47 °C), agreed with ASTM D 6751 standard, the biodiesel had an acid value (0.9 ± 0.28 mg KOH/g) that was greater than the specified limitations (Ching-Velasquez et al. 2020).

A past study looked into biodiesel production based on the composition of fatty acids. The fish waste sample was drained of extra water and completely dried in an oven at 55–60 °C. Using the Soxhlet extraction process, crude oils were extracted in petroleum ether. The crude fish oil that was extracted underwent methylation to produce fatty acid methyl esters (FAME). The GCMS system was utilized for analyzing FAME, and the NIST library was consulted in order to identify the fatty acids that were present. There were in total 21 fatty acids, of which 53.53% were saturated fatty acids (SFA), 22.1% were monounsaturated fatty acids (MUFA), and 24.37% were polyunsaturated fatty acids (PUFA). The significant fatty acids contained in fish oil, including myristic acid (C14:0), palmitic acid (C16:0), palmitoleic acid (C16:1), oleic acid (C18:1), linoleic acid (C18:2), linolenic acid (C18:3), docosapentaenoic acid (C22:5), and docosahexaenoic acid (C22). The biodiesel made from fish waste has these characteristics because it has a greater viscosity, cetane number, and density at the greatest percentage of SFA. As a result, the transesterification process of the fish waste has a significant potential to make biodiesel from the fatty acids in FAME (Azad et al. 2019b).

In order to produce biodiesel from Rohu processing waste (RPW), its oil was examined for its characterization and optimization requirements. The wet reduction method generated the highest oil yield (95.2%) by volume when compared to other solvent extraction techniques. The physical and chemical characteristics of the oils were investigated prior to transesterification reaction to produce biodiesel. For optimal biodiesel conversion of oil, different parameters such as temperature, time, catalysts, and alcohols were considered. The highest biodiesel yield (95.9%) was obtained with 0.75% KOH as the catalyst, reaction time was found to be 60 min, while reaction temperature was 55 °C, with 1:0.5 (v/v) oil to methanol ratio. The triglycerides in RPW oil were converted into biodiesel, according to laboratory analyses, palmitic acid methyl ester and α-linolenic acid methyl ester being the two most dominant esters found in the final product. Moreover, it was noted that biodiesel had a melting point temperature range of 52.56–4.11 °C, indicating good cold flow properties and having enhanced oxidative stability. These results indicate the possibility of fish waste as a feedstock for making biodiesel (Kudre et al. 2017).

Previous literature showed that biodiesel produced from fish (Cyprinus carpio) waste can be considered as a potential source of biodiesel. The finished biodiesel fuel was further gone through purification, washing, and drying processes. In order to eliminate contaminants such as colors, excess alcohol, excess catalysts, glycerol, and soaps, the washing phase uses biodiesel in a 1:1 ratio with warm distilled water for at least four times. The drying stage is completed by utilizing sodium sulphate anhydrous and drying the biodiesels in an oven set to 48 °C. After being dried and filtered, biodiesel was characterized using GC–MS analysis, which revealed seven peaks that represented all major fatty acid methyl esters and 100% of the triglycerides that had been methylated. The chain length of the fatty acids present in Cyprinus carpio ranged from 12 to 18 carbon with 18C fatty acids as the major fatty acids of 51.94%. Characterization studies on total acid number, total base number, kinematic viscosity, and multi-element analysis showed the quality of biodiesel exhibiting good fuel properties comparable to mineral diesel met with ASTM D6751 standard except for the phosphorus content and total base number value (Saifuddin and Boyle 2017).

4 Engine Performance and Emissions from Fish Waste Biodiesel

The seven peaks in the biodiesel made from fish (Cyprinus carpio) waste are fatty acid methyl esters, proving that all of the triglycerides were successfully converted to methyl esters. Under engine loads of 15 (Nm), fish-based biodiesel significantly reduced carbon monoxide (CO) and hydrocarbon (HC) emissions without modifying the engine. The produced biodiesel's kinematic viscosity was within the acceptable range for international standard (ASTM D6751). As compared to conventional diesel, the biodiesel was shown to have a low base number and a reduced specific fuel usage. The performance of engine using biodiesel at 2000 rpm with a load of 15 Nm, observed that specific fuel consumption (SFC) of 0.581 mL/s and 0.588 mL/s from biodiesel and diesel respectively. While fish-based biodiesel has a higher calorific value, having a significant impact on biodiesel with lower SFC value. Moreover, conventional diesel produces more carbon monoxide (0.029% by vol) than biodiesel (0.027% by vol). This is due to the higher kinematic viscosity and lower cetane number, which result in poor auto-ignition and significant carbon monoxide emissions. Moreover, biodiesel emits (18 ppm) and produces less unburned hydrocarbon than regular diesel (28 ppm) during engine combustion. This happens because high cetane number causes auto-ignition and faster evaporation. In addition to this, biodiesel was found to release more nitrogen oxide emissions (525 ppm) as compared to conventional diesel (515.40 ppm), due to higher exhaust temperature during combustion (Saifuddin and Boyle 2017).

Research study conducted in the past investigated the performance of engine was also analyzed through the pyrolysis of waste plastic. Waste fish byproducts were used to make fish oil biodiesel, and fish oil methyl ester (FOME) was obtained through transesterification. B20FOME is created by combining 20% FOME with 80% mineral diesel, and B20FOME10WPO is created by combining 20% FOME with 10% fuel made from waste plastic. To carry out the various engine tests, the 10 HP single cylinder 4 stroke direct injection water-cooled diesel engine was run at 1500 rpm with a load that steadily increased from 0 to 100%. Blends of biodiesel observed an increase in brake thermal efficiency from 3.82 to 10.02%. Comparable results were observed for unburned emissions of hydrocarbon and carbon monoxide; however, NOx emissions and smoke opacity were found to be slightly higher in biodiesel in comparison to conventional diesel. According to the literature, the fuel produced by converting plastic trash is also a bright promise because it operates on the recycling principle and is crucial for both the environment and the economy

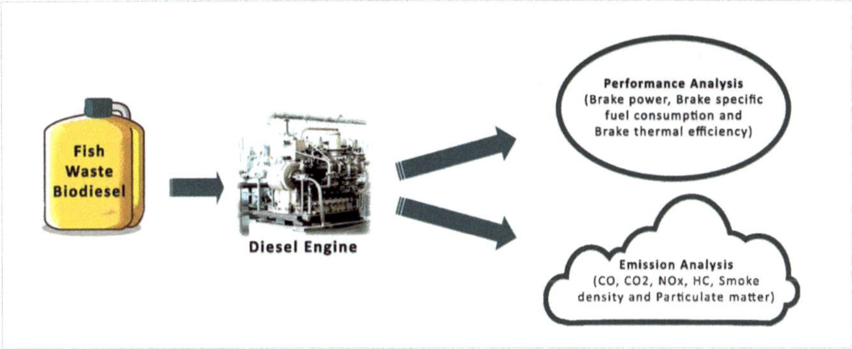

Fig. 2 Schematic representation of a fish waste biodiesel powered engine, its performance, and emission analysis

(Deepak and Manjunatha 2020). Experimental comparisons of the combustion properties, performance, and exhaust emissions of conventional diesel fuel with biodiesel produced from waste fish oil (WFO) and its blends (B25, B50, and B75) were made. The engine performance experiments were conducted using a single cylinder E6 Ricardo engine under steady state conditions at a range of engine loads. The results showed that waste fish oil biodiesel performs better than regular diesel fuel in terms of in-cylinder pressure and shorter heat release rate duration. Additionally, the use of biodiesel and its mixes could result in more consistent combustion with smaller cycle-to-cycle variations. In comparison to regular diesel, biodiesel has more thermal efficiency (2.92%) and with variable loads, the combustion loss is around 1.1% lower than that of diesel fuel. Biodiesel and its blends showed a declining trend in the amount of carbon monoxide emissions (5.2–27%), but unburned HC observed a large decline (11.6–70%). Due to more effective combustion produced by the oxygen component of the fuel, using biodiesel and its mixes causes increased CO_2 emissions (average increase of 7.2%) and NOx emissions (range of 1.9–12.8%) (Gharehghani et al. 2017). Figure 2, depicts the fish waste made biodiesel's operated engine, its performance, and emission analysis.

5 Biogas Production from Fish Waste

The combustible gas produced through renewable source such as biomass containing carbohydrates, fats, and proteins can be used as the fuel in the vehicle and energize the power plant. This renewable gas helps in building the circular economy of the country as it improves the utilization of biomass. The biogas generation helps to reduce the consumption of fossil fuels, in mitigating climate change, and ensure that everyone has access to cheap, green energy. A published report in 2020, by the International Energy Agency (IEA) highlighted the significance of untapped biogas produced through organic waste (Kasinath et al. 2021). Fish waste comprises liquid and solid waste that is rich in lipids and proteins, which helps to produce biogas. Fish waste also releases ammonia, long-chain fatty acids, sodium, calcium, magnesium, and potassium, which influences anaerobic decomposition. Fish waste can be combined with sludge, cow dung, water hyacinth, and sisal pulp as a co-digestate (Bücker et al. 2020). The co-digestion of fish waste and sisal pulp has been conducted as part of a comparative investigation. According to the compiled data, sisal pulp and fish waste were combined in mixture ratios of 50:50, 37:67, 25:75, and 20:80, respectively, to yield 0.31, 0.62, 0.48, and 0.44 (m^3/kg VS) of biogas (Mshandete et al. 2004). In addition to co-digestion of fish waste and cow manure has been conducted with varying ratios, with the highest yield of biogas being attained at a mixed ratio of 1:1.2 for the two wastes. Using the water displacement method, the gas was collected and measured in a displacement tank (Salam et al. 2009). The biogas is formed through anaerobic digestion (AD) because under this condition carbohydrates, proteins, and fats are converted into biogas through a favorable reaction at mesophilic temperature condition (35 °C). Depending upon the type of fish waste, the yield of biogas is increased in AD if inoculum is added. Furthermore, the high yield of biogas production indicates a high amount of hydrogen and carbon. During AD, fish waste releases enormous amount of ammonia, which impedes the digestion of substrates. Additionally, the high yield of biogas production suggests a high carbon and hydrogen content. Fish excrement releases a significant amount of ammonia during AD, preventing substrate digestion. Higher concentration of ammonia can lead to the growth of volatile fatty acids, and higher amounts of ammonia also inhibits the growth of methanogenic bacteria (Cadavid-Rodríguez et al. 2019). The anaerobic digestion process carried out in the absence of light, so the microorganisms used in this process such as methanogens and acetogens can have more favorable reaction conditions. The digestate, which is the output of AD, contains major proportion of methane, carbon dioxide, nitrogen, hydrogen, and hydrogen sulphide (Yuvaraj et al. 2019). The biochemical methane potential (BMP) test is carried out to determine the production of biogas, as this test involves digestion of substrate at constant temperature at varying time. Fish waste was used to produce biogas in a 1.3 L glass digester, with a reaction based on batch-scale setup. The reaction took place in a chamber with a regulated temperature that was set at 36.5 °C. A modified Gompertz equation was used to simulate the production of biogas. The variation between the model and the experiment using the modified Gompertz equation was about 16.5%. While Chen and Hashimoto's model was used to simulate the reaction's hydraulic retention time under mesophilic conditions. Also, it took about 52 days to produce one tonne of biogas. Over 90% of the biogas can be produced during this time. The best co-digestion for producing biogas was investigated using animal manure. Methane and biogas had potential of 554 and 757, mL/g VS, respectively. The fish waste was discovered to be a highly promising substrate to produce biogas by anaerobic digestion. The anaerobic co-digestion of the fish waste may be better explained by the modified Gompertz model. Fish waste using animal waste may be an effective method (Kafle and Kim 2012).

6 Possibility of Producing Bioenergy and Operating Internal Combustion Engine from Fish Waste Biogas

Waste products from the production and processing of fish constitute an important feedstock of materials for the generation of bioenergy. In Norway, fish is produced in abundance because there is so much hydropower available, and the bioenergy industry is now developing slowly. Over the past 45 years, the global per capita intake of fish has nearly doubled. Fish processing wastes may eventually be used to make biofuels considering the growth in the biofuel market. In order to achieve the desired specification, the oil obtained from fish waste needs to be converted into biodiesel before blending with diesel because fish wastes contain high fat content that is more favorable for biodiesel production. In Norway, biogas has sustainable usage for vehicle as a fuel, and it also can have preferred utilization in Sweden. While for other European and Non-European countries, biogas has the potential to be used as an alternative fuel to natural gas (Ward and Løes 2011). Since the massive usage of conventional fossil fuels has posed great threat to the environment concerns, and also the depletion of natural resources has highlighted the importance of renewable energy to meet the energy demand and also be used in internal combustion engines as fuel (Ramesha et al. 2015). Regardless of the engine operating conditions, a comparative analysis of biogas and natural gas on spark ignition engines. It was observed that natural gas fueled spark ignition engine has more power output than biogas fueled due to low calorific value of biogas. However, the usage of biogas enables significant reductions in exhaust nitrogen oxide (NO_x) emissions (Mustafi et al. 2006).

An attempt was made to adapt a 4-cylinder gasoline engine such that it could run on compressed, purified biogas instead of conventional fossil fuels. Pressured, unprocessed biogas was pumped into a transportable floating type gas holder with a volume capacity of 0.74 m^3. The purified biogas was compressed using a reciprocating compressor through a two-stage series of enrichment and moisture removal operations using activated alumina into the steel cylinder to improve the quality of the methane content. As storage for engine use, the enriched biogas was fed into the LPG tank for 20 min at 10 bars at an average concentration of 73.67% CH_4 with no traces of H_2S. LPG conversion kit installation and mounting were part of the modification. Utilizing an electro-dynamometer as a variable load, an examination of the engine's performance and combustion characteristics was conducted using pure compressed biogas and gasoline. The results demonstrated that a high percentage of CO_2 and other gas contaminants were the primary cause of the compressed biogas' declining power output. It also demonstrates that the thermal efficiency of biogas is lower than that of gasoline due to its lower calorific value. It indicates that by removing harmful gases from the mixture, the overall engine performance can be increased (Hernandez and Villanueva 2018). Research has been conducted to find out the utility of biogas on petrol-powered spark ignition engine. Also, the biofuel was blended with conventional fuel to further investigate the impact at increased compression ratio with varying load while at constant engine speed. The brake thermal efficiency and brake specification fuel consumption were reduced in biogas than conventional fuel. These parameters were obtained using 100% biogas with no blending with conventional fuel. On the contrary, other parameters, such as NO_x (48.96%), HC (63%), and CO (56.42%) emissions were on the lowest compared to conventional fuels. Furthermore, the combustion analysis showed that using biogas reduced peak pressures and changed the peak pressure's location by 2–3° crank angle. Finally, as compared to gasoline, the pressure was found to be attained at the lowest in 100% biogas, and that was around 24.69% reduction as compared to gasoline (Simsek and Uslu 2021).

7 Conclusion

Fishmeal, fish oil, dietetic products (chitosan), pharmaceuticals, animal feed, natural pigments, food-packaging applications (chitosan), cosmetics (collagen), enzyme isolation, and soil fertilizer are just a few examples of the value-added products made from fish waste and byproducts that have gained importance in recent years. Overall, the production of biogas and biodiesel from fish waste is covered in detail in this chapter. The chapter also discusses using fish waste-derived biogas and biodiesel for internal combustion engines, respectively, in addition to technical details on the manufacturing of fish waste biofuels and their characterization. When compared to mineral diesel, the performance and emission characteristics of internal combustion engines using fish waste biodiesel were determined to be more favorable. The waste materials left over after fishing can be used to make renewable biofuels (biogas and biodiesel), whilst also preventing the organic wastes from polluting the aquatic or marine environment. A potential alternative feedstock for the production of sustainable biofuels has been found as fish waste. The potential for reducing some unfavorable aspects of fossil fuel production and use, such as conventional and greenhouse gas (GHG) pollutant emissions, exhaustible resource depletion, and dependence on unreliable foreign suppliers, can be realized by substituting fossil fuels with biofuels produced from valorization of fish waste as a renewable organic material.

References

Ahmad AL et al (2022) Environmental impacts and imperative technologies towards sustainable treatment of aquaculture wastewater: a review. J Water Process Eng 46:102553

Arvanitoyannis IS, Kassaveti A (2008) Fish industry waste: treatments, environmental impacts, current and potential uses. Int J Food Sci Technol 43(4):726–745

Al Azad S, Wuen LY, Mohamad Lal MT (2019) Potential of fish wastes as feedstock for biodiesel. Adv Biosci Biotechnol 10(5):109–118

Al Azad S, Wuen LY, Mohamad Lal MT (2019) Potential of fish wastes as feedstock for biodiesel. Adv Biosci Biotechnol 10:109–118

Bücker F et al (2020) Fish waste: an efficient alternative to biogas and methane production in an anaerobic mono-digestion system. Renew Energy 147:798–805

Cadavid-Rodríguez LS, Vargas-Muñoz MA, Plácido J (2019) Biomethane from fish waste as a source of renewable energy for artisanal fishing communities. Sustain Energy Technol Assess 34:110–115

Ching-Velasquez J et al (2020) Production and characterization of biodiesel from oil of fish waste by enzymatic catalysis. Renew Energy 153:1346–1354

Choudhury A et al (2022) Anaerobic digestion challenges and resource recovery opportunities from land-based aquaculture waste and seafood processing byproducts: a review. Biores Technol 354:127144

Dauda AB et al (2019) Waste production in aquaculture: sources, components and managements in different culture systems. Aquac Fish 4(3):81–88

Deepak RK, Manjunatha D (2020) Performance, combustion and emission characteristics of B20 fish biodiesel blended with waste plastic oil on a diesel engine. In: International conference on thermo-fluids and energy systems. journal of physics: conference series, pp 1–13

Gharehghani A, Mirsalim M, Hosseini R (2017) Effects of waste fish oil biodiesel on diesel engine combustion characteristics and emission. Renew Energy 101:930–936

Hernandez NM, Villanueva EP (2018) Production, purification and utilization of biogas as fuel for internal combustion engine. In: AIP conference proceedings 2018. AIP Publishing

Jayasinghe P, Hawboldt K (2012) A review of bio-oils from waste biomass: focus on fish processing waste. Renew Sustain Energy Rev 16(1):798–821

Kafle GK, Kim SH (2012) Evaluation of the biogas productivity potential of fish waste: a lab scale batch study. J Biosyst Eng 37(5):302–313

Kasinath A et al (2021) Biomass in biogas production: pretreatment and codigestion. Renew Sustain Energy Rev 150:111509

Keshri A, Chowksey H, Girish CR (2018) Biodiesel production from fish waste and characterizing the properties. Int J Mech Prod Eng Res Dev 9(9):449–456

Khiari Z (2021) Sustainable upcycling of fisheries and aquaculture wastes using fish-derived cold-adapted proteases. Front Nurtition 9:1–10

Kudre TG, Bhaskar N, Sakhare PZ (2017) Optimization and characterization of biodiesel production from rohu (Labeo rohita) processing waste. Renew Energy 113:1408–1418

Martins GI et al (2015) Physical and chemical properties of fish oil biodiesel produced in Brazil. Renew Sustain Energy Rev 42:154–157

Mshandete A et al (2004) Anaerobic batch co-digestion of sisal pulp and fish wastes. Biores Technol 95(1):19–24

Mustafi NN, Raine RR, Bansal PK (2006) The use of biogas in internal combustion engines: a review. In: ASME 2006 internal combustion engine division spring technical conference

Ramesha DK et al (2015) Combustion, performance and emissions characteristics of a biogas fuelled diesel engine with fish biodiesel as pilot fuel. Biofuels 6(1–2):9–19

Saifuddin M, Boyle AN (2017) Biodiesel production from fish (Cyprinus carpio) waste and evaluation of engine performance. Sains Malays 46(10):1771–1778

Salam B, Islam M, Rahman MT (2009) Biogas from anaerobic digestion of fish waste. In: Proceedings of the international conference on mechanical engineering 2009. Dhaka, Bangladesh

Samat AF et al (2018) The potential of biodiesel production derived from fish waste. In: IOP conference series: materials science and engineering. Penang, Malaysia

Sanz-Lazaro C et al (2021) The environmental effect on the seabed of an offshore marine fish farm in the tropical Pacific. J Environ Manage 300:113712

Sedeeq SH, Al-Layla NMT, Fadhil AB (2019) Production, analysis, and evaluation of biodiesel from mixed castor seed oil, bitter almond oil, and waste fish oil

Simsek S, Uslu S, Simsek H (2021) Experimental study on the ability of different biogas level dual fuel spark ignition engine: Emiss Mitig Perform Combust Anal 76:74

Thirukumaran R et al (2022) Resource recovery from fish waste: prospects and the usage of intensified extraction technologies. Chemosphere 299:134361

Wan Mahari WA et al (2022) A state-of-the-art review on producing engineered biochar from shellfish waste and its application in aquaculture wastewater treatment. Chemosphere 288:132559

Ward AJ, Løes A-K (2011) The potential of fish and fish oil waste for bioenergy generation: Norway and beyond. Biofuels 2(4):375–387

Yuvaraj D et al (2019) Production of biofuels from fish wastes: an overview. Biofuels 10(3):301–307

Fish Waste: Understanding the Pollution Potential and Sustainable Mitigation Strategies

S. L. Rasmiya Begum, S. M. M. S. Himaya, M. S. M. Imthiyas, and S. M. M. S. Afreen

Abstract Fish production expansion around the world has increased waste output, which has become a stressor to the environment, despite meeting the expanding protein need and can be considered as the other side of a coin. Similar to other waste, fish waste has the potential to cause environmental pollution if not properly disposed of. This chapter provides baseline information in this regard, and the main inferences are as follows: (1) Fish waste, both solid and liquid, has the potential to cause all types of pollution, with water pollution receiving the most attention because fish waste and effluent from fish processing industries are commonly discarded into water systems (2) Fish waste pollution has drastic effects starts with water quality alteration and ends with jeopardizing biodiversity (3) Cleaner production, valorizing fish waste, raising awareness, and implementing rules and regulations are a few ways to reduce fish waste pollution, with the transformation of fish waste into marketable products being particularly attractive from an environmental and economic standpoint. Finally, while fish waste is a cause of pollution in the ecosystem, it may be reduced with the proper procedures.

Keywords Air pollution · Byproducts · Fish waste · Fish processing · Generation · Marine · Ocean dumping · Valuable products · Wastewater · Water quality

---

S. L. Rasmiya Begum (✉) · S. M. M. S. Himaya · S. M. M. S. Afreen
Department of Biosystems Technology, Faculty of Technology, South Eastern University of Sri Lanka, Oluvil, Sri Lanka
e-mail: imara@seu.ac.lk

S. M. M. S. Himaya
e-mail: himaya1989@gmail.com

S. M. M. S. Afreen
e-mail: Afreen0899@gmail.com

M. S. M. Imthiyas
Sri Lanka School of Agriculture, Palamunai, Sri Lanka
e-mail: masmy.msm@gmail.com

S. Maqsood et al. (eds.), *Fish Waste to Valuable Products*, Sustainable Materials and Technology, https://doi.org/10.1007/978-981-99-8593-7_20

1 Introduction

Due to the healthy lifestyle, fish product consumption has dramatically increased in recent years. Fisheries and aquaculture productivity have expanded significantly between 1954 and 2014, according to the Food and Agricultural Organization (FAO), because of rapid advances in aquaculture and rapid innovations in fishing technology in the United States (Coppola et al. 2021). to meet the growing demand of the increasing world population. In 2020, the world's total production of fisheries and aquaculture was 177.8 million tons out of which 157.4 million tons were used for human consumption (Fao 2022). A certain amount of fish is routinely lost during both capture and aquaculture, with the total amount expected to be 30 million tons in 2018, accounting for 23% of global catches (Guillen et al. 2018). Fish discards can make up 20 to 70 percent of the total weight of the fish, depending on the species, product, and processing methods (Jaiswal et al. 2014). In the fishing industry, "discards" refers to any organic material of animal origin in the catch that is dumped at water bodies, notably the sea. Discarding was deemed a severe problem by the FAO, and it was advised to reduce its code of conduct for ethical fishing (Guillen et al. 2018). Beyond this, intensified processing of caught fish into various products also could discard a considerable amount of waste which has little to no economic value (Dauda et al. 2019). Yorio and Caille as cited by Islam, Khan and Tanaka (2004b) stated that waste generated from processing plants was 58% of the total landings from 1989 to 2001 in coastal cities of Argentina.

When fish waste is improperly disposed of, it can endanger the environment and have an adverse effect on a larger coastal zone on a variety of ecological levels (Fig. 1). The amount of fish waste produced globally has increased dramatically; it is estimated that over two-thirds of all fish is discarded, which poses significant economic and environmental problems (Caruso et al. 2020). Small, non-food fish known as marine trash fishes are frequently abandoned in coastal environments, landfills, and neighbourhood fish markets, attracting flies and insects, emitting harmful fumes, giving off disagreeable odours at the dumps, and contaminating the soil and groundwater as a result of organic waste decomposing (Gayathri et al. 2018). Additionally, the effluent from the processing plants also causes adverse changes to the environment due to the presence of high levels of organic chemicals including salt, nitrogen, oil and fat, and the biochemical and chemical oxygen demands (BOD and COD). The release of such organic wastes affects the biodiversity and community structure of benthic assemblages (Arvanitoyannis and Kassaveti 2008). Since liquid waste from the fish processing industry is swiftly destroyed by microbes, it has the potential to be hazardous to both the environment and human health (Kurniasih et al. 2018). Consequently, the public's concern has increased regarding fish farming's detrimental effects thus necessitating that waste management must be recognized by the fish processing industry to reduce pollution whereas the industry's recovery of usable byproducts from fish wastes is known as a popular waste reduction approach because of their nutritive values. Fish waste has a higher concentration of n-3 polyunsaturated fatty acids than other organic waste products (PUFAs) (Barroso et al. 2019).

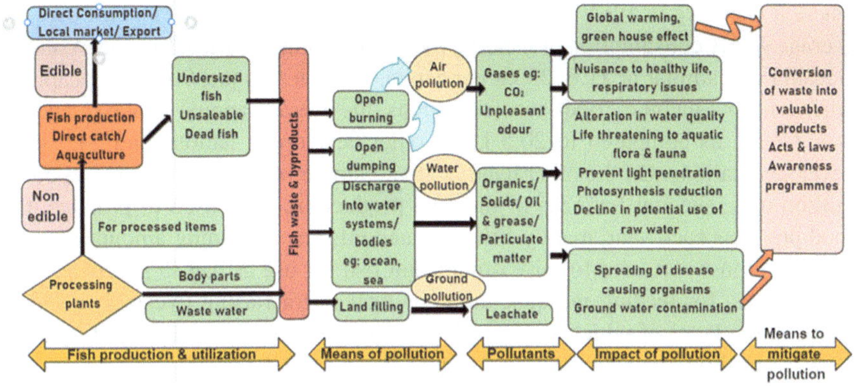

Fig. 1 Schematic presentation of fish waste as a source of pollution

The literature that is now available indicates that fish waste has been used to produce a variety of compounds, each with a potential value, including feed, biofuels, natural colours, food packaging (chitosan), cosmetics, and soil fertilizer (Arvanitoyannis and Kassaveti 2008). In this context, this chapter comprehensively overviews and discusses various means of how fish waste and byproducts act as a source of pollution in the ocean, air and land, and what is their impact on humans and the environment. Consequently, baseline information provided through this chapter would serve as a guide and reference source in the production as well as management of fish and fish waste in the future.

2 Global Fish Production, Utilization and Waste Generation

Overall 179 tons of fish were produced globally in 2018, whereas China as a leading fish producer contributed 35% of the production (Amoakwah et al. 2020). A major portion of manufacturing in 2018—outside of China—came from Asia (34%), followed by North America (14%), Europe (10%), and Africa (7%). The Food and Agriculture Organization (FAO) declared that the yearly catch of fish, crustaceans, and mollusks from inland waters around the world had surpassed 10 million tons for the first time in 2008 (Amoakwah et al. 2020). Africa's inland fisheries generate 2.1 million tons of fish annually, accounting for 24% of the world's overall production (Janko 2014). Even today, aquaculture provides over 25% of the world's seafood supply, and the FAO estimates that by 2030, that proportion will be close to 50% (Jayasankar 2001).

The caught and farmed fish can be eaten either as fresh or processed. Only about 30–40% of fisheries products are consumed fresh around the world; the rest 60–70% are processed for human utilization as well as other uses (Islam et al. 2004a). In 2016, 53% of fisheries products were intended for human use (Boyd and Davis 2020). Hence, the world's fish consumption has increased by 3.1% per year on average, where it was 9 kg in 1961 and 20.5 kg in 2018 (Coppola et al. 2021). However, the average yearly consumption of fishery products in North America, Europe, Asia, and Oceania was similar at 21.6–25 kg per capita, while it was just 9.9 kg per capita in Africa. In low-income, food-insecure nations, the average annual intake per person was 7.7 kg, as compared to 20.5 kg in other developing nations and 24.9 kg in developed nations. On the other hand, a great number of capture fisheries, about 20 million tons, account for 12% of the total production, were used to produce non-food products each year, mainly fish meal and oil. The remaining amount (roughly 4 million tons) was used to raise ornamental fish for use as bait in pharmaceutical applications, for feeding in aquaculture, as well as for raising livestock and furry animals, the latter of which was largely reduced to fish meal and fish oil by about 80% (18 million tons) (Arvanitoyannis and Kassaveti 2008).

In the fish industry, it is generally accepted that the edible flesh portion is the "main product" and that the remaining components, such as the head, skin, trimmings, scale, viscera, and bone, are known as waste (Dauda et al. 2019) (Fig. 2). The manufacture of processed fish generates enormous volumes of solid waste (30–85%) globally, reliant on the species, stage of processing (gutting, scaling, filleting), as well as the kind of fishery that is discarded (Surasani 2018). In general, more than 20 million tons of fish waste including non-target" species, byproducts, and waste from the processing of fish, are generated yearly throughout the world where about 5.2 million tons are discarded annually by European Union (Barroso et al. 2019) and this was later on used for various purposes. China's annual fish waste products from fish processing, or about half of the nation's aquaculture sector, may supply between 420,000 and 650,000 tons of fishmeal, according to Cao et al. (2015).

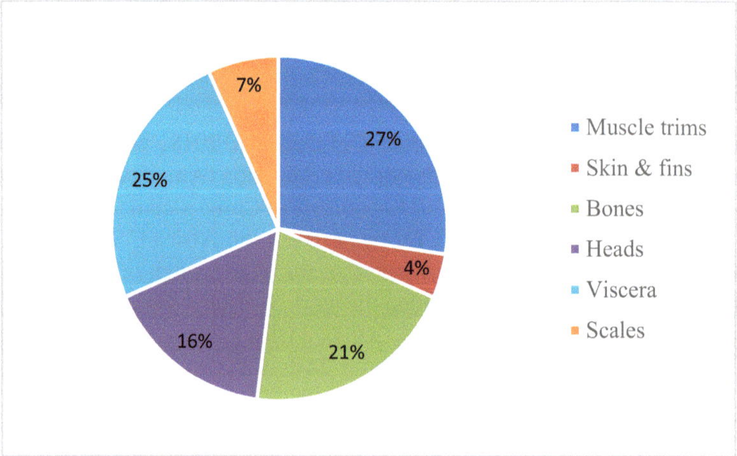

Fig. 2 Fraction of fish waste

3 Means of Pollution by Fish Waste

Fish waste either in the form of liquid or solid is known to cause environmental pollution due to the characteristics of pollutants' richness. High levels of organic substances, nitrogen, salt, oil and fat are present in fishery waste (Islam et al. 2004b). Liquid waste primarily water used during fish processing stages such as washing, thawing, cleaning, and cooking has the potential to cause water pollution when it is discharged improperly without treatment into the water systems. Small volumes of wastewater (effluent) from fish processing enterprises contaminate a large number of water bodies, leading to water pollution (Mihret et al. 2015). The pollutants in the wastewater inclusive of organics, solids, salts, oil, grease, ammonia, and cleaning agents (e.g., chlorine, surfactants) should be lesser than the standard limits released into the water bodies, otherwise, they alter the conditions at water-bodies and adversely affects aquatic living beings (Karkal and Kudre 2020). Islam et al. (2004b) well documented the characteristics of effluents of fish processing industries, for instance, the wastewater from the tuna processing industry had 125–300 m^3/day water flow and with 700, 1600, 500, 250 mg/l BOD, COD, TSS, FOG respectively. Hence, Kurniasih et al. (2018) stated that the water quality parameters of BOD and COD concentration in wastewater at the fishery industry area of the fishing port of Nizam Zachman Jakarta—North Jakarta were 7 times higher than the water quality standards. Ahumada et al. (2004) mentioned that the highest organic load received from fish processing altered the dissolved oxygen level of coastal water from Lota Bay. All these findings imply that fish waste has the potential to cause pollution. Furthermore, the volume of wastewater generated and concentration of pollutants and their consequences depend on several factors such as type and quantity of fish processed, climate, type of fish processing industry (e.g., fish meal, fish oil), hydrological condition of water systems. For example, effluents generated with processing oily fish can be rich in pollutants due to higher oil content than white fish species. Comparatively higher BOD, COD, TSS, Oil and fat were observed in effluent resulted from fish meal production than surimi production in Japan (Islam et al. 2004b). Beyond the waste water, whole fish either spoiled or non-spoiled can cause marine pollution. In Bangladesh, the practice of pitch back of by catch (unintended catch of non-target species) as well as dumping of fish parts into the sea as there are no facilities to store them in fishing boats affect the marine ecosystem (Mozumder et al. 2022). Moreover, the disposal of plastic materials (e.g., vessels, containers) and chemicals used during fish processing also can pollute the marine environment (Abdel-Shafy and Mansour 2018).

Next to water pollution, there is a possibility of air pollution from fish waste. Perishable fish waste left in the open air emits the gases like NH_3, and H_2S, giving an unpleasant odour when enzymatic and or microbial degradation occurs. As these gases have low threshold limits, for instance, 2.1×10^{-4} ppm for NH_3 and 21.4 ppm for H_2S, even a small amount of gases escaping into the surrounding atmosphere, is quiet enough to cause air pollution (Khan et al. 2017). Besides, strong odours are frequently produced from various stages of the fish meal process at fish processing factories. According to Schmidtsdorff (1974), the odour emission was high during drying (60–80%), while it was low in pneumatic meal conveying (2–5%). Hence, the processes of raw material conveying, cooking also emit fairly considerable amount of odour (10–20%). The disgusting odour of rotten fish is a major problem that can cause respiratory irritation and health issues in residential neighbourhoods. Moreover, Wu (1995) stated that sediments settled on the sea bed also can produce hazardous gases such as ammonia, methane, and hydrogen sulfide due to anoxic and reducing conditions.

On the other hand, solid wastes comprising of various fish parts like heads, bones, skin, and plastic evolved during cleaning, skinning, and packing are disposed of unsafely at an open dump site in general, which could serve as a breeding site for insects like mosquito, flies and attract animals, thus leading to the exposed land surface get polluted (Cao et al. 2007). Further, fish waste disposed of at landfill sites generates leachate upon decomposition of organic waste and its subsequent infiltration and percolation with rain water, have the potential to contaminate the groundwater and soil. Fish waste and leftovers improperly disposed of and dumped on the soil result in land pollution, thus making the environment uncomfortable (Abdel-Shafy and Mansour 2018).

Fish waste cause pollution not only as raw when it is left open in the environment but the management techniques also have a chance to cause pollution. For instance, thermal treatments, composting of fish waste and byproducts as a management approach release the greenhouse gases of CH_4 and CO_2, thus having the chance to pollute the air where the concentration of these gases is much concern. Compared to incineration, composting fish waste emitted high CO_2 and CH_4 (Hristov et al. 2017)). Meanwhile, landfilling as a management option for fish waste found with higher values of global warming potential (579.66 kg CO_2 eq/ton) and carbon footprint accounted for greater emissions of greenhouse gas CH_4 (Hristov et al. 2017). Furthermore, huge quantities of fish waste are dumped and typically burned to increase the management process's energy use, cost, and environmental impact (Arvanitoyannis and Kassaveti 2008).

4 Impact of Pollution on Living Beings and the Environment

Disposal of fish waste as solid wastes generated either at fishing ports or processing centres is a noxious and pervasive issue not only in many developing nations but even in developed countries as it has adverse effects on various components of ecosystems.

4.1 Impact on Humans and Living Beings

The abundance of animals in fish waste dumps close to human areas increases the risk of animal attack on humans and livestock theft. A large population of Hamadryas baboons *(Papio hamadryas)* coupled with fish waste and chicken meat waste dumps close to cities caused frequent clashes with people because they encroached on crops in agricultural areas (Plaza and Lambertucci 2017). Dingoes (*Canis lupus dingo*) attacked people in Australia while feeding on organic (fish) waste, and have even been known to result in deaths (Thompson 2005). In Ethiopia, hyena predation on cattle is more severe close to fish and organic waste dumps (Mihret et al. 2015). Additionally, fish waste dumps have an impact on bird populations because of the foul odours that cause serious environmental problems (Novaes and Cintra 2013) and Plaza and Lambertucci (2017) stated that the chance of an aircraft crash increases if birds are close to airports, All of these incidents of conflicts resulting from the disposal of fish wastes and other organic wastes.

Furthermore, fish waste dumped at open sites harbours flies, mosquitoes, and pathogenic microbes consequently causing health issues for people living around. Hence, the obnoxious smell from either raw or decomposing fish waste creates an unwillingness to reside closer to these sites and makes the people shift their places. Most residents believed that air disturbances caused by rotting, toxic material signified the possibility of contamination because of the overwhelming odour (Doron 2021).

4.2 Impact on Environment

On the other hand, wastewater discharge from fish processing plants attributes to negative effects on the environment. The release of wastewater primarily into the surface water bodies (e.g., sea, ocean, rivers, canals) alters the water quality, and as a result, makes it unsafe for the survival of aquatic flora and fauna. In particular, the entry of wastewater with heavy loads of organic matter from fish processing industries lowers the dissolved oxygen level, which in turn, creates an anaerobic condition which retard the growth and functioning of aquatic organisms. Oil and grease-contaminated wastewater prevents photosynthesis, inhibits direct oxygen exchange at the air–water interface, lowers light penetration, and reduces water column oxygenation (Ahumada et al. 2004). Hence, carbon dioxide resulting from organic waste conversion and its subsequent dissolution within water alters the acidity of saltwater, which has an impact on marine life (Gregory et al. 2019). The biomass, density, and variety of the benthos, plankton, and nekton are all decreased as a result of fish farm waste, and the natural food webs are also altered. This has an effect on a greater coastal zone on numerous ecological levels (Arvanitoyannis and Kassaveti 2008). In contrast, the possibility of increased algae and bacteria feeding on fish excrement also concerns scientists, where few species of algae may create toxic substances that are dangerous to humans, fish, and marine mammals (Guillen et al. 2018).

5 Means to Minimize the Pollution

One of the problems most detrimental to the environment has been the production of fish waste. The negative impacts of fish waste particularly on the marine ecosystem have drawn attention to the search for a solution to combat the pollution effect. Here are some means to minimize such adverse effects encountered with fish waste-based pollution.

5.1 Minimization of Fish Waste Generation/Cleaner Production

Prevention and or minimization of generation is the first and foremost management strategy for any type of waste. Cleaner production, which in this context is defined as the ongoing application of integrated, preventive environmental measures to processes, goods, and services, helps to reduce pollution by preventing the formation of unnecessary waste. It aims to improve the overall effectiveness and lower hazards to both people and the environment. Improved housekeeping practices, process optimization and raw material substitution are a few techniques that can be used to achieve cleaner production. Since the water and effluent from fish processing plants are the main sources of pollution, a reduction in water consumption and pollution load of effluent would help to minimize the pollution. Below listed are a few examples of strategies to be adopted to reduce water consumption and pollution load.

- Installation of devices to limit or regulate water flow for manual cleaning procedures
- Cleaning surfaces with high pressures as opposed to large volumes
- Utilizing reasonably clean wastewater for other purposes; for instance, thawing wastewater might be utilized for offal fuming or as the first stage in cleaning unclean regions
- Utilizing cooling systems with closed circuits
- Soaking equipment and floors beforehand to remove grime before the final clean
- Removing solid waste from the ground and using it as a byproduct rather than flushing it down the toilet
- Using an offal hopper instead of the effluent system to collect the blood and offal after vacuum-cleaning dressed fish
- Installing screens or traps in drains to stop solids from entering the wastewater system

- Employing dry cleaning methods, scraping equipment before cleaning, pre-cleaning using air guns, and mopping up spills on the floor with squeegees.

6 Valorization of Fish Waste and Byproducts

Rather than disposing of the waste considering the hazards associated with it, the utilization of waste known as "waste valorization" become popular in the present era as a sustainable waste management strategy, leading to reducing environmental pollution. Utilizing byproducts can help the industry produce goods more sustainably because it may be able to increase income while also lowering disposal expenses for the materials involved (Mo et al. 2018). The conversion of undesired fish wastes from aquaculture and fishing into usable products has gained increasing attention due to its intrinsic characteristics and serves as an alternative for high-cost products. For example, fish waste is recycled into fish feed and fish oil is being utilized as a feed ingredient in fish and poultry farming because of its higher protein content. Reutilization of fish waste lowers the cost of animal production and provides a useful basis of rich-quality protein and energy (Arvanitoyannis and Kassaveti 2008). Utilizing this waste helps address issues with environmental contamination while increasing food security (Surasani 2018). A hydroxyapatite adsorbent prepared from fish waste has been found effective in the removal of lead from water, which in turn reduces heavy metal pollution in the environment (Omar et al. 2019). Similarly, several value-added products have been derived from fish waste and Table 1 summarizes the widely developed products and their potential uses. As a crucial strategy for waste reduction, there is an increasing desire for the industry to recover marketable byproducts from fish waste (Guillen et al. 2018).

Besides, reusing and recycling wastewater from fish processing plants is another means of reducing pollution. The UNEP (2000) advised using vacuum pumps and relatively clean effluent from cooling systems for washing animals and fish. Recycled water could be used to wash the transport vehicles and the floors of the pens (Bailone et al. 2021).

Furthermore, the conversion of fish waste into valuable products as valorization not only indicates environmental pollution but also boosts the economy of a country. Specifically, a country like Bangladesh which is self-sufficient in fish production and generates 93 thousand tons of waste in a year, finds its way to reduce the negative effects caused by the generated huge waste via selling the raw and or dried fish waste collected from local market and processing centres either to retailers or exporters like China, Indonesia, Thailand. A kilogram of fish waste is valued at 70–90 BDT and it was estimated to earn 80 million BDT in a year by exporting 1/3 of the fish waste generated in Bangladesh (Mozumder et al. 2022). However, the production of good quality products from fish waste is challenged by the poor hygienic nature of fish waste, lack of awareness about the potential of fish waste to be utilized, lesser involvement of public and stakeholders, lack of processing facilities etc. In this context, proper segregation of degradable and non-degradable components of waste,

Table 1 Fish waste reduction or utilization approaches

Fish waste-based products		Advantage	Reference
Feed supplement	Animal feed	Cost-effective, waste reduction	Arvanitoyannis and Kassaveti (2008)
	Feedstuff	Protein substitute for swine	Gasco et al. (2020)
	Fish silage	Decreasing feeding costs, Animal protein	Ferraz De Arruda et al. (2007)
	Poultry feed	Sardine fish waste is used for poultry feeding	Shabani et al. (2018)
	Fish or mink food	Reducing the amount of salmon cannery waste solids	Gregory et al. (2019)
Fuel	Biodiesel/biogas	Alternative source of fuel, cost-effective	Arvanitoyannis and Kassaveti (2008)
Miscellaneous uses	Natural pigments (e.g., astaxanthin or βcarotene),	Additional revenue, reduce the disposal cost	Alfio et al. (2021)
Fertilizer	Fertilizer	Improve the soil fertility	Radziemska et al. (2019)
	Soil amendment - Fish offal (heads, skin, viscera and skeletons of rainbow trout	Improve the soil fertility	Arvanitoyannis & Kassaveti (2008)
Enzymes	Pepsin	Caviar production of fish	Coppola et al. (2021)
	Proteases	Pepsin soluble collagen extraction	Coppola et al. (2021)
	Lipases	Defatting of fish skin	Sae-leaw and Benjakul (2018)

awareness enhancement, and the establishment and update of processing facilities, of existing would help address these challenges.

7 Laws and Legislations

People in many developing countries like Bangladesh are of lack awareness and poor knowledge regarding the adverse impacts of waste disposal on the environment and the benefits of reutilizing the waste (Mozumder et al. 2022). Hence, in the absence of laws and regulations, any legal actions cannot be carried out against the people or industries for dumping the waste into the ocean. In this context, proper awareness among people and stakeholders through social and mass media as well as the establishment of laws and legislation would be helpful to minimize the pollution caused. Several countries have passed various acts, laws and regulations restricting the dumping of waste into water bodies to prevent and or minimize water pollution. Accordingly, the Marine Environmental Protection Law (MEPL), a fundamental law formed in 1982, has regulations on the disposal of fish wastes at sea and governs the protection of marine environments in China (Keyuan 2006). China has the East Asian region's longest coastline, and its coastal environmental laws and practices have an impact on the entire East Asian Sea region. Additionally, the Ministry of Agriculture in China issued regulations for fishery loss calculation of accidents of water pollution as well as regulations and procedures for investigating and handling pollution accidents in fishery waters in 1997 to put the investigation and management of pollution accidents in aquaculture on a legal footing (Cao et al. 2007). Similar to this, the marine ecosystem in the USA is preserved by the Ocean Dumping Ban Act of 1988, an addition to the Marine Protection, Research, and Sanctuaries Act of 1972. An international convention for the prevention of pollution from ships (MARPOL), enforced by over 150 countries around the globe, controls the marine pollution by ships including fishing vessels.

Only a few countries, including Norway, the Faroe Islands, Iceland, Chile, New Zealand, and now the EU, have discard bans in place, which outlaw the practice and mandate that all catches of species subject to the restrictions be landed (Guillen et al. 2018). The discard prohibition for European Union fisheries is a key component of the most current overhaul of the Common Fisheries Policy (Guillen et al. 2018). As a result of these acts and laws, the quantity of waste thrown away could be decreased, meanwhile, violations also occur. At this juncture, penalties and punishment should be made. People who disobey the laws, rules, and regulations governing the sea, administration of marine scientific research, and foreign affairs are punished by the ocean administrative penalty enforcement authority (Fu 2005). However, it is still unclear how cutting back on discards would affect the environment in the short- and long-term, as well as the economic and social implications that go along with them.

Besides the above-mentioned main ways, the establishment of modern and innovative processing centres closer to landing sites, installation of own wastewater treatment plants inside the processing factories, collaborative activities between public and private sectors for promoting better utilization of fish waste, conducting empirical studies and improving the existing waste management systems also would help in pollution control.

8 Conclusion

Fish waste and leftovers discarded carelessly onto the ground and waterbodies have the potential to pollute the resources of the environment, especially water because of the heavy pollutant load. Meanwhile, there are possibilities for pollution when fish wastes are managed properly via composting, landfilling and incineration due to emission of greenhouse gases such as CH_4 and CO_2. Therefore, fish waste, as a source of pollution, is a serious environmental issue and causes substantial economic losses. As a result, there is currently a push for the creation of a sustainable fish waste management system that strives to minimize waste output and maximize its potential use, whereas recycling of fish waste into usable products and reuse of wastewater with or without minimal treatments could be better means for managing the fish waste, while those minimize the pollution. Further, several factors namely awareness, social acceptability, facilities available, processing conditions and facilities and costs incurred may challenge fish waste management via a reutilization approach. These can be overcome with awareness campaigns, government and non-government sector collaboration for fish waste reutilization, establishment and update of processing facilities. Even though fish waste has been found potential for pollution, the studies conducted in this regard are limited and it should be explored more in the future.

References

Abdel-Shafy HI, Mansour MSM (2018) Solid waste issue: Sources, composition, disposal, recycling, and valorization. Egypt J Pet 27(4):1275–1290. https://doi.org/10.1016/j.ejpe.2018.07.003

Ahumada R, Rudolph A, Contreras S (2004) Evaluation of coastal waters receiving fish processing waste : lota bay as a case study Liquid industrial wastes arising from fish meal and fish oil processing industries are an environmental problem of consequence in the coastal zones of Chile and Peru. Environ Monitor Assess 90(Table I):89–99

Alfio VG, Manzo C, Micillo R (2021) 'From fishwaste to value: an overview of the sustainable recovery of omega-3 for food supplements. Molecules 26(4). doi: https://doi.org/10.3390/molecules26041002

Amoakwah E et al (2020) Assessment of heavy metal pollution of soil-water-vegetative ecosystems associated with artisanal gold mining. Soil Sediment Contam 29(7):788–803. https://doi.org/10.1080/15320383.2020.1777936

Arvanitoyannis IS, Kassaveti A (2008) Fish industry waste: Treatments, environmental impacts, current and potential uses. Int J Food Sci Technol 43(4):726–745. https://doi.org/10.1111/j.1365-2621.2006.01513.x

Bailone R et al (2021) Sustainable water management in slaughterhouses by cleaner production methods—A review. Renewable Agric Food Syst 36(2):215–224. https://doi.org/10.1017/S1742170520000083

Barroso FG et al (2019) Production of n-3-rich insects by bioaccumulation of fishery waste. J Food Compos Anal 82(June):103237. https://doi.org/10.1016/j.jfca.2019.103237

Boyd CE, Davis RP (2020) Lentic freshwater: Ponds-Aquaculture ponds. Encycl World's Biomes Elsevier. https://doi.org/10.1016/B978-0-12-409548-9.11956-6

Cao L et al (2015) China's aquaculture and the world's wild fisheries. Science 347(6218):133–135. https://doi.org/10.1126/science.1260149

Cao L et al. (2007) Aquaculture Pollution in China Subject Area 5.1: Microbial studies and technologies supporting waste disposal, management, and remediation of municipal and industrial hazardous wastes Environmental Impact of Aquaculture and Countermeasures to Aquaculture. Environ Sci Pollut Res-Int 14(7):452–462. Available at: http://crsps.net/wp-content/downloads/AquaFish/Inventoried10.29/8-2007-4-47a.pdf

Caruso G et al. (2020) Fishery wastes as a yet undiscovered treasure from the sea: biomolecules sources, extraction methods and valorization. Marine Drugs 18(12). doi: https://doi.org/10.3390/md18120622

Coppola D et al (2021) Fish waste: from problem to valuable resource. Mar Drugs 19(2):1–39. https://doi.org/10.3390/md19020116

Dauda AB et al (2019) Waste production in aquaculture: Sources, components and managements in different culture systems. Aquacult Fish 4(3):81–88. https://doi.org/10.1016/j.aaf.2018.10.002

Doron A (2021) Stench and sensibilities: On living with waste, animals and microbes in India. Aust J Anthropol 32(S1):23–41. https://doi.org/10.1111/taja.12380

Fao (2022) The International Year of Artisanal Fisheries and Aquaculture OBJECTIVES OF THE YEAR. NewsletterIslam, 2004b, (Iyafa 2022), p 2022. Available at: www.fao.org/artisanal-fisheries-aquaculture-2022

Ferraz De Arruda L, Borghesi R, Oetterer M (2007) Use of fish waste as silage—A review. Braz Arch Biol Technol 50(5):879–886. https://doi.org/10.1590/s1516-89132007000500016

Fu K-C (2005) International Energy Policy, the Arctic and the Law of the Sea

Gasco L et al (2020) Insect and fish by-products as sustainable alternatives to conventional animal proteins in animal nutrition. Ital J Anim Sci 19(1):360–372. https://doi.org/10.1080/1828051X.2020.1743209

Gayathri S et al (2018) Ultrasound-assisted microextraction of β, ε-carotene-3,3′-diol (lutein) from marine microalgae Chlorella salina: effect of different extraction parameters. Biomass Conversion and Biorefinery 8(4):791–797. https://doi.org/10.1007/s13399-018-0331-9

Gregory SP et al. (2019) Subsurface microbial hydrogen cycling: Natural occurrence and implications for industry. Microorganisms 7(2). doi: https://doi.org/10.3390/microorganisms7020053

Guillen et al. (2018) A review of the European union landing obligation focusing on its implications for fisheries and the environment. Sustainability (Switzerland) 10(4). doi: https://doi.org/10.3390/su10040900

Hristov AN et al (2017) Discrepancies and uncertainties in bottom-up gridded inventories of livestock methane emissions for the contiguous United States. Environ Sci Technol 51(23):13668–13677. https://doi.org/10.1021/acs.est.7b03332

Islam MS, Khan S, Tanaka M (2004b) Waste loading in shrimp and fish processing effluents: potential source of hazards to the coastal and nearshore environments. Mar Pollut Bull 49(1–2):103–110. https://doi.org/10.1016/j.marpolbul.2004.01.018

Islam MS, Khan S, Tanaka M (2004a) Coppola et al., 2021 Marine Pollut Bull 49(1–2):103–110. doi: https://doi.org/10.1016/j.marpolbul.2004.01.018

Janko AM (2014) Fish production, consumption and management in Ethiopia. Int J Econ Manag Sci 03(03):460–466. https://doi.org/10.4172/2162-6359.1000183

Jayasankar R (2001) Integrated farming of seaweeds with finfish and shellfish

Karkal SS, Kudre TG (2020) Valorization of fish discards for the sustainable production of renewable fuels. J Clean Prod 275:122985. https://doi.org/10.1016/j.jclepro.2020.122985

Keyuan Z (2006) Joint development in the south china sea: a new approach

Khan MA et al (2017) Recent trends in electrochemical detection of NH3, H2S and NOx gases. Int J Electrochem Sci 12(3):1711–1733. https://doi.org/10.20964/2017.03.76

Kurniasih SD, Soesilo TEB, Soemantojo R (2018) Pollutants of fish processing industry and assessment of its waste management by wastewater quality standards. E3S Web Conf 68:6–9. doi: https://doi.org/10.1051/e3sconf/20186803006

Mihret G, Tsegaye G, Gidey Y (2015) Livestock loss by the spotted hyena (Crocuta crocuta) in and around a waste dumping site in Northern Ethiopia. Int J Biodivers Conserv 7(1):50–53. https://doi.org/10.5897/ijbc2014.0791

Mo WY, Man YB, Wong MH (2018) Use of food waste, fish waste and food processing waste for China's aquaculture industry: Needs and challenge. Sci Total Environ 613–614:635–643. https://doi.org/10.1016/j.scitotenv.2017.08.321

Mozumder MMH et al (2022) Sustainable utilization of fishery waste in Bangladesh—a qualitative study for a circular bioeconomy initiative. Fishes 7(2):1–21. https://doi.org/10.3390/fishes7020084

Novaes WG, Cintra R (2013) Factors influencing the selection of communal roost sites by the black vulture coragyps atratus (Aves: Cathartidae) in an urban area in central amazon. Zoologia 30(6):607–614. https://doi.org/10.1590/S1984-46702013005000014
Omar S, et al. (2019) A review on lead sources, occurrences, health effects, and treatment using hydroxyapatite (HAp) adsorbent made from fish waste. Water Air Soil Pollut 230(12). doi: https://doi.org/10.1007/s11270-019-4312-9
Plaza PI, Lambertucci SA (2017) How are garbage dumps impacting vertebrate demography, heath, and conservation? Global Ecol Conserv 12:9–20. https://doi.org/10.1016/j.gecco.2017.08.002
Radziemska M et al (2019) Valorization of fish waste compost as a fertilizer for agricultural use. Waste Biomass Valoriz 10(9):2537–2545. https://doi.org/10.1007/s12649-018-0288-8
Sae-leaw T, Benjakul S (2018) Lipase from liver of seabass (*Lates calcarifer*): Characteristics and the use for defatting of fish skin. Food Chem 240:9–15. https://doi.org/10.1016/j.foodchem.2017.07.089
Schmidtsdorff W (1974) Symposium, international exhibition, Jonkoping, Denmark
Shabani A et al (2018) Preparation of fish waste silage and its effect on the growth performance and meat quality of broiler chickens. J Sci Food Agric 98(11):4097–4103. https://doi.org/10.1002/jsfa.8926
Surasani VKR (2018) Acid and alkaline solubilization (pH shift) process: a better approach for the utilization of fish processing waste and by-products. Environ Sci Pollut Res 25(19):18345–18363. https://doi.org/10.1007/s11356-018-2319-1
Thompson S (2005) Modeling waste management options for greenhouse gas reduction. J Environ Inf 6(1):16–24. https://doi.org/10.3808/jei.200500051
Wu RSS (1995) The environmental impact of marine fish culture: towards a sustainable future. Mar Pollut Bull 31(4–12):159–166. https://doi.org/10.1016/0025-326X(95)00100-2

Strategies to Reduce/Manage Fish Waste

Fatima Zahra Aboudamia, Fatima El Amerany, and Abderrahim Jaouad

Abstract Recovery of fish waste has taken the priorities of industrialist and scientists, given its richness in high-value products. A better management of this marine resource could increase the profit margin and reduce environmental pollution. During last decade, fish wastes were used as raw material of some marine biopolymers such as chitin/chitosan and its derivatives, gelatin and collagen, mineral compound as hydroxyapatite, and vitamins. Thus, fish waste could transform into animal feed (fish meal or pet feed), fish silage, protein hydrolysates, bioactive peptide, omega-3, biodiesel/biogas, and soil fertilizer. However, their application is limited to the laboratory scale. This chapter will highlight fish waste as feedstock of marine compounds and different strategies to reduce and manage fish wastes. The best management of fish wastes could enhance a circular economy and zero marine waste; consequently, it could open new avenues for natural marine compounds.

F. Z. Aboudamia (✉) · A. Jaouad
Laboratory of Material Science and Process Optimization, Department of Chemistry, Cadi Ayyad University, BP2390, 40000 Marrakesh, Semlalia, Morocco
e-mail: Aboudamia.fatimazahra@gmail.com

F. E. Amerany
Laboratory of Sustainable Development and Health Research, Department of Chemistry, Faculty of Science and Technology of Marrakech, Cadi Ayyad University, PO Box 549, 40000 Marrakech, Morocco

Department of Biology, Environment and Materials, Higher Normal School, Interdisciplinary Laboratory in Bio-Resources, Cadi Ayyad University, PO Box 575, 40000 Marrakech, Morocco

S. Maqsood et al. (eds.), *Fish Waste to Valuable Products*, Sustainable Materials and Technology, https://doi.org/10.1007/978-981-99-8593-7_21

Graphical abstract for fish waste management

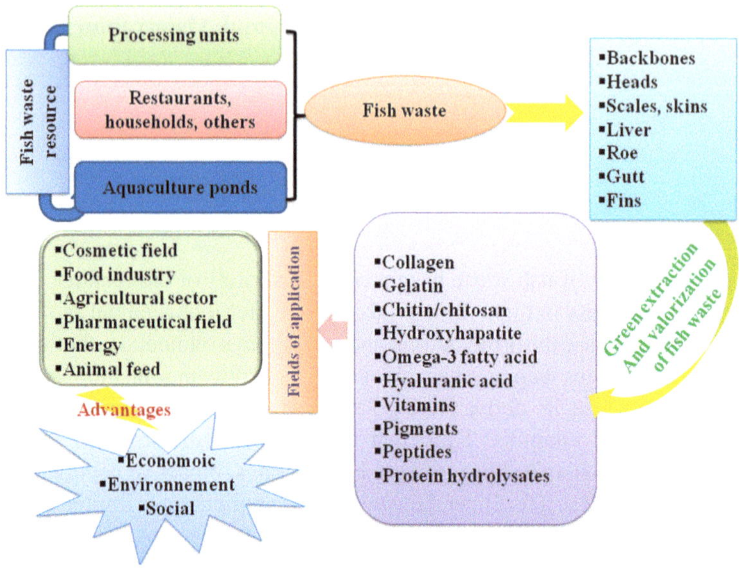

Keywords Fish waste · Recovery · Chitin/chitosan · Omega-3 · Collagen · Hydroxyapatite · Gelatin

1 Introduction

With COVID-19, the population is oriented towards the consumption of marine products. Certainly, this high consumption needs supply food demand. In the other side, and despite its primary role in the international and national economy, the fishing sector generates marine litter from processing units, aquaculture, restaurants, households, others. The sustainable management of fish litter produced from seafood processing becomes one of the major worldwide environmental issue. Fish waste could define as various fish species or by-catch products without commercial value and dwarfed or deteriorated commercial species (Caruso 2016). All over the world, the flesh is the most valuable part of the fish. However, after processing, scales, fins, viscera, heads, skeletons, and others are considered fish waste or the non-edible part of the fish. Those parts are less valuable economically and account for approximately 40–60% of the fish weight. The non-valorization of these marine by-products and by-catch could increase the pollution because they contain high contents of organic matter with a very high degradation rate, which influences marine ecosystems and generates greenhouse gas emissions, and offensive odors.

Internationally, half of the catches are consumed directly by humans. Nevertheless, the other half covers by-catch and filleting waste, which is little or no valued, like heads and abdominal parts, which are not edible. Morocco is characterized with vast coastal area, and the fishing sector occupies a primary place in the national economy. In 2022, the Moroccan kingdom provided significant leadership in Africa with a national fish production of 1.55 million tons, which is around 1% of world fish production. The sector mobilizes 19,064 operational units, including 334 active in offshore fishing, 1,800 in coastal fishing and 17,130 in artisanal fishing. The various activities carried out in this sector are freezing, canning, semi-canning, packaging of fresh fish, packaging of shellfish, fish meal, oil, shelling of shrimp, storage, and others. In addition, due to its inland water resources, Morocco is recognized in aquaculture as producer and the leader exporter among the North African countries (Hülya and Tahir 2021). Nationally, despite the significant wealth and the important variety of fishery products and sector activities, the recovery of fish by-products or by-catch is almost rare, and it is related to fishmeal and oil, however, those activities consumed a lot of energy, and exhausted marine resources by producing low-quality of the obtained product.

Fish processing industries, aquaculture ponds, restaurants, and households generate large amounts of solid and liquid wastes; that cause environmental pollution. However, better valorization of those fish waste could produce valuable high products such as protein, lipid fractions, minerals, enzymes, peptides, polysaccharides, biogas, biodiesel, omega-3, animal feed, fertilizer, etc. The principal aim of this present chapter is to highlight the necessary strategies to manage or reduce the fish waste impact on the environment.

2 Fish Waste: From Marine Litter to Valuable Compounds

2.1 Fish Waste in Worldwide

The increase in the world's population was followed by an increase in demand for healthy foods, especially seafood, which is the top recommended food for meeting nutritional needs. Seafood can be divided into 25 taxonomic groups (Table 1), and these aquatic groups can be found in 21 fishing areas (Fisheries 2022). Seafood farming is considered among the fastest-growing food sector over the last two decades (Clawson et al. 2022). In addition, this naturally occurring product has different benefits for human health due to its containing high levels of vitamins, especially vitamin D, polyunsaturated fatty acids (i.e., Omega-3), amino acids (i.e., glutamic acid), and minerals (i.e., iodine) (Reames 2012). The composition of seafood can vary between species. For example, high levels of oil can be found under the skin layer of some sea mammals (i.e., seals, dolphins, and whales), in comparison to

Table 1 Taxonomic groups of aquatic species (Fisheries 2022). (https://www.fao.org/fishery/en/aqspecies/search)

Taxonomic group	Species	Taxonomic group	Species
Delphilidae	31	Pleuronectidae	11
Clupeidae	27	Hemiscylliidae	10
Acipenseridae	20	Merlucciidae	10
Scombridae	20	Engraulidae	7
Squalidae	20	Balaenopteridae	6
Penaeidae	19	Orectolobidae	6
Phocidae	19	Parascylliidae	6
Ziphiidae	18	Salmonidae	6
Gadidae	16	Sciaenidae	6
Carangidae	14	Carcharhinidae	5
Otaridae	14	Lamnidae	5
Sparidae	13	Phocoenidae	5
Squatinidae	13		

other seafood species (Saadoun et al. 2015). This oil is stored as a mass of fat cells, which is also called blubber. In fish, Bogard et al. (2015) showed that the levels of fatty acids in *Tenualosa ilisha, Hypophthalmichthys molitrix,* and *Glossogobius giuris* were 183 g kg^{-1}, 41 g kg^{-1}, and 4 g kg^{-1}, respectively.

From 2010 to 2020, an increase in global fish production amounted to over 18% (Boyd et al. 2022). A recent work showed that only 70% of the total fish caught every year is consumed or processed, while the remaining is discarded, on the land surface or water sources, due to its rapid spoilage and storage problems (Nirmal and Maqsood 2022). Also, only one third or half the quantity of the fish is comestible, and the remaining portion, which represents a mixture of scales, skins, vertebrae, dorsal fins, intestines, livers, heads, stick water, and bacteria, is considered waste. In warm conditions, fish waste can be degraded quickly and release carbon dioxide because of the increase in the activity of bacteria that feed on the organic compounds that existed in the raw materials. Therefore, if fish waste is not treated properly it can lead to atmospheric pollution.

Different studies reported that fish waste contains different valuable by-products such as collagen, peptides, chitin, oils, and enzymes (Fernandes 2016; Muthumari et al. 2016; Ivanovs and Blumberga 2017; Abuine et al. 2019; Aboudamia et al. 2020a, b; Aboudamia et al. 2021). Thus, in order that the impact of fish waste on the environment reduced, it would be necessary to reuse it or to extract its by-products for further use in various fields, including agronomy, pharmaceutical engineering, and water treatments.

2.2 *Composition of Fish Waste*

The isolation of useful substances (collagen, chitin, lipids, fatty acids, enzymes, and hydroxyapatites), as well as the production of other compounds (gelatin, peptides, and chitosan) from discarded parts of fish processing, would have a great interest, especially in the application for boosting plant growth as well as for food and drugs biotechnology (Friess 1998; El Amerany 2020, 2021, 2022).

2.2.1 Collagen, Gelatin, and Peptides

Collagen is a Greek word, which means producing gum (Wang 2021). Also, it is a protein (Fig. 1) that is presented in all human, animal, and fish organs, such as skin, tendons, and ligaments as well as cartilaginous and connective tissues (León-López et al. 2019). It is synthesized after vigorous exercise in order to provide support to connective tissue (Langberg et al. 1999). This protein is characterized by the dominance and the combination of amino acids, especially glycine, proline, and hydroxyproline (Li and Wu 2018).

Collagen can have different structures; it can be amino acid triplet, α-helix, triple helix, or fibrils (Wang 2021). Regarding collagen size, its molecule can be less than 1.3 nm, whereas the thickness of its fibril can vary from 50 to 500 nm (Fratzl 2008).

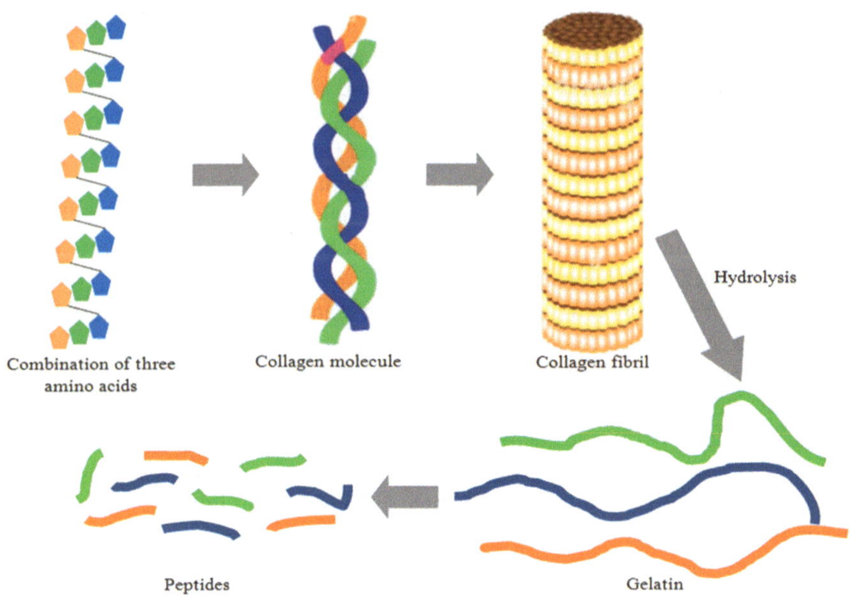

Fig. 1 Collagen structure

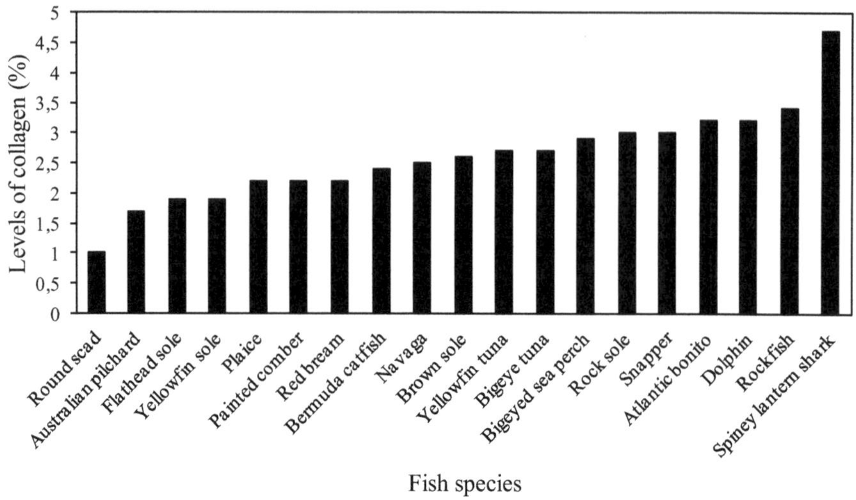

Fig. 2 Collagen levels in fish species

Collagen is often used in the medical field, especially for tissue engineering, wound healing, treating bone defects, and rebuilding teeth (Li et al. 2021). In addition, collagen has other biological properties, including a high capacity in absorbing water and in forming emulsions, which make them a desired biomaterial in the agri-food industry. The levels of collagen in fish muscles are ten-fold lower than in red meat (Sikorski et al. 1984) and as shown in Fig. 2, it can vary between species.

Gelatin is a transparent or slightly yellow, and solid substance, almost tasteless and odorless. It is made from fish waste (epidermis, scales, and bones) (Usman et al. 2022). Even though the level of gelatin obtained from pig skin is up to 80% in the market, in comparison to others made from bovine, this product could not meet the demand for food, especially for Muslim people, who accept only Halal food. Therefore, Gelatin obtained from fish waste can be used as a substitute for those made from other animals like pigs.

Gelatin can be applied in agri-food and drug manufacturers either as a stabilizer or as a thickener to modify the viscosity of a solution without altering the whole its characteristics (Saha and Bhattacharya 2010). Gelatin is a polymer of amino acids, produced after the denaturation of collagen (Vázquez et al. 2021) (Fig. 1). This gel is characterized by a lower gelation property than mammalian gelatin due to low levels of proline and hydroxyproline. Therefore, it is recommended to add ionic polysaccharides to gelatin, such as alginate and chitosan to improve its functional properties (Derkach et al. 2020).

To induce the rate of gelatin, collagen can be degraded at pH 4 (Usman et al. 2022). In addition, a high quantity of gelatin can be obtained by using enzymes (i.e., pepsin enzyme) and non-thermal technics such as high pressure and ultrasonication treatment, to break the non-covalent bonds (Usman et al. 2022).

Fish peptides are polymers, characterized by short sequences of amino acids ($\leq$20) (Fig. 1) (Ucak et al. 2021). They can be produced under the action of several enzymes, including alcalase and trypsin (Brandelli et al. 2015). Various studies found that strongly charged peptides, isolated from fish waste, had beneficial effects on human health, like building blocks for proteins and fighting cancers, pathogenic agents, and other diseases (Gildberg 2004; Masso-Silva and Diamond 2014). These substances can be found in different fish parts such as muscle protein, bones, and mucous layer (Senevirathne and Kim 2012).

2.2.2 Chitin and Chitosan

Chitin is a material composed of two units' types: N-acetylglucosamine (more than 50%) and N-glucosamine (less than 50%), which are connected by β(1–4) glycosidic bonds (Fig. 3) (Pellis et al. 2022). Chitin can not be dissolved in water and organic solvents because of its crystal structure, which therefore limits its use in several areas (Austin 1973).

Chitin can be isolated from the inner layer of the cell wall of insects (i.e., black soldier fly), microorganisms (i.e., *fusarium oxysporum*), and spores (i.e., arbuscular mycorrhizal fungi), exoskeletons of arthropods (i.e., shrimps), and scales of fish (Elieh-Ali-Komi and Hamblin 2016; Lagat et al. 2021; Aboudamia et al. 2020a, b). This polymer is characterized by three polymorphic crystalline structures, such as alpha (α), beta, and gamma, which differ according to chitin's chains arrangement (Martínez et al. 2001). Regarding the first type, its chains are put in an antiparallel order, however, for the second type, the chains are all parallel (Martínez et al. 2001). However, for the type γ chitin, every one chain is arranged in the opposite direction of two chains that have the same polarity (Martínez et al. 2001). In all chitin's types, the chains are linked together by hydrogen bridge between either the amide group or the carbonyl group of the adjacent chain (Darmon and Rudall 1950). As Aboudamia et al. (2020a, b) demonstrated in their last research, chitin extracted from fish scales has a crystal structure of type β.

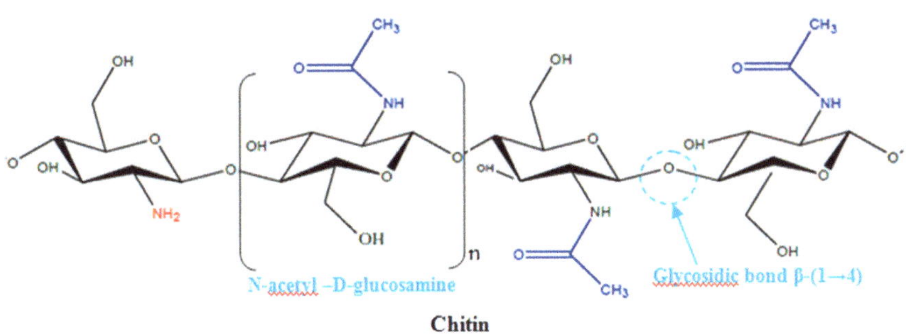

Fig. 3 Chitin structure

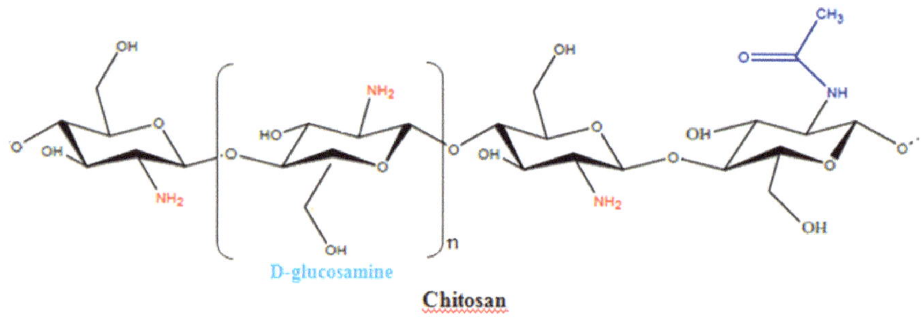

Fig. 4 Chitosan structure

From the hard material covering the skin or the scales of various fish, such as bocachico fish (*Prochilodus magdalenae*), parrotfish (*Chlorurus sordidus*), red snapper (*Lutjanus argentimaculatus*), rohu (*Labeo rohita*), nile tilapia (*Oreochromis niloticus*), and European sardine (*Sardina pilchardus*) the yield of chitin can be 45.2%, 45%, 33%, 22%, 20%, and 10%, respectively (Muslim et al. 2013; Boarin-Alcalde and Graciano-Fonseca 2016; Rumengan et al. 2017; Molina-Ramírez et al. 2021; Aboudamia et al. 2020a, b).

Chitosan, an aminopolysaccharide, is derived from the copolymerization of two monomers: glucosamine (or 2-amino-2-deoxy-beta-d-glucopyranose) and N-acetylglucosamine (or 2-acetamide-2-deoxy-beta-d-glucopyranose) (Fig. 4), which are joined together by glycosidic bonds (Domínguez-Delgado et al. 2014). This biopolymer is characterized by the dominance of the D-glucosamine unit (more than 50%) in the polymeric chains (Fig. 4), which allows them to be part of different organic acids, including lactic acid, chloride acid, and acetic acid (Rinaudo 2006). In 1859, chitosan was discovered for the first time by Rouget (Ruiz and Corrales 2017) who found out that heating chitin in an alkaline solution produces a soluble material.

Chitosan is a substance that is not widely available in nature. It can be isolated only from the cell wall of some microorganisms, such as fungi (e.g., Mucor Rouxii) (White et al. 1979; Zamani et al. 2007; Tayel et al. 2011) and yeasts (e.g., Candida Albicans) (Pochanavanich and Suntornsuk 2002) and in the abdominal wall of the termite queen (Pillai and Ray 2012). It is, moreover, obtained through the complete or incomplete deacetylation of chitin that is seen anywhere and available in large quantity (Baxter et al. 1992).

Chitosan's yield produced from fish sources can be varied depending on the extraction method applied. For instance, Ooi et al. (2021) showed that chitosan's yield produced from the scales of red snapper (*Lutjanus johnii*) can be 49% and 33% if the raw material was treated with 0.05% and 3% HCl, respectively.

2.2.3 Lipids and Fatty Acids

Lipids are considered the main components of the cell membranes. They can act as regulators of membrane layer permeability, cellular signaling and responses, and gene expression (Calder 2012). Many researchers have demonstrated that fish oil, especially Omega-3 fatty acid, can be beneficial for patients with high cholesterol and heart disease (Chaddha and Eagle 2015), and this oil can be found in fish meat as well as its waste. A study has shown that the oil isolated from the waste of the *Atlantic salmon*, especially heads, and soft tissues, contains high levels of saturated (11% palmitic acid), monounsaturated (47% oleic acid), and polyunsaturated (15% linoleic acid) fatty acids and antibiotics (i.e., nafcillin, oxacillin, and penicillin). This oil has as well as bactericidal activity against *Pseudomonas aeruginosa* (Inguglia et al. 2020). Regarding *Sardinella lemuru* waste, the highest level of lipids, which was up to 5%, can be found either in heads, intestines, and livers (Khoddami et al. 2009). In addition, this waste is composed of saturated (36% palmitic acid and 9% steric acid), monounsaturated (22% oleic acid), and polyunsaturated (16% docosahexaenoic acid) fatty acids.

In addition, Suriani and Komansilan (2019) have reported that increasing the yield of mono- and poly-unsaturated fats and reducing the levels of saturated fats, which lead to cancer and gaining weight, is possible through using the process of urea crystallization.

2.2.4 Enzymes

Fish wastes contain diverse kinds of enzymes (i.e., collagenases, peptidases, proteases) (Table 2) that can modify macromolecules and accelerate the deterioration of waste (Venugopal 2016). The isolation of these enzymes required four main steps: (i) collection of the enzyme extract, (ii) metabolite fractionation, (iii) enzyme purification, and (iv) enzyme activation through biochemical modification of their shape (Daboor et al. 2012).

To produce different and high-quality products for human and animal consumption, enzymes are becoming a crucial element of the production processes applied by the food and medical industry (Shahidi and Kamil 2001). For example, enzymes isolated from cold fish species can be used to protect pharmaceutical and food products (Venugopal 2016).

2.2.5 Hydroxyapatite

Hydroxyapatite is a mineral species that can be found in bone, teeth, and scales, and it is mainly composed of two elements, such as calcium (Ca) and phosphate (P) (Mustafa et al. 2015). This material is considered stable only if Ca/P ratio is more than 1.67 (Fiume et al. 2021).

Table 2 Enzymes isolated from fish species

Family enzyme	Compounds	Fish species	Properties	References
Collagenases	Collagenases	Smooth weakfish (*Cynoscion leiarchus*)	Activity increased when the temperature is around 55 °C whereas the pH is equal to 8	de Melo Oliveira et al. (2017)
Peptidases	Chymotrypsin A	Atlantic salmon (*Salmo salar* L.)	Activity reduced when temperature is more than 38 °C Activity increased in low pH	Zhou et al. (2011)
	Chymotrypsin B			Rungruangsak-Torrissen et al. (2006)
	Trypsin		Not shown	
Proteases	Caspase-3	White amur (*Ctenopharyngodon idella*)	Activity decreased during cold temperatures	Jiang et al. (2022)
	Calpain			
	Cathepsin B		Activity increased only during the first day of ice storage	
	Cathepsin L		Activity increased only during 3 days of ice storage	
	Cathepsin	Yellow pike (*Congresox talabon*)	Activity increased when the temperature is around 50 °C and the pH is equal to 5	Pertiwi et al. (2020)
	Chymosin	Atlantic tuna (*Thunnus obesus*)	Activity of enzyme is less sensitive at pH values above 6.4	Tavares et al. (1997)
	Pepsin	Catfish and milkfish	Activity of enzyme is more important when the temperature range of 20–40 °C, whereas the pH is less than 3	Nurhayati et al. (2020)

Hydroxyapatite that existed in nature is better than synthetic one because of its higher metabolic activity and stability as well as its biocompatibility and non-toxicity (Granito et al. 2018). Hydroxyapatite from fish bone and scales is cheap and easy to produce. Based on many different protocols, the production of hydroxyapatite from fish waste can be done through 4 main steps: (i) heating the raw material to remove lipids and meat surrounding the material used, (ii) drying and crushing, and (iii) calcinations (Kerian 2019). Hydroxyapatite isolated from natural resources is mainly used for bone tissue engineering after controlling its microstructure to allow blood flow through the material.

2.2.6 Vitamins and Pigments

Carotenoid pigments, specifically carotene and astaxanthin, are the cause of the appearance of yellow, orange, and red stains in the meat and skin of fish (Simpson 2007). In addition, fish liver stores high levels of vitamins such as vitamin D (Lock et al. 2010).

3 Strategies to Reduce and Manage Fish Wastes

The Moroccan government inaugurate various sustainability initiatives as the National Roadmap for Biomass Energy Valorization (BEV) to 2030, the National Energy Efficiency Strategy to 2030, the Green Hydrogen Roadmap, the Low Emission Development Strategy for 2050 (LEDS), and others. In order to achieve the objectives of those initiatives, the National Ministry of the Environment, national laboratories, and municipalities are looking for green solutions to decrease the negative impact of fish by-products in the environment, while applying eco-friendly technologies. The choice and the application of these strategies require a great challenge, and the collaboration of several actors (environmental, scientific, industrial, financial), and civil society. During the last decade, several governments have oriented and encouraged fish companies to valorize their marine by-products. Circular business models aim to find a solution to environmental problems by incorporating new scientific knowledge and green technologies into a new economic system (Rosa et al. 2019). To decrease fish waste generated, various eco-friendly solutions could be adopted. Fish companies and aquaculture should apply the traditional hierarchy of waste management from the best choice to the least favorable one. The model of the inverted pyramid or 5R gives optimal solutions for human and environment from prevention to disposal (Fig. 5).

Romero-Hernández and Romero (2018) reported that the circular economy is established on sale, disposal, and prevention compared with waste management. Thus, it is recommended to make a framework for each processing plant and aquaculture in order to integrate the economic, environmental, and social. In addition, the application of a circular bioeconomy which is an integral part of the circular

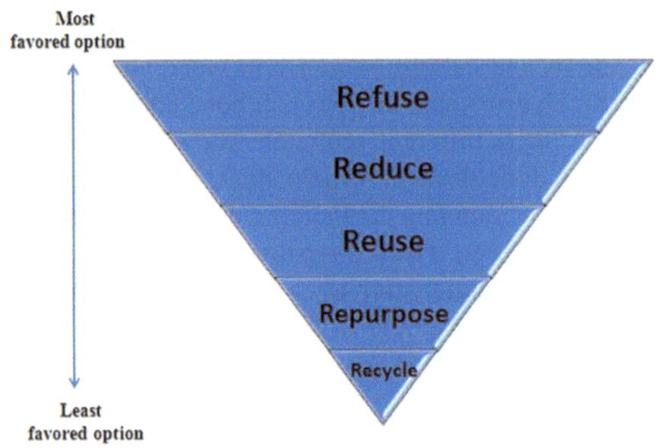

Fig. 5 The inverted pyramid for fish waste management

economy is primordial to achieve sustainability in terms of resources and environmental sustainability (Coppola et al. 2021). The European Commission, defined bioeconomy as "the production of renewable biological resources and the conversion of these resources and waste streams into value-added products, including food, feed, bio-based products and bioenergy" (Commission, E. A Sustainable Bioeconomy for Europe). The application of circular bioeconomy in the fisheries sector could increase the awareness by society and companies, sustain production and consumption, increase valorization of resources and zero waste, contribute to stakeholders and policymakers, and support of politics (Fig. 6).

As mentioned in the first axis of this chapter, the fish marine waste contains a valuable compound of high-commercial added as minerals, polysaccharides, proteins,

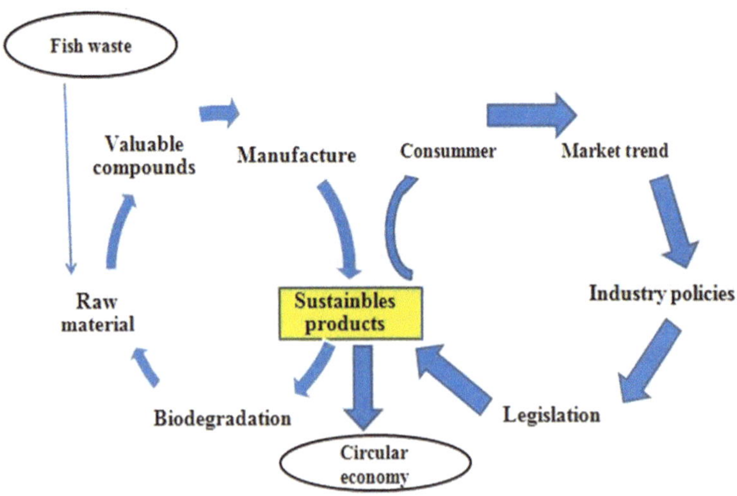

Fig. 6 The inter- and multi-disciplinary process for fish wastes

lipids, and others. In general, fish by-products and by-catch might be applied in various fields from agriculture to pharmaceutic. In this section, we will highlight various scenarios that might be used to reduce or manage fish waste in energy, and agriculture as vital sectors.

3.1 Energy

3.1.1 Biodiesel

Fossil fuels are considered as non-renewable energy, which causes the depletion of natural resources by the emission of greenhouse gases, as well as atmospheric contaminants, some product of incomplete combustion such as NOX, SOX, CO, volatile organic elements with particles of atmospheric matter (Kannahi and Arulmozhi 2013). Recently, biodiesel might produce from fish litter, and among the most used parts are the head, spine, skin, stomach, and tail. Fish by-products as marine litter might be the inexpensive resource of biodiesel (Yahyaee et al. 2013) or bio-oil (Wisniewski et al. 2010). Fish oil obtained from marine fish contains higher rate of omega-3 polyunsaturated fatty acids and its use as source of biofuels will minimize greenhouse gases and preserve biofuel feedstock for the future generation. According to the study of Mata et al. (2014), it is demonstrated that residual biodiesel properties as density, flash point, heat of combustion, acide value depends on waste fat (ester profile). In addition, fish oil could enhance fuel flow fluidity at low operating temperatures, due to its composition which consists mainly from unsaturated fatty acids as PUFAs (Karkal and Kudre 2020). However, the high rate of mineral and moisture in fish by-products requires adequate technologies for energy conversion (Fig. 7).

The biodiesel production could be obtained using various methods such as pyrolysis (Wisniewski et al. 2010), transesterification (Jung et al. 2019), mico-emulsion, and direct blending (Behçet 2011) with diesel fuel. Bhaskar (2018) reported that transesterification chemical process is one of the best processes to decrease the viscosity of the vegetable or animal fat oil and to increase the cetane number of biodiesel. In addition, a simple transesterification of fish oil was done by Sharma et al (2014) who converted fish waste into biodiesel of high FAME content using a single-step transesterification by alkaline catalyst (CH_3ONa). The production of biodiesel from marine fish litter as an alternative of fossil fuels could solve environmental and political problems.

3.1.2 Biogas

Biogas could be obtained from various organic sources and wastes during anaerobic digestion. Fish waste is rich source of proteins considered as biodegradable organics, making it a proper substrate for anaerobic digestion (Yulisa et al. 2022). The production of biogas from fish by-products and by-catch is among the simplest processes for

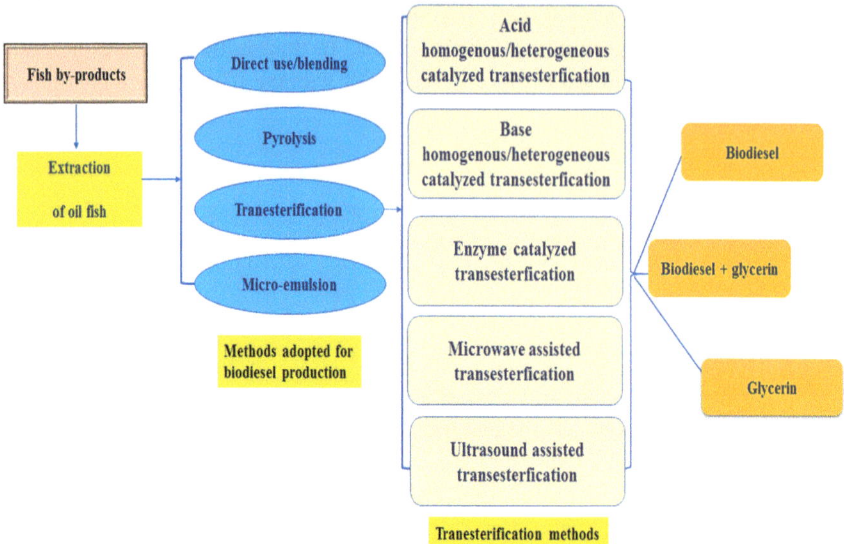

Fig. 7 Outline of various methods applied for the production of biodiesel from fish by-products and its perspective of use (Karkal and Kudre 2020)

energy conversion, and it could be viable and sustainable feedstock and alternative biofuels. However, methane as accumulated intermediate from protein degradation could affect negatively system stability and even process failure (Solli et al. 2018). The study of Cadavid-Rodríguez et al. (2019) reported that fish waste could be an alternative source of biomethane production, the produced biogas was used as an alternative energy for artisanal fishing communities in Tumaco city, Colombia. According to Bücker et al. (2020), industrial fish processing waste can be efficiently converted to methane in a mono-digestion process and produced methane at 540.5 CH_4 mL gVS_1. In addition, the fish waste had the best daily methane production performance and produced biogas with higher methane content than fish crude oil waste in less time.

As already mentioned in Sect. 3.1, the production of biogas from fish waste could minimize the greenhouse gases and increase economic incomes.

3.2 *Biofertilizer*

Biofertilizers are defined as the microorganisms (bacteria/fungi) that help the plants to grow by ameliorating the nutrient supply to the host plant when given to seeds, plants, or the soil (Daniel et al. 2022). According to the FAO 2016, the application of biofertilizers enhances soil fertility and sustains plant growth, thus enhancing crop yields. Biofertilizers are classified within various classifications according to their functions and their mechanisms of action. The study of Nosheen et al. (2021) showed that nitrogen-fixers (N-fixers), potassium solubilizers (K solubilizers), phosphorus solubilizers (P solubilizers), and plant growth-promoting rhizobacteria (PGPR) are the most favorable nutrient for the plant. Fish by-products and by-catch are the among source for the production of biofertilizers, due to their richness of Ca, N, P (Illera-Vives et al. 2015). The production requires the blending of acids to neutralize the pH to eliminate fungal growth, and the process of maturation demands between 3 and 4 days to achieve maturation. Moreover, organic matter ameliorates the soil fertility. It plays as a natural fixation to the soil physical, chemical, and biological properties which sustain the use of soil for long term (Jubin and Radzi 2022). Previous studies revealed that fish waste as abdominal parts, heads, outer shells, scales, or skins have a higher influence on growth and crop yields. (Hepsibha and Geetha 2021; Shahsavani et al. 2017). The application of bio-fertilizer from fish by-products and by-catch could mitigate the environmental pollution caused by using chemical fertilizer.

Finally, we can mention that several constraints may confront the industrials in the sector of the recovery of fish by-product /by-catch and among them are the absence of knowledge, lack of finance, lack of awareness, low skilled workers. In addition, the recovery of fish waste will be a new sector that is in line with the principles of the blue economy using environmentally friendly and sustainable technologies, and while producing high-value substances applicable in different fields.

4 Conclusion and Future Perspective

For the greater benefits of our environment, it is mandatory to elaborate strong collaborations between scientists and industrials to learn from their depth experiences. In addition, the government should support the industries that make use of fish by-products; because they have great economic and environmental potential in the country Furthermore, it is essential to include civil society and the mass media to show the major interest of fish by-products. In schools and universities, students must be involved in workshops or congresses by showing different documentaries about valuables products that can be made using from fish by-products, thus by mentioning the major interest of recovery in the depollution. It is also necessary to organize workshops for professionals to increase their knowledge in the field of marine waste recovery. Finally, the recovery of fish by-products is a great deal with important advantages for our society, economy, and environment.

References

Aboudamia FZ, Kharroubi M, Neffa M, Aatab F, Hanoune S, Bouchdoug M, Jaouad A (2020a) Potential of discarded sardine scales (*Sardina pilchardus*) as chitosan sources. J Air Waste Manag Assoc 70(11):1186–1197

Aboudamia F, Kharroubi M, Neffa M, Jaouad A, Bouchdoug M (2020b) Extraction and characterization of ß-chitin from sardine's scales Sardina pilchardus (Walbaum, 1792). Moroc J Chem 8(1):8–1
Aboudamia FZ, Aatab F, Jaouad A, Bouchdoug M, Kharroubi M (2021) Sardine scales: a promising source of marine biomaterials. Lett Appl NanoBioScience 11:3954–3969
Abuine R, Rathnayake AU, Byun HG (2019) Biological activity of peptides purified from fish skin hydrolysates. Fish Aquat Sci 22(1):1–14
Austin PR (1973) Solvents for and purification of chitin, pp 2–4. United States Patent
Baxter A, Dillon M, Taylor KA, Roberts GA (1992) Improved method for i.r. determination of the degree of N-acetylation of chitosan. Int J Biol Macromol 14(3):166–169
Behçet R (2011) Performance and emission study of waste anchovy fish biodiesel in a diesel engine. Fuel Process Technol 92(6):1187–1194
Bhaskar SV (2018) A comprehensive review on waste fish oil as feed-stock of biodiesel. Int J Res Appl Sci Eng Technol 6(3):1374–1378
Boarin-Alcalde L, Graciano-Fonseca G (2016) Alkali process for chitin extraction and chitosan production from Nile tilapia (*Oreochromis niloticus*) scales. Lat Am J Aquat Res 44(4):683–688
Bogard JR, Thilsted SH, Marks GC, Wahab MA, Hossain MA, Jakobsen J, Stangoulis J (2015) Nutrient composition of important fish species in Bangladesh and potential contribution to recommended nutrient intakes. J Food Compos Anal 42:120–133
Boyd CE, McNevin AA, Davis RP (2022) The contribution of fisheries and aquaculture to the global protein supply. Food Secur 1–23
Brandelli A, Daroit DJ, Corrêa APF (2015) Whey as a source of peptides with remarkable biological activities. Food Res Int 73:149–161
Bücker F, Marder M, Peiter MR, Lehn DN, Esquerdo VM, de Almeida Pinto LA, Konrad O (2020) Fish waste: an efficient alternative to biogas and methane production in an anaerobic monodigestion system. Renew Energy 147:798–805
Cadavid-Rodríguez LS, Vargas-Muñoz MA, Plácido J (2019) Biomethane from fish waste as a source of renewable energy for artisanal fishing communities. Sustain Energy Technol Assess 34:110–115
Calder PC (2012) Mechanisms of action of (n-3) fatty acids. J Nutr 142(3):592S-599S
Caruso G (2016) Fishery wastes and by-products: a resource to be valorized. J Fish Sci 10(1):0–0
Chaddha A, Eagle KA (2015) Omega-3 fatty acids and heart health. Circulation 132(22):pe350–e352
Clawson G, Kuempel CD, Frazier M, Blasco G, Cottrell RS, Froehlich HE, Metian M, Nash KL, Tobben J, Verstaen J, Williams DR, Halpern BS (2022) Mapping the spatial distribution of global mariculture production. Aquaculture 553:738066
Commission E. A sustainable bioeconomy for Europe: strengthening the connection between economy, society and the environment. https://ec.europa.eu/research/bioeconomy/pdf/ec_bioeconomy_strategy_2018.pdf. Last Accessed 17 Sept 2020
Coppola G, Gaudio MT, Lopresto CG, Calabro V, Curcio S, Chakraborty S (2021) Bioplastic from renewable biomass: a facile solution for a greener environment. Earth Syst Environ 5:231–251
Daboor SM, Budge SM, Ghaly AE, Brooks MS, Dave D (2012) Isolation and activation of collagenase from fish processing waste
Daniel AI, Fadaka AO, Gokul A, Bakare OO, Aina O, Fisher S, Burt AF, Mavumengwana V, Keyster M, Klein A (2022) Biofertilizer: the future of food security and food safety. Microorganisms 10(6):1220
Darmon SE, Rudall KM (1950) Infra-red and X-ray studies of chitin. Discuss Faraday Soc 9:251–260
de Melo D, Oliveira V, Assis CRD, Herculano PN, Cavalcanti MTH, de Souza Bezerra R, Figueiredo AL (2017) Collagenase from smooth weakfish: extraction, partial purification, characterization and collagen specificity test for industrial application. Bol Inst Pesca 43(1):52–64
Derkach SR, Voron'ko NG, Kuchina YA, Kolotova DS (2020) Modified fish gelatin as an alternative to mammalian gelatin in modern food technologies. Polymers 12(12):3051

Domínguez-Delgado CL, Rodríguez-Cruz IM, Fuentes-Prado E, Escobar-Chávez JJ, Vidal-Romero G, García-González L, Puente-Lee RI (2014) Drug carrier systems using chitosan for non parenteral routes. Pharmacology and Therapeutics

El Amerany F, Rhazi M, Wahbi S, Taourirte M, Meddich A (2020) The effect of chitosan, arbuscular mycorrhizal fungi, and compost applied individually or in combination on growth, nutrient uptake, and stem anatomy of tomato. Sci Hortic 261:109015

El Amerany F, Taourirte M, Wahbi S, Meddich A, Rhazi M (2021) Use of metabolomics data analysis to identify fruit quality markers enhanced by the application of an aminopolysaccharide. RSC Adv 11(56):35514–35524

El Amerany F, Rhazi M, Balcke G, Wahbi S, Meddich A, Taourirte M, Hause B (2022) The effect of chitosan on plant physiology, wound response, and fruit quality of tomato. Polymers 14(22):5006

Elieh-Ali-Komi D, Hamblin MR (2016) Chitin and chitosan: production and application of versatile biomedical nanomaterials. Int J Adv Res 4(3):411

Fernandes P (2016) Enzymes in fish and seafood processing. Front Bioeng Biotechnol 4:59

Fisheries FAO (2022) Aquaculture Division. Rome: Food and Agriculture Organization of the United Nations

Fiume E, Magnaterra G, Rahdar A, Verné E, Baino F (2021) Hydroxyapatite for biomedical applications: a short overview. Ceramics 4(4):542–563

Fratzl P (2008) Collagen: structure and mechanics, an introduction. In: Fratzl P (ed) Collagen. Springer, Boston, pp 1–13

Friess W (1998) Collagen–biomaterial for drug delivery. Eur J Pharm Biopharm 45(2):113–136

Gildberg AR (2004) Enzymes and bioactive peptides from fish waste related to fish silage, fish feed and fish sauce production. J Aquat Food Prod Technol 13(2):3–11

Granito RN, Renno ACM, Yamamura H, de Almeida MC, Ruiz PLM, Ribeiro DA (2018) Hydroxyapatite from fish for bone tissue engineering: a promising approach. Int J Mol Cell Med 7(2):80

Hepsibha BT, Geetha A (2021) Effect of Biofertilizer (Fermented fish waste–Gunapaselam) on structure and biochemical components of Vigna radiata leaves. Res J Chem Environ 25:7

Hülya S, Tahir SYA (2021). Aquaculture production of North African countries in the year 2030. Surv Fish Sci 8(1):107–118

Illera-Vives M, Seoane Labandeira S, Brito LM, López-Fabal A, López-Mosquera ME (2015) Evaluation of compost from seaweed and fish waste as a fertilizer for horticultural use. Sci Hortic 186(2018):101–107

Inguglia L, Chiaramonte M, Di Stefano V, Schillaci D, Cammilleri G, Pantano L, Mauro M, Vazzana M, Ferrantelli V, Nicolosi R, Arizza V (2020) Salmo salar fish waste oil: fatty acids composition and antibacterial activity. PeerJ 8:e9299

Ivanovs K, Blumberga D (2017) Extraction of fish oil using green extraction methods: a short review. Energy Procedia 128:477–483

Jiang Q, Yang F, Jia S, Yu D, Gao P, Xu Y, Xia W (2022) The role of endogenous proteases in degrading grass carp (*Ctenopharyngodon idella*) myofibrillar structural proteins during ice storage. LWT 154:112743

Jubin JJ, Radzi NM (2022) Application of fish waste fertilizer on the growth of maize (Zea mays). In: IOP conference series: earth and environmental science, vol 1059, no 1, p 012070). IOP Publishing. Food and Agriculture Organization of the United Nations FAO 2015, World State of Soil Resources

Jung JM, Oh JI, Park YK, Lee J, Kwon EE (2019) Biodiesel synthesis from fish waste via thermally-induced transesterification using clay as porous material. J Hazard Mater 371:27–32

Kannahi M, Arulmozhi R (2013) Production of biodiesel from edible and non-edible oils using Rhizopus oryzae and Aspergillus niger. Asian J Plant Sci Res 3(5):60–64

Karkal SS, Kudre TG (2020) Valorization of fish discards for the sustainable production of renewable fuels. J Clean Prod 275:122985

Kerian K (2019) Synthesis and characterization of hydroxyapatite powder from fish bones and scales using calcination method. Synthesis 28(18):82–87

Khoddami A, Ariffin AA, Bakar J, Ghazali HM (2009) Fatty acid profile of the oil extracted from fish waste (head, intestine and liver) (*Sardinella lemuru*). World Appl Sci J 7(1):127–131
Lagat MK, Were S, Ndwigah F, Kemboi VJ, Kipkoech C, Tanga CM (2021) Antimicrobial activity of chemically and biologically treated chitosan prepared from black soldier fly (*Hermetia illucens*) Pupal Shell Waste. Microorganisms 9(12):2417
Langberg H, Skovgaard D, Petersen LJ, Bülow J, Kjær M (1999) Type I collagen synthesis and degradation in peritendinous tissue after exercise determined by microdialysis in humans. J Physiol 521(1):299–306
León-López A, Morales-Peñaloza A, Martínez-Juárez VM, Vargas-Torres A, Zeugolis DI, Aguirre-Álvarez G (2019) Hydrolyzed collagen—sources and applications. Molecules 24(22):4031
Li P, Wu G (2018) Roles of dietary glycine, proline, and hydroxyproline in collagen synthesis and animal growth. Amino Acids 50(1):29–38
Li Y, Liu Y, Li R, Bai H, Zhu Z, Zhu L, Zhu C, Che Z, Liu H, Wang L, Huang L (2021) Collagen-based biomaterials for bone tissue engineering. Mater Des 210:110049
Lock EJ, Waagbø R, Wendelaar Bonga S, Flik G (2010) The significance of vitamin D for fish: a review. Aquac Nutr 16(1):100–116
Martínez JP, Falomir MP, Gozalbo D (2001) Chitin: a structural biopolysaccharide. eLS
Masso-Silva JA, Diamond G (2014) Antimicrobial peptides from fish. Pharmaceuticals 7(3):265–310
Mata TM, Mendes AM, Caetano NS, Martins AA (2014) Properties and sustainability of biodiesel from animal fats and fish oil. Chem Eng Trans 38:175–180
Molina-Ramírez C, Mazo P, Zuluaga R, Gañán P, Álvarez-Caballero J (2021) Characterization of chitosan extracted from fish scales of the Colombian endemic species Prochilodus Magdalenae as a novel source for antibacterial starch-based films. Polymers 13(13):2079
Muslim T, Rahman MH, Begum HA, Rahman MA (2013) Chitosan and carboxymethyl chitosan from fish scales of Labeo rohita. Dhaka Univ J Sci 61(1):145–148
Mustafa N, Ibrahim MHI, Asmawi R, Amin AM (2015) Hydroxyapatite extracted from waste fish bones and scales via calcination method. Appl Mech Mater 287–290
Muthumari K, Anand M, Maruthupandy M (2016) Collagen extract from marine finfish scales as a potential mosquito larvicide. Protein J 35(6):391–400
Nirmal NP, Maqsood S (2022) Seafood waste utilization: isolation, characterization, functional and bio-active properties, and their application in food and nutrition. Front Nutr 1351
Nosheen S, Ajmal I, Song Y (2021) Microbes as biofertilizers, a potential approach for sustainable crop production. Sustainability 13:1868
Nurhayati T, Ambarsari L, Nurilmala M, Abdullah A, Rakhmawati IAI (2020) Pepsin activity from gastric of milkfish and catfish in Indonesian Waters. In: IOP conference series: earth and environmental science, vol 404, no 1, p 012060. IOP Publishing
Ooi HM, Munawer MH, Kiew PL (2021) Extraction of chitosan from fish scale for food preservation and shelf-life enhancer
Pellis A, Guebitz GM, Nyanhongo GS (2022) Chitosan: sources, processing and modification techniques. Gels 8(7):393
Pertiwi RM, Pamungkas ID, Prastyo DT, Trisdiani DU, Pasaribu E, Nurhayati T (2020) Cathepsin characterization from crude extract of yellow pike (*Congresox talabon*). In: IOP conference series: earth and environmental science, vol 404, no 1, p 012013). IOP Publishing
Pillai SK, Ray SS (2012) Chitosan-based nanocomposites. Nat Polym 2:33–68
Pochanavanich P, Suntornsuk W (2002) Fungal chitosan production and its characterization. Lett Appl Microbiol 35(1):17–21
Reames E (2012) Nutritional benefits of seafood. Southern Regional Aquaculture Center
Rinaudo M (2006) Chitin and chitosan: properties and applications. Prog Polym Sci 31(7):603–632
Romero-Hernández O, Romero S (2018) Maximizing the value of waste: from waste management to the circular economy. Thunderbird Int Bus Rev 60(5):757–764
Rosa P, Sassanelli C, Terzi S (2019) Towards circular business models: a systematic literature review on classification frameworks and archetypes. J Clean Prod 236:117696

Ruiz GAM, Corrales HFZ (2017) Chitosan, chitosan derivatives and their biomedical applications. Biol Act Appl Mar Polysacch 87
Rumengan IFM, Suptijah P, Wullur S, Talumepa A (2017) Characterization of chitin extracted from fish scales of marine fish species purchased from local markets in North Sulawesi, Indonesia. In: IOP conference series: earth and environmental science, vol 89, no 1, p 012028. IOP Publishing
Rungruangsak-Torrissen K, Moss R, Andresen LH, Berg A, Waagbø R (2006) Different expressions of trypsin and chymotrypsin in relation to growth in Atlantic salmon (Salmo salar L.). Fish Physiol Biochem 32(1):7–23
Saadoun IM (2015) Impact of oil spills on marine life. In: Larramendy ML, Soloneski S (eds) Emerging pollutants in the environment-current and further implications. IntechOpen, London
Saha D, Bhattacharya S (2010) Hydrocolloids as thickening and gelling agents in food: a critical review. J Food Sci Technol 47(6):587–597
Senevirathne M, Kim SK (2012) Development of bioactive peptides from fish proteins and their health promoting ability. Adv Food Nutr Res 65:235–248
Shahidi F, Kamil YJ (2001) Enzymes from fish and aquatic invertebrates and their application in the food industry. Trends Food Sci Technol 12(12):435–464
Shahsavani S, Abaspour A, Parsaeeyan M, Yonesi Z (2017) Effect of fish waste, chemical fertilizer and biofertilizer on yield and yield components of bean (Vigna sinensis) and some soil properties. Iran J Pulses Res 8(1):45–59
Sharma YC, Singh B, Madhu D, Liu Y, Yaakob Z (2014) Fast synthesis of high quality biodiesel from 'waste fish oil' by single step transesterification. Biofuel Res J 1(3):78–80
Sikorski ZE, Scott DN, Buisson DH, Love RM (1984) The role of collagen in the quality and processing of fish. Crit Rev Food Sci Nutr 20(4):301–343
Simpson BK (2007) Pigments from by-products of seafood processing. In: Shahidi F (ed) Maximising the value of marine by-products. Woodhead Publishing
Solli L, Schnurer A, Horn SJ (2018) Process performance and population dynamics of ammonium tolerant microorganism during co-digestion of fish waste and manure. Renew Energy 125:529–536
Suriani NW, Komansilan A (2019) Enrichment of omega-3 fatty acids, waste oil by-products canning tuna (*Thunnus* sp.) with urea crystallization. J Phys Conf Ser 1317:012056
Tavares JFP, Baptista JAB, Marcone MF (1997) Milk-coagulating enzymes of tuna fish waste as a rennet substitute. Int J Food Sci Nutr 48(3):169–176
Tayel AA, Moussa SH, Wael F, Elguindy NM, Opwis K (2011) Antimicrobial textile treated with chitosan from *Aspergillus niger* mycelial waste. Int J Biol Macromol 49(2):p241-245
Ucak I, Afreen M, Montesano D, Carrillo C, Tomasevic I, Simal-Gandara J, Barba FJ (2021) Functional and bioactive properties of peptides derived from marine side streams. Mar Drugs 19(2):71
Usman M, Sahar A, Inam-Ur-Raheem M, Rahman UU, Sameen A, Aadil RM (2022) Gelatin extraction from fish waste and potential applications in food sector. Int J Food Sci Technol 57(1):154–163
Vázquez JA, Hermida-Merino C, Hermida-Merino D, Piñeiro MM, Johansen J, Sotelo CG, Pérez-Martin RI, Valcarcel J (2021) Characterization of gelatin and hydrolysates from valorization of farmed salmon skin by-products. Polymers 13(16):2828
Venugopal V (2016) Enzymes from seafood processing waste and their applications in seafood processing. Adv Food Nutr Res 78:47–69
Wang H (2021) A review of the effects of collagen treatment in clinical studies. Polymers 13(22):3868
White SA, Farina PR, Fulton I (1979) Production and isolation of chitosan from Mucor rouxii. Appl Environ Microbiol 38(2):323–328
Wisniewski A Jr, Wiggers VR, Simionatto EL, Meier HF, Barros AAC, Madureira LAS (2010) Biofuels from waste fish oil pyrolysis: chemical composition. Fuel 89(3):563–568
Yahyaee R, Ghobadian B, Najafi G (2013) Waste fish oil biodiesel as a source of renewable fuel in Iran. Renew Sustain Energy Rev 17:312–319

Yulisa A, Chairattanawat C, Park SH, Jannat MAH, Hwang S (2022) Effect of substrate-to-inoculum ratio and temperatures during the start-up of anaerobic digestion of fish waste. Ind Domest Waste Manage 2(1):17–29

Zamani A, Edebo L, Sjöström B, Taherzadeh MJ (2007) Extraction and precipitation of chitosan from cell wall of zygomycetes fungi by dilute sulfuric acid. Biomacromol 8(12):3786–3790

Zhou L, Budge SM, Ghaly AE, Brooks MS, Dave D (2011) Extraction, purification and characterization of fish chymotrypsin: a review. Am J Biochem Biotechnol 7(3):104–123

Fish Waste and By-Product Utilization: A Circular Economy

Raman Jasrotia, Seema Langer, and Menakshi Dhar

Abstract Significant increase in population growth across the globe followed by the rapid rise in industrialization and urbanization has augmented aquaculture production; consequently, the amount of fish waste generated has also seen a remarkable rise around the globe. The management of waste is one of the main problems that have a great effect on the environment. The utilization of fish waste and fish by-products permits the waste reduction that else would lead to the pollution of both the terrestrial and aquatic ecosystems in the future. The considerable stress of the fishery wastes on the ecosystem puts forward the importance of using it as a potential source compound that helps to promote good health. The nutritive value of fish waste products is almost similar to the edible parts of fish. Collagen, gelatin, bioactive peptides, protein hydrolysates, enzymes, anti-microbial peptides, pigments, chitosan, chitin, lipids, and minerals that have high nutritional value, good flavor, and are suitable for storage can be generated by re-processing the fish waste. The use of this new biological source for generation of the compounds having high value will also prove beneficial for the sustainable use of biotic sources. Utilization of fish by-products symbolizes an important tool in lowering the problem of hunger and food shortage in developing countries. In addition to this, the multifariousness of the productive chains promotes the generation of employment opportunities and as a result, turns out to be advantageous both for the environment and for the upliftment of the socio-economic conditions of human society. The utilization of fish wastes and fishery by-products will also help in improving the economy of fish-processing industries.

Keywords Economy · Fish-processing · Food shortage · Fishery wastes · Sustainable use

---

R. Jasrotia (✉) · S. Langer · M. Dhar
Department of Zoology, University of Jammu, Jammu 180006, India
e-mail: ramanjasrotia18@gmail.com

R. Jasrotia
Department of Zoology, Cluster University of Jammu, Jammu, India

S. Maqsood et al. (eds.), *Fish Waste to Valuable Products*, Sustainable Materials and Technology, https://doi.org/10.1007/978-981-99-8593-7_22

1 Introduction

Fish farming for commercial purposes such as food products is known as aquaculture production. It includes the rearing of fish in tanks or any other enclosures on a commercial scale, generally for the procurement of food. The target of aquaculture management is to increase the output production. The aquaculture sector has shown a fast expansion in recent years as compared to any other livestock sector. This sector has depicted exponential growth over the last few decades as it serves as a fundamental source for fulfilling the rising demand for animal protein, but with the traditional linear model, aquaculture may become unsustainable. This sector generates huge quantities of wastes that have adverse impact on environment. Such effects may be lessened by the utilization of these waste products with implementation of circular economy strategies (Dauda et al. 2019). The circular economy model suggests that wastes generated from the fisheries sector have applicability in agriculture, horticulture, pharmacology, food, and feed industries. Such approaches valorize the fish waste products and reduce the effect on the environment and provide additional benefits.

2 Fish Productivity

At present, the fisheries production is approximately 160 million tonnes across the globe. Aquaculture production is dependent on net primary productivity, the way this yield navigates through food chain network in an aquatic ecosystem and enters the human food chain (Iverson 1990). In 2002, 76% of the production in fisheries sector was used directly for consumption by humans, and the leftover 24% was used for the fish oil and fish meal production (FAO 2004). In 2004, out of the total 164 million tonnes of aquatic production at the global level, about 77% was from the marine ecosystem, and only 23% was contributed from the inland waters. Out of the total production, 32% of molluscs, and fish comes from the aquaculture sector, 66% was from capture production. With the rapid rise in aquaculture production, it has been estimated that by 2030, production from aquaculture will be very close to capture production (Brander 2007). According to the studies done by Jamu and Brummet (2004) and FAO (2011), it has been found that aquaculture received great importance on the global level due to food shortage problems.

Fish farming can be considered as an approach for the development of the economy and to mollify to economic development and mollifying poverty (Mwaijande and Lugendo 2015). Aquaculture production has great potential as it generates employment and establishes food security by providing highly nutritive proteins and other essential micronutrients of animal origin (FAO 2012). A report developed by FAO, in 2012 has also predicted that fish is the source of food for about 3 billion people across the world. Fish farming helps to improve the income of different communities and boost food security at the global level (Shava and Gunhidzirai 2017). In the past few years, the aquaculture industries are rapidly growing in the food industry sector across the globe (Tsani and Koundouri 2018). Production from the aquaculture sector even provides economic benefits to the society and thus forms an important pillar of the Blue Growth targets. Worldwide, 40% of the fish used for consumption by humans are obtained from commercial farms (Goldburg and Naylor 2005). Thus, the fisheries sector has an immense contribution to the economy of any country. In India, fish production has shown an increasing trend and in 2006, it has attained a record of 6.4 million tonnes (Ayyappan and Diwan 2006). According to their study, 1.1% of the total GDP is contributed by the fisheries sector, and India stands at fourth place in terms of total fish production all across the world. Sustainable Development Goals in its Agenda for the year 2030 has set aim for the contribution of fish production towards the nourishment of the world's population. Costello et al. (2020) concluded that 30% of the protein consumed across the globe is derived from edible fish.

Fish production for food has shown a significant increase during the last few years. The total fish trade across the globe increased from 8 billion US dollars in 1976 to 109 billion US dollars in 2010. In 2014, aquaculture production was recorded to be 3.4 million tons across the globe (FAO 2016). A remarkable rise in the growth of global fish supply was observed in the period between 1961 and 2016, and during this period, the average annual growth rate of the fisheries sector was 3.2%. A similar increase has been observed in the global fish consumption per capita, in which it was reported that in 1916, the consumption per capita was 9 kg, which rose to 20.2 kg in 2015 (FAO 2018). As a result, a dramatic rise in fish waste and by-products has been seen across the globe.

3 Fish By-Products

A substantial quantity of effluents such as feed waste, feces, medications, etc. is discharged from the fish production units, which have a serious impact on the environment. There is a growing need to integrate fish food production with a suitable action for the fish waste and by-products processing. It is need of time to manage fish waste in a better way to control the various environmental-related issues and for the commercial use of fish waste. Fish waste is used for the production of fertilizers, fish oil, and fish meal and some of the fish waste may be used as a feed in aquaculture production practices (Mo et al. 2018; Stevens et al. 2018). Fertilizers derived from the fish waste are suitable for organic farming, generally for the plants grown in gardens. Based on the kind of fish and processing level, 30–70% of the fish is considered as waste (Ahuja et al. 2020). Researchers in aquaculture field have evaluated that the huge amount of waste released from the fishery sector is thrown away which besides having a serious impact on the environment also leads to the loss of valuable commodities, which can be obtained from the fish waste. Production of fish waste has been validated as a pressing issue that has stimulated the research for finding out a feasible solution (Sardà et al. 2013). It has been evaluated that the

Fig. 1 Different fish tissues such as scales, air bladder, flesh, viscera, fins, operculum

different fish tissues such as gut, skin, ovaries, head, operculum, fin, liver, and viscera are disposed of and are categorized as fish waste (Fig. 1).

The composition and the quantity of fish waste vary with the fishing areas and thus the percentage of discarded material is not consistent (Davies et al. 2009). According to the report published by the FAO (2019); it has been estimated that every year approximately nine million tonnes of fish waste are discarded. Kandyliari et al. (2020) carried out a study to find out the percentage of waste products of the meager and gilthead sea bream. They found that the fish waste obtained from these fishes includes 5–9% bones, 2–3% scales, 5–7% intestines, 17–19% heads, 6–7% skin, and 1–2% trimmings. They calculated the nutritional value of such fish by-products and found it similar to the fish fillet.

During the fish filleting, only 30–50% of the meat is obtained while the 45% of the body of a fish includes 4–5% skin, 24–34% bones, and 21–25% head remains unutilized and is discarded (Ghaly et al. 2013). According to Mo et al. (2018), large quantities of fish waste are generated from fish processing industries and this constitutes about 25% of the total productivity. Huge quantities of the scales removed from the body of a fish are part of fish by-products and being less biodegradable their management is difficult and they become a problem for the environment. Waste generated from aquaculture varies that depends upon the fish species and region from where the fish is collected. Fish farms release huge amounts of organic waste and the management of this waste is the need of an hour as it poses a great risk to the environment. Appropriate technologies should be used for the management of fish waste.

In 2018, fishery production reached the volume of 210.9 million tonnes (FAO 2020), which in turn increased the productivity of huge amounts of fish waste and by-products. These fish wastes such as skin, bones, trimmings, head, and intestines are excellent sources of protein, calcium, and lipids. It has been found that fish waste contains approximately 49.22–57.92% protein content, 7.16–19.10% fat content, and 21.79–30.16% ash content (Vidotti et al. 2003; Abbey et al. 2017). Such wastes with good biochemical composition are dumped in the environment, and this leads to various environment-related issues.

Wastes derived from the aquaculture sector can be grouped into four different categories according to Dauda et al. (2019):

(a) Solid waste materials are derived from unconsumable part of the animal.
(b) Dissolved organic matter includes elements such as nitrogen and phosphorus.
(c) Dissolved chemical compounds.
(d) Pathogens.

Recent studies by Lee et al. (2019) have classified the fish wastes into two categories:

(a) Biological that includes the leftover food and fecal matter.
(b) Waste discharge consists of nitrates and organic waste.

Thus, it is imperative to use the fish by-products to obtain certain value-added products like minerals, chitin, enzymes, collagen, and polyunsaturated fatty acids, thus reducing the adverse impact of fish by-products on the environment and maximizing the economic benefits (Shahidi et al. 2019; Parvathy et al. 2018).

4 Utilization of Fish Waste

Various products can be processed out of the fish waste, which can contribute to the economic growth of any nation and even lessen the problems generated due to the accumulation of the fish waste. Composition of the by-products of fish is similar to fish fillet that is consumed as a food product. The fish by-products are an essential source of fatty acids, proteins, and minerals. Studies carried out by Kandyliari et al. (2020) had disclosed that revealed that head, viscera, and bones are significant sources of lipids and calcium, skin is the rich source of protein. Fish by-products are an important source of proteins and fatty acids, which are known to have antioxidant, anti-tumor, anti-hypertensive, and anti-bacterial activities. The use of fish by-products is procuring significant attention these days as different bio-compounds are extracted from such by-products as oil, enzymes, peptides, collagen, chitosan, biofuel, biogas, fertilizers, fish insulin, fish meal, and fish sauce. Valorization of various fish by-products has been depicted in Fig. 2.

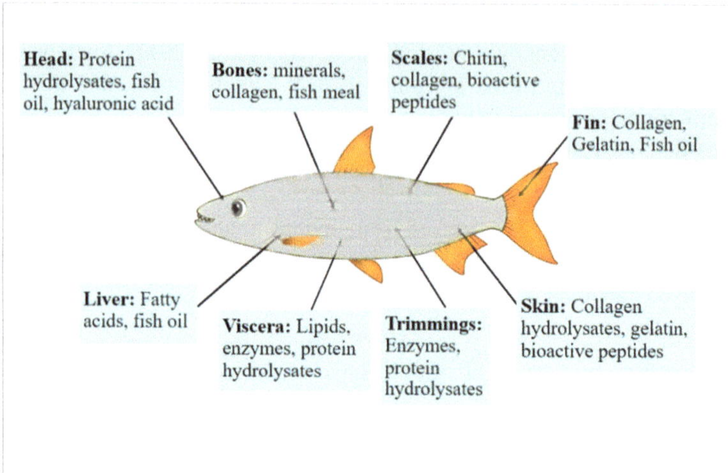

Fig. 2 Valorization of various fish by-products

4.1 Collagen

Collagen extraction from fish skin ameliorates usefulness of the byproducts derived from fish. Extraction of the collagen from the scales of fish has numerous applications in the medical sciences (Rodríguez et al. 2018). It has been used in various cosmetic products and medicines due to its property of decreased immunogenicity (Sionkowska et al. 2020). Based on structural features, Type 1 and Type 2 collagen are found in the fish, which have wide applications in the field of cosmetics and wound healing medicines (Karim and Bhat 2009). By weight, approximately, 30% collagen is present in the bones, skin, and fins of fish. Fish waste symbolizes the inexpensive and huge sources of collagen (Jafari et al. 2020). Nagai and Suzuki (2000) proposed that the bones, fins, and skin of fish are responsible for the extraction of 36–54% of collagen. Scales of the sardines are important sources of collagen (Belouafa et al. 2018). Scales of various species of fish such as *Pagrus major, Catla catla, Oreochromis nilotics, Hypophthalmichthys nobilis, Cirrhinus mrigala* (Ikoma et al. 2003; Fengxiang et al. 2011; Mahboob 2015) have been used for the collagen extraction.

4.2 Gelatin

Fish gelatin consists of both non-essential and essential building blocks of proteins; thus, it can be used in various food industries. In food sources, bovine and porcine gelatin acts as alternative source for fish gelatin. Gelatin obtained from fish waste can be utilized in bakeries, dairy, cosmetics, shampoos, photography, tablets, capsules, and syrups (Lin et al. 2017; Ishaq et al. 2020). Collagen present in the fish skin is an important source of gelatin production (Jamilah and Harvinder 2002). According to Ahmad et al. (2017); five different steps viz., cleaning, isolation, purification, concentration, and drying are used for converting fish waste to gelatin. Both the acid and alkali processes are involved in the hydrolyzation of collagen into gelatin. At acidic pH, gelatin is isolated from the skin of fish by hydrolyzation of collagen at acidic (Baziwane and He 2003). Because of the high concentration of collagen in the bones and skin of fish, the production of gelatin from the waste of the fish is done on extensive level these days (Alfaro et al. 2015).

4.3 Chitin

Fish waste especially the scales of the fish are excellent source of chitin and its derivatives. Chitin was isolated for the first time by Zaku et al. (2011) from the common carp scales. They dried, macerated, washed the scales with acid, rinsed them with de-mineralized water, and then de-proteinized. This way they obtained the chitin from the fish waste. Chitin and chitosan were obtained from the scales of a fish, *Tilapia nilotica* by Uawonggul and Ruksakulpiwat (2002) and Alcalde and Fonseca (2016). Approximately, 20% of the chitin was obtained from 50 gm of the scales of fish by them. Similarly, 45% chitin yield was obtained from *Chlorurus sordidus* by Rumengan et al. (2017). Chitin obtained from the waste of fish products was utilized in numerous fields such as nutraceuticals, pharmaceutics, bioremediation, and cosmetic industry. The chitosan derived from the *Papuyu* fish scales has been used for ameliorating process of removal of iron from groundwater (Irawan et al. 2018). Chitin finds its huge applications in the field of pharmaceutical industries as they have anti-bacterial, anti-oxidant, anti-cholestremic, and anti-thrombogenic properties. Several authors have also reported the antibacterial activities of chitosan (Zheng and Zhu 2003; Benhabiles et al. 2012). Chitosan is ideal for the storage of easily perishable commodities such as vegetables, meat, eggs, dairy products, and sea food as the films developed from the chitosan can form oxygen barriers (Coppola et al. 2021). Chitosan is isolated from the fish scale waste mainly by following three steps, i.e., deprotonation, demineralization, and deacetylation (Aichayawanich and Saengprapaitip 2019). Suresh et al. (2022) revealed that from the 100 gm of fish scale; 12% of the chitosan can be extracted. Deacetylation of the chitin using the chitin deacetylase produces chitosan (Santos et al. 2020).

4.4 Fish Oil

The volume of fish waste augments with an increase in the production of fish. In 2016, out of the fish oil produced in total, 26% was obtained from the waste products of the fish (Jackson and Newton 2016). Fish oil is obtained from the head, viscera, fins, tail, and skin of the fish. The viscera of the fish waste contains a considerable quantity of fat, oil, and proteins (Kudre et al. 2017). Yasin et al. (2021) extracted the fish oil from waste products of a fish, *Pangasius hypophthalmus* using the modified Soxhlet method. The nutrients present in the fish waste especially fatty acids increase its demand in the market. Various techniques are utilized for the extraction of fish oil from fish by-products and the selection of the technique for the fish oil extraction (Kerton et al. 2013; Bonilla-Méndez and Hoyos-Concha 2018). Khoddami et al. (2012) utilized the various organs of the fish, *Euthynnus affinis* for the extraction of oil. They concluded that the predominant fatty acids in the tuna wastes were stearic acid, docosahexaenoic acid, and oleic acid. According to their studies, the head of the tuna contains a suitable amount of omega-3 fatty acids, ratio of ω-3/ω-6, and lipid content so it is suitable for consumption by humans. Different methods utilized for the extraction of the fish oil from the waste products of the fish include Goldfisch, chloroform–methanol, Bligh and Dyer, and acid digestion (Shahidi 2003). Fish oil contains a long chain of fatty acids and can be used in the food and feed industry, nutraceuticals, and aquaculture. Fish oil has high value as it consists of essential polyunsaturated fatty acids viz., eicosapentaenoic acid, docosahexaenoic acid, and omega-3 fatty acids, which are advantageous for the health of humans (Khoddami et al. 2009). The omega-3 fatty acids are known to have positive impacts such as reducing blood pressure, symptoms of asthma, improving learning ability and the survival of a person suffering from cancer, prevent arrhythmias and atherosclerosis (Kim and Park 2006; Tawfik 2009). Extraction of the fish oil was done from the liver waste of ray species by Sellami et al. (2018) and the fatty acid profiling revealed that the unsaturated fatty acids were present in the major concentration in all the samples. Besides this, various carotenoids and phenolic compounds were also found in these oils that are known to have antioxidant activity. Inguglia et al. (2020) evaluated that the waste products of *Salmo salar* can be used to treat various pathogenic infections.

4.5 Biofuels

Biofuels can be used as an alternative to conventional energy sources as they are free from pollution and so are advantageous over conventional fossil fuels. Oil extracted from fish waste acts as a basic material for biodiesel production (Samat et al. 2018). Research studies have shown that more than 50% of the fish is transformed into fish waste. Extracted fish fat from the fish waste was subjected to saponification and then transesterification. This process resulted in the production of biodiesel (Girish et al. 2017). Results of studies done by Zhang et al. (2020) revealed that *insitu* transesterification of the waste of fish can be done for the production of biodiesel. Lin and Li (2009) suggested that the biodiesel extracted from the marine fish has high hydrogen and carbon content, low oxygen content, high heating value, and cetane index. They obtained a high yield (>97%) from the fish waste. The various physico-chemical parameters of biodiesel obtained from waste of *Tilapia* were evaluated by Martins et al. (2015), and it was found to have all the characteristics of good fuel. Moroccan fish oil waste was used as a raw material by Kara et al. (2018) for the production of biodiesel, and this fuel was found to be free from glycerol, and in accordance with international standards. Wastes of the fish can be utilized for the production of biodiesel as they are non-toxic and biodegradable. When compared with conventional fuel, biodiesel is less toxic and produces fewer amounts of carbon dioxide and other particulates. The efficiency of biofuels can be confirmed by the usage of various techniques such as gas chromatography-mass spectrometry (GCMS), Fourier-transform nuclear magnetic resonance (FT-NMR), and proton nuclear magnetic resonance (H-NMR) (Yuvaraj et al. 2019).

4.6 Enzymes

Enzymes have numerous applications and are utilized in agriculture, textile, pharmaceuticals, various other manufacturing units. Fish wastes are good sources of enzymes such as elastase, alkaline phosphatases, lipases, hyaluronidase, proteases, transglutaminases, acetylglucosaminidase, etc. Enzymes isolated from the wastes of the fish inhabiting cold waters are more useful as they work at low temperatures and thus save energy and provide protection to the food items (Venugopal 2016). Various proteolytic enzymes, viz., elastase, collagenase, and pepsin have been isolated from the viscera of fish. Gildberg (2004) utilized the process of autolysis for the isolation of pepsin from the fish silage. Generally, protease enzyme is found in fish waste. Fish viscera is the rich source of various enzymes, viz., hyaluronidase, chitinase, and alkaline phosphatase (Shahidi and Kamil 2001). Pepsin isolated from the viscera of fish is highly active under acidic conditions (Morrissey and Okado 2007). Sriket (2014) isolated collagenase enzyme from the muscles of *Scomber japonicas, Paralichthys olivaceous*, common carp, and rainbow trout. Myrnes and Johansen (1994) isolated the lysozyme from the waste of Arctic scallops, which was found to inhibit the growth of both the Gram negative and Gram positive bacteria (Guérard et al. 2005). Harikrishna et al. (2017) reported that the fish scale waste is used as a raw material for the alkaline protease production by *Bacillus altitudinis* GVC II. Pepsin described from the viscera of fish waste can be used as an alternative for the hog-derived pepsin (Kim and Dewapriya 2014).

4.7 Bioactive Peptides

Fish waste contains high-quality protein, thus representing an important source of biofunctional peptides. Peptides isolated from the fish wastes depict HIV protease inhibitory, anti-microbial, and calcium binding activity. Biopeptides and protein hydrolysates derived from the scales of seabream, the skin of pollack, bones, and scales of yellowtail are known to inhibit the action of the main component of renin-angiotension system under *invitro* conditions. Thus, it acts as a natural inhibitor and helps in the treatment of hypertension (Harnedy and FitzGerald 2012). Biopeptides produced by the process of protein hydrolysis of fish waste possess anti-oxidant activity, besides having various functional and nutritional activities (Tacias-Pascacio et al. 2021). Peptides derived from the skin waste of *Ctenopharyngodon idella* were found to have anti-oxidant properties (Cai et al. 2015). Vázquez et al. found that fish hydrolysates derived from the head, trimmings, and viscera of *Scophthalmusmaximus* have anti-hypertensive and anti-oxidant properties. Peptides derived from skin of *Magalapis cordyla* and *Otolithes ruber* reduce the peroxidation of polyunsaturated fatty acids (Kumar et al. 2012).

4.8 Fertilizers

Fertilizers can be manufactured from the fish waste, which are then used in the horticultural fields, and they help to ameliorate the quality of soil and plants. Fertilizers prepared from fish waste can be used in gardens, field crops, and vegetable production as they increase the nutrient content of the soil, decrease the incidence of diseases in plants and eliminate unwanted plants (Jayvardhan and Arvind 2020). Kusuma et al. (2019) prepared a compost by placing 2 kg of fish waste in a porous container having 15.08% of moisture content and kept it as such for 2 weeks. This fish waste was then mixed with the remaining fruits and vegetables and compost was prepared. They recorded high levels of carbon, nitrogen, potassium, and phosphorus in the compost made from fish waste. The results of the experiment carried out by Radziemska et al. (2019) revealed that compost made from fish waste is suitable for use in agricultural fields. They observed that compost made from fish waste increased the dry and fresh matter of the leaves of *Lactuca sativa.* Significant rise in concentration of different elements like nitrogen, potassium, calcium, sodium, and phosphorus was observed in the leaves of this plant. Ranasinghe et al. (2021) explained that liquid organic fertilizers can be produced by the hydrolysis of fish waste, which can be utilized in agricultural fields.

5 Role of Fish Waste in the Circular Economy Era

Every year, extraction of 90 billion tonnes of primary materials is done and only 9% material is recycled (UNEP 2019). This process upsets the ecological balance and has an adverse effect on climate, ecosystem, and the health of humans. By enhancing the utility of any resource and incorporating the circularity concept during the process of production and utilization, the circular economy elevates. A circular economy is pivotal in securing future prosperity and economic benefits. It is an alternative economic model that aims to reduce the waste to minimum levels by reusing and recycling the waste products. Across the globe, our economy is 8.6% circular and the remaining 91.4% is wasted according to the Circularity Gap Report, 2021. The main aim behind the circular economy is to design out the waste products and the environmental problems caused by them by keeping the products in use and aid in regenerating the natural systems. This involves the maintenance of the value of the products as far as possible by giving them back into the nature at the end of their utilization and this way helps in decreasing the waste generation. And thereby declines the over-use of raw materials, greenhouse gas (GHG) emission, and the loss of biodiversity. Management of fish waste in an efficient manner should be aligned with the concept of circular economy. Recycling and reusing of the fish waste into new products can lessen the dependence on virgin resources. The model of circular economy in aquaculture is based on re-utilization of fish waste and by-products, in which the outputs of one cycle become inputs for any another cycle. For example, processing of fish produces left over wastes such as scales, viscera, fins, gills, skin, trimmings, etc. These waste products generated from the fish processing can be used in feedstock, as a fertilizer for growing crops, and this way helps to support the other forms of biodiversity. This way the circular economy helps in the valorization of the fish waste and by-products that may otherwise cause various environmental problems. The principles of circular economy revolve around the reduction of waste, increase of efficiency and favoring of more sustainable ecosystems. As far as the perspective of circular economy is concerned, the reutilization of the products in the foodservice industry is considered as most popular in declining the waste (Tola et al. 2023).

The utility and maximum value of products are maintained in the circular economy by increasing the use of resources and extending the lifetime of any product. The main idea of circular economy is based on regenerative development, which means that the cycling of earth's resources restores and enhances the economy instead of depleting it. In aquaculture, the adoption of a circular economy addresses the problem of fish waste by the creation of value-added products from these waste products, thus contributing to the development of a healthy ecosystem and providing employment opportunities. The proper management of fish wastes by the use of a circular economy has a positive influence on the economy and reduces the generation of waste, thereby, protecting both the health and the environment (Fraga-Corral et al. 2022). The most efficacious approach to reusing the fisheries by-products and wastes is the recovery of important biomolecules such as gelatin, lipids, collagen, and pigments, which can be used in the food industry, pharmaceuticals, and cosmetics. The circular economy helps in the sustainable development and the maintenance of the resources and products for a long period of time and this way minimizes the generation of waste (Carus and Dammer 2018). According to the studies conducted by Angouria-Tsorochidou et al. (2021), the circular economy helps in the conservation of environment and alleviated poverty so as to ameliorate the well-being of individuals and significantly decline the risks posed by the fish waste for the environment. The main challenge behind the blue economy is to generate the power of products obtained from the sea, which can then be utilized as a raw material for various applications. Circular economy focuses on the preparation of value-added product from the waste so that it gets recycled and has no harmful effect on the environment. The waste generated from the different parts of the fish supply chain can be processed and utilized in the pharmaceutical units.

The circular economy focuses on expanding the rate of reuse of fish waste products and this way helps in the reduction of pressure on the various natural resources. As far as sea food is concerned, large amount of biomass is being lost through the discard or by-catch from aquaculture such as sludge, and by-products that are produced by the fish processing such as offal and trimmings. All these can be valorized and used for the generation of value-added products, production of food and feed generation for the pets. Awareness, communication, and interaction with the collaborators play a key role in development of circular economy as education and acceptance among the local people will increase the demand for the valorization of fish waste (Cooney et al. 2023). In general, the circular production system will have a great impact on the production efficiency and it will lessen the demand of natural resources and food production and all this will address the different goals of the United Nations Sustainable Development Programme and thereby contributing a lot to the food security and maintenance of the life on the Earth.

6 Conclusion

The fish by-product exploitation will be helpful in the sustainability of aquaculture. The valorization of fish waste, which is considered to be useless, is known to play a key role in resolving impact of waste on environment. The principle aim of fishery management is to convert the fish wastes and recover the significant product before their disposal. Conversion of the biodegradable fish waste into different bio compounds is the most effective method for the processing of fish waste. Fish waste finds its prominent use in cosmetics, pharmaceutical, and textile manufacturing units. Processing of fish waste products is essential for conservation of natural resources. Sustainable use of fishery waste expands the potential of the aquaculture sector by providing employment and thus generates income for the local communities. It is necessary to manage the fish wastes in a better way to overcome the environment-related issues and concomitantly ensure the use of by-products. For this, the government should build up several regulations and policies along with requisite infrastructure and facilities. The present paper illustrates the production process of different value-added products with several applications that can be extracted from fish wastes. The utilization and the minimization of the fish waste and its conversion into numerous useful products are beneficial both from an economic and environmental point of view. Fishery wastes have numerous uses, and this industry can play an important role in ameliorating the livelihoods, creation of employment, increasing the environmental sustainability, and this way it can contribute a lot in the economy of a nation (Mozumder et al. 2022).

References

Abbey L, Glover-Amengor M, Atikpo MO, Atter A, Toppe J (2017) Nutrient content of fish powder from low value fish and fish byproducts. Food Sci Nutr 5:374–379. https://doi.org/10.1002/fsn3.402

Ahmad T, Ismail A, Ahmad SA, Khalil KA, Kumar Y, Adeyemi KD, Sazili AQ (2017) Recent advances on the role of process variables affecting gelatin yield and characteristics with special reference to enzymatic extraction: a review. Food Hydrocoll 63:85–96

Ahuja I, Dauksas E, Remme JF, Richardsen R, Løes AK (2020) Fish and fish waste-based fertilizers in organic farming—with status in Norway: a review. Waste Manage 115:95–112. https://doi.org/10.1016/j.wasman.2020.07.025

Aichayawanich S, Saengprapaitip M (2019) Isolation and characterization of chitosan from fish scale waste. IOP Conf Ser Earth Environ Sci 301:012051

Alcalde LB, Fonseca GG (2016) Alkali process for chitin extraction and chitosan production from Nile tilapia (*Oreochromis niloticus*) scales. Lat Am J Aquat Res 44:683–688. https://doi.org/10.3856/vol44-issue4-fulltext-3

Alfaro AD, Balbinot E, Weber CI, Tonial IB, Machado-Lunkes A (2015) Fish gelatin: characteristics, functional properties, applications and future potentials. Food Eng Rev 7(1):33–44

Angouria-Tsorochidou E, Teigiserova DA, Thomsen M (2021) Limits to circular bioeconomy in the transition towards decentralized biowaste management systems. Resour Conserv Recycl 164:105207

Ayyappan S, Diwan AD (2006) Fisheries research and development in India. Fish Chimes 26(1):19–23

Baziwane D, He Q (2003) Gelatin: the paramount food additive. Food Rev Intl 19(4):423–435

Belouafa S, Bourja L, Villain S, Tayane S, Bennamara A, Abourriche A (2018) Biocomposite based on collagen/calcium salts extraction from sardine scales. In: Smart application and data analysis for smart cities (SADASC'18)

Benhabiles MS, Salah R, Lounici H, Drouiche N, Goosen MFA, Mameri N (2012) Antibacterial activity of chitin, chitosan and its oligomers prepared from shrimp shell waste. Food Hydrocoll 29:48–56

Bonilla-Méndez JR, Hoyos-Concha JL (2018) Methods of extraction, refining and concentration of fish oil as a source of omega-3 fatty acids. Corpoica Cienc Tecnol Agropecu 19:645–668

Brander KM (2007) Global fish production and climate change. Proc Natl Acad Sci USA 104(50):19709–19714. https://doi.org/10.1073/pnas.0702059104

Cai L, Wu X, Zhang Y, Li X, Ma S, Li J (2015) Purification and characterization of three antioxidant peptides from protein hydrolysate of grass carp (Ctenopharyngodon idella) skin. J Funct Foods 16:234–242

Carus M, Dammer L (2018) The circular bioeconomy—concepts, opportunities, and limitations. Ind Biotechnol 14:83–91

Cooney R, de Sousa DB, Fernández-Ríos A, Mellett S, Rowan N, Morse AP, Hayes M, Laso J, Regueiro L, Wan AHL, Clifford E (2023) A circular economy framework for seafood waste valorisation to meet challenges and opportunities for intensive production and sustainability. J Clean Prod 392:136283

Coppola D, Lauritano C, Esposito FP, Riccio G, Rizzo C, de Pascale D (2021) Fish waste: from problem to valuable resource. Mar Drugs 19(2):116

Costello C, Cao L, Gelcich S, Cisneros-Mata MÁ, Free CM, Froehlich HE, Golden CD, Ishimura G, Maier J, Macadam-Somer I, Mangin T (2020) The future of food from the sea. Nature 588(7836):95–100

Dauda AB, Ajadi A, Tola-Fabunmi AS, Akinwole AO (2019) Waste production in aquaculture: sources, components and managements in different culture systems. Aquac Fish 4(3):81–88

Davies RWD, Cripps SJ, Nickson A, Porter G (2009) Defining and estimating global marine fisheries bycatch. Mar Policy 33(4):661–672

Fengxiang Z, Anning W, Zhihua L, Shengwen H, Lijun S (2011) Preparation and characterisation of collagen from freshwater fish scales. Food Nutr Sci 02:818–823
Food Agriculture Organization of the United Nations (2004) Fisheries Department. The State of World Fisheries and Aquaculture. FAO, Rome
Food and Agriculture Organization (FAO). Kassam L, Subasinghe R, Phillips M (2011) Aquaculture farmer organizations and cluster management: concepts and experiences. FAO Fish Aquac Tech 563:90
Food and Agriculture Organization FAO (2012) The state of world fisheries and aquaculture. FAO Fisheries and Aquaculture Department Publications, Rome, Italy
Food and Agriculture Organization FAO (2016) The state of world fisheries and aquaculture. Contributing to food security and nutrition for all, Rome, p 200
Food and Agriculture Organization FAO (2018) The state of world fisheries and aquaculture 2018. http://www.fao.org/state-of-fisheries-aquaculture/2018/en
FAO (2019) The State of Food and Agriculture 2019
Food and Agriculture Organisation FAO (2020) El estado mundial de la pesca y la acuicultura. La sostenibilidad en acción
Fraga-Corral M, Ronza P, Garcia-Oliveira P, Pereira AG, Losada AP, Prieto MA, Quiroga MI, Simal-Gandara J (2022) Aquaculture as a circular bio-economy model with Galicia as a study case: how to transform waste into revalorized by-products. Trends Food Sci Technol 119:23–35. https://doi.org/10.1016/j.tifs.2021.11.026
Ghaly AE, Ramakrishnan VV, Brooks MS, Budge SM, Dave DJ (2013) Fish processing wastes as a potential source of proteins, amino acids and oils: a critical review. Microb Biochem Technol 5(4):107–129. https://doi.org/10.4172/1948-5948.1000110
Gildberg A (2004) Enzymes and bioactive peptides from fish waste related to fish silage, fish feed and fish sauce production. J Aquat Food Prod Technol 13:3–11
Girish CR, Gambhir MM, Deshmukh T (2017) Production of biodiesel from fish waste and characterization of produced biodiesel. Int J Civ Eng Technol 8(9):1–6
Goldburg R, Naylor R (2005) Future seascapes, fishing, and fish farming source. Front Ecol Environ 3(1):21–28
Guérard F, Sellos D, Gal YL (2005) Fish and shellfish upgrading, traceability. Adv Biochem Engin/Biotechnol 96:127–163
Harikrishna N, Mahalakshmi S, Kumar K, Reddy G (2017) Fish scales as potential substrate for production of alkaline protease and amino acid rich aqua hydrolyzate by *Bacillus altitudinis* GVC11. Indian J Microbiol 57(3):339–343. https://doi.org/10.1007/s12088-017-0664-2
Harnedy PA, FitzGerald RJ (2012) Bioactive peptides from marine processing waste and shellfish: a review. J Funct Foods 4(1):6–24
Ikoma T, Kobayashi H, Tanaka J, Walsh D, Mann S (2003) Physical properties of type I collagen extracted from fish scales of *Pagrus major* and *Oreochromis niloticas*. Int J Biol Macromol 32(3–5):199–204
Inguglia L, Chiaramonte M, Di Stefano V, Schillaci D, Cammilleri G, Pantano L, Mauro M, Vazzana M, Ferrantelli V, Nicolosi R et al (2020) *Salmo salar* fish waste oil: fatty acids composition and antibacterial activity. PeerJ 8:e9299. https://doi.org/10.7717/peerj.9299
Irawan C, Nata I, Putra M, Marisa R, Asnia M, Arifin Y (2018) Biopolymer of chitosan from fish scales as natural coagulant for iron—contaminated groundwater treatment. J Rekayasa Kim Lingkung 13:93–99
Ishaq A, Rahman U, Sahar A, Perveen R, Deering AJ, Khalil AA, Aadil RM, Hafeez MA, Khaliq A, Siddique U (2020) Potentiality of analytical approaches to determine gelatin authenticity in food systems: a review. Leb Wiss Technol 121:108968
Iverson RL (1990) Control of marine fish production. Limnol Oceanogr 35:1593–1604
Jackson A, Newton RW (2016) Project to model the use of fisheries by-product in the production of marine ingredients, with special reference to the omega 3 fatty acids EPA and DHA. University of Stirling, Stirling, UK

Jafari H, Lista A, Siekapen MM, Ghaffari-Bohlouli P, Nie L, Alimoradi H, Shavandi A (2020) Fish collagen: extraction, characterization, and applications for biomaterials engineering. Polymers 12:2230

Jamilah B, Harvinder KG (2002) Properties of gelatins from skins of fish—black tilapia (*Oreochromis mossambicus*) and red tilapia (*Oreochromis nilotica*). Food Chem 77(1):81–84

Jamu D, Brummett R (2004) Opportunities and challenges for African aquaculture. World Fish Centre, Malawi

Jayvardhan BV, Arvind BC (2020) Preparation of organic fertilizer from fish waste: a new avenue in employment generation. Int Res J Sci Eng A10:21–24

Kandyliari A, Mallouchos A, Papandroulakis N, Golla JP, Lam TT, Sakellari A, Karavoltsos S, Vasiliou V, Kapsokefalou M (2020) Nutrient composition and fatty acid and protein profiles of selected fish by-products. Foods 9:190

Kara K, Ouanji F, Lotfi EM, Mahi ME, Kacimi M, Ziyad M (2018) Biodiesel production from waste fish oil with high free fatty acid content from Moroccan fish-processing industries. Egypt J Pet 27(2):249–255. https://doi.org/10.1016/j.ejpe.2017.07.010

Karim A, Bhat R (2009) Fish gelatin: properties, challenges, and prospects as an alternative to mammalian gelatins. Food Hydrocoll 23:563–576

Kerton FM, Liu Y, Omari KW, Hawboldt K (2013) Green chemistry and the ocean-based biorefinery. Green Chem 15:860–871

Khoddami A, Ariffin AA, Bakar J, Ghazali HM (2009) Fatty acid profile of the oil extracted from fish waste (Head, intestine and liver) (*Sardinella lemuru*). World Alied Sci J 7:127–131

Khoddami A, Ariffin AA, Bakar J, Ghazali HM (2012) Quality and fatty acid profile of the oil extracted from fish waste (head, intestine and liver) (*Euthynnus affinis*). Afr J Biotech 11(7):1683–1689

Kim SK, Dewapriya P (2014) Enzymes from fish processing waste materials and their commercial applications. In: Kim SK (eds) Seafood processing by-products. Springer, New York. https://doi.org/10.1007/978-1-4614-9590-1_10

Kim JS, Park JW (2006) Mince from seafood processing by-products and surimi as food ingredients. In: Maximising the value of marine by-products, pp. 246–255. Woodhead Publishing Cambridge, London, UK

Kudre TG, Bhaskar N, Sakhare PZ (2017) Optimization and characterization of biodiesel production from rohu (*Labeo rohita*) processing waste. Renew Energy 113:1408–1418

Kumar NS, Nazeer RA, Jaiganesh R (2012) Purification and identification of antioxidant peptides from the skin protein hydrolysate of two marine fishes, horse mackerel (*Magalaspis cordyla*) and croaker (*Otolithes ruber*). Amino Acids 42(5):1641–1649

Kusuma IM, Syafrudin, Yulianto B (2019) Utilization of fish waste processing as compost raw material in Tambak Lorok market. E3S Web Conf 125:07004

Lee J, Kim I, Emmanuel A, Koh S (2019) Microbial valorization of solid wastes from a recirculating aquaculture system and the relevant microbial functions. Aquacult Eng 87:102016

Lin CY, Li RJ (2009) Engine performance and emission characteristics of marine fish-oil biodiesel produced from the discarded parts of marine fish. Fuel Process Technol 90:883–888

Lin L, Regenstein JM, Lv S, Lu J, Jiang S (2017) An overview of gelatin derived from aquatic animals: properties and modification. Trends Food Sci Technol 68:102–112

Mahboob S (2015) Isolation and characterization of collagen from fish waste material-skin, scales and fins of *Catla catla* and *Cirrhinus mrigala*. J Food Sci Technol 52(7):4296–4305

Martins GI, Secco D, Rosa HA, Bariccatti RA, Dolci BD, De Souza SNM, Santos RF, Da Silva TRB, Gurgacz F (2015) Physical and chemical properties of fish oil biodiesel produced in Brazil. Renew Sustain Energy Rev 42:154–157. https://doi.org/10.1016/j.rser.2014.10.024

Mo WY, Man YB, Wong MH (2018) Use of food waste, fish waste and food processing waste for China's aquaculture industry: needs and challenge. Sci Total Environ 613:635–643. https://doi.org/10.1016/j.scitotenv.2017.08.321

Morrissey MT, Okada T (2007) Marine enzymes from seafood by-products. Maximizing the value of marine by-products. In: Shahidi F (ed) CRC Press. FL, Boca Raton, pp 374–396

Mozumder MMH, Uddin MM, Schneider P, Raiyan MHI, Trisha MGA, Tahsin TH, Newase S (2022) Sustainable utilization of fishery waste in Bangladesh—a qualitative study for a circular bioeconomy initiative. Fishes 7:84

Mwaijande FA, Lugendo P (2015) Fish-farming value chain analysis: Policy implications for transformations and robust growth in Tanzania. J Rural Community Dev 10(2):47–62

Myrnes B, Johansen A (1994) Recovery of lysozyme from scallop waste. Prep Biochem 24:69–80

Nagai T, Suzuki N (2000) Isolation of collagen from fish waste material-skin, bone and fins. Food Chem 68:277–281

Parvathy U, Nizam KM, Zynudheen AA, Ninan G, Panda SK, Ravishankar CN (2018) Characterization of fish protein hydrolysate from red meat of *Euthynnus affinis* and its application as an antioxidant in iced sardine. J Sci Ind Res 77:111–119

Radziemska M, Vaverková MD, Adamcová D, Brtnický M, Mazur Z (2019) Valorization of fish waste compost as a fertilizer for agricultural use. Waste Biomass Valor 10:2537–2545. https://doi.org/10.1007/s12649-018-0288-8

Ranasinghe R, Kannangara BTSDP, Rathnayake RMCS (2021) Hydrolysis of fish waste using fruit wastes of *Ananas comosus* and *Carica papaya* for the formulation of liquid fertilizers. J Environ Prot Sustain Dev 5(3):96–106

Rodríguez MIA, Barroso LGR, Sánchez ML (2018) Collagen: a review on its sources and potential cosmetic applications. J Cosmet Dermatol 17:20–26

Rumengan I, Suptijah P, Wullur S, Talumepa A (2017) Characterization of chitin extracted from fish scales of marine fish species purchased from local markets in North Sulawesi, Indonesia. IOP Conf Ser Earth Environ Sci 89:012028

Samat AF, Muhamad NAS, Rasib NAA, Hassan SAM, Sohaimi KSA, Iberahim NI (2018) The potential of biodiesel production derived from fish waste. IOP Conf Ser Mater Sci Eng 318:012017

Santos VP, Marques N, Maia P, Lima M, Franco LO, Campos-Takaki GM (2020) Seafood waste as attractive source of chitin and chitosan production and their applications. Int J Mol Sci 21(12):4290. https://doi.org/10.3390/ijms21124290

Sardà F, Coll M, Heymans JJ, Stergiou KI (2013) Overlooked impacts and challenges of the new European discard ban. Fish Fish 16:175–180

Sellami M, Ben Rebah F, Gargouri Y, Miled N (2018) Lipid composition and antioxidant activity of liver oils from ray species living in Tunisian coasts. Arab J Chem 11:233–239. https://doi.org/10.1016/j.arabjc.2014.07.010

Shahidi F, Varatharajan V, Peng H, Senadheera R (2019) Utilization of marine by-products for the recovery of value-added products. J Food Bioact 6:10–61. https://doi.org/10.31665/JFB.2019.6184

Shahidi F, Kamil JYVA (2001) Enzymes from fish and aquatic invertebrates and their application in the food industry. Trends Food Sci Technol 12:435–464

Shahidi F (2003) Extraction and measurement of total lipids. Current Protocols in Food Analytical Chemistry

Shava E, Gunhidzirai C (2017) Fish farming as an innovative strategy for promoting food security in drought risk regions of Zimbabwe. Jamba (Potchefstroom, S Afr) 9(1):491. https://doi.org/10.4102/jamba.v9i1.491

Sionkowska A, Adamiak K, Musiał K, Gadomska M (2020) Collagen based materials in cosmetic applications: a review. Mater (Basel) 13(19):4217. https://doi.org/10.3390/ma13194217

Sriket C (2014) Proteases in fish and shellfish: role on muscle softening and prevention. Int Food Res J 21(2):433

Stevens JR, Newton RW, Tlusty M, Little DC (2018) The rise of aquaculture by-products: increasing food production, value, and sustainability through strategic utilisation. Mar Policy 90:115–124. https://doi.org/10.1016/j.marpol.2017.12.027

Suresh S, Umesh M, Sarojini S (2022) Valorization of fish waste for chitosan production: a sustainable approach. ECS Trans 107(1):16259

Tacias-Pascacio VG, Castañeda-Valbuena D, Morellon-Sterling R, Tavano O, Berenguer-Murcia Á, Vela-Gutiérrez G, Rather IA, Fernandez-Lafuente R (2021) Bioactive peptides from fisheries residues: a review of use of papain in proteolysis reactions. Int J Biol Macromol 184:415–428. https://doi.org/10.1016/j.ijbiomac.2021.06.076
Tawfik MS (2009) Proximate composition and Fatty acids profiles in most common available fish species Saudi market. Asian J Clin Nutr 1:50–57
Tola F, Mosconi EM, Branca G, Natali F, Gianvincenzi M, Nososva B, Colantoni A (2023) Analysis of wastage mechanisms in the supply chain of fish products in a circular economy perspective: empirical research. Heliyon 9(6):E17449
Tsani S, Koundouri P (2018) A methodological note for the development of integrated aquaculture production models. Front Mar Sci 4:406. https://doi.org/10.3389/fmars.2017.00406
Uawonggul N, Ruksakulpiwat C (2002) Study on dye-binding interactions of chitosan obtained from the fish scale of tilapia (*Tilapia nilotica*). In: Proceedings of the 28th congress on science and technology of Thailand, Bangkok, Thailand, 24–26 October 2002
UNEP (2019) Gender equality and preventing plastic pollution; Coordinating body on the seas of East Asia: Bangkok, Thailand
Venugopal V (2016) Enzymes from seafood processing waste and their applications in seafood processing. Adv Food Nutr Res 78:47–69. https://doi.org/10.1016/bs.afnr.2016.06.004
Vidotti RM, Viegas EMM, Carneiro DJ (2003) Amino acid composition of processed fish silage using different raw materials. Anim Feed Sci Technol 105:199–204. https://doi.org/10.1016/S0377-8401(03)00056-7
Yasin NM, Ahmad NAN, Hanapi MM (2021) Extraction of FAME from fish waste by using modified soxhlet method. IOP Conf Ser Mater Sci Eng 1092(1):012015
Yuvaraj D, Bharathiraja B, Rithika J, Dhanasree S, Ezhilarasi V, Lavanya A, Praveenkumar R (2019) Production of biofuels from fish wastes: an overview. Biofuels 10(3):301–307
Zaku S, Emmanuel S, Aguzue O, Thomas S (2011) Extraction and characterization of chitin; a functional biopolymer obtained from scales of common carp fish (*Cyprinus carpio* L.): a lesser known source. Afr J Food Sci 5:478–483
Zhang T, Du B, Lin Y, Zhang M, Liu Y (2020) Production of biodiesel and high-protein feed from fish processing wastes using in situ transesterification. Molecules 25(7):1650
Zheng LY, Zhu JF (2003) Study on antimicrobial activity of chitosan with different molecular weights. Carbohyd Polym 54:527–530

MIX
Papier aus verantwortungsvollen Quellen
Paper from responsible sources
FSC® C105338

If you have any concerns about our products,
you can contact us on
ProductSafety@springernature.com

In case Publisher is established outside the EU,
the EU authorized representative is:
Springer Nature Customer Service Center GmbH
Europaplatz 3, 69115 Heidelberg, Germany

Printed by Libri Plureos GmbH
in Hamburg, Germany